# International Postgraduate Conference on Mechanical Engineering (IPCME 2021)

IOP Conference Series: Materials Science and Engineering
Volume 1078

Pekan, Malaysia
19 - 20 January 2021

**Editors:**

**Kumaran Kadirgama**
**Wan Sharuzi Wan Harun**
**Devarajan Ramasamy**
**Norazlianie Sazali**

ISBN: 978-1-7138-3394-9
ISSN: 1757-8981

**Printed from e-media with permission by:**

Curran Associates, Inc.
57 Morehouse Lane
Red Hook, NY 12571

CURRAN ASSOCIATES INC. proceedings.com

**Some format issues inherent in the e-media version may also appear in this print version.**



Institute of Physics
Dirac House, Temple Back
Bristol BS1 6BE UK

Phone: 44 1 17 929 7481
Fax: 44 1 17 920 0979

techtracking@iop.org

**Additional copies of this publication are available from:**

Curran Associates, Inc.
57 Morehouse Lane
Red Hook, NY 12571 USA
Phone: 845-758-0400
Fax: 845-758-2633
Email: curran@proceedings.com
Web: www.proceedings.com

# TABLE OF CONTENTS

The International Postgraduate Conference on Mechanical Engineering (IPCME 2021) is the second international conference organized by the Faculty of Mechanical & Automotive Engineering Technology, Universiti Malaysia Pahang (UMP), Malaysia. This conference aimed to provide a forum for discussion, sharing and disseminating ideas and expertise among researchers within and outside the UMP, especially postgraduate students across Malaysia. The conference is also an effort to foster the implementation of the latest research with a culture of innovation in rapidly growing mechanical engineering technology. The conference focused on interdisciplinary and multidisciplinary research topics concerning advanced manufacturing, automotive engineering, biomechanics, heat transfer, vibration, and other mechanical engineering topics.

Most interestingly, the conference was fully managed by the postgraduate students of UMP with guidance from the institutional academic and non-academic staff. This is to nurture the leadership and decision-making skills of UMP's postgraduate students in managing a conference on their future working career. The first IPCME conference was successfully organized on October 31, 2018, with 110 participants. Initially, the biennial IPCME conference was planned to be held in April 2020 with a face-to-face presentation. However, due to the uncertain circumstances of the COVID-19, the conference was postponed to January 19, 2021, with the video-conferencing presentation. Nevertheless, we still managed to receive 69 paper submissions, including international participants from China, Indonesia, and Iraq.

Therefore, it is an honor to make the IPCME 2021 a success despite the pandemic. We would like to express our appreciation to Perbadanan Usahawan Nasional Berhad (PUNB) is sponsoring our conference. Special thanks to our keynote speaker Prof. Datin. Ts. Dr. Mimi Sakinah binti Abdul Munaim and Ir. Ts. Wan Mohd Ikhwan bin Wan Yusoff for the motivational and interesting speech. We would also like to express our gratitude to Prof. Dr. Mahadzir bin Ishak and Assoc. Prof. Ir. Dr. Mohd Hafizi bin Zohari as the patron and chairman of IPCME 2021.

Most importantly, thank you to the IOP Conference Series: Materials Science and Engineering conference proceeding in supporting the publication of IPCME 2021. Last but not least, thank you to all participants, committees and all those who supported IPCME 2021. We sincerely apologize for any inconvenience caused and look forward to seeing you in our next IPCME conference.

**The Editors**

1. Assoc. Prof. Ir. Ts. Dr. Kumaran Kadirgama
2. Assoc. Prof. Ir. Ts. Dr. Wan Sharuzi Wan Harun
3. Assoc. Prof. Ts. Dr. Devarajan Ramasamy
4. Ts. Dr. Norazlianie Sazali

**The Committees**

1. Mohd Farid Zainudin
2. Norshalawati Mat Yusof
3. Dara Affydah Mohd Affendi Lim
4. Suhaida Mohamed Salleh
5. Amanina Farhana Ahmad
6. Mahdhir Mohd Yusof
7. Muhammad Hanafi Yusop
8. Muhammad Naqiuddin Mat Salleh
9. Nurul Fatimah Abdullah
10. Maliessa Nabilah Mazelan
11. Muhammad Aminuddin Pi Remli
12. Atik Faysal
13. Mohd Haslizam Abdullah
14. Munir Afnan Mukhtar
15. Muhammad Aizzuddin Abdullah
16. Vorathin Epin
17. Zetty Akhtar Abd Malek
18. Sakinah Muhammad Hisham
19. Ahmad Jamil Sulaiman Hanifah
20. Lim Sheh Hong
21. Wajiha Tasnim Urmi
22. Ma Quan Jin
23. Li Mao Qi
24. Ashwindran Naidu Sanderasagran
25. Amira Qistina Syamimi Zaifuddin
26. Sreenath Sukumarran
27. Siti Rohaida Mohamed
28. Agus Nugroho
29. Nurul Anati Sulaiman
30. Aina Najwa Nadzri
31. Nur Anis Nabila Mokhtar
32. Nur Khaleeda Romli

IOP Conf. Series: Materials Science and Engineering 1078 (2021) 011002 doi:10.1088/1757-899X/1078/1/011002

# Peer review declaration

All papers published in this volume of *IOP Conference Series: Materials Science and Engineering* have been peer reviewed through processes administered by the Editors. Reviews were conducted by expert referees to the professional and scientific standards expected of a proceedings journal published by IOP Publishing.

- **Type of peer review: Double-blind**

- **Conference submission management system: Google form and email at ipcme@ump.edu.my**

- **Number of submissions received: 69**

- **Number of submissions sent for review: 68**

- **Number of submissions accepted: 46**

- **Acceptance Rate (Number of Submissions Accepted / Number of Submissions Received X 100): 67%**

- **Average number of reviews per paper: 2**

- **Total number of reviewers involved: 35**

- **Any additional info on review process: We received a total of 69 paper submissions for the IPCME 2021. All the papers were first screened for related topics in mechanical engineering which were divided into 7 main focus groups (advanced fluid, automotive engineering, energy sustainability, manufacturing, advanced structural integrity and vibration, structural performance materials and human engineering). Plagiarism checking was also carried for all the papers by using Turnitin software with recommended similarity index of no more than 25%. The papers were then sent to 35 reviewers with expertise in the 7 focus groups. A total of 68 papers were sent for review and 1 paper was withdrawn by the authors prior to the review process. From the first review, only 46 papers were accepted and 22 papers were rejected due to poor quality. In order to maintain the quality of the published paper, 39 papers were given major revision and only 7 papers received minor revision. All the major revision papers were then re-evaluated after the authors have submitted their revised manuscript. The second review process was to make sure that the authors have made all the necessary corrections as recommended by the reviewers.**

- **Contact person for queries: Assoc. Prof. Ir. Ts. Dr. Kumaran Kadirgama (kumaran@ump.edu.my)**

# Water quality monitoring via cloud application: Brief review

**A N Abdul Muta'ali[1,2], N Sazali[1,*], S A C Ghani[1] and J Walter[2]**

[1] Faculty of Mechanical & Automotive Engineering Technology, Universiti Malaysia Pahang, 26600 Pekan, Pahang, Malaysia
[2] Hochschule Karlsruhe-Technik und Wirtschaft, Karlsruhe University of Applied Sciences, 76012 Karlsruhe, Germany

*Corresponding author's e-mail: azlianie@ump.edu.my

**Abstract.** With the growing issue of water pollution and resources, the need for a remote water monitoring system has never been more crucial. The implementation of remote monitoring system has a growing interest in the current technology development. Evolution of wireless sensor networks, coding automation and control systems led to the introduction of Internet of Things. This paper aims to review a monitoring system via cloud application to make known of current available system. Cloud application system was studied extensively by various researcher, but research study focusing for water quality monitoring is still scarce. Two main objectives in the studies were designing low-cost system and ease of operation. Most of researchers designed their system using ZigBee, one of IoT system that is currently trending. Currently, cloud system was not widely applied in real industrial application due to lack of experimental evidence. Several enhancements that can be made were proposed.

**Keywords:** Monitoring System; Cloud Application; Distance Test; System Design.

## 1. Introduction

Water quality is mainly assessed to standards related to ecosystem's health, human contact safety and drinking water. Natural bodies of water (i.e. lake, river, sea and etc.) vary due to the environment around them. The International Organization for Standardization (ISO) has released guidelines for water quality. From the European Environment Agency (EEA) assessment, only 38% of the surface body water in the European Union (EU) are considered to have good chemical status [1]. As urban population increases, multiple cities have resorted to drawing fresh water from distant watersheds as local surface and groundwater sources has either become depleted or polluted. Water quality monitoring has been a primary way of identifying water pollution problems. Monitoring has also provided data that shows the trends over time. Water monitoring for large water bodies has always been one that requires a lot of time and effort. Real time data collecting from the water source is necessary to ensure its quality. The data collection requires professional personnel to go out to the field and collect the data during that time. The data collected covers only as of that time. There is no continuous data collection done. This makes it difficult to detect contamination in the water as there is no way to update on the current water quality. Several studies have demonstrated remotely sensed data can be used to map water quality parameters that meet the application needs [2]. However, the use of remotely sensed data on an operational basis for monitoring water quality is limited to areas that are within the range of cellular coverage. The environment that is off grid is generally located in areas that

IOP Conf. Series: Materials Science and Engineering 1078 (2021) 012001 doi:10.1088/1757-899X/1078/1/012001

are far from civilization. It also makes it difficult to return and collect data after several months. One method implemented to retrieve the data is using satellite applications.

Most deployed systems store the data collected internally. The data would then need to be obtained every few weeks or months where personnel would have to physically return to the deploy site. This would not be an issue for deployment within a close and safe area. However, it becomes a problem when the system is deployed further. The journey itself would add up to the cost of the deployment. Water quality monitoring for isolated and remote areas can be quite an expensive task. In the very least it would require personnel to visit the site and manually monitor the parameters using probes. The visits would add up to become a costly operation. There is no real time data collection method available for an affordable price. Available products in the market are designed for professional uses thus carry a heavy price tag with it. With it comes with the charge of setting it up.

## 2. Past studies on water quality monitoring

Among previous research and projects done on the topic of long-distance monitoring, there has been quite a number covering the design and development of LoRa implementation. However, most of these papers only cover the theory of the system. At present, most domestic water quality monitoring is still using manual methods. The detection process involved sampling, sample transportation and preservation, laboratory data measured. The Internet of Things (IoT) brings the promise of constant monitoring of objects with wireless devices to improve everyday aspects of life. Wireless sensor network (WNS) is an evolving field with an extensive range of applications. It packs sensing, computation, and communication into a single sensor node [3]. This typically consists of large numbers of sensor devices with processing capability, sensor and/or actuator, a power source, multiple types of memory and a radio frequency (RF). The sensors are deployed over a large field and inter-networked together, monitoring physical or environmental conditions that generate readings and deliver them to a sink node to be further processed [4]. The wireless network industry has a growing interest towards low power wide area wireless network (LPWAN) [5]. LPWAN technologies have in fact successfully proposed wide area connectivity to tens of kilometers for low data rate, low power and low throughput applications. The market is expected to be huge. About a quarter of 30 billion IoT/M2M devices are predicted to be connected to the Internet using LPWAN [6].

### *2.1. Long Range (LoRa)*

LoRa is an LPWAN protocol for IoT applications. Semtech Corporation introduced extensive utilization of advanced spread spectrum technologies with the LoRa product line [7]. LoRa runs with the spread spectrum modulation technique which offers an increase link budget and better immunity to network interferences. LoRa has multiple orthogonal spreading factors (SF); numbered between 7 to 12. The SF provides tradeoff between data rate and range. Lower SF means more chirps are sent per second; therefore, more data can be encoded per second. Having a higher SF can increase the range but decreases the data rate and vice versa [8]. Relationship between spreading factor and airtime for LoRa modulations can be referred in Figure 1 below. LoRa uses the forward error correction (FEC) to increase the receiver sensitivity. Code rate (CR) defines the amount of FEC which ranges between 0 and 4. CR=0 means that there are no FEC. The code rates are as shown in the Table 1 below. The redundancy provides the receiver with information to detect and often correct errors in the message. The downside of it is that it also reduces the effective data rate. As the CR value increases, the data rate decreases in each bandwidth spectrum. LoRa offers the usage of scalable bandwidth; 125 kHz, 250 kHz or 500 kHz [9]. Wider band usage in LoRa makes it resistant to channel noise, Doppler effects and fading. Transmitter sends spreaded data at a chip rate equal to the systems' bandwidth in chips per-second-per-Hertz [8].

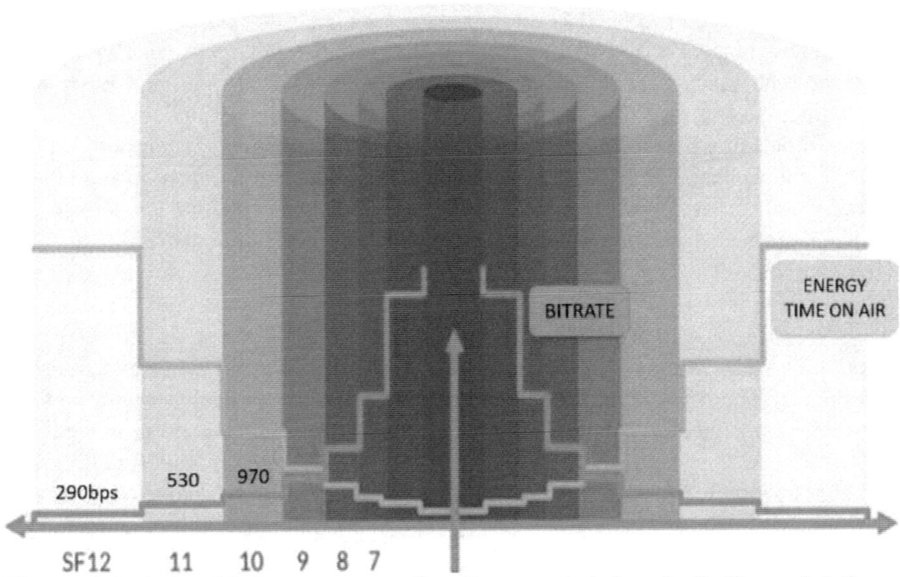


**Figure 1.** Relationship between spreading factor and airtime for LoRa modulations.

**Table 1.** Code rates in LoRa.

| CR Value | No. of redundant bits | Coding rate |
|---|---|---|
| 1 | 1 | 4/5 |
| 2 | 2 | 2/3 |
| 3 | 3 | 4/7 |
| 4 | 4 | 1/2 |

LoRa has a maximum packet size of 256 bytes. The packet consists of 4 fields; preamble field, header field, payload field and CRC [7]. The preamble field is used to synchronize the receiver with the incoming data from the receiver. Depending on the choice of operation modes, the header field could specify the number of bytes for code rate and payload or have it fixed. In the latter, the frame does not contain this field thus reducing the transmission time. The header also carries 2 bytes of cyclic redundancy check (CRC) field, allowing the receiver to ignore packets of invalid header. The payload field varies from 2 to 255 bytes. Within this field are 3 sub-field; Message Authentication Code (MAC) header, MAC payload and Message Integrity Code (MIC). MAC header defines the frame type, protocol version and its direction. The actual data is located in MAC payload. MIC is the digital signature of the payload. The last field which is 2 bytes, CRC, is optional. It comprises cyclic redundancy check bytes for error protection of the payload. Free space path loss (FSPL) is the attenuation of radio energy between the feedpoints of two antennas that results from the combination of the receiving antenna's capture area plus the obstacle free, line-of-sight path through the free space. FSPL is used to predict the strength of an RF signal at a particular distance. It is crucial to find the balance between battery performance and long-range communication. Higher SF requires longer airtime to send data compared to lower SF thus consuming more battery power. However, the extended airtime allows for better sensitivity.

*2.2. Literature review survey*
We have reviewed some of researchers which has designed system that implemented wireless sensor network (WSN). Wang et al. [10] in their publication has designed a monitoring system for sewerage

based on ZigBee. To overcome communication issue between sensor and base station and to increase the communication distance, multi-distributed nodes was proposed by them which resulted in 80% of reliability. On the other hand, Geetha and Gouthami [11] presented a smart IoT-based water quality monitoring system in their paper. It was less complex and is cost-effective. Cloete et. al. [12] provided a thorough system design which is also IoT-based for water monitoring. The system can detect water contamination in the water bodies. ZigBee module was used as communication system between the nodes in which it will gives off alarm when unsafe levels of water quality is detected. The farthest distance that the system can work in non-line-of-sight is 13m, a very big achievement for small scale study. Similar with Cloete et. al. [12], Rekhi et. al. [13] published their research work on monitoring pollution in water system.

Jiang et. al. [14] took into account the water temperature in their water monitoring system. They reported that their system can monitor the temperature and pH of water with high accuracy. Similar with other reported researches, they also used ZigBee protocol to design the monitoring system software. However, they did not report on how far their wireless communication will be able to work. Table 2 below outlined the summary of researches done by various authors. The trend that can be observed was that all the systems aims to monitor the data from longer distance with different settings input. Based on the table, most researchers reported on using ZigBee transmitter and receiver due to its long-range distance and ease of operation. It was known that ZigBee usage in the automation (monitoring) system possess many advantages such as low power consumption, battery that can lasts longer and can be easily tailored according to needs [15]. When there is low power consumption, overall cost can be lowered which is good for industrial purpose. There is another module that currently is gaining attention, named XBee ZB [16, 17].

**Table 2.** Summary of past research by various authors.

| Method | Result | Ref. |
|---|---|---|
| • Wireless sensor network based on ZigBee | A monitoring system of WSN and GPRS technology was successfully designed. | [10] |
| • UART protocol<br>• Data stored in Ubidots cloud | A low cost and less complex monitoring were proposed. | [11] |
| • ZigBee receiver and transmitter modules | Wireless communication of 13 m was achieved. | [12] |
| • Platformed on WSNs and RFID systems | Tags to be deployed on the waterway banks to allow sensor to detect position of pollution. | [13] |
| • Main processor: MSP430F1611, co-processor: CC2430<br>• Zigbee Radio Frequency Module<br>• LCD display | The system successfully performed online auto-monitoring of an artificial lake. | [14] |
| • 3G network and Winsock MySQL database | Data transmitted using 3G network to the server | [18] |
| • Microcontroller: Arduino Uno R3<br>• LabVIEW as a front panel for the user interface | The parameters are observed through direct connection of the Arduino and LabVIEW successfully. | [19] |
| • ZIC2410 single-chip microprocessor<br>• ZigBee protocol system | Designed a real-time monitoring of inpatient physiological signal within hospital. | [20] |

## 3. Future perspectives

Water parameters to be measured include water salinity, conductivity and temperature. A cloud server is to be configured as a data repository. In the future, data collected should be displayed in a remote PC. Also, the system needs to be cost effective and requires minimum maintenance. This system is still new in terms of its development. There are still room for improvements to be done on it. Some of the suggestion includes PCB fabrication of the electrical components in used. Eventhough the use of electronic parts is considered minimum, a PCB fabrication would provide better and more secure connection between the components as opposed to jumper cables. There are four main components that needs more focus which are mechanical, electrical, software and documentation. The software component consists of the Arduino IDE and a user interface in which the Arduino IDE serves as a platform to code the program. The user interface should act as a front end to monitor the water parameters without having to read through lines of data. The electrical component of the system includes the PCB design, microcontroller, power source, sensors and display should be capable of long-distance wireless communication. The power source must be able to provide for long periods of time. The display is not a necessary part but would be helpful for early detection of any issue. In the mechanical part, there is the system housing. Existing research provides a better understanding of what the market needs. The products that are currently available in the market either store their data in a local storage, in their brands' cloud system or both. There is a redundancy of data stored and this results in waste of data space. For domestic uses, there is no data collection for observation as it is only available for products of professional uses. Therefore, a system that allows users to monitor current and past readings and obtain the data for their own analysis should be designed.

## 4. Conclusion

Currently, usage of LoRa is very limited despite of volume of research done. Most deployed systems stored the data collected internally and can only be obtained every few weeks or months where personnel would have to physically return to the deploy site. In this paper, we have briefly reviewed the current available system for water quality monitoring. LoRa system is one of the promising technology due to its competitiveness, technical advantage and its potentiality for IoT application which is using wireless sensor networks. Two main objectives in the existing studies reported were designing low-cost system and ease of operation. Future improvements that can be made include extensive study on the components that can provide better and secure connection such as PCB fabrication, able to work in longer periods and able to monitor current and past reading easily.

## Acknowledgments

The authors would like to acknowledge the financial support from Universiti Malaysia Pahang and Hochschule Karlsruhe-Technik und Wirtschaft (HsKA) under grant number RDU192703 and UIC191514.

## References

[1] European Environment Agency (EEA) 2018 *Environmental Indicator Report 2018* (Luxembourg: Publications Office of the European Union)
[2] Avdan Z Y, Kaplan G, Goncu S and Avdan U 2019 *ISPRS Int. J. Geo-Inf.* **8** 553
[3] Cho S, Akyildiz I F, Bender M D and Uzunalioglu H 2002 *Wireless Networks* **8** 403-415
[4] Shen H 2001 *Applied Mathematics and Mechanics* **22** 270-281
[5] Slauson J 2016 *The Bloomsbury Encyclopedia of Design Volume 2* (London: Bloomsbury Academic)
[6] Mikhaylov K, Petajajarvi J, Makelainen M, Paatelma A and Hanninen T 2015 *IEEE Sensors*
[7] Semtech Corporation 2014 *Annual Reports* (Washington D.C.: United States Securities and Exchange Commission)
[8] Noreen U, Bounceur A and Clavier L 2017 *3rd International Conference on Advanced Technologies for Signal and Image Processing (ATSIP2017)* pp. 1-6

[9] Reynders B, Meert W and Pollin S 2016 *23rd International Conference on Telecommunications (ICT)*
[10] Wang S, Horoshenkov K, Tait S, Abd-Alhameed R, Hu Y, Elkhazmi E and Gardiner J 2009 *Asia Pacific Microwave Conference*
[11] Geetha S and Gouthami S 2016 *Smart Water* **2** 1-19
[12] Cloete N A, Malekian R and Nair L 2016 *IEEE Access* **4** 3975-3990
[13] Rekhis S, Ellouze N and Boudriga N 2012 *8th ACM Symposium on QoS and Security for Wireless and Mobile Networks*
[14] Jiang P, Xia H, He Z and Wang Z 2009 *Sensors* **9** 6411-6434
[15] Tomar A 2011 *Element 14: the knode* 1-24
[16] Piyare R, Lee S R 2013 *International Journal of Scientific & Engineering Research* **4**
[17] Kuzminykh I, Snihurov A, Carlsson A 2017 *14th International Conference The Experience of Designing and Application of CAD Systems in Microelectronics* pp. 133-136
[18] Huang Y, Chou C, Jau J and Sandnes F E 2010 *7th International Conference on Autonomic and Trusted Computing*
[19] Taru Y K and Karwankar A 2017 *International Conference on Computing Methodologies and Communication*
[20] Song A, Si G 2017 *29th Chinese Control and Decision Conference* pp. 2618-2621

# Water monitoring system design for data collection at specific intervals via cloud application

**A N Abdul Muta'Ali[1,2], N Sazali[1,*], S A C Ghani[1] and J Walter[2]**

[1] Faculty of Mechanical & Automotive Engineering Technology, Universiti Malaysia Pahang, 26600 Pekan, Pahang, Malaysia
[2] Hochschule Karlsruhe-Technik und Wirtschaft, Karlsruhe University of Applied Sciences, 76012 Karlsruhe, Germany

*Corresponding author's email: azlianie@ump.edu.my

**Abstract.** With the growing issue of water pollution and resources, the need for a remote water monitoring system has never been more crucial. The implementation of remote monitoring systems has a growing interest in the current technology development. This project aims to design a water monitoring system via cloud application. Based on previous research and papers written on the topic, a few system designs have been theorized but not implemented. In this project, the data is transmitted through the combination of LoRa and Wi-Fi connection. The system is also powered by rechargeable battery and solar panels. The communication using long range achieved a distance test of 898 m. The full system has the capability to be deployed in a distance and have the data collected uploaded to a cloud. A web server accessed through the local IP address serves the user interface that provides current data of the water parameters.

**Keywords:** Monitoring System; Cloud Application; Distance Test; System Design.

## 1. Introduction

Water is arguably the most important resource on Earth. With the current environmental situation, water management has never been more crucial. There are 148 countries that share 267 international river basins [1,2]. 60% of the water from this flow around the world. As urban population increases, some cities have resorted to drawing fresh water for distant watersheds due to the depletion and pollution of local surface and groundwater sources [3]. Monitoring water quality allows for data collection needed to safeguard the environment against contamination from point sources. Water quality is a description of the water conditions, the chemical, physical and biological characteristics, with respect to its suitability of its purpose. Water quality is usually sampled and analyzed in laboratories. Since the late 20th century, systems were developed to collect real-time data of the water source. In 2009, with the increasing attention to the controversial drilling technique for natural gas production near Susquehanna river basin, Nexsens Technology came up with a solution of monitoring the water quality and made it available to the public to keep tabs on their local rivers [4]. Over 40 remote monitoring stations were installed as water monitoring became more crucial as day passed. Remote monitoring refers to the specification that helps monitor network operational activities using remote devices known as probes or monitors. In the early 21st century, water remote monitoring began to be deployed to monitor parameters [5-8]. Remote monitoring systems are possible through various methods of data transmission such as satellite, cellular, radio and even telephone modem. Each of these methods carries their own pros and cons, as documented by predecessors before. The methods vary as to meet different needs. Water monitoring for large water bodies has always been one that requires a lot of time and effort. Real time data collecting from the water source is necessary to ensure its quality. The data collection requires professional personnel to go out to the field and collect the data during that time. The data collected covers only as of that time. There is no continuous data collection done. This makes it difficult to detect contamination in the water as there is no way to update on the current water quality. Several studies have demonstrated remotely sensed data can be used to map water quality parameters that meet the application needs. However, the use of remotely sensed data on an operational basis for monitoring water quality is limited to areas that are within the range of cellular coverage. The environment that is off grid is generally located in areas that are far from civilization and sometimes inaccessible, providing many challenges to maintain wireless networks [9]. It also makes it difficult to return and collect data after several months. One method implemented to retrieve the data is using satellite applications.

Most deployed systems store the data collected internally. The data would then need to be obtained every few weeks or months where personnel would have to physically return to the deployed site [10]. This would not be an issue for deployment within a close and safe area. However, it becomes a problem when the system is deployed further. The journey itself would add up to the cost of the deployment. Water quality monitoring for isolated and remote areas can be quite an expensive task. In the very least it would require personnel to visit the site and manually monitor the parameters using probes. The visits would add up to become a costly operation. There is no real time data collection method available for an affordable price. Available products in the market are designed for professional uses thus carry a heavy price tag in addition to the charge of setting it up. The objective of this project is to design an efficient and cost-effective water monitoring system via cloud. The system is to be able to collect data at specific time intervals, link the microcontrollers in use for long range communication while allowing for the microcontrollers to connect to a cloud system.

**Table 1.** Components specifications.

| Materials | Specification |
|---|---|
| Microcontroller | • SX1276 LoRa ESP32<br>• OLED Screen<br>• Semtech SX1276 radio |
| pH sensor | • Module Power 5V<br>• Measuring range 0-14 pH<br>• Accuracy ± 0.1 pH (25 °C)<br>• Response time $\leq$ 1 min |
| Temperature sensor | • DS18B20 Digital Temperature Sensor<br>• Power supply 3- 5 V<br>• Accuracy: ± 0.5 °C |
| TDS Sensor | • Input voltage 3.3-5V<br>• Waterproof probe<br>• TDS range 0-1000ppm<br>• Accuracy ± 10 F.S. (25 °C) |
| Power supply | • Mini Solar Panel 5V<br>• Rechargeable Li-ion Batteries 3.6V<br>• TP4056 Lithium Battery charger |

## 2. Materials and methods

### *2.1. System flowchart and components*

The system consists of two ESP32 modules, one to collect data and send it to the other, where the data is then uploaded to the cloud server. The ESP32 connected to the sensors is powered by a rechargeable battery, charged by solar panels. This allows it to be deployed for a longer period. As shown in Figure 1, the general view of how the system works. The two ESP modules are connected using long range connectivity (LoRa). This allows for data to be transmitted over a large range. The second ESP functions as a gateway to connect to the Internet. The data collected is sent through IFTTT (If This Then That). IFTTT is a free web-based service that creates chains of simple conditional statements, called applets. This applet is a result of triggers and actions to send the reading obtained to the Google Sheet assigned. From the gateway, the data is uploaded over to Google Sheets with the help of IFTTT. The user interface is created in a web browser, accessible through the IP address obtained from the receiver ESP. This allows for the data to be accessed immediately from the gateway. Components specifications are as listed in Table 1 whilst flowchart can be referred in Figure 1.

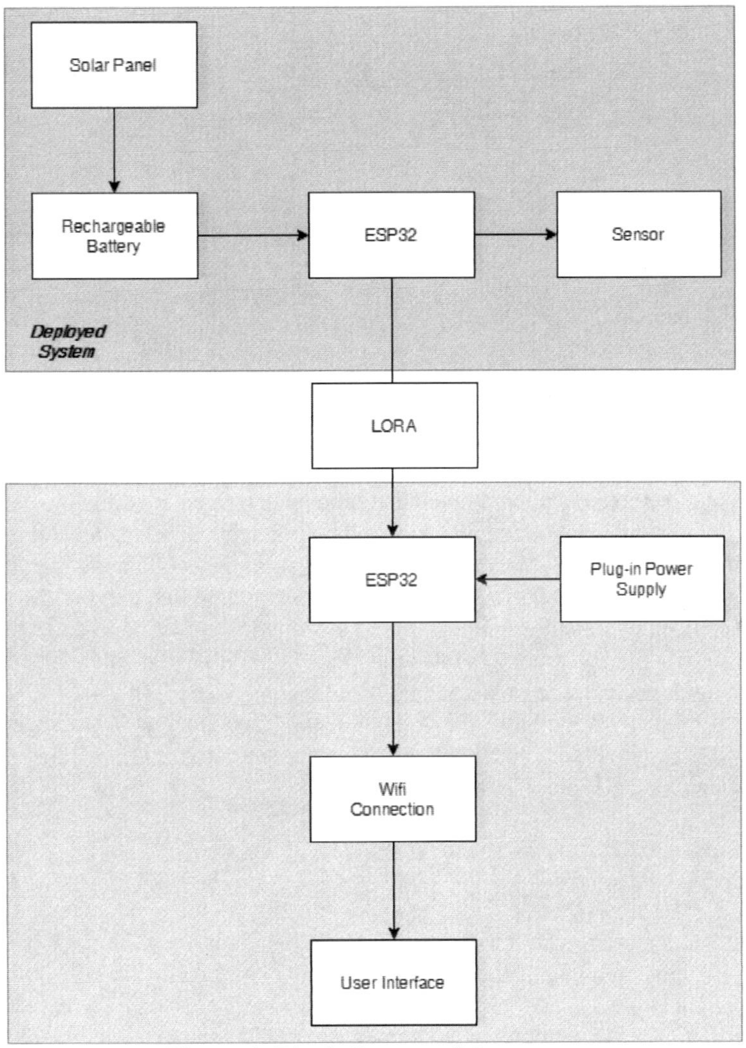


**Figure 1.** Block diagram of the system.

### *2.2. Pin allocation plan*

The connection of the sensor to the ESP32 is planned as shown in the figure below. The TDS, temperature and pH sensors are connected to GPIO 36, 38 and 34 respectively. All the sensors are powered through the ESP board. With the use of the SX1276 LoRa ESP32 microcontroller, the radio and OLED module are built in. No further additional module needs to be added. As shown in the Figure 2 below, the sensor modules are connected to the 5V supply from the ESP32. The same goes for all the grounding. There is a 4.7 kΩ pull-up resistor between the Vcc and Data pin of the temperature sensor. This would ensure a more reliable communication.

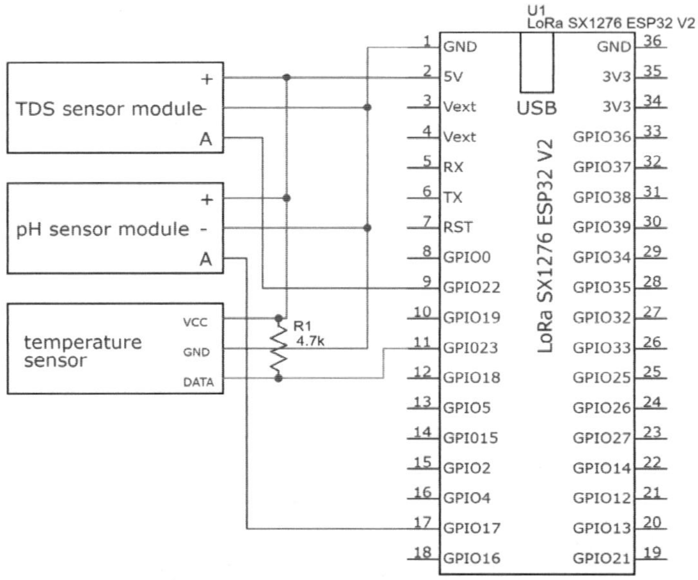


**Figure 2.** Schematic diagram of ESP32.

### *2.3. Power source*

As the aim of the project is to have the monitoring system be deployed, it would be useful for the system to have its own renewable source of energy. Powering it with solar panels is an alternative method that is put to test in this assembly. The solar panels used have an output voltage of 5V. Connected directly to the solar panels is a TP4056 battery charger module. This module functions as the battery charger. It prevents the battery from overcharging and reverse polarity connection. The battery is a li-ion 18650 battery. The battery outputs 3.6V when it is fully charged. As the ESP32 works with 3.3V, a low dropout voltage regulator circuit needs to be implemented. For this, an MCP1700-3302E is chosen as the low-dropout regulator (LDO). To smooth out the voltage peaks, capacitors are connected in parallel to the LDO. For this, a 100nF and 100uF capacitors are used. The output voltage will then power the ESP32 by connection through the 3.3V pin.

## 3. Results and discussion

### *3.1. Sender code*

The coding for the Sender requires it to be synced with the Receiver, contains the necessary libraries for the sensors and the ability to transmit the data obtained. For the two ESPs to be able to communicate with each other, a sync word is set. The sync word allows the Receiver to only read the data from a specific source instead of any random radio transmission. The sensors each have their own sensor libraries and settings. As each sensor reading is separate, it is compiled into a single string value, LoRaMessage, which is sent through the LoRa connection. The string consists of a reading ID, the temperature reading, pH value and TDS reading. In between each of the data, a symbol character is added to accommodate to the Receiver's coding. A snippet of the code is shown below in Figure 3.

```
/*-----------------Send Data----------------------------*/

void sendData(){

  LoRaMessage = String(readingID) + "/" + String(temperature) + "&" + String(phValue) + "#" +
String(tdsValue);

 LoRa.beginPacket();

 LoRa.setTxPower(14,RF_PACONFIG_PASELECT_PABOOST);

 LoRa.print(LoRaMessage);

 LoRa.endPacket();

 readingID++;

}
```

**Figure 3.** Code to string data for lora transmission.

### *3.2. Receiver code*

There are a few additional libraries required for the Receiver compared to the Sender. The ESPAsyncWebServer library provides the web server to request data from the Receiver. The Wi-Fi library is also necessary for the Receiver. The wifi connection needs to be initialized before connection to the server. The SSID and password for the router is inputted in the coding directly as shown in Figure 4. The data received are sent to the IFTTT server with the 'makeIFTTTRequest' function.

```
/*=============set up wifi connection===========================*/

const char* ssid     ="bridge";

const char* password ="password";

/*==================IFTTT server connection=====================*/

const char* resource ="/trigger/ESP32_water/with/key/c8vajtrYEGq1_Ejhhy4Yt_";

const char* server1  ="maker.ifttt.com";

/*===========Create AsyncWebServer on port80===================*/

AsyncWebServer server(80);
```

**Figure 4.** Setup for the Receiver code.

For the web server, the HTML code is written in the same file. It is also possible to have the HTML be called from a different file however as the coding is not too complicated, it is unnecessary. The readings are connected to the web server with the coding shown in Figure 5.

```
//Route for root / web page
server.on("/", HTTP_GET, [](AsyncWebServerRequest *request){
 request->send(SPIFFS, "/index.html", String(), false, processor);
});
server.on("/style.css", HTTP_GET, [](AsyncWebServerRequest *request){
 request->send(SPIFFS, "/style.css", "text/css");
});
server.on("/temperature", HTTP_GET, [](AsyncWebServerRequest *request){
 request->send(SPIFFS, "/index.html", temperature.c_str(), false, processor);
});
server.on("/ph", HTTP_GET, [](AsyncWebServerRequest *request){
 request->send(SPIFFS, "/index.html", ph.c_str(), false, processor);
});
server.on("/tds", HTTP_GET, [](AsyncWebServerRequest *request){
 request->send(SPIFFS, "/index.html", tds.c_str(), false, processor);
});
```

**Figure 5.** Sending the data collected to the web server.

### *3.3. User interface*

The web interface of the system should provide a current overview of the condition of the water. Shown on the interface is the latest reading of the temperature, pH and TDS. Along with that, users can also download the data collected in the xls file for further analysis from the Google Sheets. As shown in Figure 6, the data are tabulated automatically through the IFTTT setting. Tabs for each parameter where the graph of the reading collected is shown in Figure 7.

| | A | B | C | D |
|---|---|---|---|---|
| 1 | Date | temperature | pH value | tds |
| 2 | 01-01-2020 | 22.68 | 6.87615933 | 51.0 |
| 3 | 02-01-2020 | 22.68 | 6.85577404 | 53.0 |
| 4 | 03-01-2020 | 22.68 | 6.49871328 | 59.0 |
| 5 | 04-01-2020 | 22.69 | 6.18471519 | 64.0 |
| 6 | 05-01-2020 | 22.69 | 6.21029986 | 76.0 |
| 7 | 06-01-2020 | 22.69 | 6.77514099 | 78.0 |
| 8 | 07-01-2020 | 22.69 | 6.12274251 | 80.0 |
| 9 | 08-01-2020 | 22.69 | 6.82312351 | 80.0 |
| 10 | 09-01-2020 | 22.69 | 6.12659479 | 90.0 |
| 11 | 10-01-2020 | 22.69 | 6.67174834 | 93.0 |

**Figure 6.** Sample of data obtained from the Google Sheets.

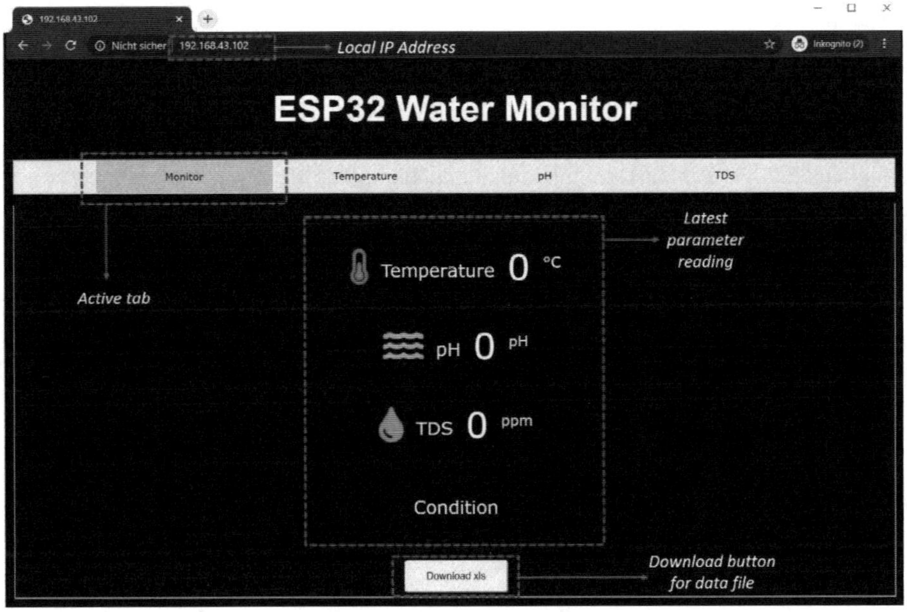


**Figure 7.** Parameter tab view.

### *3.4. Distance test*

The distance capability of the LoRa connection was done multiple times. The Sender is set up at a fixed location as the Receiver is carried further away from it. The location of the two microcontrollers are then mapped onto Google Maps based on their coordinate to measure the distance. The initial distance test on the ESP had a result of approximately 300m. The sender was located near a window in a building on the fourth floor. Then, the second test was done away from any building interference. The receiver could collect data at 645.18 m. Both ESPs were at ground level. The final test was done to confirm the result from the first and second test. The location test was done around the same location to recreate the same result. The distance observed was approximately 200 m further than that of the predecessor with 896.66 m as shown in Figure 8.

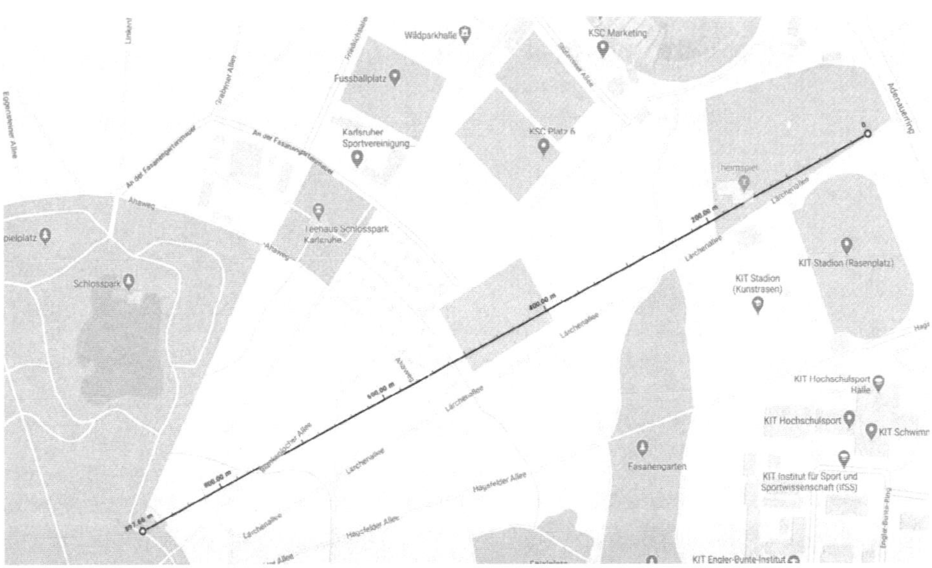


**Figure 8.** Distance test recorded using Google Map.

## 4. Conclusion

As a conclusion, a monitoring system that can collect data at set time intervals was successfully created. Any sudden or progressive changes thanks to the water parameters are traceable thanks to the system. Data collected are uploaded directly to Google Sheets via Wi-Fi connection. The system allow for easy data analysis in the long run as the data set is accessible. A web interface is also created to monitor the latest data collected from the system and download the data directly. The connection between the microcontrollers are achieved through LoRa connection for data transmission. The data transfer presented a different way of connection to traditional water monitoring system that are in the market. It is still a relatively new method in its implementation and technologies are ever so growing. There is bound to be new upgrades that the LoRa could offer in the future.

**Acknowledgments**

The authors would like to acknowledge the financial support from Universiti Malaysia Pahang and Hochschule Karlsruhe-Technik und Wirtschaft (HsKA) under grant number RDU192703 and UIC191514.

## References

[1] Masundire H M 2008 *Achieving Sustainable Development and Promoting Developent Cooperation – Dialogues at the ECOSOC* (New York: United Nations)
[2] Biswas A K and Tortajada C 2019 *International Journal of Water Resources Development* **35** 913-916
[3] Jha M K, Sah R K, Rashmitha M S and Sinha R 2018 *International Conference on Inventive Research in Computing Applications*
[4] NexSens Technology 2019 Accessed at https://www.nexsens.com/blog/water-monitoring-buoy- systems.htm
[5] Antonini K, Langer M, Farid A and Walter U 2017 *Acta Astronautica* **140** 10-17
[6] Wang X, Ma L and Yang H 2011 *Proceedia Engineering* **15** 2680-2684
[7] Kim Y, Schmid T, Charbiwala Z M, Friedman J and Srivastava M B 2008 *Proceedings of the 6th ACM Conference on Embedded Network Sensor System* 309-322
[8] Jiang P, Xia H, He Z and Wang Z 2009 *Sensors* **9** 6411-6434
[9] Toledano-Ayala M, Herrera-Ruiz G, Soto-Zarazua G M, Rivas-Araiza E A, Bazan Trujillo R D and Porras-Trejo R E 2011 *Sensors* **11** 71414-7161
[10] Schreiber M E, Schwartz B F, Orndorff W, Doctor D H, Eagle S D and Gerst J D 2015 *Advances in Watershed Science and Assessment* 205-231

# Optimization of processing conditions on mechanical properties of Polypropylene nanocomposite reinforced with Gigantochloa Scortechinii via ANOVA

**S Z Khamis[1,*], O Mamat[1] and M H Othman[2]**

[1]Faculty of Mechanical Engineering, Universiti Teknologi Petronas, Malaysia
[2]Faculty of Mechanical engineering and manufacturing Universiti Tun Hussien Onn Malaysia

[*]Corresponding author's email: siti_19001006@utp.edu.my

**Abstract.** Defining the level and value of the appropriate parameters is very significant in seeing the quality in mechanical properties of the injection molding products, particularly affecting the advanced material such as nanocomposites. Consequently, this study was aimed to optimize the processing conditions to the mechanical properties of the sample in nanocomposite polypropylene reinforced with *Gigantochloa Scortechinii* fiber. The tests to be tested were flexural strength, shrinkage and warpage, and the factor chosen were melting temperature, filling time, packing pressure and screw speed. The bamboo fiber types *Gigantochloa Scortechinii* in form nano powder, and preheat at 120°C after alkaline treatment, and so blended with polypropylene, Maleic anhydride, and cloisite 20A using two step of twin-screw extruder brabender. According the result from ANOVA, the flexural strength and shrinkage for 0wt% and 3wt% of *Gigantochloa Scortechinii* fiber, the most influence parameter is filling time, while the melt temperature for 6wt% of gigantochlia scortechinii fiber. For the result of warpage, for 0wt% and 3wt% of fiber *Gigantochloa Scortechinii*, the melting temperature is the highest factor prevailing warpage. Whereas for 6wt% of the fiber *Gigantochloa Scortechinii*, the packing pressure and screw speed are the highest factor. The conclusions of this study will be invaluable for quality dominance and benchmarking for future researchers.

**Keywords:** Injection Molding, Cloisite 20A, Gigantochlia Scortechinii, ANOVA.

## 1. Introduction

Today, the injection molding in polymer processing is to the highest degree in manufacturing. This is because, this competence can produce for series production and can also save time and costs. Nevertheless, the culmination of the proficiency is the optimized processing conditions to ascertain that the in demand product properties can be accomplished [1]. As far as the design of the experiment (DOE) is concerned, Taguchi Method has introduced numerous statistical methods to define the optimum parameter in addition to the greatest contributory factor in mind for the pick out reactions. In the study of the injection molding process, many researchers performed excessively using Taguchi method. Previous researchers have obviously defined how the responses and materials are affected by the parameter in processing [2,3,4,5]. Several previous searches related to the scope of this study also been carried out [7][8]. The result will be used to determine the level and the factors of the study.

An examination noted that the nanoclay, polymer, nanocomposite and natural fiber have become more unremarkably referable to the polymer and advance material are widely used and naturally plentiful with a high aspect ratio [10].With respect to additional reference, the submitter reviewed the properties, application and the preparation of polymer nanocomposites [11]. Nevertheless, the effects of clay types, and percent distribution in the composite paying the primary role when blended with any fillers and the choice of seizing parameters have been highlighted as important to build sure the whole material is fitted to the matrix polymer in order to prevent problems properties of injected mould [12].

## 2. Experimental

Figure 1 illustrates the main process of experience.

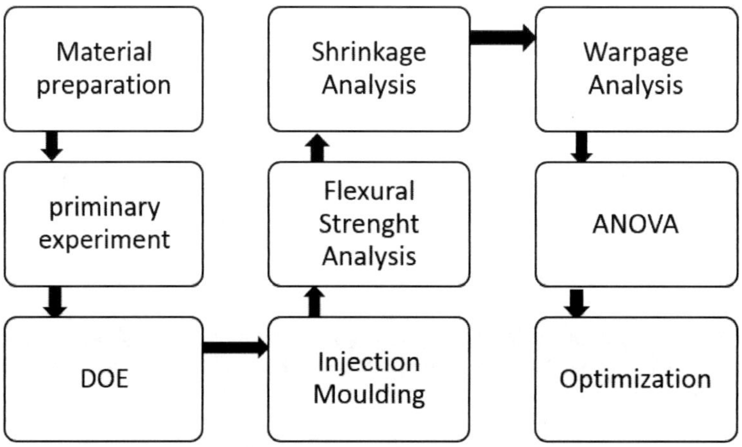


**Figure 1**. The frame process of the experiment.

*2.1 Material preparation*

The composition of compounding were divided into three different composites as shown in table 1 below. The material was the raw polypropylene (PP), polypropylene grafted-maleic anhydride (PPgMA) as compatibillizer, cloisite 20A and *Gigantochlia Scortechinii* (bamboo fiber).

**Table 1**. The composition of the sample.

| Composition | PP | PPgMA | Cloisite 20A | *Gigantochlia Scortechinii* |
|---|---|---|---|---|
| 1 | 84 wt% | 15 wt% | 1 wt% | 0 wt% |
| 2 | 81 wt% | 15 wt% | 1 wt% | 3 wt% |
| 3 | 78 wt% | 15 wt% | 1 wt% | 6 wt% |

The figure 2 shows the preparation of the fibers starts with clean up the bamboo using paper. The fiber was use is the whole bamboo include the skin of bamboo. Then the bamboo was chipped into small size (10cm), then the bamboo was soaked in sodium hydroxide for the treatment, this treatment was commonly used is alkaline treatment. After the treatment, the bamboo is dry under the sun in naturally before using dry oven at 120°C for 6 hours. The Granulator machine was used to crush the bamboo into powder (nano size) as shown below.

**Figure 2**. The flow processing to prepared the *gigantochlia scortechinii* fibers.

*2.2 Design of the experiment*

The injection molding machine was used is types Nessei NP7-1F which is the education scale machine. By using the mould for standard ISO D790 for flexural strength, the sample was prepared regarding the formulation respectively. By using Taguchi Method, the effect of several parameters most effectively to determine in using the orthogonal array ($L_93^4$) and the details was prepared at table 2. This method was best used to find optimal process values to improve product quality. Four factors of processing condition were chosen with 3 different levels and bring out 27 samples for the 9 trials of sample have been tested.

**Table 2**. The orthogonal array by Taguchi Method in Minitab

| Melting Temperature (%) | Screw Speed (%) | Packing Pressure (%) | Filled Times (s) |
|---|---|---|---|
| 165 | 25 | 30 | 1 |
| 165 | 30 | 35 | 2 |
| 165 | 35 | 40 | 3 |
| 170 | 25 | 30 | 3 |
| 170 | 30 | 35 | 1 |
| 170 | 35 | 40 | 2 |
| 175 | 25 | 30 | 2 |
| 175 | 30 | 35 | 3 |
| 175 | 35 | 40 | 1 |

*2.3 Measuring of flexural strength*

The flexural strength of the specimen was measured using a universal test machine (10KN) based on three bending points according to ISO 178 standard with the speed of 2.04317 mm/min.

*2.4 Determination of warpage and shrinkage.*

Warpage and shrinkage were bound together. The different shrinkage is occurred by the deformation. The defect happens due to numerous factors and parameters which were high screw pressure, melting temperature and filled time rate is overly low and partially turnout is excessively hot [1]. The shrinkage and the warpage were selected for optimisation by checking the injection molding setting. These superiority features were measured in samples following injection molding. The sample that was ejected from the mould must be cleaned of the runner and flashing prior to measuring the sample. The shrinkage was measured by means of the equations (1) and (2), the coefficient thermal expansion (α) is the value of tool steel by value ($6.45\text{x}10^{-6/°F}$) and the symbol $T_{\text{mould}}$ is refered to temperature of mould. While, $T_{\text{ambient}}$ is referred the ambient temperature.

$$L_C = L[1 + \alpha\,(T_{mould} - T_{ambient})] \tag{1}$$

$$S = \frac{L_C - L_M}{L_C} \tag{2}$$

Subsequently, to obtain the data needed to deform a specimen, the equation (3) is used. Relates to the plate maximum and is the average thickness of the plate.

$$Z = h - t_a \tag{3}$$

*2.5 Measuring the ANOVA*

ANOVA has been used to provide the variance measure level. It determines the variability (variance) of the data using the Minitab software.

## 3. Result and discussion

*3.1 Flexural strength*

Table 3 shows the ANOVA result of the flexural strength for 0wt.%, 3wt.%, and 6wt% of *Gigantochlia Scortechinii fiber*. The calculation is based on the outcome of the S/N ratio.

**Table 3.** The flexural strength results of ANOVA for 0wt.% ,3wt.% and6 wt.% of *Gigantochlia Scortechinii* fiber.

| Parameter | Flexural Strength 0 wt.% GS fiber | | | Flexural Strength 3 wt.% GS fiber | | | Flexural Strength 6 wt.% GS fiber | | |
|---|---|---|---|---|---|---|---|---|---|
| | Seq SS | Adj SS | P% | Seq SS | Adj SS | P% | Seq SS | Adj SS | P% |
| Melting Temperature (A) | 0.0374 | 0.0374 | 15.5 | 0.05863 | 0.05863 | 35.84 | 0.261 | 0.261 | 31.84 |
| Packing Pressure(B) | 0.0073 | 0.0073 | 3.04 | 0.00873 | 0.00873 | 5.34 | 0.1628 | 0.1628 | 19.86 |
| Screw Speed(C) | 0.0073 | 0.0073 | 3.04 | 0.00873 | 0.00873 | 5.34 | 0.1628 | 0.1628 | 19.86 |
| Filled Times (D) | 0.1891 | 0.1891 | 78.42 | 0.08751 | 0.08751 | 53.48 | 0.2332 | 0.2332 | 28.45 |
| Total | 0.2412 | | 100 | 0.1636 | | 100 | 0.8198 | | 100 |

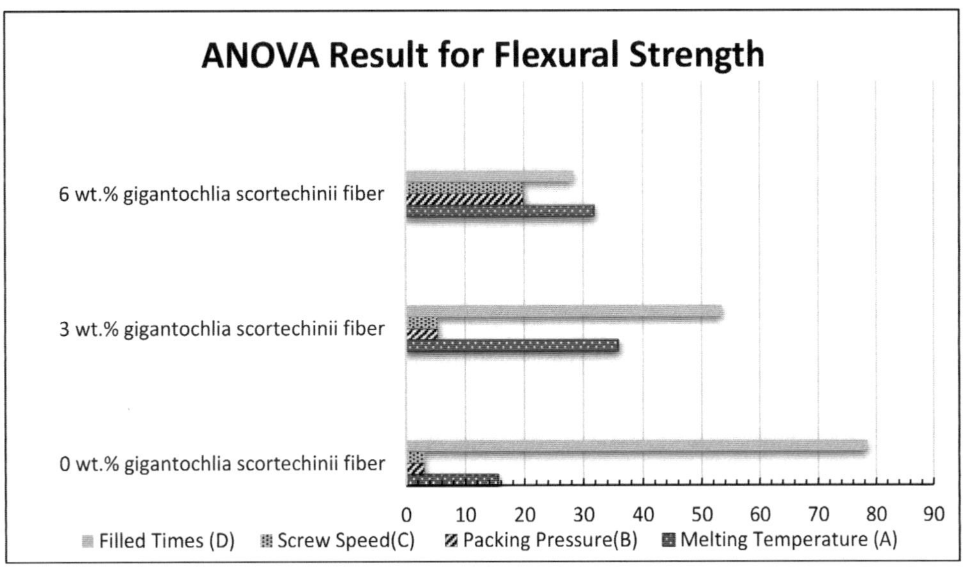


**Figure 3.** The graph of the flexural strength for all wt.% of GS.

According to Table 3, In the result of 0 wt.% of *Gigantochlia Scortechinii fiber*, the largest percent contribution was (D) filling time to 78.42% and followed by (A) melting temperature to 15.5%. The lowest percent was (C) screw speed and (B) packing pressure was 3.04%. At 3%wt.% of *Gigantochlia Scortechinii* fiber, the highest filling time setting (D) of 53.48%, the lowest setting was packing pressure (B ) and filled time (C) of 5.34% for both parameters. Other parameters representing the melting temperature (A) at 35.84% were the most important to effect the flexural strength of the sample.The highest factor of 6 wt% sample of the flexural strength result is the melting temperature (A) followed by the Screw speed (C) 28.45%, filling time setting (D) at 31.84% and the the lowest percent contribution are packing pressure setting (B) at 19.86%. As regards the percentage of the value, it appears that all injection molding processing conditions have contributed to the findings of this research. According to Elsabbagh, *et al.* [13]the collective of the holding pressure of injection molding could improve the composites mechanical properties. This was in line for to the resulting crystalline the molecular chains in the polymer and the orientation of the fibers. In contrast, the increase in temperature in injection and filling times during the processing contributed to strength less of the mechanical performance in composite.

*3.2 Shrinkage*

Table 4 shows the ANOVA of shrinkage result for all *Gigantochlia Scortechinii fiber*s by wt%. The calculation is based on the result of the S/N ratio.

**Table 4.** The shrinkage results of ANOVA for 0wt.% , 3wt.% and 6wt.% of *Gigantochlia Scortechinii* fiber.

| Parameter | Shrinkage 0.wt% *Gigantochlia Scortechinii* fiber | | | Shrinkage 3.wt% *Gigantochlia Scortechinii* fiber | | | Shrinkage 6.wt% *Gigantochlia Scortechinii* fiber | | |
|---|---|---|---|---|---|---|---|---|---|
| | Seq SS | Adj SS | P% | Seq SS | Adj SS | P% | Seq SS | Adj SS | P% |
| Melting Temperature (A) | 0.2109 | 0.2109 | 6.19 | 2.934 | 2.934 | 8.08 | 5.8958 | 5.8958 | 55.1 |
| Packing Pressure(B) | 0.8779 | 0.8779 | 25.72 | 6.995 | 6.995 | 19.25 | 0.8611 | 0.8611 | 8.03 |
| Screw Speed(C) | 0.8779 | 0.8779 | 25.72 | 6.995 | 6.995 | 19.25 | 0.8610 | 0.8610 | 8.03 |
| Filled Times (D) | 1.4464 | 1.4464 | 42.37 | 19.411 | 19.411 | 53.42 | 3.1011 | 3.1011 | 28.93 |
| Overall | 3.4132 | | 100 | 36.335 | | 100 | 10.719 | | 100 |

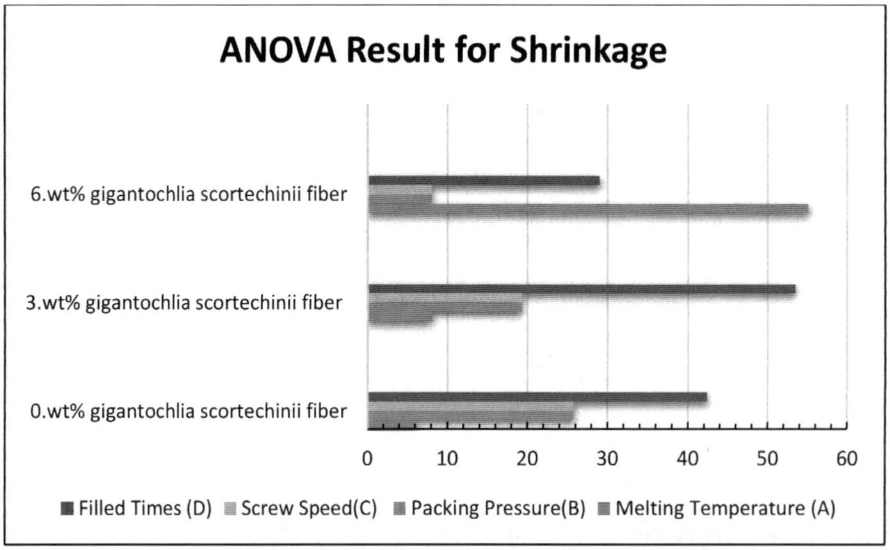


**Figure 4.** The graph of the shrinkage for all wt.% of GS.

Table 4 presents the result of the shrinkage for all wt.% of *Gigantochlia Scortechinii* fiber. At 0 wt%, the highest percentage are (D) filling times by 42.37%, whereas the (A) melt temperature at 6.19%. Other parameter appears to contribute moderately to these percent values for (C) screw speed and (B) packing pressure by 25.72%. At 3 wt% *Gigantochlia Scortechinii* fiber, the percent contribution for injection molding parameters indicates the (D) filling times the highest percent contribution of 53.42%. The (A) melt temperature was the lowest figure at 8.08% and for packing pressure (B) and screw speed (C) appear to be a moderate contributor was 19.25%.Then for the resulting sample 6 wt% *Gigantochlia Scortechinii* fiber, (A) melt temperature is the highest percentage contribution at 55.1% and followed by (D) filling times at 28.93%. The lowest value is 8.03% of (C) screw seed and (B) packing pressure was the lowest at. As regards the percent contribution for all parameters are interrelated with each other in the injection molding process. From the hilmi *et al* [7] was agreeing the factor of melt temperature is rather less in effect on the shrinkage. Nevertheless, the researcher Altan *et.al* specified that the melt temperature was considered more significant than the holding pressure to minimizing the defect of shrinkage [14]. According to the researcher, to get the optimum result for the shrinkage is difficult to discover, this is because it depends on the types and the composition of materials has been used.

*3.3 Warpage*

Table 5 shows the warpage ANOVA result for all wt.% of *Gigantochlia Scortechinii* fiber. The calculation is based on the result of the S/N ratio.

**Table 5.** ANOVA of Warpage results for 0,3 and 6 wt.% of Gigantochlia Scortechinii fiber.

| Parameter | Warpage 0.wt% *Gigantochlia Scortechinii* fiber | | | Warpage 3.wt% *Gigantochlia Scortechinii* fiber | | | Warpage 6.wt% *Gigantochlia Scortechinii* fiber | | |
|---|---|---|---|---|---|---|---|---|---|
| | Seq SS | Adj SS | P% | Seq SS | Adj SS | P% | Seq SS | Adj SS | P% |
| Melt Temperature (A) | 173.14 | 173.14 | 72.21 | 5.195 | 5.195 | 39.09 | 1.9330 | 1.9330 | 10.67 |
| Packing Pressure (B) | 12.79 | 12.79 | 5.33 | 2.766 | 2.766 | 20.81 | 8.0540 | 8.0540 | 44.44 |
| Screw Speed (C ) | 12.79 | 12.79 | 5.33 | 2.766 | 2.766 | 20.81 | 8.0540 | 8.0540 | 44.44 |
| Filling Times (D) | 41.06 | 41.06 | 17.13 | 2.564 | 2.564 | 19.29 | 0.0810 | 0.0810 | 00.45 |
| Total | 239.78 | | 100 | 13.291 | | 100 | 18.122 | | 100 |

In accordance with table 5, at 0 wt% *Gigantochlia Scortechinii* fiber, the (A) melt temperature parameter contributes to 72.21% as the highest percentage, while the (C) screw speed and (B) packing pressure were the bottommost at 5.33%. The (D) filling time appears to contribute moderately at 17.13% intakes. The result at 3 wt% of *Gigantochlia Scortechinii* fiber indicates that the uppermost percentage with 39.09% was the (A) melt temperature and the lowest is (D) filling time at 19.29%. The parameter of (C) screw speed and (B) packing pressure appears to make a moderate contribution at 20.81%.The last is 6 wt% of *Gigantochlia Scortechinii* fiber, the highest percentage was (C) screw speed and (B) packing pressure at 44.44%, the (D) filling time was the lowest at 0.45% and the (A) melt temperature is the amongst at 10.67%. Based on the result, 6 wt% of *Gigantochlia Scortechinii* fiber notes that the value of packing pressure for warpage also has the highest in the result of ANOVA. According to researcher oktem.H in his finding was stated the packing pressure is the most important parameter need to control to minimize the warpage.

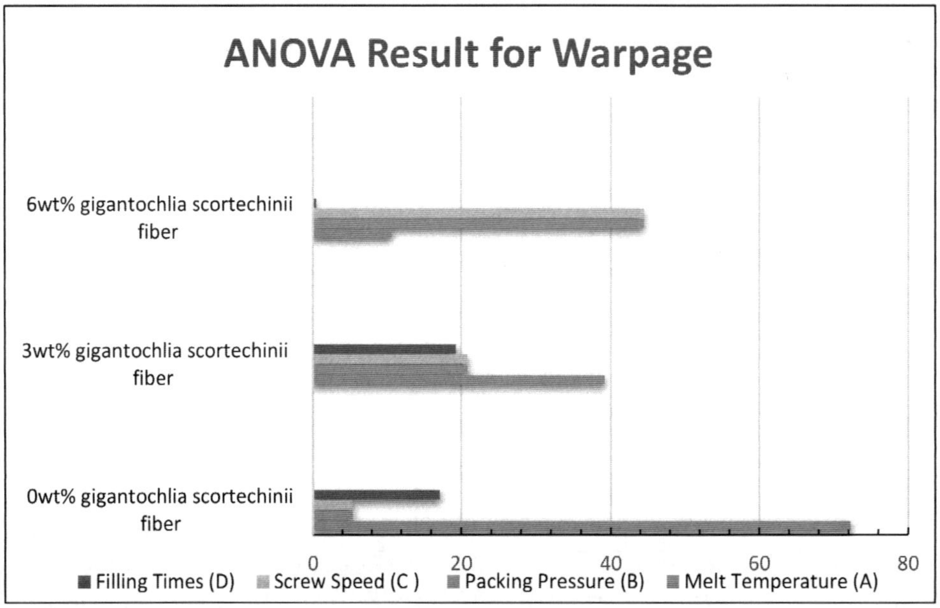


**Figure 5.**The graph of the warpage for all wt.% of GS.

## 4. Conclusion

ANOVA was applied to find deviations in the middling performance of each material tested. Consequently, the melting temperature is the most significant settings for 0% wt.% of *Gigantochlia Scortechinii* fiber on the minimum the warpage. Whereas for flexural strength and shrinkage is filling time. Next, for 3 wt% of *Gigantochlia Scortechinii* fiber shows the melt temperature parameter is the significant factor. While for flexural strength and shrinkage is filling time. For 6 wt.% of *Gigantochlia Scortechinii* fiber, the most important parameters for warpage were packing pressure and screw speed. While, for the flexural strength and shrinkage is melting temperature. Based on hilmi *et al.,* The major cause of deformation is commonly known as the variance of shrinkage to the injection process in the case of thin walled parts of plastic. The properties of material, the geometry of the parts and the conditions of the injection molding process are classified as factors influencing variations in the withdrawal of parts during injection molding.The ANOVA analysis, with the highest percentage of the quality endpoint, provided the best sample. In a conclusive way, the objective if this study were achieved for the parameters that influence flexural strength, shrinkage and warpage. Other than that, to make the effect of the increasing in the quality of bamboo fiber (*Gigantochlia Scortechinii* fiber) in the composition also be accomplished by changing the optimal setting with respected to the parameters that result from the combination of the best setting.After completing this experiment, valuable experiences from different aspects are acquired. These have been field tested on the injection moulding machine and also include the running cycle of the injection moulding machine. This will provide better information on how to use other machines that have not been used.

**Acknowledgments**
This research was funded by the Ministry of Higher Malaysia through FRGS (Fundamental Research Grant Scheme).

**References**

[1] S. Kashyap and D. Datta, "Process parameter optimization of plastic injection molding: a review," *Int. J. Plast. Technol.*, vol. 19, no. 1, pp. 1–18, 2015.

[2] D. Kumar, "with Taguchi Approach - A Review Plastic Injection Molding with Taguchi Approach - A Review Kalpit Jain," pp. 1–4, 2016.

[3] J. O. Osarenmwinda and D. D. Olodu, "Optimization of injection molding process parameters in the molding of High Density Polyethylene (HDPE)," *J. Appl. Sci. Environ. Manag.*, vol. 22, no. 2, p. 203, 2018, doi: 10.4314/jasem.v22i2.8.

[4] E. Farotti and M. Natalini, "Injection molding. Influence of process parameters on mechanical properties of polypropylene polymer. A first study.," *Procedia Struct. Integr.*, vol. 8, pp. 256–264, 2018, doi: 10.1016/j.prostr.2017.12.027.

[5] O. Mohd Hilmi, S. Hasan, W. N. A. W. Muhammad, and Z. Zuramashana, "Optimising injection molding parameter setting in processing polypropylene-clay composites through Taguchi method," in *Applied Mechanics and Materials*, 2013, vol. 271, pp. 272–276.

[6] M. H. Othman, S. Shamsudin, and S. Hasan, "The Effects of Injection Molding Processing Parameters and Mould Gate Size towards Weld Line Strength," *Adv. Mech. Mater.*, vol. 229–231, pp. 2536–2540, 2012.

[7] W. N. a. Wan Muhammad, M. H. Othman, S. Hasan, and M. F. Ruslee, "A Study of Injection Molding Optimum Parameters to Control Shrinkage and Warpage of Polypropylene-Nanoclay Hinges Samples," *Appl. Mech. Mater.*, vol. 465–466, pp. 1011–1015, 2014, doi: 10.4028/www.scientific.net/AMM.465-466.1011.

[8] M. H. Othman, "Polypropylene-Nanoclay Nanocomposites: The Optimisation of Parameter Settings fiber Towards Quality of Snap Fit Samples," 2014.

[9] Y. Zare, "Recent progress on preparation and properties of nanocomposites from recycled polymers: A review," *Waste Manag.*, vol. 33, no. 3, pp. 598–604, 2013, doi: 10.1016/j.wasman.2012.07.031.

[10] V. Sessini, M. P. Arrieta, J. M. Raquez, P. Dubois, J. M. Kenny, and L. Peponi, "Thermal and composting degradation of EVA/Thermoplastic starch blends and their nanocomposites," *Polym. Degrad. Stab.*, vol. 159, pp. 184–198, 2019, doi: 10.1016/j.polymdegradstab.2018.11.025.

[11] S. Z. Khamis, M. H. Othman, S. Hasan, and M. H. I. Ibrahim, "Characterizetion of Flexural Strength, Warpage and Shrinkage of Polypropylene-Nanoclay-Nanocomposites Blend with Gigantochloa Scortechinii.," in *IOP Conference Series: Materials Science and Engineering*, Aug. 2017, vol. 226, no. 1, doi: 10.1088/1757-899X/226/1/012163.

[12] S. Z. Khamis *et al.*, "Multiple Responses Optimisation in Injection Molding Parameter for Polypropylene-Nanoclay-Gigantochloa Scortechinii via Taguchi Method.," in *Journal of Physics: Conference Series*, Mar. 2019, vol. 1150, no. 1, doi: 10.1088/1742-6596/1150/1/012062.

[13] A. Elsabbagh, A. Ramzy, L. Steuernagel, and G. Ziegmann, "Models of flow behaviour and fibre distribution of injected moulded polypropylene reinforced with natural fibre composites," *Compos. Part B Eng.*, vol. 162, no. April 2018, pp. 198–205, 2019, doi: 10.1016/j.compositesb.2018.10.105.

[14] M. Altan, "Reducing shrinkage in injection moldings fiber via the Taguchi, ANOVA and neural network methods," *Mater. Des.*, vol. 31, no. 1, pp. 599–604, 2010, doi: 10.1016/j.matdes.2009.06.049.

# A review on bonding of Polyamide reinforced carbon fibre via additive manufacturing

**Nisa Naima Khalid[1,*], Nabilah Afiqah Mohd Radzuan[1], Abu Bakar Sulong[1] and Farhana Mohd Foudzi[1]**

[1] Faculty of Engineering and Built Environment, Universiti Kebangsaan Malaysia, 43600 UKM Bangi, Selangor, Malaysia.

*Corresponding author's email: P108540@siswa.ukm.edu.my

**Abstract.** Continuous carbon fibre offers a huge potential which is a lightweight, high strength, low density, a low coefficient of thermal expansion, and high thermal conductivity and stiffness. In addition, polyamide-6 (PA6) has good mechanical and physical properties thus, it makes PA6 as effectiveness between matrix and reinforcement of continuous carbon fibre. 3D printing as a cost effective in additive manufacturing was covered in a complex part building using CAD software based on CAD model makes fisible finish the elements. The main objectives of this paper are to review on interlayer bonding of polyamide reinforced carbon fibre by using 3D printing method on their mechanical properties. This paper scope was a review on the bonding of polymeric composites which is polyamide reinforced carbon fibre as the material used and their printing parameter which is important for their orientation during printing. The printing parameter in 3D printing was set to vary the quality and mechanical behavior of the sample. The sample was reviewed on their bead orientation consist of 0°, 45°, and 90° directional angle of the sample in form of their bonding. Mechanical properties of the samples were tests using tensile test and resulted at 0° it achieves higher tensile strength and modulus when tensile loading direction is parallel to deposition direction. At 90° direction, the fibre was perpendicular with the tensile loading direction and resulted in a decreased tensile property.

**Keywords:** Continuous Carbon Fibre; 3D Printing; Bonding; Orientation.

## 1. Introduction

The complex part building, cost-effectiveness, and sustainability have shown that additive manufacturing (AM) technologies have a huge potential. Subsequently, AM gets quite possibly in most encouraging fields in manufacturing components and gains considerable attention. AM technologies have numerous methods [1]. One of the AM technologies is layer manufacturing or commonly referred to as 3D printing in additive manufacturing is a very dynamically developing direction in manufacturing technology. The article presents considerations on the possibilities of using 3D printing in the development of manufacturing technology by using a polymer composite [2]. Carbon fibre strengthened polymer composites (CFRP) offer huge preferences over metals in which it gives a lightweight, high strength and stiffness, and they offer to resists corrosion and fatigue. Thermosetting epoxy matrices which are widely used in CFRPs have a stiffness to weight ratio and high in strength which can be used in aviation applications. Surface treatment in carbon fibre and sizing technologies in materials have been created aviation in the pass of years which resulted in achieving high interfacial fibre-matrix bond strength and favorable mechanical properties [3].

The superiority of continuous carbon fibre can be fabricated using these printing methods such as 3D printed of continuous fibre reinforced parts awake researchers' attention. Traditional manufacturing uses the same method which needs a large number of printing parameters to be all considered in AM process. Impact of printing parameters of 3D-printing composites using continuous carbon fibres on mechanical properties was studied. In addition, the literature on polyamide (PA) based composites reinforced with continuous carbon fibres in 3D printing on mechanical properties was also explored. The mechanical behaviors of 3D printed parts with various build orientations which are horizontal and vertical were additionally evaluated [4]. It shows with the use of 3D printing makes it feasible to finish the elements. It has using ready-made metal elements. The specific features in 3D printing which have been created by CAM software based in CAD model have been brought out to adapts to the capabilities of this method [2].

3D printing technology in fused deposition modeling (FDM) which is widely used for utilized low-cost was utilized adhesive properties of thermoplastic materials and hot melt. Polyamide (PA) in engineering thermoplastic polymer materials is important because it has an excellent comprehensive performance. FDM technique is used in 3D printing as it exponentially evolved manufacturing technology covering a process of joining materials to make objects from 3D model data, usually layer-by-layer, which has received significant attention in recent years by offering numerous advantages over traditional subtractive methods since it was first described in 1986 by Charles Hull [3]. The 3DP technology provides the ability to directly create objects with intricate geometrical features in a cost-effective way, which requires no mold tool and offers near-net-shape manufacturing in a relatively short period of time[5]. Moreover, the 3DP is generally advantageous in products and fabrication of customized parts with custom-fitted design and complex while being equipped for harnessing digital information for the realization of a robust and decentralized 3D manufacturing system. A major factor that will limit the performance of PACF in 3D printing is interlayer delamination and is exacerbated owing to the introduction of continuous fibres in this process [6].

## 2. Polymer composite feeler

In engineering thermoplastic, polyamide 6 (PA6) has mostly used in the automotive industry, electronic components, and mechanical parts [7]. In recent studies, the crystalline polyamide 6 (PA6) was modified by introducing amorphous polymer to prevent severe warpage so that it could be applied to FDM technology. Polyamide-6 (PA6) or can be called polycaprolactam have good mechanical and physical properties with it is a biodegradable and synthetic polymeric substance [8]. The effectiveness between matrix and reinforcement of continuous carbon fibre (CCF) reinforced polyamide was chosen in this study due to their superior mechanical properties. From previous studies, CCF tows were not a pure cluster of continuous carbon fibres. Specifically, CCF tows were found in continuous carbon fibre reinforced polyamide, contains 48 wt % pure continuous carbon fibres. Also, 3D printed were manufacture based on CCF with the temperature of the fibre and matrix nozzle was 270°C, while the printing speeds of the continuous carbon fibre reinforced polyamide (CCFRPA) was 15.8 mm/s. Besides that, the average CCF width was 1.2 mm [1].

### *2.1. Design compaction of PACF using CAD software*

The process starts with a 3D model created with CAD (Computer Aided-Design) *programs* [9]. The designed part is exported to STL (Surface Tessellation Language) format. The sliced model was sent to a 3D-Printer and created as a physical part as Figure 1 below.

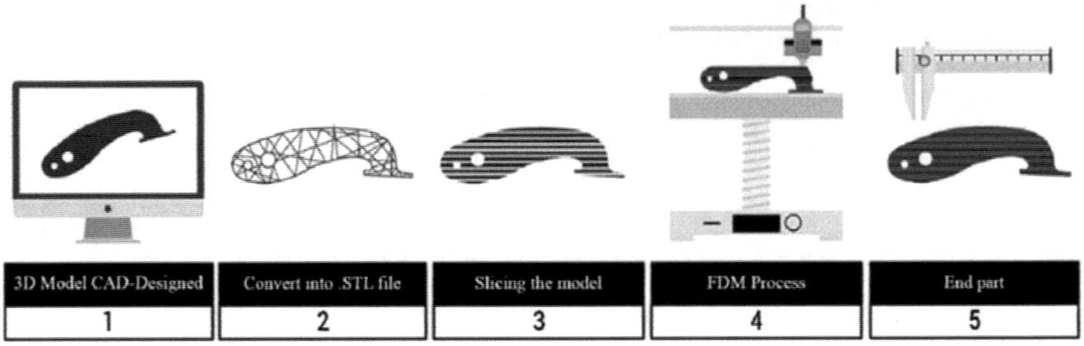


**Figure 1.** Schematic representation of the 3D printing process [10].

Fused Deposition Modeling (FDM) sometimes also called 3D printing technologies were used as a method in this study. A wide range of polymeric materials using this technology can be produced. Currently, fibre reinforced thermoplastic composites have become more important in this technique. The technologies consist of driving polymers of low melting point into a semi-liquid state and extruding them in a controlled way through the nozzle until the desired layers are deposited. Various printing parameters which includes of the height of the layer, printing pattern, placement of the part in the build plate, nozzle diameter, and infill density can be set in order to vary the quality, appearance, and mechanical behavior of the sample [10]. Figure 2 shows rectangular plates were produces in the xy plane which contains three printing architectures that differ in terms of bead orientation within the stacked layers. The samples were fabricated according to the printing parameters [11].

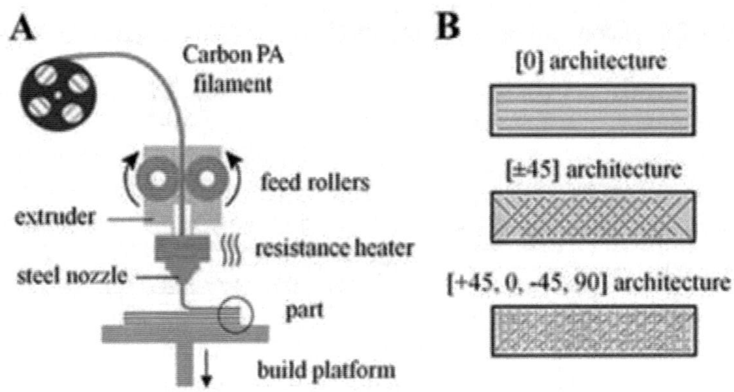


**Figure 2.** (a) Conceptual sketch of the FDM manufacturing process and (b) schematic representation of 3D-printed specimens with different printing architectures [6].

## 3. Characterization

Mechanical test of FDM-printed specimens on the tensile property was tested by using GT-7001-HC6, GOTECH TESTING MACHINE INC, Taiwan China at speed 2mm/min. A further test was observed on fracture surface by field-emission scanning electron microscope FESEM, QUANTA 450, FEI. Test specimens dimension of 250 × 15 × 1 mm (length × width × thickness) for the tensile test was followed ISO 527-5 [12]. In addition, the re-entrant honeycomb structure is subsequently manufactured, and the in-plane compression test is conducted at a constant strain rate of 2 mm/min [13].

**Table 1.** FDM printing parameter for PACF specimens [13].

| Parameter | Value |
|---|---|
| Printing speed | 30 mm/s |
| Infill density | 100% |
| Nozzle diameter | 0.4 mm |
| Printing raster angle | 0°, 45°/-45°, 90° |
| Layer thickness | 0.15 mm |
| Nozzle temperature | 260 °C |
| Environment temperature | 22 °C |
| Built-plate temperature | 30°C, 55°C, 80°C, 105°C |

Since fiber orientation is directly related to the molten polymer of polyamide fluid during the printing, three-directional angles consist of 0°, 45°/-45°, and 90° are adopted as illustrated in Figure 3. Figure 3 shows the specimen was been produced according to the printing parameter of Table 1 and the build-plate temperature was 105 ◦C. It shows all three kinds of specimens have exhibit a flat surface. Less material agglomerates were observed due to a tendency of fewer flow rate fluctuations, and carbon fibre is well infiltrated in PA6 matrix. Along the printing direction, fibres tend to be distributed to the orientation, which may influence the tensile property. To acquire good mechanical properties in PACF composites, the effect of deposition path during 3D printing on tensile strength was investigated [13].

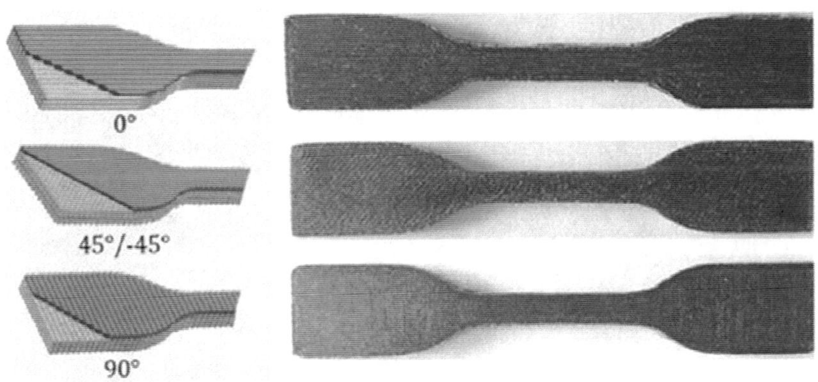


**Figure 3.** PACF specimen with different directional angles [13].

The interface adhesion between carbon fibre and polyamide was improved, which are beneficial for stress transfer from the matrix to fiber. In addition, a tensile strength of fiber-reinforced thermoplastics is also connected to distribution of fiber orientations [14]. When the tensile loading direction is parallel to the deposition direction, fibers can bear more loadings, achieving higher tensile strength and modulus [13]. When the printing direction along 90◦, although fibers are still oriented along printing direction, they are perpendicular with tensile loading direction, which fails in carrying a load and thus result in decreased tensile property.

Polyamide-6 was characterized by Fourier-transform infrared spectroscopy (FTIR). Fourier-transform infrared (FTIR) spectroscopy is the most well-knowns methodology currently used for the identification and quantification of microplastics. Focal plane array–based µ-FTIR imaging gave a considerable decrease in analysis time and rise the accuracy of the results by impartial measurement of the entire filter surfaces [15]. FTIR was resulted on PA6 have a weak peak at 3060 $cm^{-1}$ on which represents a stretching vibration of crystalline hydrogen-bonded -N-H [16]. Flexural behaviors of 3DP-CCFRPA with different raster angles and stacking sequences, in different loading directions (parallel or perpendicular to the thickness directions of specimens) or namely as horizontal and vertical were reviews. Note that 3DP-CCFRPA had three mutually perpendicular planes of symmetry. The intersection of these planes defines three axes that coincide with the 0° continuous fibre raster direction (X axis), the thickness direction (Z axis), and the width direction (Y axis) perpendicular to the other twos [17]. 0° angle exhibit a higher flexural strength but compared to 45° and 90°. At 90° angle, flexural strength has lower flexural strength and lowest energy absorption among the other specimen [1].

## 4. Conclusion

In conclusion, the mechanical properties of polyamide reinforced carbon fibre (PACF) composite used in 3D printing method were review. Interlayer bonding on the specimen was influence by the printing parameter through the manufacturing process which affected their mechanical properties on the sample. Orientation and interlayer bonding are found to own a close relationship with the tensile properties of interlayer adhesion. In orientation, continuous fibers with 0° raster angle could effectively improve the bear-loading capability of composites. Besides, 0° to have a higher tensile strength according to their tensile property parallel to deposition direction. The flexural strength at 0° angle also shown have higher energy absorption among the other angles. Can be concluded, the bonding of polyamide reinforced carbon fibre at 0° angle have higher mechanical properties among the other raster angle of 45° and 90°.

## Acknowledgments

The authors wish to gratefully thank and acknowledge the Center for Research and Instrumentation Management (CRIM), University Kebangsaan Malaysia for their financial support to complete this study under grant number GGPM-2020-002.

## References

[1] Y. Peng, Y. Wu, S. Li, K. Wang, S. Yao, and Z. Liu, "Tailorable rigidity and energy-absorption capability of 3D printed continuous carbon fiber reinforced polyamide composites," *Compos. Sci. Technol.*, vol. 199, no. February, p. 108337, 2020, doi: 10.1016/j.compscitech.2020.108337.

[2] L. Moldovan, H. Fuwen, I. Ciobanu, and I. Ciobanu, "ScienceDirect ScienceDirect ScienceDirect Costing models for Tomasz capacity optimization in Industry CAM System System for for Manufacturing Manufacturing Innovative Innovative Hybrid Hybrid Design Design Using Using 3D Printing between used capacity and," *Procedia Manuf.*, vol. 32, pp. 22–28, 2019, doi: 10.1016/j.promfg.2019.02.178.

[3] N. van de Werken, H. Tekinalp, P. Khanbolouki, S. Ozcan, A. Williams, and M. Tehrani, "Additively manufactured carbon fiber-reinforced composites: State of the art and perspective," *Addit. Manuf.*, vol. 31, p. 100962, 2020, doi: 10.1016/j.addma.2019.100962.

[4] K. Raney, E. Lani, and D. K. Kalla, "ScienceDirect Experimental characterization of the tensile strength of ABS parts manufactured by fused deposition modeling process," *Mater. Today Proc.*, vol. 4, no. 8, pp. 7956–7961, 2017, doi: 10.1016/j.matpr.2017.07.132.

[5] S. C. Daminabo, S. Goel, S. A. Grammatikos, H. Y. Nezhad, and V. K. Thakur, "Fused deposition modeling-based additive manufacturing (3D printing): techniques for polymer material systems," *Mater. Today Chem.*, vol. 16, p. 100248, 2020, doi: 10.1016/j.mtchem.2020.100248.

[6] L. Meng, T. Xiaoyong, S. Junfan, Z. Weijun, L. Dichen, and Q. Yingjie, "Impregnation and interlayer bonding behaviours of 3D-printed continuous carbon- fi ber-reinforced poly-ether-ether-ketone composites," vol. 121, no. February, pp. 130–138, 2019, doi: 10.1016/j.compositesa.2019.03.020.

[7] Y. Ma *et al.*, "Enhanced interfacial properties of carbon fiber reinforced polyamide 6 composites by grafting graphene oxide onto fiber surface," *Appl. Surf. Sci.*, vol. 452, pp. 286–298, 2018, doi: 10.1016/j.apsusc.2018.04.274.

[8] B. M. Razavizadeh and R. Niazmand, "Characterization of polyamide-6/ propolis blended electrospun fibers," *Heliyon*, vol. 6, no. 8, p. e04784, 2020, doi: 10.1016/j.heliyon.2020.e04784.

[9] T. Yao, J. Ye, Z. Deng, K. Zhang, Y. Ma, and H. Ouyang, "Tensile failure strength and separation angle of FDM 3D printing PLA material: Experimental and theoretical analyses," *Compos. Part B Eng.*, vol. 188, no. January, p. 107894, 2020, doi: 10.1016/j.compositesb.2020.107894.

[10] E. Verdejo, D. Toro, J. Coello, A. Matínez, E. Técnica, and S. De Ingenieros, "ScienceDirect ScienceDirect ScienceDirect Analysis of the influence of the variables of the Fused Deposition Modeling ( FDM ) process on the variables mechanical of a carbon Analysis of the influence of the of properties the Fused Deposition polyamide Mod," *Procedia Manuf.*, vol. 41, pp. 731–738, 2020, doi: 10.1016/j.promfg.2019.09.064.

[11] F. Lupone, E. Padovano, A. Veca, L. Franceschetti, and C. Badini, "Innovative processing route combining fused deposition modelling and laser writing for the manufacturing of multifunctional polyamide/carbon fiber composites," *Mater. Des.*, vol. 193, p. 108869, 2020, doi: 10.1016/j.matdes.2020.108869.

[12] M. Ueda, S. Kishimoto, M. Yamawaki, R. Matsuzaki, and A. Todoroki, "3D compaction printing of a continuous carbon fi ber reinforced thermoplastic," *Compos. Part A*, vol. 137, no. April, p. 105985, 2020, doi: 10.1016/j.compositesa.2020.105985.

[13] X. Peng, M. Zhang, Z. Guo, L. Sang, and W. Hou, "Investigation of processing parameters on tensile performance for FDM-printed carbon fiber reinforced polyamide 6 composites," *Compos. Commun.*, vol. 22, no. August, p. 100478, 2020, doi: 10.1016/j.coco.2020.100478.

[14] J. M. O. Rourke And V. D. Scott, "Evaluation Of Al-Cu-Mg-Ag Alloys Containing Discontinuous Alumina Fibres : I . Microstructural Evolution," Vol. 56, 1996.

[15] Y. Chen *et al.*, "Identification and quantification of microplastics using Fourier-transform infrared spectroscopy: Current status and future prospects," *Curr. Opin. Environ. Sci. Heal.*, vol. 18, pp. 14–19, 2020, doi: 10.1016/j.coesh.2020.05.004.

[16] R. J. Tapper, M. L. Longana, I. Hamerton, and K. D. Potter, "A closed-loop recycling process for discontinuous carbon fibre polyamide 6 composites," *Compos. Part B Eng.*, vol. 179, p. 107418, 2019, doi: 10.1016/j.compositesb.2019.107418.

[17] J. S. Shim, J. Kim, H. Jeong, and J. Choi, "Printing accuracy , mechanical properties , surface characteristics , and microbial adhesion of 3D-printed resins with various printing orientations," *J. Prosthet. Dent.*, pp. 1–8, doi: 10.1016/j.prosdent.2019.05.034.

# Acoustic properties of different blending ratios of empty fruit bunch and oil palm frond

**L S Ewe[1,*], Mohammad Nazhan Nasir[1], W K Yew[2], R Mageswaran[1] and Zawawi Ibrahim[3]**

[1] College of Engineering, Universiti Tenaga Nasional (UNITEN), Putrajaya Campus, Jalan IKRAM-UNITEN, 43000 Kajang, Selangor, Malaysia
[2] School of Engineering and Physical Science, Heriot-Watt University Malaysia, No 1, Jalan Venna P5/2, Precint 5, 62200, Putrajaya, Malaysia.
[3]Engineering and Processing Division, Malaysian Palm Oil Board (MPOB), No. 6, Persiaran Institusi, Bandar Baru Bangi, Kajang 43000, Selangor, Malaysia

*Corresponding author's email: laysheng@uniten.edu.my

**Abstract.** The society understands the importance of using sound-absorbing panels to protect our health against the noise pollution. However, the repercussions of fibreglass as sound-absorbing panels on human health and the environment have increased year by year. Natural fibers are more environmentally friendly and have significantly lower production cost. This research emphasizes the acoustic properties of different blending ratios of Empty Fruit Bunch (EFB) and Oil Palm Frond (OPF at four different thickness. Their acoustic properties were tested by using the Impedance Tube Method (ITM) from frequency range of 0 - 6400 Hz. The findings show the Sound Absorption Coefficient (SAC) value increases as the frequency increases for all samples. Both blending ratios have good SAC results. For blending ratio of 30% EFB – 70% OPF, the SAC of the panels achieved 80% and above at frequency between 3500 - 6400 Hz for samples with thickness of 14 mm, 16 mm and 18 mm. However, for blending ratio of 20% EFB – 80% OPF, the SAC of all samples are found 0.8 and above at frequency between 3500 - 6400 Hz. Notably, the SAC values for both blending ratio are increased with increasing in thickness. It might be attributed to the effect of tortuosity. The SAC values for all samples are also found to achieve unity ( ~ 0.96 ) at different frequency range. Furthermore, the frequency range for SAC to reach 0.8 and above has been increased from 4500 – 6400 Hz to 3500 – 6400 Hz with an increased in OPF contents.

**Keywords**: Empty Fruit Bunch (EFB), Oil Palm Frond (OPF), Sound Absorption Coefficient (SAC), Blending Ratio.

## 1. Introduction

It has already been more than a decade, that we are facing noise pollution and the harmful consequences of using fibre glass acoustic panels. These noise pollutions are getting out of hand day by day. However, due to the ignorance of human beings, this noise pollution is treated lightly [1]. Based on the statistics done by the Occupational Safety and Health Administration (OSHA), 85 dB is maximum limit of sound that humans able withstand, anything higher than that is injurious to health [2]. Well, one of the precaution methods that could reduce noise pollution is to follow the limit of noise level, control noise level near sensitive area and etc. However certain matters are uncontrollable such as construction sites, transportation (airports and railways), etc. Thus, the best solution is put in sound-absorbing panels to absorb unwanted noise [3]. Ordinarily these sound-absorbing panels are used in cinema theatres, recording studios, offices, meeting rooms, factory and etc.

The common selections of sound-absorbing panel used in the market are the fiberglass, cotton, mineral fiber, polyester, etc. However, fiberglass as a sound absorbing panel pose severe health threat towards humans due to the shedding of fiberglass [4]. Scientist and researcher found lung tumour in an animal which was triggered by fiberglass, the same material used in making of sound-absorbing panels. In addition, fiberglass can also potentially cause lung infection, irritation to skin, eyes and also pulmonary diseases [5] [6]. This information was shared by the US government when a research was shown on fiberglass. At the same time, the industrial companies that produce this synthetic fiberglass also pollute the air and water with their waste products.

Researcher and scientist have strived forward for an environmentally friendly solution which uses natural fiber as sound-absorbing material and to test their acoustic properties. Natural fiber has more advantageous properties compared to synthetic fibers. They are renewable, biodegradable, and most important of all, it is safe to humans and surrounding [7]. There are many other natural fibers that have been tested for their acoustic properties such as bamboo fiber [8], pineapple [9], tea-leaf [10] etc. Till today oil palm trunk (OPT) has shown prominent acoustic result according to Kalaivani et. al [11]. OPT natural fiber was able to reach sound absorption coefficient (SAC) of 0.99, which meant that 99% of the sound is absorbed and 1% of the sound is reflected to surrounding as heat energy. This was achieved at a frequency range of 3000 - 6000 Hz for sample thickness of 12 mm [11]. Moreover, arenga pinnata and kenaf fiberboard had their acoustic properties tested out. The SAC results from arenga pinnata and kenaf were good as well [12] [13].

As can be seen from the previous research, it shows that natural fibers have a very promising acoustical property, therefore this research is captivating and engaging your attention. This research emphasizes on the acoustic properties of natural fibers of empty fruit bunch (EFB) and oil palm frond (OPF) at different thickness. With this research, we can solve the environmental pollution and concurrently would be able to replace the synthetic made fiberglass sound-absorbing panels with natural fiber sound-absorbing panels.

## 2. Methodology

Dry method process has been used to fabricate the low density fiberboard (LDF). The amounts of EFB and OPF natural fibers are required in preparing a total of 8 fiberboards have been computed. Urea Formaldehyde (UF) glue used as resin to blend the fibers. The mass of the natural fibers and moisture content are calculated using a pre programme excel sheet prepared by MPOB, where it can achieve the targeted blending ratio (30% EFB – 70% OPF, 20% EFB – 80% OPF), density ($\rho$ = 120 kg/m$^3$) and thickness (12 mm, 14 mm, 16 mm and 18 mm). The fiberboards were fabricated at Research Station MPOB / UKM located at Pekan Bangi Lama, Kajang.

The fabrication process involved chipping, refining, glue bending, mat-forming, pre-press, hot press and cool down as shown in Figure 1. For further interpretation can refer to our previous publications [3] [7] [14]. For Acoustic testing, the SAC value were obtained using the Bruel & Kjaer (B&K) Impedance Tube Type 4206 apparatus as shown in Figure 2. SAC was measured using 2 microphones, where the impedance tube attached to the analyser. Measurement of 2 microphones transfer function method is based on the international standards of ASTM E1050-2 for sound absorption coefficient The labview software was used to calculate the SAC.

## 3. Results and discussion

The acoustic properties of the materials is determined by analyzing its sound absorption coefficient, SAC ($\alpha$). The ratio of sound transmitted through a material and the incident sound that emitted from a material is the concept behind SAC ($\alpha$) [15]. For instance, when a material achieved SAC ($\alpha$) value of 0.98 at a certain frequency. It indicates that this material can absorb 98% of the noise and only 1% of the remaining noise is reflected to the surrounding as heat energy at this frequency. The SAC, ($\alpha$) is

IOP Conf. Series: Materials Science and Engineering 1078 (2021) 012005 doi:10.1088/1757-899X/1078/1/012005

obtained and measured from impedance tube method (ITM), which is following the ASTM E1050-95 (ISO 10534-2) test technique [16].

Figure 3 displays the SAC, $\alpha$ of four different thickness with the blending ratio of 30% EFB – 70% OPF. SAC, $\alpha$ values are found to rise as the frequency increases from 0 - 6400 Hz for all samples. It is notable that, at the lower region medium frequency (~1500 Hz), sample thickness 18 mm has the highest SAC, $\alpha$, which is 0.52 if compared to the samples with thickness of 12 mm, 14 mm, and 16 mm. Where the SAC, $\alpha$ values are ranging from 0.24 - 0.38. Above 4500 Hz, it is noticeable that the SAC for samples with thickness of 14 mm and 16 mm surpass the sample with thickness of 18 mm. The SAC for these two samples (14 mm and 16 mm) almost reached unity (≥ 0.98) above 4500 Hz too. While, sample with thickness of 14 mm reached unity ($\alpha$ = 1.00) at the frequency range of 5500 - 6000 Hz. Furthermore, the $\alpha$ for all samples, except for sample with thickness of 12 mm possessed the $\alpha$ of 0.8 and above from a frequency range of 3250 - 6400 Hz. Whereas, the sample thickness of 12 mm only achieved α more than 0.80 at frequency above 4750 Hz. On average, all the samples performed splendidly at high frequency (4000 – 6400 Hz) with maintaining more than 90% sound absorption rate except for sample with thickness of 12 mm. If compared with the research of Wood Fiber Board (WFB), at high frequency of 4000 Hz, the SAC value for WFB was only 0.68 while the average SAC for EFB and OPF mixture was 0.89 at high frequency range [17].

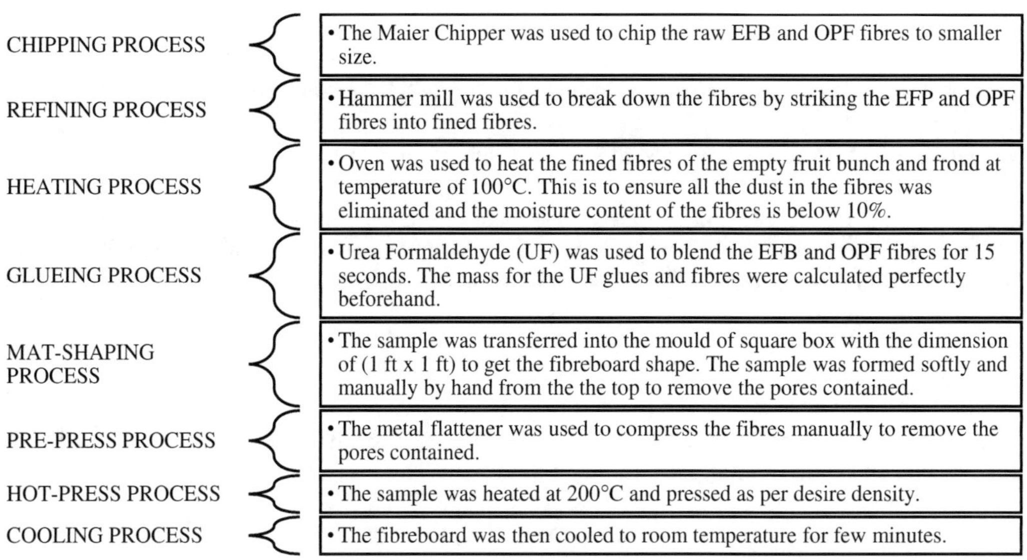


**Figure 1.** Low density fiberboard (LDF) fabrication process.

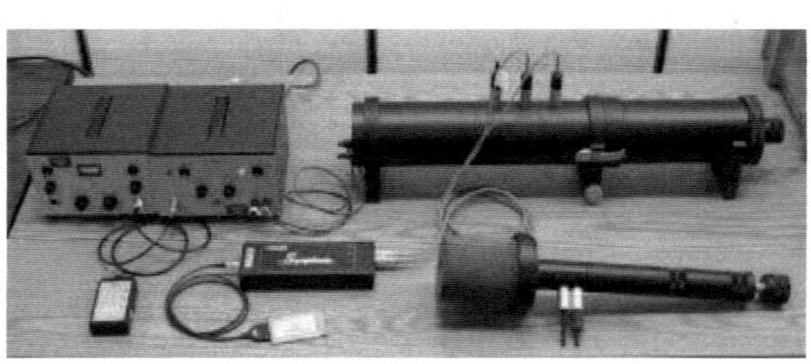

**Figure 2.** Setup of B&K Tube Type 4206.

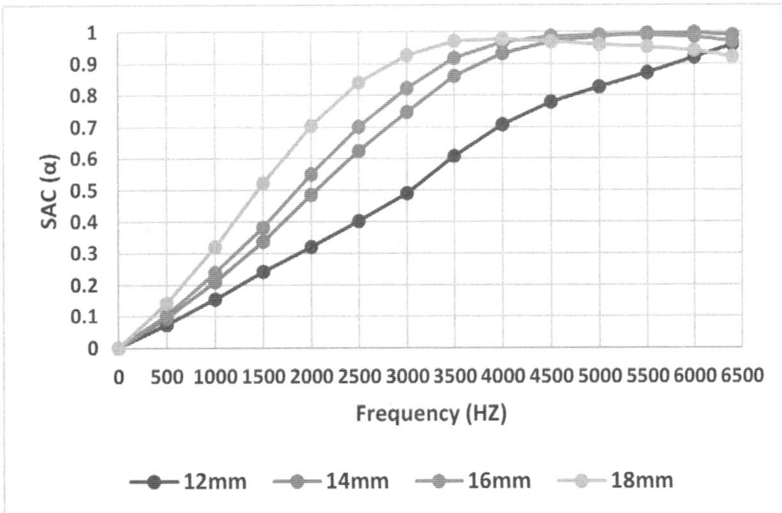


**Figure 3.** SAC, $\alpha$ versus frequency (Hz) in four different thickness with blending ratio of 30% EFB – 70% OPF.

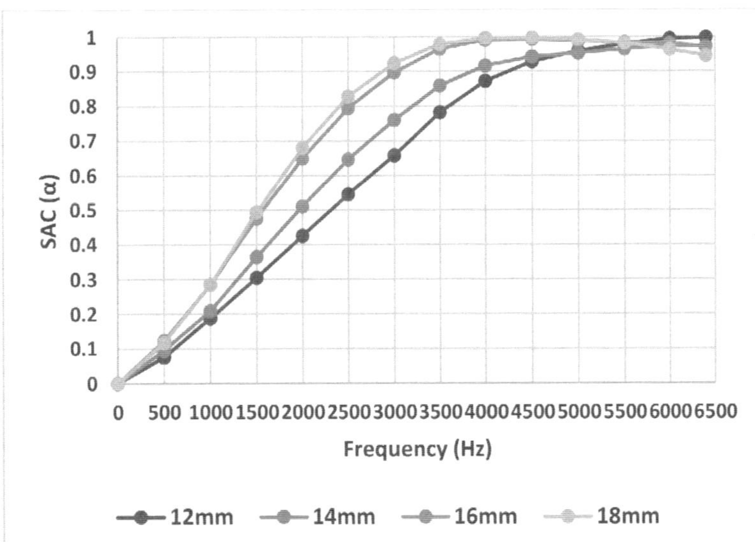


**Figure 4.** SAC $\alpha$ versus frequency (Hz) in four different thickness with blending ratio of 20% EFB – 80% OPF.

Figure 4 exhibits the SAC, $\alpha$ of four different thickness with the blending ratio of 20% EFB – 80% OPF. SAC, α values are found to increase as the frequency increase from 0 - 6400 Hz for all samples. From the figure above, samples with thickness of 16 mm and 18 mm performed better at mid-range frequency compare to samples with thickness of 12 mm and 14 mm. Based on past research, thickness plays a vital role in acoustic performance [18]. Sample with thickness of 12 mm was able to perform very well at a high frequency range of 6000 Hz to 6400 Hz with achieving 100 % sound absorption rate. Moreover, all samples exhibit the SAC more than 0.80 at the frequency range above 3500 Hz to 6400 Hz and SAC of 0.90 and above at frequency above 4000 Hz except for sample with a thickness of 12 mm, where it only achieves SAC of 0.87 at the same frequency. Based on sound absorption values and classes over a range of standard test frequencies according to BS EN ISO 354, the SAC of 0.8 and above is considered as A-rated products having the highest rated sound absorption.

It is remarkable that the SAC values increase with increasing in thickness from 12 mm to 18 mm for these two blending ratio at different frequency range. It can be explained by the tortuosity effect. The tortuosity is a measure of the shape of the air void passages and it brings a direct effect to the material sound absorption properties. The amount of fibers used is increasing with the sample's thickness increased. For porous material, an increase in the amount of fibers used will lead to a more complex and tortuous path and hence more energy dissipation due to high friction loss and more absorbing capability. The thicker the fiberboard, the more straight or twisted path that prolongs the time taken for the wave in contact with the fibers. Hence, it leads to better sound absorption for each of the reflection with the fibers. Thus, tortuosity plays a vital role in determining the sound absorption of the fiberboard.

## 4. Conclusion

In overall, the acoustic properties of blended OPF & EFB natural fibres were studied according to their different blending ratio and thickness. The experimental results above imply that the frequency range for SAC to reach 0.8 and above has been increased from 4500 – 6400 Hz to 3500 – 6400 Hz with an increased in OPF contents. This research has been shown the tortuosity plays a crucial role in determining the acoustic properties from low to medium frequency range. The increased of sinuous path in the fibers due to the increase in sample's thickness led to more sound absorption for the sample.

**Acknowledgments**

The authors would like to thank The Energy University's BOLD grant with project code: RJO 105178441060 and the Malaysian Oil Palm Board (MPOB) for supporting this research work.

## References

[1] Singh, N. & Davar, S.C., "Noise pollution-sources, effects and control," *J. Hum. Ecol.*, vol. 16, no. 3, pp. 181–187, 2004.

[2] P. A. Niquette, "Noise exposure: Explanation of OSHA and NIOSH safe exposure limits and the importance of noise dosimetry," *Can. Hear. Rep.*, vol. 9, no. 3, pp. 22–29, 2012.

[3] R.Mageswaran, E. L. Sheng, and Z. Ibrahim, "The Thickness Effects On Acoustic Properties Of Oil Palm Trunk Natural Fiber In Density Of 170 Kg/m3," *Sci Int*, vol. 30, no. 6, pp. 845–849, 2018.

[4] Navy Environmental Health Center, "Man-Made Vitreous Fibers," vol. 12, no. October, p. 110, 1997.

[5] Iarc, "Iarc Monographs on the Evaluation of Carcinogenic Risks To Humans," *Iarc Work. Gr. Eval. Carcinog. Risks To Humans*, vol. 96, pp. i-ix+1-390, 2002.

[6] K. H. Kilburn, D. Powers, and R. H. Warshaw, "Pulmonary effects of exposure to fine fibreglass : irregular opacities and small airways obstruction," pp. 714–720, 1992.

[7] R. Mageswaran, L. S. Ewe, W. K. Yew, and Z. Ibrahim, "Acoustic Properties of Mixing Empty Fruit Bunch and Oil Palm Frond Natural Fibres," *Int. J. Recent Technol. Eng.*, vol. 8, no. 4, pp. 6347–6349, 2019.

[8] T. Koizumi, N. Tsuijuchi, and A. Adachi, "The development of sound absorbing materials using natural bamboo fibers and their acoustic properties," *INTER-NOISE NOISE-CON Congr. Conf. Proc.*, vol. 2002, no. 5, pp. 713–718, 2002.

[9] A. Putra, K. H. Or, M. Z. Selamat, M. J. M. Nor, M. H. Hassan, and I. Prasetiyo, "Sound absorption of extracted pineapple-leaf fibres," *Appl. Acoust.*, vol. 136, no. November 2017, pp. 9–15, 2018.

[10] S. Ersoy and H. Küçük, "Investigation of industrial tea-leaf-fibre waste material for its sound absorption properties," *Appl. Acoust.*, vol. 70, no. 1, pp. 215–220, 2009.

[11] R. Kalaivani, L. S. Ewe, B. K. Yap, N. A. Talik, and Z. Ibrahim, "The effects of density on microstructure and acoustic properties of OPT natural fibers," *J. Fundam. Appl. Sci.*, vol. 10, no.

3s, pp. 434–445, 2018.
[12] M. Saad and I. Kamal, "Kenaf Core Particleboard and Its Sound Absorbing Properties," *J. Sci. Technol.*, vol. 4, no. 2, pp. 23–34, 2013.
[13] L. Ismail, M. I. Ghazali, M. Shahruddin, and M. A. . Ahmad, "Sound Absorption of Arenga Pinnata Natural Fiber," *Int. J. Mater. Metall. Eng.*, vol. 4, no. 7, pp. 438–440, 2010.
[14] Z. Ibrahim, "Production of Medium Density Fibreboard (MDF) FROM Oil Palm Trunk," *J. Appl. Sci.*, vol. 11, 2014.
[15] F. D'Alessandro, F. Asdrubali, and N. Mencarelli, "Experimental evaluation and modelling of the sound absorption properties of plants for indoor acoustic applications," *Build. Environ.*, vol. 94, pp. 913–923, 2015.
[16] S.Jung, Y. Kim, Y. Lee, S. Cho and J. Lee, "Measurement of sound transmission loss by using impedance tubes", Journal of the Korean Physical Society, vol. 53, no. 2. pp. 596-600, 2008.
[17] A. Nandanwar, M. C. Kiran, and K. C. Varadarajulu, "Influence of Density on Sound Absorption Coefficient of Fibre Board," *Open J. Acoust.*, vol. 07, no. 01, pp. 1–9, 2017.
[18] E. M. Samsudin, L. H. Ismail, A. A. Kadir, and I. Norfaslia, "Thickness, density and porosity relationship towards sound absorption performance of mixed palm oil fibers," *24th Int. Congr. Sound Vib.*, pp. 0–8, 2017.

# The effect of graphene addition on the Young's modulus and tensile strength of kenaf fibre composites

**Nabilah Afiqah Mohd Radzuan[1,*], Mihirssen Gunasegran[1] and Nisa Naima Khalid[1]**

[1]Department of Mechanical and Manufacturing Engineering, Faculty of Engineering & Built Environment, Universiti Kebangsaan Malaysia, Malaysia

[*]Corresponding author's email: afiqah@ukm.edu.my

**Abstract.** Natural fibres such as kenaf fibres may be used to reinforce polymer matrices such as polypropylene (PP). Such composite materials are in demand for the production of interior components for automotive vehicles due to their low density (1.2 g/cm³) and good mechanical properties (Young's modulus of 11 GPa and tensile strength of 780 MPa). The main objective of this study is to determine the effect of the addition of graphene fillers to the tensile strength and Young's modulus of kenaf fibre composites. For this kenaf/graphene/polypropylene (PP) composites are to be studied via computer simulations using Abaqus CAE software and to compare the experimental data of Young's modulus and tensile strength of the composite from previous researchers with the results of the simulations. For the Abaqus CAE simulations, general static and dynamic explicit analyses were conducted and set at 20 wt. % kenaf fibres, 0, 1, 3, to 5 wt. % graphene. The highest value of Young's modulus was 1 600 MPa achieved with the material with composition of PP/kenaf/graphene 5 wt. % while the highest value of tensile strength was achieved by the composition of PP/kenaf/graphene 1 wt. % at 23.07 MPa. On the other hand, the values of tensile strength increase with the addition of graphene (1 wt. %) at first due to the improvement of interface adhesion between the polymer matrix and non-organic particles. Further addition of graphene content (3 – 5 wt. %) results in a decrease of tensile strength due to the presence of agglomeration and defects such as voids and fibre pull-outs which weakens the adhesion between the fibres and matrix. The addition of graphene to PP/kenaf composites was proved to improve the Young's modulus and tensile strength of the materials. This improves the suitability of the materials in the production of interior components for automotive vehicles.

**Keywords:** Natural Fibre Composites; Kenaf; Graphene; Young's Modulus; Tensile Strength.

## 1. Introduction

Pollution can be defined as the presence or introduction of pollutants (solid, liquid, or gaseous) or energy (heat, light, radiation, or sound) into an environment, be it directly or indirectly caused by human activities or natural causes [1]. This includes pollution of the air, water and land, which may cause changes to the climate and weather systems, which in turn may cause disastrous effects such as famine, drought floods and erratic weather patterns [2,3]. With the worsening of such a phenomenon, there could be significant negative effects on human life, agricultural yield, and the economy in the long term [4]. Everyday occurrences such as spills, accidents, negligence, and vandalism all contribute to pollution that leads to harmful effects to our health and environment. Although, human activities have inadvertently been causing pollution for thousands of years, this condition has worsened due to the relatively recent human modernisation and industrialisation, advancement of transportation technology, and increase in population since the 20th century [5].

Various efforts have been taken to overcome the problem of pollution due to human activities. One of the efforts pertains to the processing of composite materials used in daily life. A composite material is made up of two or more different materials; a matrix component and the reinforcing components, usually in the form of fibres [6]. The applications of composite materials include medicine, aerospace, automotive components, and sporting goods, conventionally, synthetic materials such as such as fibreglass and aramids are used. However, these materials are non- biodegradable, thus not environmentally friendly; they may lead to pollution upon disposal [7,8]. Natural fibres such as jute, sisal, and kenaf are gaining more attention for this application. The benefits of using natural fibres in composites are they are available in large quantities, have low cost, and have mechanical properties up to par with synthetic fibres. Most importantly, these fibres are biodegradable and do not contain toxic substances [9]. Specifically, kenaf fibres are favourable as composite fillers as they have good tensile strength of around 780 MPa, tensile strength of 11 GPa, and a low density; approximately 1.2 g/cm³ [8,10]. An application of a kenaf composite, such as a kenaf/polypropylene (PP) composite would be to fabricate interior components for automotive vehicles. European car manufacturers such as BMW and Audi have already begun fabricating interior parts from natural fibre composites [11]. The low density material would help in reducing the vehicle's mass, fuel consumption, and emissions, while having sufficient mechanical properties and being environmentally friendly [12].

To further improve the suitability of the kenaf composite, its mechanical properties can be improved, as the automotive components must be able to withstand loads in use [13]. One way to achieve this would be to add fillers to increase the Young's modulus and tensile strength of the composite. Previous research has shown that the addition of graphene as a filler to kenaf composites is able to improve the mechanical properties [14]. However, as of yet, there is not many comprehensive studies regarding the effect of graphene addition on the mechanical properties of PP/kenaf composites, as well as a lack of computer simulations that could aid in making estimations regarding the material's properties. Thus, the main objective of this study is to find the effect of the addition of graphene fillers to the tensile strength and Young's modulus of kenaf/PP composites at room temperature.

## 2. Methodology

The research was started with preparation of the raw materials. Polypropylene (PP) powder of grade HM20/70P with particles less than 90 micrometres in size was used as the composite matrix. C-500 graphene nanoplatelets were used as fillers, with surface areas of 500 m²/gram. Micron sized kenaf fibres (size 20 mesh) were firstly treated with sodium hydroxide solution with a concentration of 6 wt. %. The fibres were soaked for three hours in the alkali solution before being rinsed with distilled water until it free from the alkali substance. This was determined by testing the fibres with litmus paper. The fibres were then dried in a heating oven for 24 hours at 80 °C. The graphene nanoplatelets and PP powder were physically mixed by hand for 5 minutes, followed by ball milling at 200 rpm for one hour using the Fritsch Pulverisette 6 machine. Next, the materials were mixed together using a Sigma Blade process at 190 °C and 45 rpm. The composition of the composite samples were 40 wt. % kenaf, with graphene content of 0, 1, 3, and 5 wt. % and the PP powder making up the rest of the composition (55, 57, 59 wt. %). The PP/graphene mixtures were poured into the machine compartment for 5 minutes to allow the polymer to melt evenly, followed by the kenaf fibres for another 10 minutes. The materials were removed and allowed to cool and harden. The hardened materials were crushed into a coarse powder using a crushing machine after that. The materials were then placed onto a mould with dimensions 175 mm × 175 mm × 2 mm. The material was then shaped with a hot press machine at 190 °C and 5 MPa for a 5-minute holding time. The moulds were then removed and allowed to cool for three hours before the composite material plate was removed. The samples were then shaped using a hand saw according to specifications of ASTM D 3039 for tensile tests, which is 250 mm x 25 mm x 3 mm. This shaped was repeated do for all the compositions of samples.

After the sample is drawn, new material should be produced in the 'Property' module. Values for mechanical properties (Young modulus, tensile strength, Poisson ratio) and physical properties of the sample (density). It should be noted that the program does not use units for those values, and the values entered must be consistent throughout the simulation process. After that, a new part needs to be produced with the material, and the part is assigned to the drawn component (Assign Section). Next to the Assembly module, the sample component given the material must be selected (Create Instance). After that, in the Step module, a new step needs to be created, by selecting Static, General. The time period is reduced from one second to 1.0 seconds to shorten the simulation time, and 'Nlgeom' needs to be turned on (On). Thereafter, in the Constrain module, a new 'coupling' type barrier is set at the top of the sample. Next, a mesh for the sample will be generated on the Mesh module. The net size used is 2 mm was chosen because the results of the simulation will be more accurate. However, a size that is too small will take a lot of time to complete the analysis. The type of mesh chosen is tetrahedral type, as other types of mesh do not fit such components, and many steps need to be taken to use them. The number of elements is 1240 and the number of nodes is 2079. For explicit dynamic analysis, the values for the material should be included in the 'Property' module. Values for mechanical properties such as Young modulus and Poisson ratio are the same as in general static analysis. Values for plastic deformation are taken from previous studies on the strength of PP / kenaf composite mechanical properties. The value for ductile damage is (Ductile Damage) taken from previous studies as well. The same steps for the Assembly are taken, followed by the Step module. The Explicit Dynamic Type is selected, and the following values are entered for Mass Scaling with the reference point is also set at 2 cm from the top center of the sample. In the interaction module, the constraint (Constrain) type Coupling (coupling) is selected. The same steps for the Mesh and Job modules as in general static analysis are taken. Results will be generated later in the Visualization module.

## 3. Result and discussion

### *3.1. PP/Kenaf/Graphene composite samples*

After remove the mould from the hot press machine, the material was left to be cooled for 3 hours to avoid defects such as cracks and warping before being removed from the mould and shaped into test samples for tensile testing as per the standards of ASTM D3039 using a hand saw. It had found that the addition of graphene changes the physical appearance of the composites, causing them to become considerably darker in shade, even with only 1 wt. % graphene addition. The samples were shown in Figure 1 (a) – (d). found that the PP/kenaf sample could not be formed as the composite material plate formed by the hot press process shattered upon opening the mould. This may be due to weaker adhesion between the PP matrix and kenaf fibres, causing the material to have lower mechanical properties. Previous studies show that the composite comprised of just PP and kenaf have a lower values of both Young's modulus and tensile strength compared to PP/kenaf/graphene composites [14]. Besides that, the defect may have been due to insufficient amounts of wax applied on the moulds, causing the composite material to adhere strongly to the mould upon opening [15]. Furthermore, it is possible that the material did not have enough time to cool down and harden completely. Premature opening while the material was still in 2 phases (solid and semi-solid) prevented the plate from staying in one piece [16].

From FESEM (Field Emission Scanning Electron Microscopy) analyses from previous research, excellent dispersion of materials with no sign of agglomeration was observed for the PP/kenaf/graphene 1 wt. % sample (Idumah & Hassan 2016). The same was observed for the PP/kenaf/graphene 3 wt. % sample, which also showed the intact, embedded graphene nanoplatelet sheets protruding from the fracture surface studied. The increase in mechanical properties (tensile strength from 17.5 MPa to 21.3 MPa) reported for the sample with the 3 wt. % graphene content is attributed to the homogenous distribution of the fillers within the polymer matrix. On the other hand, the PP/kenaf/graphene 5 wt. % sample shows a decrease in tensile strength to 17.3 MPa. the cause for the drop in tensile strength was attributed to the presence of agglomerates and fibre pull-outs, and defects such as voids in the PP matrix.

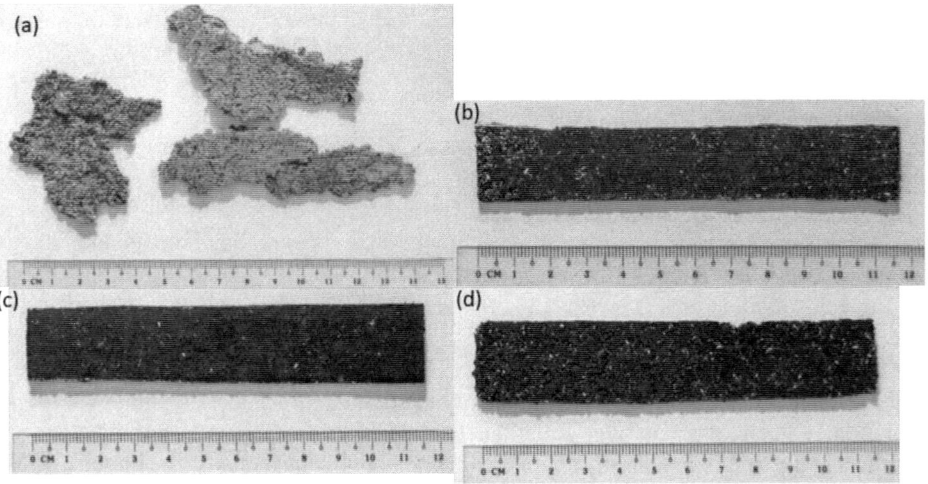


**Figure 1.** (a). Tensile testing sample: PP/kenaf composite, (b). Tensile testing sample: PP/kenaf/graphene 1 wt. % composite, (c). Tensile testing sample: PP/kenaf/graphene 3 wt. % composite, (d). Tensile testing sample: PP/kenaf/graphene 5 wt. % composite.

### *3.2. Computer simulations: General static analysis*

For the general static analysis, the tensile test was set up in a fashion similar to a physical test following the ASTM D3039 standards, with a sample of dimensions 250 mm × 25 mm × 3 mm, with the sample clamped at both ends. For the PP/kenaf composite without graphene addition, the highest values of stress and strain during tensile testing were 22.955 MPa and 0.02404 respectively. The values of Young's modulus for the samples was determined by calculating the gradient of the stress-strain graphs generated by the software for each simulation and the value of tensile strength was the highest stress obtained from the stress- strain graphs. For the PP/kenaf composite, the Young's modulus was determined to be 954 MPa while the tensile strength was 22.95 MPa. For the PP/kenaf/graphene 1 wt. % sample, the highest value of stress obtained was 23.07 MPa, corresponding to the material's tensile strength, while the highest value for its strain was 0.021165. The Young's modulus value obtained was 1090 MPa. this shows a 0.12 MPa or 0.5 % increase in tensile strength compared to the PP/kenaf sample, while the Young's modulus increased by 135.34 MPa or 14.18 %.

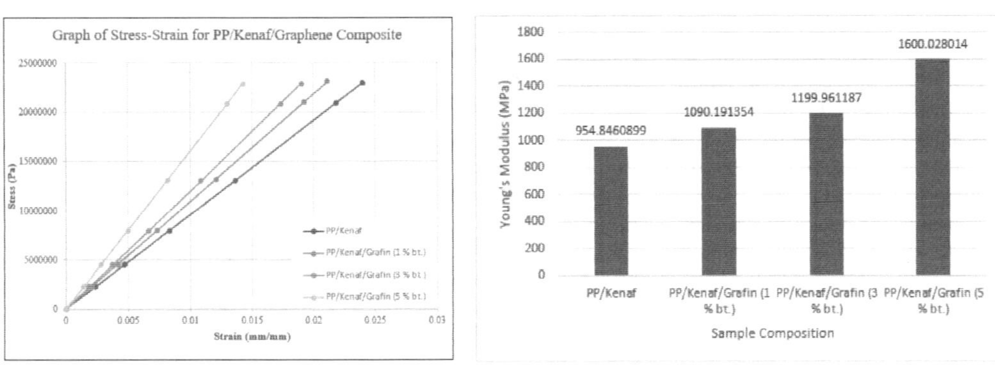


(a) (b)

**Figure 2.** (a) Stress-strain graphs of PP/kenaf/graphene samples, (b) Young's modulus values of samples (general static analysis).

Next, the PP/kenaf/graphene 3 wt. % sample had shown on (tensile strength) value of 22.878 MPa, with a strain value of 0.019066, and Young's modulus of 1200 MPa; a decrease of 0.076 MPa (0.33 %) in tensile strength and increase of 245.15 MPa (25.67 %) in Young's modulus when compared to the PP/kenaf composition. Lastly, the PP/kenaf/graphene 5 wt. % sample had a tensile strength of 22.846 MPa, highest strain value of 0.014279, and Young's modulus value of 1600 MPa. This corresponds with a 0.109 MPa (0.47 %) decrease in tensile strength and 645.16 MPa (67.57 %) increase in Young's modulus value compared to the sample without graphene addition. The stress-strain curves for all the sample compositions is displayed on Figure 2. It can be seen that increasing graphene content in the sample composition causes steeper gradients in the graphs. This indicates that the higher the graphene content, the higher the Young's modulus of the sample. The values of Young's modulus are plotted and compared on a bar chart in Figure 2.

From Figure 2(b), it is clear to see that the values of Young's modulus increase with the addition of graphene fillers. This may be attributed to a uniform and homogenous distribution of the graphene filler in the PP matrix [17]. Graphene has very high mechanical properties and stiffness; its Young's modulus is up to 1 TPa and maximum tensile strength is 130 GPa, and it contributes to the increase of mechanical properties of the composites [18]. Furthermore, the addition of graphene may have improved the adhesion between the PP matrix and kenaf fibres. This reduces and restricts the deformation of the matrix during the elastic region of tensile testing phase, thus increasing the Young's modulus. Other than that, the possible alignment of the graphene nanoplatelets parallel to the direction of flow of the polymer matrix during the hot press process further improves the composites' Young's moduli. Moreover, the dispersion of inorganic materials, such as graphene, within the polymer material plays a role in the structure of the matrix. The movement of molecular chains will be restricted as cross- links between the graphene and PP matrix are formed. This increases the composite system's stiffness. As PP is a crystalline polymer, the addition of graphene as a filler will have an effect on the heterogenous nucleation, which leads to an increase in the degree of crystallinity or may even alter the crystal structure. This also increases the stiffness of the composites.

The tensile strength of the composites is laid out and compared using a bar chart as seen in Figure 3. Generally, the values do not undergo any significant changes, with a range of only 0.228 MPa between the lowest value of 22.85 MPa for the PP/kenaf/graphene 5 wt. % sample and the highest value of 23.08 MPa for the PP/kenaf/graphene 1 wt.% sample. It can be seen that the trend of tensile strength is an increase in the value from 22.95 MPa (PP/kenaf sample) to 23.08 MPa (PP/kenaf/graphene 1 wt. % sample), before gradually decreasing to 22.88 MPa (PP/kenaf/graphene 3 wt. % sample), to 22.85 MPa (PP/kenaf/graphene 5 wt. % sample). The initial increase in tensile strength can be attributed to the interface adhesion between the inorganic particles and the polymer matrix. For particles in the form of thin sheets, the effect of adhesion depends significantly on the surface area of the particle, whereby a particle with higher specific surface area results in a composite with higher mechanical properties due to better interface adhesion [19]. Graphene is a 2- dimensional material with a high surface area relative to its weight (500 $m^2/g$ surface area and 12.01 g/mol molecular weight for C-500 graphene nanoplatelets) and is capable of increasing the interface adhesion with the PP matrix [18]. The subsequent decline in tensile strength could be due to filler-filler interactions (that of the graphene and kenaf fibres) caused by agglomeration which increases with the increase in filler content. The effect of agglomeration will weaken the adhesion between the fillers and matrix, and thus weakening the whole composite system [17]. Besides that, the presence of other defects such as pores and voids, non-uniform dispersion of fillers, and fibre pull-outs weakens the composite system by causing stress concentrations and a non-homogenous sample [20].

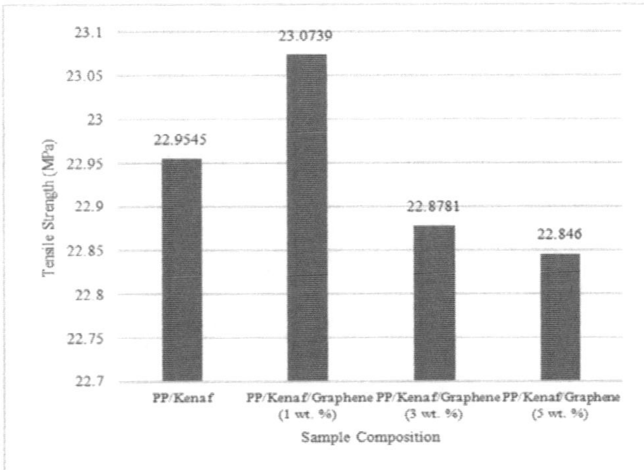


**Figure 3**. Tensile strength values of samples (general static analysis).

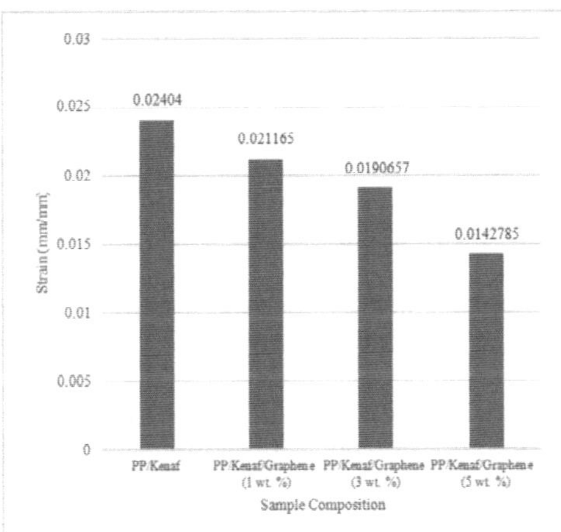


**Figure 4**. Strain values of samples (general static analysis).

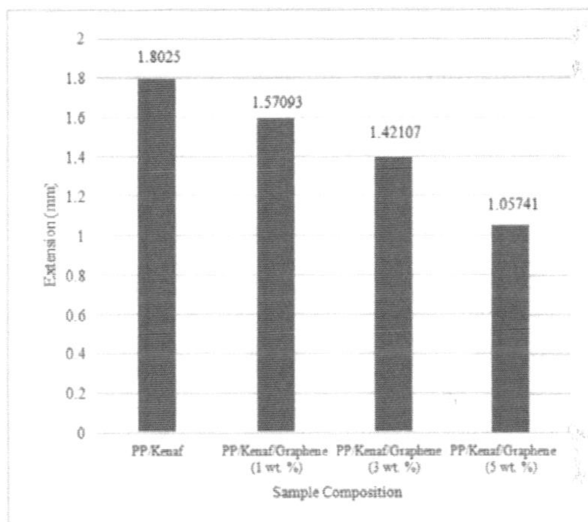


**Figure 5**. Extension values of samples (general static analysis).

The strains and extensions for the composite samples after tensile testing are displayed on bar charts on Figure 4 and Figure 5 respectively. The values of strain show a consistent decrease with increase in graphene content, from 0.024 for the PP/kenaf sample to 0.014 for the PP/kenaf/graphene 5 wt. % sample as seen in Figure 4. This results in a decrease of 0.01 (mm/mm) or 41.67 %. This indicates that the increase in graphene content increases the brittleness of the composite material. The same decreasing trend is observed for the extension of the samples as shown in Figure 5; from 1.80 mm for the PP/kenaf sample to 1.06 mm for the PP/kenaf/graphene 5 wt. % sample, which is a 41.33 % decrease (0.76 mm). These decreases in strain and extension are possibly due to the graphene particles initiating β-PP nucleation, or the incompatibility between the matrix and the fillers. Besides that, the graphene addition increases the composites' densities and causes them to become more brittle [17]. The consistent decrease in strain values justifies the increase in Young's modulus with increasing graphene content for the samples, even though there is a decrease in the tensile strength of the samples. The value of Young's modulus is obtained by calculating the gradient of the stress-strain graphs for each composition that is to divide the highest value of stress with the highest value of strain on the straight-line graph. Since the strain values decrease more drastically (41.67 %) compared with the decrease in tensile strength values (0.47 %), the values of Young's modulus increase with the increase of graphene content.

## 4. Conclusion

The effect of addition of graphene to the Young's modulus and tensile strength of PP/kenaf composites has been investigated throughout this research. Test samples of the composite materials were successfully fabricated using hot press method according to ASTM D3039 specifications for tensile testing. Static general and dynamic explicit computer simulations using Abaqus CAE software were conducted to study the aforementioned mechanical properties of the composite materials. From the general static analysis, the highest value of Young's modulus was 1 600 MPa achieved with the material with composition of PP/kenaf/graphene 5 wt. % while the highest value of tensile strength was achieved by the composition of PP/kenaf/graphene 1 wt. % at 23.07 MPa compared to a Young's modulus of 954 MPa and tensile strength of 22.95 MPa for the PP/kenaf composite without addition of graphene. The values of Young's modulus were similar when compared to previously conducted studies, while the values of tensile strength and the effect of graphene addition on the properties of the composites were congruent to the previous studies as well. It can be concluded, with the additional of graphene in PP/Kenaf have shown in decreasing of their tensile strength. Meanwhile, addition of graphene in PP/Kenaf have resulted increases in their Young's Modulus.

**Acknowledgements**

The authors wish to gratefully thank and acknowledge the Center for Research and Instrumentation Management (CRIM), Universiti Kebangsaan Malaysia for their financial support to complete this study under grant number GGPM-2020-002.

## References

[1] W. Wang, Y. Zhang, S. Wang, C.Q. Fan, H. Xu, Distributions of phthalic esters carried by total suspended particulates in Nanjing, China, Environ. Monit. Assess. 184 (2012) 6789–6798. https://doi.org/10.1007/s10661-011-2458-z.

[2] T.E. Lovejoy, L. Hannah, Biodiversity and climate change: Transforming the biosphere, 2019.

[3] C. Lu, S. Grundy, Urban Agriculture and Vertical Farming, Elsevier, 2017. https://doi.org/10.1016/B978-0-12-409548-9.10184-8.

[4] J. Bell, L. Paula, T. Dodd, S. Németh, C. Nanou, V. Mega, P. Campos, EU ambition to build the world's leading bioeconomy—Uncertain times demand innovative and sustainable solutions, N. Biotechnol. 40 (2018) 25–30. https://doi.org/10.1016/j.nbt.2017.06.010.

[5] Y. Zare, K.Y. Rhee, Effects of carbon nanotubes and interphase properties on the interfacial conductivity and electrical conductivity of polymer nanocomposites, Polym. Int. 69 (2020) 413–422. https://doi.org/10.1002/pi.5969.

[6] R. Taherian, M.J. Hadianfard, A.N. Golikand, Manufacture of a polymer-based carbon nanocomposite as bipolar plate of proton exchange membrane fuel cells, Mater. Des. 49 (2013) 242–251. https://doi.org/http://dx.doi.org/10.1016/j.matdes.2013.01.058.

[7] M.H. Cetin, B. Ozcelik, E. Kuram, E. Demirbas, Evaluation of vegetable based cutting fluids with extreme pressure and cutting parameters in turning of AISI 304L by Taguchi method, J. Clean. Prod. 19 (2011) 2049–2056. https://doi.org/10.1016/j.jclepro.2011.07.013.

[8] M. Ramesh, Kenaf (Hibiscus cannabinus L.) fibre based bio-materials: A review on processing and properties, Prog. Mater. Sci. 78 (2016) 1–92. https://doi.org/http://dx.doi.org/10.1016/j.pmatsci.2015.11.001.

[9] R.A. Ilyas, S.M. Sapuan, M.R. Ishak, E.S. Zainudin, Development and characterization of sugar palm nanocrystalline cellulose reinforced sugar palm starch bionanocomposites, Carbohydr. Polym. 202 (2018) 186–202. https://doi.org/10.1016/j.carbpol.2018.09.002.

[10] H.M. Akil, M.F. Omar, A.A.M. Mazuki, S. Safiee, Z.A.M. Ishak, A. Abu Bakar, Kenaf fiber reinforced composites: A review, Mater. Des. 32 (2011) 4107–4121.

https://doi.org/10.1016/j.matdes.2011.04.008.
[11] J.A. García Sánchez, J.M. López Martínez, J. Lumbreras Martín, M.N. Flores Holgado, H. Aguilar Morales, Impact of Spanish electricity mix, over the period 2008-2030, on the Life Cycle energy consumption and GHG emissions of Electric, Hybrid Diesel-Electric, Fuel Cell Hybrid and Diesel Bus of the Madrid Transportation System, Energy Convers. Manag. 74 (2013) 332–343. https://doi.org/10.1016/j.enconman.2013.05.023.
[12] N.A.M. Radzuan, A.B. Sulong, M. Rao Somalu, Electrical properties of extruded milled carbon fibre and polypropylene, J. Compos. Mater. 51 (2017) 3187–3195. https://doi.org/10.1177/0021998316688075.
[13] P. Wambua, J. Ivens, I. Verpoest, Natural fibres: Can they replace glass in fibre reinforced plastics?, Compos. Sci. Technol. 63 (2003) 1259–1264. https://doi.org/10.1016/S0266-3538(03)00096-4.
[14] C.I. Idumah, J.E. Ogbu, J.U. Ndem, V. Obiana, Influence of chemical modification of kenaf fiber on xGNP-PP nano-biocomposites, SN Appl. Sci. 1 (2019) 1–11. https://doi.org/10.1007/s42452-019-1319-1.
[15] M.K.F. MdRadzi, A.B. Sulong, N. Muhamad, M.A. MohdLatiff, N.F. Ismail, Effect of Filler Loading and NaOH Addition on Mechanical Properties of Moulded Kenaf/Polypropylene Composite, Pertanika J. Trop. Agric. Sci. 38 (2015) 583–590.
[16] C.G. Kang, J.S. Choi, K.H. Kim, Effect of strain rate on macroscopic behavior in the compression forming of semi-solid aluminum alloy, J. Mater. Process. Technol. 88 (1999) 159–168. https://doi.org/10.1016/S0924-0136(98)00383-5.
[17] N. Afiqah, M. Radzuan, A.B. Sulong, M. Irwan, M. Firdaus, T. Husaini, E.H. Majlan, Fabrication of multi-filler MCF / MWCNT / SG-based bipolar plates, Ceram. Int. 45 (2019) 7413–7418. https://doi.org/10.1016/j.ceramint.2019.01.028.
[18] R.B. Mathur, S.R. Dhakate, D.K. Gupta, T.L. Dhami, R.K. Aggarwal, Effect of different carbon fillers on the properties of graphite composite bipolar plate, J. Mater. Process. Technol. 203 (2008) 184–192. https://doi.org/http://dx.doi.org/10.1016/j.jmatprotec.2007.10.044.
[19] W.F. Santos, M. Quattrone, V.M. John, S.C. Angulo, Roughness, wettability and water absorption of water repellent treated recycled aggregates, Constr. Build. Mater. 146 (2017) 502–513. https://doi.org/10.1016/j.conbuildmat.2017.04.012.
[20] N.A.M. Radzuan, A.B. Sulong, M.R. Somalu, Influence the filler orientation on the performance of bipolar plate, Sains Malaysiana. 48 (2019) 669–676. https://doi.org/10.17576/jsm-2019-4803-21.

# Additive manufacturing in medical applications: A brief review

**Bo Sun**[1,2,*], **Quanjin Ma**[1,2], **Xinfu Wang**[1], **Jinyan Liu**[3] **and M R M Rejab**[1,2]

[1]School of Mechanical Engineering, Ningxia University, 750021 Yinchuan, China
[2]Structural Performance Materials Engineering (SUPREME) Focus Group, Faculty of Mechanical & Automotive Engineering Technology, Universiti Malaysia Pahang, 26600 Pekan, Pahang, Malaysia
[3]School of Fine Arts, Shandong Yingcai University, 370115, Shandong, China

*Corresponding author's email: superbsunbo@163.com

**Abstract.** In the medical field, due to the obvious individual differences between patients, the complexity of human tissues and the fact that patients are too price-sensitive, it is desperate to search for a new biological material adjuvant therapy, which far more closely matches the pathological characteristics of patients. It is noted that additive manufacturing has the ability to overcome those limitations by virtue of its exclusive characteristics of personalized customization, small-scale production and high precision. This paper aims to summarize the characteristics and applications of 3D printing technology used at four levels in the medical field, including adjuvant therapy and anatomical model, implants and scaffolds, biological 3D printing, and pharmaceuticals, as well as, to present the recent innovation and discuss existing issues with 3D printing that related to medical applications. It is highlighted that 3D printing technology will be further built on medical application.

**Keyword:** Additive Manufacturing; Adjuvant Therapy; Implants; Biological Material; Pharmaceutical.

## 1. Introduction

3D printing technology used to be born in the 1990s and has been extensively used in a variety of fields such as industry, pharmacy, architecture, clothes, training [1-4]. Thanks to its blessings of precision, quick prototyping and customization, it is now increasingly imperative to realize an individualized and accurate therapeutic cure [5, 6]. This technology separates a 3D model into multiple layers of slices alongside a certain set path and types a 3D object by layer by layer creation and layer by layer stacking. Its specific flow includes acquisition of statistics, image analysis and development of models [7].

Current research on 3D printing technology for medical purposes concentrates on the following four important areas: 1. Study on personalized manufacturing of bioactive permanent prosthesis implants and scaffolds; 2. Study on manufacturing pathological organ models to assist preoperative planning and surgical treatment analysis [8]; 3. Study on directly printing tissues and organs with complete life functions [9-11]; 4. Study on versatile platform for drug delivery [12]. Although such applications remain a long way from widespread clinical applications as several key technical and fundamental scientific issues that are yet to be overcome, notable scientific advancements and applications have been achieved in these areas.

## 2. 3D printing fundamentals

### *2.1. 3D printing technology*

3D printing technology is known to be an Additive Manufacturing (AM) technique used for the manufacture of products in a consecutive layering series. Normally, this technology is categorized into solid, liquid and power-based approaches based entirely on input materials [13]. The solid-based pattern is Fused Deposition Modeling (FDM), the power-based pattern is composed of Selective Laser Sintering (SLS) and Selective Laser Melting (SLM). Liquid-based trends include Stereolithography (SLA), Optical Light Processing (DLP), Direct Ink Writing (DIW), and Inkjet [14, 15].

### *2.2. 3D printed medical model methods*

There are forms of 3D part printing processes [16-18]. All these processes in 3D solid object printing, using the aggregation layer-by-layer technique and reducing manufacturing costs and improving the printing precision and efficiency of 3D printers, 3D printing technology allows significant advancements in medical equipment, implant material and cell printing [19], the processes of 3D medical models by additive manufacturing.: 1. Creating a 3D CAD model of the individual patient by the use of Computed Tomography (CT), Magnetic Resonance Imaging (MRI), 3D scanner or any design software; 2. Converting Digital Imaging and Communications in Medicine (DICOM)/3D CAD generated from these scanning technologies into STL format; 3. Printing of patient-specific implants vie using AM technologies; 4. Post-processing of implants; 5. Testing of implants; 6. Final implementation of the implant in the patient body [20]. Figure 1 indicates the method of 3D printing techniques to create an anthropomorphic thorax phantom for clinical imaging purposes.

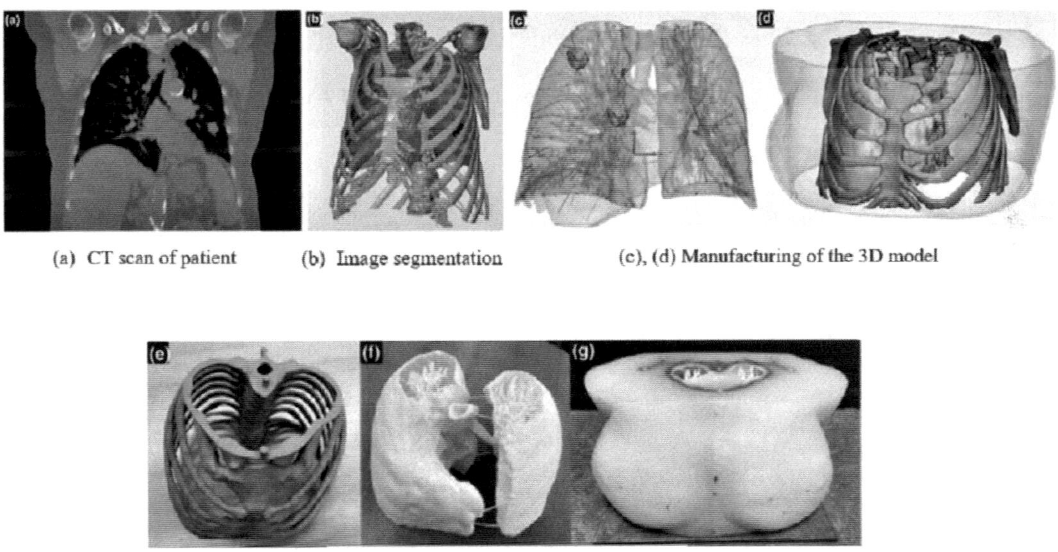


**Figure 1.** Overview of 3D printing/molding process [21].

### *2.3. 3D printed material types*

Additive Manufacturing (AM) presents flexibility for the production of any shape, and must therefore be one of the main innovations that lead to medical therapy and the coming industrial revolution [22]. Additive Manufacturing (AM) makes precise versions of patient-specific software for complex procedures. Exclusive heart valves and scaffolds are used in cardiology and are also useful for preoperative planning [23]. This will manufacture surgical instruments at a reduced cost relative to traditional technology. It creates personalized prostheses that are effortless to adjust and match the wearer [24]. Additive Manufacturing (AM) technology helps to enhance human life by removing the human organ, making artificial organs, surgical instruments and accelerating surgery. Efficient contact and substances are required in the field of medical treatment [25]. This technology is used for bioprinting of suitable medical material. A living cell and synthetic living tissue can be developed for the medical laboratory [26]. This makes it easier to build a surgical mesh that protects tissues and organs during surgery. In short, a vast range of materials are used to build 3D models for the medical industry. The various types of materials used in AM technologies are shown in table 1.

**Table 1.** The summary of materials used by Additive Manufacturing in the medical field.

| No | Materials | Application | Refs |
|---|---|---|---|
| 1 | Silicone | • Medical implants like tracheal scaffolds<br>• Single-use and reusable components, diagnostic instruments and other surgical tools<br>• Medical wires and cable components | [18, 27] |
| 2 | Plastic | • Thin tube which helps to unblock blood vessels<br>• Heart valve, intravenous blood bags, disposable syringes, eyeglass frame and lenses | [28, 29] |
| 3 | Metal | • Metals implants to fix of a fractured bone<br>• Artificial joints, scaffolds, plate, screw and valves<br>• Long term implantable devices | [30-32] |
| 4 | Tissue and cell | • Improving whole organ/damaged tissue<br>• Creating desired cell type<br>• Organs for transplantation, Human tissue repair products<br>• Improving damaged tissue and whole organs | [33-35] |
| 5 | Polylactic-acid (PLA) | • Tissue engineering application like bone<br>• Wound management of dental extraction, surgical sutures and preventing postoperative adhesions<br>• Scaffolds | [36-38] |
| 6 | Carbon Fiber | • Patient's lost limb, Medical tools | [39, 40] |
| 7 | Powder | • Presurgical planning for complex surgery<br>• Education of the medical student | [41, 42] |
| 8 | Nitinol | • Medical implants, Personalization | [43, 44] |

## 3. Medical applications

### *3.1. Bioactive permanent prosthesis implants and scaffolds*

At present, prosthesis implantation is frequently used in orthopedics and dentistry [45], which requires that the prosthesis has a desirable biocompatibility, that it does not biodegrade in the human body and that it may exist for a long time to take over the previous body tissue to function, and that the related technology is relatively mature at present. In comparison to traditional manufacturing techniques, 3D printing allows for the custom production of goods in a limited period and is used to the manufacture of implantable prostheses of ideal scale, form and mechanical properties.

Permanent implants can replace human body parts and play a parallel function in the rest of the patient's life, such as in hip joint, knee joint, shoulder joint, elbow joint or wrist prostheses [46]. Titanium has a high strength/weight ratio, non-magnetic and high corrosion resistance due to the dense titanium oxide ($TiO_2$) protective film on its surface, so permanent bone tissue prostheses are usually made of metallic titanium or other materials with a gel coating on their surface, which are enable to enhance biocompatibility, promote cell growth around the implant prosthesis and reduce the risk of inflammation and infection that may be caused by titanium or other permanent materials.

Winder et al [47]. combined CT imaging and 3D printing technology to achieve a customized titanium plate by making the patient's 3D-printed skull model in order to repair the patient's skull defect. The tailored titanium plate was better fitted to the skull and enhanced the fitting effect. Figure 2 shows a customized cranial titanium (Ti) implant and Ti hip implant for an Australian patient using selective laser melting (SLM). Scaffolding is an integral aspect of tissue engineering. The advance of the scaffold helps the weakened regions of the human body to be repaired as an alternative rather than removed, which not only fixes the issue of inadequate transplant donors, it also avoids the incompatibility of immune rejection of grafts.

In clinical practice, non-permanent scaffolding is more difficult to incorporate than permanent implants. In particular, it should not only conform with the requirements of mechanical and structural properties, but also enable the materials used to be degradable, capable of providing a surface environment for the attachment and spread of living cells, providing a pore network for survival conditions, and also directing cells via growth factors to divide into certain lines.

Lee et al. [48] used indirect 3D printing to use gelatin particles to print the mold with the desired shapes using gelatin shapes, and then a scaffold was generated from the printing. In addition, the polycaprolactone and chitosan were used to accurately control the internal morphology of the scaffold, and the inert 3D printing material applied to the scaffold was modified with a bioactive apatite coating. The scaffold was used to simulate the human mandibular condyle. The prepared scaffold material therefore had strong cell compatibility, and after surface treatment, its bioactivity was further increased. While several similar research and clinical implementations have been performed on scaffolds, complex organs with more specific multicellular structures involving vascular networks are still difficult to produce.

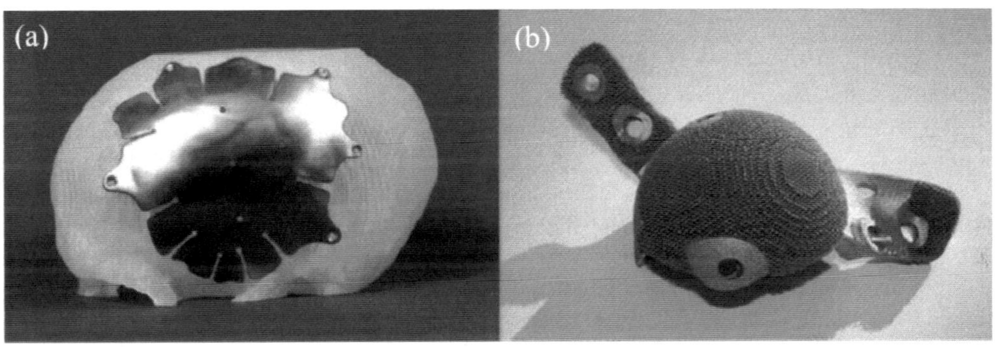


**Figure 2.** (a) Stereolithographic resin model of a full skull with a custom Ti plate in place; (b) Hip implant fabricated by SLM Solutions in Germany [47].

*3.2. Pathological organ models to assist preoperative planning and surgical treatment analysis*

For clinical treatment and preoperative planning, high-fidelity physical organ models are of significant importance, and can also be used in medical education in colleges and universities, offering more intuitive stereoscopic vision and tactile sensation for students [49]. The patient condition can only be understood from the 2D plane by means of magnetic resonance imaging (MRI) or computed tomography (CT) technologies due to the individual variations between patients, which cannot provide more accurate preoperative guidance for surgery, whereas the traditional model production procedure is not only difficult, but also not appropriate for rapid prototype modeling.

Whereas the advent of 3D printing can overcome these problems easily, prior to surgery, physicians are able to conduct surgery preparation and diagnosis using 3D printed models. In addition, these 3D printed models include procedure repeatability, automate steps and make the whole operation phase more precise, thus reducing the operation's anesthesia time and infection risk and eventually improving the operation's success rate. Cardiovascular disease is a chronic and regularly occurring disease in hospitals, with elevated clinical death and injury rates. Having a proper diagnosis of coronary disease prompt and reliable is conducive to effective management of the disease, thus reducing the mortality risk of the disease.

Farooqi et al. [50] conducted a cardiac magnetic resonance imaging of a woman with tricuspid valve hypoplasia, and printed a 3D model on this basis. They could quickly and plainly view the severely dilated and deformed right ventricular from the outside of the 3D printed model, and the tricuspid valve leaflet could be seen to be relocated from the inside to the right ventricular outflow, which was useful in preparing the positioning of inflow and outflow sleeves for ventricular assistance systems. Figure 3 displays the left and right ventricles models of the heart, the myocardial models, the cardiac fibrous skeleton and the various solid valve models created by 3D printing.

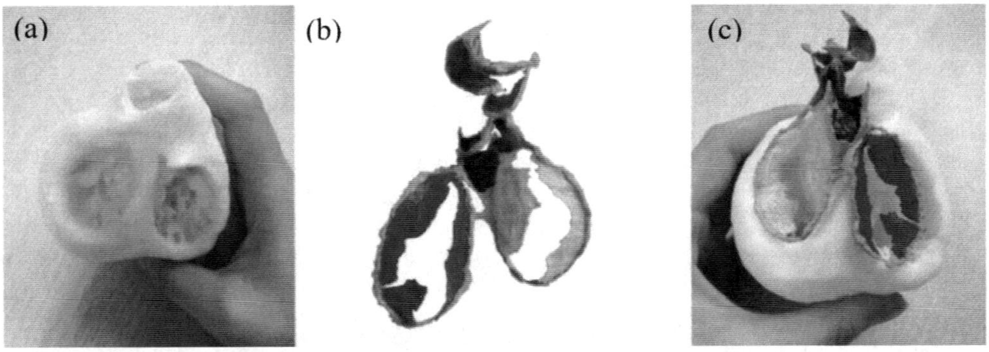


**Figure 3.** 3D printed models of heart: (a) The left and right ventricles models of the heart and myocardial models; (b) The cardiac fibrous skeleton and various valve solid models; (c) The assembly of components [51].

By comparing and studying the cardiac fibrous skeleton and the left and right ventricles of the heart and their myocardial versions, we can better understand the anatomical configuration of the valve and its spatial association with the adjacent cardiac chambers. The nervous system comprises primarily the central nervous system and the peripheral nervous system. There are a host of disorders in the central nervous system, including multiple pathogenesis. In clinical practice, the occurrence of illness may have a significant effect on the health of patients and may also jeopardize the wellbeing of patients.

Chen et al. [52] used the photographs collected from magnetic resonance scanning to print the knee joint disease model in order to provide orthopedic postgraduates with a more vivid and understandable anatomical type of knee joint events, so that students can not only learn more about the causes, diagnosis and treatment of the disease, but also better understand the anatomical parts, the spatial arrangement, g. Around the same time, they can follow the surgical criteria and execute simulation surgery experiments so that students can discover information that are not technically accessible and improve their knowledge of the disease and anatomy..

There are some drawbacks with respect to the use of realistic models for adjuvant surgery. Person refining models have high data quality standards, so hospitals need to configure high-resolution imaging equipment to collect relevant photographs from patients in order to generate 3D printed models. Physicians, meanwhile, need to expend a lot of time on simulating and planning surgery before surgery, which prolongs treatment time and is not appropriate for emergency surgery. Therefore, the implementation is only at the initial level of discovery.

### *3.3. Directly printed tissues and organs*

In 2000, Professor Thomas Boland of Clemson University, USA, put forward a new idea called 'cell and organ printing,' which marks the roots of current 3D bioprinting technology. 3D Biological Printing is a manufacturing process that uses rapid prototyping (RP) for the printing of biomaterials and biological units individually or in series and directly produces three-dimensional tissues or organs layer by layer according to the specifications of bionic morphology, biological function, microenvironmental cell growth, etc. Direct cell printing is an extension of tissue engineering, relative to scaffolding, bioprinting can result in the deposition of various types of cells of different densities at different locations of the scaffold and directly printing tissues or organs [53, 54].

Bioprinting is generally primarily limited to the printing of blood vessels, skin and organs, although the complexity of printing tissues and organs focuses mostly on the internal fluid vascular network [19]. Laronda et al. [55] suggested a system for the rehabilitation of ovarian mice using gelatin hydrogel-printed artificial bio-microporous scaffolding (Figure 4). As a 3D bioprinting hydrogel material, it is also important to balance its mechanical behavior so that it can fulfill the requirements of both printing and cell growth. To fix this problem, the team used the partial cross-linking properties of gelatin at a certain temperature (Figure 4a) to print smooth and continuous hydrogel filaments and completely cross-linked the hydrogel to the cooling table to accurately print complex microporous scaffolds (Figure 4b) suitable for follicle growth. The scaffold was used to grow follicles and to further produce the artificial biological ovaries that could be prepared and inserted in the missing sections of the mice ovaries. This technology will rebuild the ovaries of the mouse and restore their function (Figure 4c).

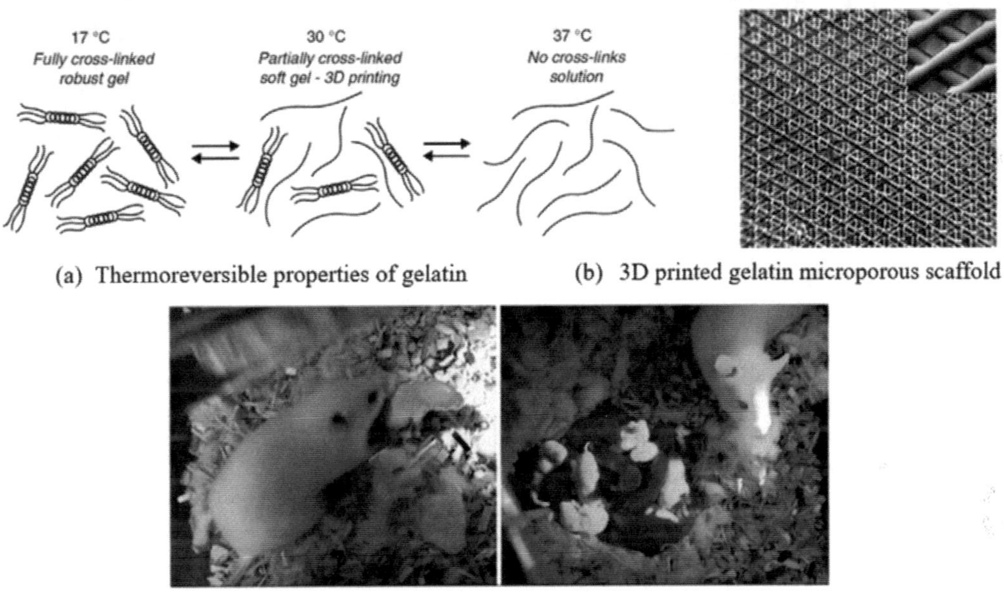


(a) Thermoreversible properties of gelatin (b) 3D printed gelatin microporous scaffold

(c) Artificial biological ovaries restore the ovarian function

**Figure 4.** Bio-prosthetic ovary created using 3D printed gelatin microporous scaffolds restores ovarian function in mice [55].

Dolati et al. [56] used MWCNT-reinforced alginate and typical alginate to make vascular catheters successful. The manufactured vascular catheters have a length of more than 1 m and have a strong perfusion capacity and high permeability, and can allow the media to diffuse radially, close to natural blood vessels. Bioprinting remains a cutting-edge approach to the growing international scarcity of organs, and the opportunity for transplant tissues and organs with a low risk of immune response is expected.

### *3.4. Versatile platform for drug delivery*

A number of 3D printing technologies are commercially viable, but detailed studies have been performed showing that Fused Deposition Modeling (FDM) has the greatest potential for the manufacture of different types of pharmaceutical products due to the large range of polymeric matrices available [57, 58]. FDM 3D printing can be described as a "flexible-dose dispenser" [59] that permits the manufacture of dosage forms with varying doses, release profiles ranging from immediate [60] to continuous [61] and changed release [62] without the need for additional processing, such as coating

[63]. This technology therefore provides a more sophisticated method for the engineering of pharmaceutical devices.

Channeled systems have been investigated to optimize the release of medications from immediate release formulations (Figure 5a) [64]. By controlling the different aspects of the channels (e.g. width, length and configuration), the release of drugs could be tailored to meet the required specification. The development of such complex geometries using traditional manufacturing methods is a challenge. Similarly, caplets containing internal holes, referred to as gaplets, are developed to facilitate separation of fast-release tablets and to prevent the need for disintegrants (Figure 5b) [65]. In contrast, 3D printed networks with controlled release properties and differing densities were also manufactured [66].

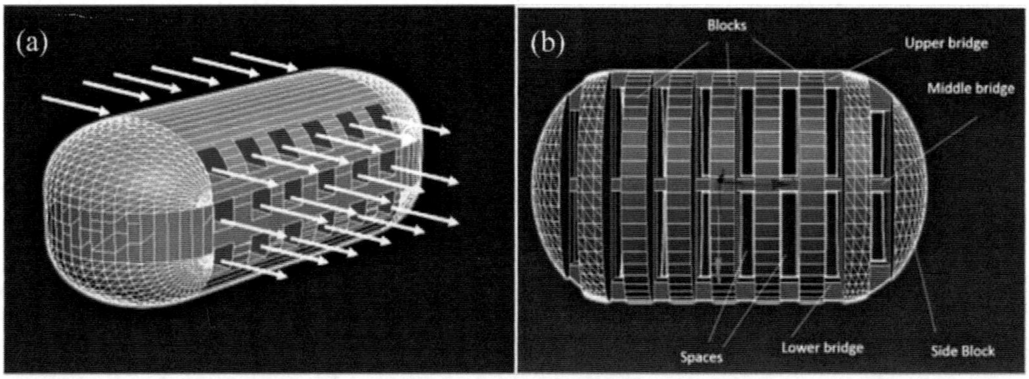


**Figure 5.** Schematic illustration: (a) Channelled caplets; (b) Gaplets prepared using FDM 3D printing [64, 65].

FDM 3D printing technology can also be used to build hollow capsule shells with separate dual compartments [67]. Each compartment may be constructed of the same or different materials and/or thicknesses. Since these compartments are precisely sealed, they can be filled with powders or even liquids [68]. This research is important for the development of personalized drug delivery systems combining various medications (e.g. metformin and glimepiride [69]) or doses of drugs, in particular because drugs can be co-formulated even though they are chemically incompatible. In comparison to traditional manufacturing technologies, FDM 3D printing technology allows on-demand manufacture of personalized patient-specific dose drug products with reliable dosing, high reproducibility and good mechanical properties.

In addition, the processing of complex dosage forms may be accomplished, such as through accommodating a broad variety of polymeric matrices and active ingredients, or through designing dosage forms that show personalized effects or target particular organs. The technology has further progressed, making it easier to integrate it with other innovations in order to build more dynamic and smart distribution mechanisms.

## 4. Conclusion

Additive Manufacturing (AM) technology is at a flourishing level, and its uses in adjuvant surgery, permanent prosthesis replacements and rehabilitation medical equipment are becoming more and more mature. Applying 3D printing to adjuvant therapy can help physicians optimize surgery preparation, promote diagnostic and preoperative instruction, develop doctor-patient contact, and assist in teaching research to increase the success rate of surgery. The individualized implant is more consistent with the initial anatomical form, which increases patient satisfaction and aesthetics following surgery and is helpful for postoperative rehabilitation. Medical use of direct printing of tissues and organs may resolve the problems of loss of organ donor and immune rejection.

While 3D printing technology has already been realized in clinical applications, the expense of 3D printing technology is still limited. The material support is insufficient, the types of printable materials are few, the price of biocompatible materials is high, and the price of 3D printing equipment and high-resolution imaging equipment are equally high, making research and development (R&D) and verification in clinical applications costly. In comparison, contrary to traditional approaches, there are a range of relevant comprehensive laws and regulations to guide this technology in getting it closer to the consumer. In fact, there are still many obstacles in the technological sector, while the cells can be directly printed at this point, it is still a long way to extend the concept of vitro tissues to clinical applications. Many similar studies have concentrated on the issue of processing, but there is a lack of breakthrough in the area of functionalization and implementation, and there is still a bottleneck in the development of complex organs. As an evolving technology, 3D printing continues to be implemented in the medical sector and has made a major contribution to clinical work, but there are also many problems that need to be tackled, such as the ongoing advancement of science and technology, continuous investment in relevant research, upgrading and changing management and regulatory frameworks, their success and application in practice.

**Acknowledgments**
This research work is strongly supported by the Structural Performance Materials Engineering (SUPREME) Focus Group and the Human Engineering (HEG) Focus Group, Universiti Malaysia Pahang, and School of Mechanical Engineering, Ningxia University, which provided the research materials and equipment.

**References**

[1] Vanderploeg A, Lee S-E and Mamp M 2017 The application of 3D printing technology in the fashion industry *International Journal of Fashion Design, Technology and Education* **10** 170-9
[2] Ventola C L 2014 Medical applications for 3D printing: current and projected uses *Pharmacy and Therapeutics* **39** 704
[3] Li D Z, Chen Z Y and Zhong J X 2019 Review on the Research and Practice of Construction 3D Printing in China from the Perspective of Multiple Participation *Journal of Civil Engineering and Mangement* **36** 1-7
[4] Zhang J L and Zhang J B 2016 The Application of 3D Printing Technology in the Innovation and Entrepreneurship Education in Colleges and Universities *Journal of Zhangjiakou Vocational and Technical College* **29** 60-3
[5] Iannotti J P, Weiner S, Rodriguez E, Subhas N, Patterson T E, Jun B J and Ricchetti E T 2015 Three-dimensional imaging and templating improve glenoid implant positioning *JBJS* **97** 651-8
[6] Li S H, Deng C, Zhou M W and Fu Z B 2018 Concepts, characteristics and values of precision medicine in the field of orthopedics *Chinese Journal of Tissue Engineering Research* **22** 2407
[7] Rengier F, Mehndiratta A, Von Tengg-Kobligk H, Zechmann C M, Unterhinninghofen R, Kauczor H-U and Giesel F L 2010 3D printing based on imaging data: review of medical applications *International journal of computer assisted radiology and surgery* **5** 335-41
[8] Butscher A, Bohner M, Doebelin N, Hofmann S and Müller R 2013 New depowdering-friendly designs for three-dimensional printing of calcium phosphate bone substitutes *Acta biomaterialia* **9** 9149-58
[9] Derby B 2012 Printing and prototyping of tissues and scaffolds *Science* **338** 921-6
[10] Saunders R E, Gough J E and Derby B 2008 Delivery of human fibroblast cells by piezoelectric drop-on-demand inkjet printing *Biomaterials* **29** 193-203
[11] Xu T, Zhao W, Zhu J-M, Albanna M Z, Yoo J J and Atala A 2013 Complex heterogeneous tissue constructs containing multiple cell types prepared by inkjet printing technology *Biomaterials* **34** 130-9
[12] Berman B 2012 3-D printing: The new industrial revolution *Business horizons* **55** 155-62
[13] Quanjin M, Rejab M, Idris M, Kumar N M, Abdullah M and Reddy G R 2020 Recent 3D and 4D intelligent printing technologies: A comparative review and future perspective *Procedia*

*Computer Science* **167** 1210-9
[14] Petrovic V, Haro J V, Blasco J R and Portolés L 2012 Additive manufacturing solutions for improved medical implants *Biomedicine* 147-80
[15] Rastogi P and Kandasubramanian B 2019 Breakthrough in the printing tactics for stimuli-responsive materials: 4D printing *Chemical Engineering Journal* **366** 264-304
[16] Boyard N, Christmann O, Rivette M, Kerbrat O and Richir S 2018 Support optimization for additive manufacturing: application to FDM *Rapid Prototyping Journal*
[17] e Costa E C, Duarte J P and Bártolo P 2017 A review of additive manufacturing for ceramic production *Rapid Prototyping Journal*
[18] Vickers N J 2017 Animal communication: when i'm calling you, will you answer too? *Current biology* **27** R713-R5
[19] Yan Q, Dong H, Su J, Han J, Song B, Wei Q and Shi Y 2018 A review of 3D printing technology for medical applications *Engineering* **4** 729-42
[20] Wang J and Zhang X D 2019 Magnetic resonance imaging-based 3D printing technology: advantages and prospects in clinical application *Chinese Journal of Tissue Engineering Research* **23** 4897
[21] Hazelaar C, van Eijnatten M, Dahele M, Wolff J, Forouzanfar T, Slotman B and Verbakel W F 2018 Using 3D printing techniques to create an anthropomorphic thorax phantom for medical imaging purposes *Medical physics* **45** 92-100
[22] Javaid M and Haleem A 2019 Industry 4.0 applications in medical field: a brief review *Current Medicine Research and Practice* **9** 102-9
[23] Hussain T, Gomez-Cia T and Valverde I 2015 Potential of 3D-printed models in planning structural interventional procedures *Interventional Cardiology* **7** 343-50
[24] Negi S, Dhiman S and Sharma R K 2014 Basics and applications of rapid prototyping medical models *Rapid Prototyping Journal*
[25] Awasthi S, Verma T, Agarwal M, Singh J V, Srivastava N M and Nichter M 2017 Developing effective health communication messages for community acquired pneumonia in children under five years of age: A rural North Indian qualitative study *Clinical Epidemiology and Global Health* **5** 107-16
[26] Ramola M, Yadav V and Jain R 2019 On the adoption of additive manufacturing in healthcare: a literature review *Journal of Manufacturing Technology Management*
[27] Bajpai P and Bajpai M 2010 Development of a high performance hybrid epoxy silicone resin for coatings *Pigment & Resin Technology*
[28] O'Neill J K, Laina V and Wilson S 2008 Referrals between consultants. The wait to be seen in a specialist plastic surgery hand clinic *Clinical Governance: An International Journal*
[29] Singh R 2014 Process capability analysis of fused deposition modelling for plastic components *Rapid Prototyping Journal*
[30] Jin Y, Li X, Campbell R I and Ji S 2018 Visualizing the hotspots and emerging trends of 3D printing through scientometrics *Rapid Prototyping Journal*
[31] Paul S C, van Zijl G P, Tan M J and Gibson I 2018 A review of 3D concrete printing systems and materials properties: Current status and future research prospects *Rapid Prototyping Journal*
[32] Wang R, Shang J, Li X, Wang Z and Luo Z 2018 Novel topological design of 3D Kagome structure for additive manufacturing *Rapid Prototyping Journal*
[33] Khalil S, Nam J and Sun W 2005 Multi- nozzle deposition for construction of 3D biopolymer tissue scaffolds *Rapid Prototyping Journal*
[34] Vasireddi R and Basu B 2015 Conceptual design of three-dimensional scaffolds of powder-based materials for bone tissue engineering applications *Rapid Prototyping Journal*
[35] Zheng Y, Wang Y, Chen R K, Deshpande S, Nelson N S, Buchman S R and Shih A J 2017 Tissue transformation mold design and stereolithography fabrication *Rapid Prototyping Journal*
[36] Drummer D, Cifuentes- Cuéllar S and Rietzel D 2012 Suitability of PLA/TCP for fused

deposition modeling *Rapid Prototyping Journal*
[37] Grasso M, Azzouz L, Ruiz-Hincapie P, Zarrelli M and Ren G 2018 Effect of temperature on the mechanical properties of 3D-printed PLA tensile specimens *Rapid Prototyping Journal*
[38] Hamad K, Kaseem M, Yang H, Deri F and Ko Y 2015 Properties and medical applications of polylactic acid: A review *Express Polymer Letters* **9**
[39] Haleem A, Javaid M and Vaishya R 2018 Industry 4.0 and its applications in orthopaedics *Journal of clinical orthopaedics and trauma* **10** 615-6
[40] Yang C, Tian X, Liu T, Cao Y and Li D 2017 3D printing for continuous fiber reinforced thermoplastic composites: mechanism and performance *Rapid Prototyping Journal*
[41] Dreßler M, Röllig M, Schmidt M, Maturilli A and Helbert J 2010 Temperature distribution in powder beds during 3D printing *Rapid Prototyping Journal*
[42] Lanzetta M and Sachs E 2003 Improved surface finish in 3D printing using bimodal powder distribution *Rapid Prototyping Journal*
[43] Bibb R, Eggbeer D and Evans P 2010 Rapid prototyping technologies in soft tissue facial prosthetics: current state of the art *Rapid Prototyping Journal*
[44] Haleem A and Javaid M 2019 Industry 5.0 and its applications in orthopaedics *Journal of clinical orthopaedics and trauma* **10** 807-8
[45] Jiang J, Zhu L Y, Yang J F and Yang J Q 2014 Review of 3 dimensional printing technology in the medical field *Machine Design and Manufacturing Engineering* **43** 5-9
[46] Jardini A L, Larosa M A, Maciel Filho R, de Carvalho Zavaglia C A, Bernardes L F, Lambert C S, Calderoni D R and Kharmandayan P 2014 Cranial reconstruction: 3D biomodel and custom-built implant created using additive manufacturing *Journal of Cranio-Maxillofacial Surgery* **42** 1877-84
[47] Winder J, Cooke R, Gray J, Fannin T and Fegan T 1999 Medical rapid prototyping and 3D CT in the manufacture of custom made cranial titanium plates *Journal of medical engineering & technology* **23** 26-8
[48] Lee J-Y, Choi B, Wu B and Lee M 2013 Customized biomimetic scaffolds created by indirect three-dimensional printing for tissue engineering *Biofabrication* **5** 045003
[49] Wang K, Ho C-C, Zhang C and Wang B 2017 A review on the 3D printing of functional structures for medical phantoms and regenerated tissue and organ applications *Engineering* **3** 653-62
[50] Farooqi K M, Cooper C, Chelliah A, Saeed O, Chai P J, Jambawalikar S R, Lipson H, Bacha E A, Einstein A J and Jorde U P 2019 3d printing and heart failure: The present and the future *JACC: Heart Failure* **7** 132-42
[51] Zhang X Q, Li Y, TAN L W and Zhang S X 2018 Construction of three-dimensional models of cardiac fibrous skeleton and valves and application of the 3D printed models *Journal of Third Military Medical University* **40** 1251-57
[52] Chen W J, Chen Y X and Chen H H Application of 3D printing knee joint disease model based on MRI reconstruction technology in orthopaedics teaching *China Modern Medicine* **25** 159-61
[53] Ozbolat I T and Yu Y 2013 Bioprinting toward organ fabrication: challenges and future trends *IEEE Transactions on Biomedical Engineering* **60** 691-9
[54] He Y, Gao Q, Liu A, Sun M and Fu J Z 2019 3D bioprinting: from structure to function *Journal of Zhejiang University (Engineering Science)* **53** 407-19
[55] Laronda M M, Rutz A L, Xiao S, Whelan K A, Duncan F E, Roth E W, Woodruff T K and Shah R N 2017 A bioprosthetic ovary created using 3D printed microporous scaffolds restores ovarian function in sterilized mice *Nature communications* **8** 1-10
[56] Dolati F, Yu Y, Zhang Y, De Jesus A M, Sander E A and Ozbolat I T 2014 In vitro evaluation of carbon-nanotube-reinforced bioprintable vascular conduits *Nanotechnology* **25** 145101
[57] Alhnan M A, Okwuosa T C, Sadia M, Wan K-W, Ahmed W and Arafat B 2016 Emergence of 3D printed dosage forms: opportunities and challenges *Pharmaceutical research* **33** 1817-32
[58] Cunha-Filho M, Araújo M R, Gelfuso G M and Gratieri T 2017 FDM 3D printing of modified

drug-delivery systems using hot melt extrusion: A new approach for individualized therapy *Therapeutic Delivery* **8** 957-66

[59] Pietrzak K, Isreb A and Alhnan M A 2015 A flexible-dose dispenser for immediate and extended release 3D printed tablets *European journal of pharmaceutics and biopharmaceutics* **96** 380-7

[60] Solanki N G, Tahsin M, Shah A V and Serajuddin A T 2018 Formulation of 3D printed tablet for rapid drug release by fused deposition modeling: screening polymers for drug release, drug-polymer miscibility and printability *Journal of pharmaceutical sciences* **107** 390-401

[61] Verstraete G, Samaro A, Grymonpré W, Vanhoorne V, Van Snick B, Boone M, Hellemans T, Van Hoorebeke L, Remon J P and Vervaet C 2018 3D printing of high drug loaded dosage forms using thermoplastic polyurethanes *International journal of pharmaceutics* **536** 318-25

[62] Goyanes A, Fina F, Martorana A, Sedough D, Gaisford S and Basit A W 2017 Development of modified release 3D printed tablets (printlets) with pharmaceutical excipients using additive manufacturing *International journal of pharmaceutics* **527** 21-30

[63] Goyanes A, Buanz A B, Hatton G B, Gaisford S and Basit A W 2015 3D printing of modified-release aminosalicylate (4-ASA and 5-ASA) tablets *European Journal of Pharmaceutics and Biopharmaceutics* **89** 157-62

[64] Sadia M, Arafat B, Ahmed W, Forbes R T and Alhnan M A 2018 Channelled tablets: An innovative approach to accelerating drug release from 3D printed tablets *Journal of Controlled Release* **269** 355-63

[65] Arafat B, Wojsz M, Isreb A, Forbes R T, Isreb M, Ahmed W, Arafat T and Alhnan M A 2018 Tablet fragmentation without a disintegrant: A novel design approach for accelerating disintegration and drug release from 3D printed cellulosic tablets *European Journal of Pharmaceutical Sciences* **118** 191-9

[66] Korte C and Quodbach J 2018 3D-printed network structures as controlled-release drug delivery systems: dose adjustment, API release analysis and prediction *AAPS PharmSciTech* **19** 3333-42

[67] Maroni A, Melocchi A, Parietti F, Foppoli A, Zema L and Gazzaniga A 2017 3D printed multi-compartment capsular devices for two-pulse oral drug delivery *Journal of Controlled Release* **268** 10-8

[68] Markl D, Zeitler J A, Rasch C, Michaelsen M H, Müllertz A, Rantanen J, Rades T and Bøtker J 2017 Analysis of 3D prints by x-ray computed microtomography and terahertz pulsed imaging *Pharmaceutical research* **34** 1037-52

[69] Gioumouxouzis C I, Baklavaridis A, Katsamenis O L, Markopoulou C K, Bouropoulos N, Tzetzis D and Fatouros D G 2018 A 3D printed bilayer oral solid dosage form combining metformin for prolonged and glimepiride for immediate drug delivery *European Journal of Pharmaceutical Sciences* **120** 40-52

# Design of a fixture for single incremental forming (SPIF) process based on design for assembly (DFA) methods

**Kamarul Al-Hafiz Abdul Razak[1], Ahmad Baharuddin Abdullah[1,*], Zahurin Samad[1] and Muhammad Farizuan Muhammad Fauzi[2]**

[1]Metal Forming Research Lab, School of Mechanical Engineering, Universiti Sains Malaysia, Engineering Campus, 14300 Penang, Malaysia
[2]Department of Mechatronic, ADTEC Kulim, 09000 Kulim, Kedah, Malaysia

*Corresponding author's email: mebaha@usm.my

**Abstract.** This paper discusses the application of Design for Assembly (DFA) Method, Boothroyd Dewhurst to select the best design of a fixture for Single Point Incremental Forming (SPIF). The 3D design of fixtures was created using SOLIDWORKS 2016 software. This research will focus on the manual handling and insertion of the fixtures by reducing the number of parts and comparing between two designed fixtures. The chosen design has been made based on the best design efficiency. The result found that the application of DFA improved the design efficiency, where welded-based Design B is 35.62% greater than design efficiency of fastener-based fixture Design A which is 23.04%.

**Keywords:** SPIF; Fixture Design; Design for Assembly.

## 1. Introduction

Sheet metal forming is one of most popular process in manufacturing sector. Due to high flexibility in demands, industry nowadays more focus on producing low production quantities, design customization, low cost, complex geometries and the most importantly the quality of the finish part [1]. One of the processes that can meet all these criteria is incremental sheet forming (ISF). Frequently, the ISF process can be determined based on the size of the fixture and the geometry of the part. Typically, the fixture consists of top plate and base plate by clamped the sheet by series of threaded fasteners [2].

### *1.1. Single point incremental forming*

Incremental sheet forming can be divided into single point incremental forming (SPIF) and two-point incremental forming (TPIF) [3]. One of the major advantages of SPIF is the process is die-less, more flexible and less tooling cost. Basically, in the SPIF, a forming tool movement controlled by a CNC machine or robot arm, push downward incrementally the clamped sheet metal and the process is continue until final profile obtained [4]. This process involves concentration force resulting from spinning of forming tool act on the sheet metal and friction forces occurred between the contacted surfaces produced the heat, thus, it facilitates forming of the required shape [5]. Figure 1 show the schematic diagram of the mechanism and basic parts involved in SPIF. Considering the understanding about SPIF process, the design of the fixture should have several criteria such as the blank space to downward force from rotating forming tools [6]. In addition, level of the fixture depends on the maximum depth of SPIF process and capability of the frame structure of the fixture to mount at worktable due to process stability. All this aspect needs to be considered in the design of the fixture for SPIF process [7].

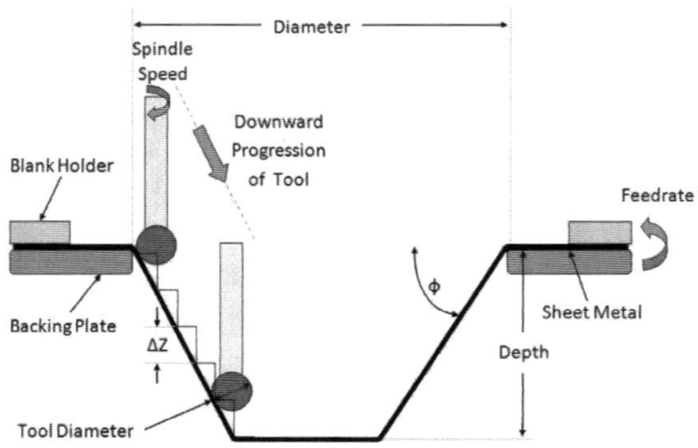


**Figure 1.** Working principle of SPIF.

### *1.2. Fixture design requirements*

In manufacturing processes, applications of fixture are wide and therefore requires interchangeable and repeatable processes to produce same part but maintain the process position [8]. In ISF process. several parameters are crucial to control the magnitude of localized force applied to the sheet [9].

### *1.3. Purpose of fixture*

Generally main purpose of fixture is to ensure the mounting of the workpiece is securable during the process and support to produce highly accuracy for the finished part [10]. Some basic purpose to develop the fixture is to quick, easy and consistently during locating the work piece. Besides, is to maintain the product quality consistently and guidance for cutting or forming process especially act as an assistant for unskilled operator and the most considerable using fixture is high interchangeability for mass production [11].

### *1.4. Elements of fixture*

Basically, all fixtures consist of the following elements:
*1.4.1.* Locators: Locators is a fixed place on the fixture to insert the workpiece. It is used to ensure the position of workpiece to maintain during the process.
*1.4.2.* Clamps: Clamps is mechanism an applied force onto the workpiece either internal (mount at the fixture itself) or external (G clamp) clamp, to secure the workpiece is firmly fixed.
*1.4.3.* Fixture Body: Main frame or structure of the fixture. Fixture body try to maintain the relationship with the other fixtures elements such as locators, clamps and supporting parts.
*1.4.4.* Support device can be fixed or adjustable in order to improve some weaknesses of the fixture.

## 2. Methodology

### *2.1. Design for assembly (DFA)*

Design for Manufacture and Assembly (DFMA) is a type of design approach that emphasizes on ease and efficiency of assembly process. By reducing the unnecessary part of a design, it possible to optimize the cost of manufacture and minimize an assemble time. DFMA terms is combination between Design for Manufacture (DFM) and Design for Assembly (DFA) [12]. DFA is a tool used to optimized the components in product design while DFM is a tool to optimize the manufacturing process in product design [13]. By referring DFA guidelines, it will assist the designer and manufacturer to understand the design rules that need to be follow when developing a product design.

According to Boothroyd Dewhurst method, the manual assembly process consists of two aspect. First is handling by means of orientating the parts and secondly is about the inserting by means of mating the parts to another group of parts [14]. The following experiment will be referred concept of alpha-beta orientation and applied digit codes from Manual Handling Classification and Manual Inserting Classification table [15]. Design efficiency can be calculated from the formula.

$$\text{Design Efficiency} = (\mathbf{3s}\ \boldsymbol{N_m}/\boldsymbol{T_m}) \times \mathbf{100}\% \tag{1}$$

Where Nm is Theoretical Minimum Number and Tm is Total Part of the part to develop the fixture.

*2.2. 3D design SOLIDWORKS*

SOLIDWORKS 2016 has been used to design the 3D fixtures. The fixtures design emphasized on the basic technique to draw on 2D with specific dimensions, then convert to 3D by extrude features. Single part was designed in "Part" section. While combining of all parts to become one complete product will be carried out in "Assembly" section in SOLIDWORKS. The design of the both fixtures are based on the dimensions of the workpiece. Aluminium sheet type of AA6061-T6, AA5052-H32 and AA1100-H14 has been used with standard size 200mm x 170mm with 3mm thickness. Material structure for both design of fixtures used is aluminium profile 40mm x 40mm with same length for the Design Fixture A and different length for the Design Fixture B. Figure 2 and Figure 3 show the envelope dimensions for both fixtures.

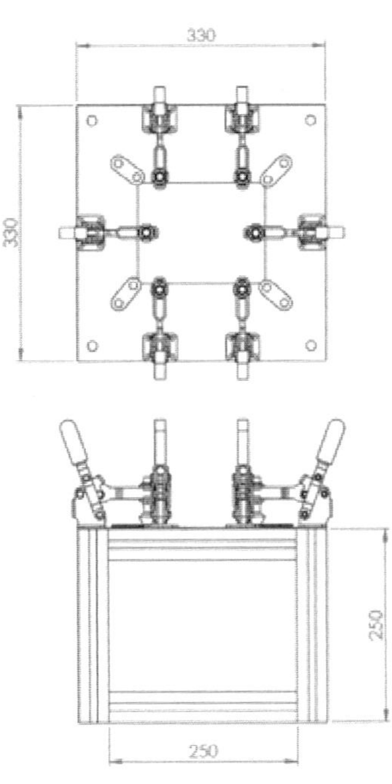


**Figure 2.** Dimension Design Fixture A.

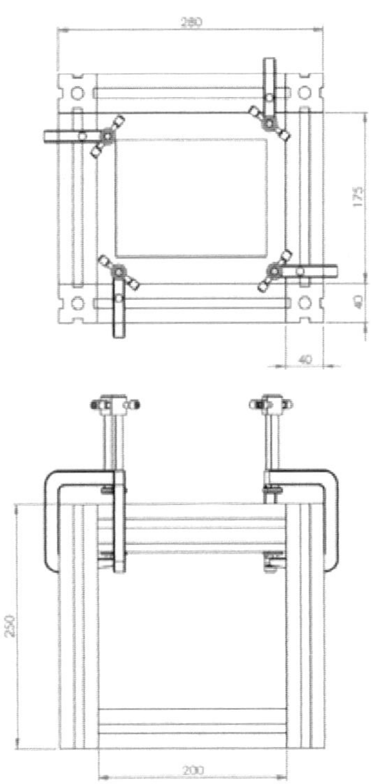


**Figure 3.** Dimension Design Fixture B.

## 3. Result and discussions

This section will focus on design analysis to determine the best design efficiency from both fixtures to achieve the objective of this research.

*3.1. Design fixture A*

This study will emphasis on of Design Fixture A. The design of the fixture creates by SOLIDWORKS 2016 software. Figure 4 show the Design Fixture A and Figure 5 show the exploded view of Design Fixture A.

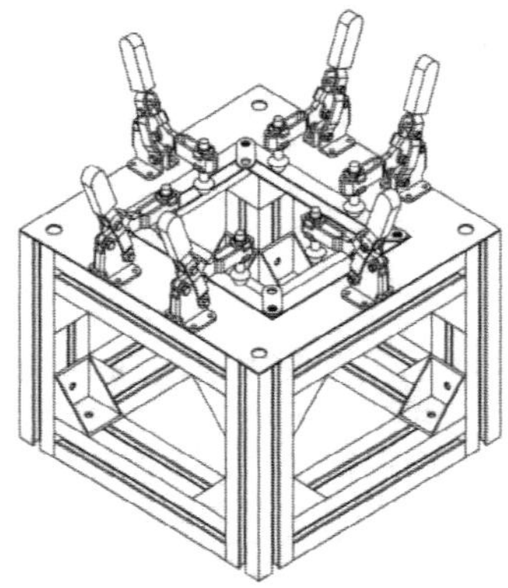

**Figure 4.** Design Fixture A.

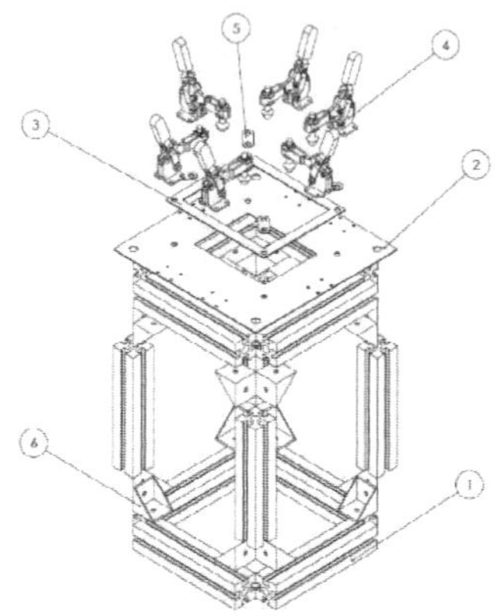

**Figure 5.** Exploded view of Design Fixture A.

Before determine the Theoretical Minimum Number (Nm), Design Fixture A need to disassemble to count the Total Part (Tm). Table 1 show the details part for Design Fixture A.

**Table 1.** Total parts of Design Fixture A.

| Part no | Name | Quantity | Material | Function |
|---|---|---|---|---|
| 1 | Frame | 12 | Aluminium Profile | Fixture main frame. |
| 2 | Layer | 1 | Mild Steel | Base for the workpiece |
| 3 | Cover | 1 | Aluminium Plate | Workpiece Locator |
| 4 | Toggle Clamp | 6 | Stainless Steel | Clamping the workpiece |
| 5 | Tighten | 4 | Aluminium Plate | Tighten the workpiece |
| 6 | Bracket | 16 | Aluminium | Connector between Frame structure |

*3.1.1.* Theoretical minimum number

Based on the Table 1, Total Parts for Design Fixture A are 40. To determine the Theoretical Minimum Number (Nm) of Design Fixture A, the parts list will be evaluating to obtain the Nm by mark 1 for essential part and 0 for non-essential part as shown in Table 2. Based on Table 2, total of Theoretical Minimum Number (Nm) of Design Fixture A are 36.

*3.1.2.* Operation time

Operation Time will be calculating based on the concept of $\alpha + \beta$ orientation and to estimate the handling and inserting time can be refer from Manual Handling Chart and Manual Inserting Chart. Table 3 shows data for the Assembly time for Design Fixture A.

**Table 2.** Theoretical Minimum Number (Nm) of Design Fixture A.

| Part no | Name | Quantity | Theoretical Minimum Number, Nm |
|---|---|---|---|
| 1 | Frame | 12 | 1 |
| 2 | Layer | 1 | 1 |
| 3 | Cover | 1 | 1 |
| 4 | Toggle Clamp | 6 | 1 |
| 5 | Tighten | 4 | 0 |
| 6 | Bracket | 16 | 1 |

**Table 3.** Assembly Time of Design Fixture A.

| Operation | Name | No of items | α | β | α+β | Handling Code | Handling Time, s | Insertion Code | Insertion Time, s | Total Operating Time, s |
|---|---|---|---|---|---|---|---|---|---|---|
| 1 | Beam | 12 | 0 | 180 | 180 | 1 5 | 2.57 | 5 9 | 12 | 174.84 |
| 2 | Layer | 1 | 0 | 180 | 180 | 0 4 | 2.18 | 0 0 | 1.5 | 3.68 |
| 3 | Cover | 1 | 0 | 180 | 180 | 0 4 | 2.18 | 0 0 | 1.5 | 3.68 |
| 4 | Toggle Clamp | 6 | 180 | 180 | 360 | 1 8 | 3 | 3 8 | 6 | 54 |
| 5 | Tighten | 4 | 0 | 180 | 180 | 0 2 | 1.88 | 3 8 | 6 | 31.52 |
| 6 | Bracket | 16 | 0 | 180 | 180 | 1 3 | 2.06 | 4 9 | 10.5 | 200.96 |
| | | | | | | | Total Operation Time | | | 468.68 |

*3.1.3.* Design efficiency

Based on the Table 3, the total operation time is 468.68 second. Design Efficiency can be calculated by using equation (1)

$$\begin{aligned} \text{Design Efficiency} &= (3s\ N_m/T_m) \times 100\% \\ &= ((3s\ x\ 36)/468.68) \times 100\% \\ &= 23.04\% \end{aligned}$$

*3.2. Design fixture B*

The design of the Design Fixture B also created by SOLIDWORKS 2016 software. Figure 6 show the Design Fixture B and Figure 7 show the exploded view of Design Fixture B. Before determine the Theoretical Minimum Number (Nm), Design Fixture B need to disassemble to count the Total Part (Tm). Table 4 show the details part for Design Fixture B.

*3.2.1.* Theoretical minimum number

Based on the Table 4, Total Parts for Design Fixture B are 17. To determine the Theoretical Minimum Number (Nm) of Design Fixture B, the parts list will be evaluating to obtain the Nm by mark 1 for essential part and 0 for non-essential part as shown in Table 5. Based on Table 5, total of Theoretical Minimum Number (Nm) of Design Fixture A are 17.

*3.2.2.* Operation Time

Operation Time will be calculating based on the concept of α + β orientation and to estimate the handling and inserting time can be refer from Manual Handling Chart and Manual Inserting Chart. Table 6 shows data for the Assembly time for Design Fixture B.

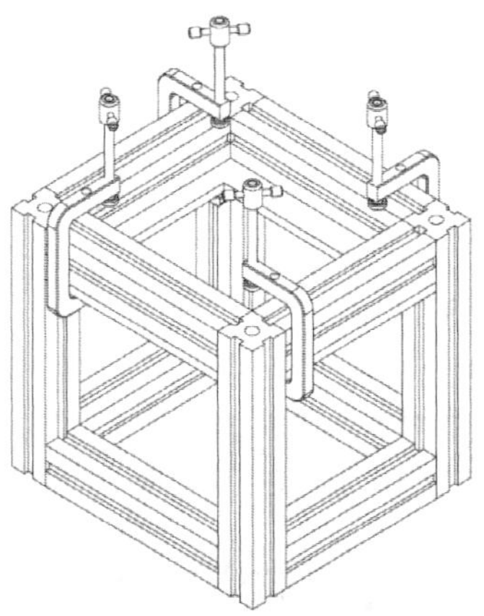

**Figure 6.** Design Fixture B.

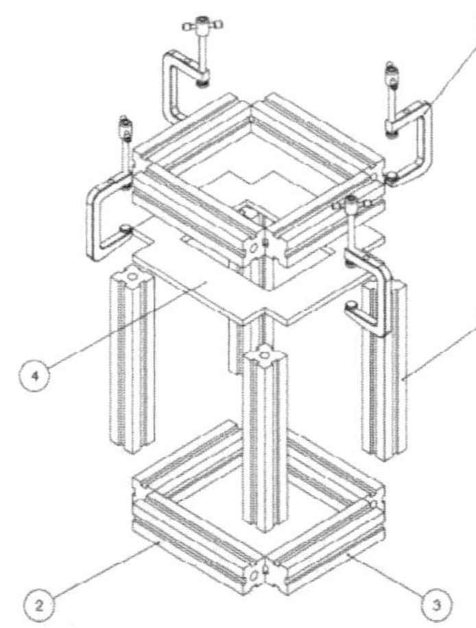


**Figure 7.** Exploded view of Design Fixture B.

**Table 4.** Total parts of Design Fixture B.

| Part no | Name | Quantity | Material | Function |
|---|---|---|---|---|
| 1 | Frame A | 4 | Aluminium Profile | Fixture main frame. |
| 2 | Frame B | 4 | Mild Steel | Base for the workpiece |
| 3 | Frame C | 4 | Aluminium Plate | Workpiece Locator |
| 4 | Base | 1 | Stainless Steel | Clamping the workpiece |
| 5 | G Clamp | 4 | Aluminium | Connector between Frame structure |

**Table 5.** Theoretical Minimum Number (Nm) of Design Fixture B.

| Part no | Name | Quantity | Theoritical Minimum Number, Nm |
|---|---|---|---|
| 1 | Frame A | 4 | 1 |
| 2 | Frame B | 4 | 1 |
| 3 | Frame C | 4 | 1 |
| 4 | Base | 1 | 1 |
| 5 | G Clamp | 4 | 1 |

**Table 6.** Assembly Time of Design Fixture B.

| Operation | Name | No of items | α | β | α+β | Handling Code | Handling Time, s | Insertion Code | Insertion Time, s | Total Operating Time, s |
|---|---|---|---|---|---|---|---|---|---|---|
| 1 | Frame A | 4 | 0 | 180 | 180 | 0 7 | 2.65 | 3 8 | 6 | 34.6 |
| 2 | Frame B | 4 | 0 | 180 | 180 | 0 7 | 2.65 | 3 8 | 6 | 34.6 |
| 3 | Frame C | 4 | 0 | 180 | 180 | 0 7 | 2.65 | 3 8 | 6 | 34.6 |
| 4 | Base | 1 | 0 | 180 | 180 | 0 2 | 1.88 | 0 6 | 5.5 | 7.38 |
| 5 | G clamp | 4 | 180 | 180 | 360 | 1 8 | 3 | 9 2 | 5 | 32 |
| | | | | | | | Total Operation Time | | | 143.18 |

*3.2.3.* Design Efficiency
Based on the Table 6, the total operation time is 143.18 second. Design Efficiency can be calculated by using equation (1)

$$\text{Design Efficiency} = (3s\, N_m/T_m) \times 100\%$$
$$= ((3s\ x\ 17)/143.18) \times 100\%$$
$$= 35.62\%$$

### *3.3. Comparison Data*

**Table 7.** Comparison Data between Design Fixture A and Design Fixture B.

| | Design Fixture A | Design Fixture B |
|---|---|---|
| Total Parts | 40 | 17 |
| Total Operation Time | 468.68 s | 143.18 |
| Design Efficiency | 23.04% | 35.62% |

Table 7 show the significant improvement on Design Fixture B compare than Design Fixture A in terms of Total Parts, Total Operation Time and Design Efficiency.

## 4. Conclusions

Based on this research, the study about Design for Assembly (DFA) is very useful in order to achieve the best design by focusing in manual handling and inserting method. The design efficiency value is very significant in order to achieve the objective this research. Results show the relationship between the total operation time is inverse proportional with design efficiency. Which means, low total operation time will increase design efficiency. Selection codes on manual handling and inserting charts also may affect the result. Even the value of design efficiency is not very high, but by compared between two design of fixture, as a conclusion, Design Fixture B is better than Design Fixture A. For future work, this research will be continued by considering the practicality of the design. Furthermore, failure analysis of structure using Finite Elements Analysis (FEA) software during the SPIF process. This research can contribute some knowledge in the jig and fixture development field.

**Acknowledgements**
The author would like to acknowledge the support from Universiti Sains Malaysia (USM) thru RUI Grant (1001/PMEKANIK/8014031), trainers from ADTEC Kulim and KISMEC Kulim for the advices and equipment facilities.

IOP Conf. Series: Materials Science and Engineering 1078 (2021) 012008 doi:10.1088/1757-899X/1078/1/012008

**References**

[1] Reddy N V and Cao J (2008). Incremental Sheet Metal Forming: A Review. *Indo-US Workshop on Smart Machine Tools*, 1–9.

[2] Siddiqi M U R, Corney J R, Sivaswamy G, Amir M and Bhattacharya R (2018). Design and validation of a fixture for positive incremental sheet forming. *Proceedings of the Institution of Mechanical Engineers, Part B: Journal of Engineering Manufacture*, ***232*(4)**, 629–643.

[3] Nimbalkar D H and Nandedkar V M (2013). Review of Incremental Forming of Sheet Metal Components. *International Journal of Engineering Research and Applications*, ***3*(5)**, 39–51.

[4] Koh, K H, Kang J G and Jung J Y (2016). The analysis of forming forces in single point incremental forming. *MATEC Web of Conferences*, ***81***, 10–13.

[5] Patel J, Rangrej S, Sharma U and Mahavidyalaya B V (2015). *Design and Analysis of Flexible Fixture for Single Point*. ***2*(5),** 2–6.

[6] Crenganiș M, Bârsan A, Racz S G and Daniela Iordache M (2018). *Single Point Incremental Forming Using Kuka Kr6-2 Industrial Robot-a Dynamic Approach*. (November), 133–140.

[7] Uheida E H (2016). *Development and optimisation of incremental sheet forming of titanium grade 2—Process mapping*. (December), 192.

[8] Radhwan H, Effendi M S M, Farizuan Rosli M, Shayfull Z and Nadia K N (2019). Design and Analysis of Jigs and Fixtures for Manufacturing Process. *IOP Conference Series: Materials Science and Engineering*, ***551*(1).**

[9] Dwivedy M and Kalluri V (2019). The effect of process parameters on forming forces in single point incremental forming. *Procedia Manufacturing*, ***29***, 120–128.

[10] Karad P A A, Wagh B, Shukla A, Gujar C and Pyata N (2016). *A Review on Design Consideration and Need of Fixture in Manufacturing Industries*. ***4*(2),** 428–432.

[11] More N K, Buktar R B, Ali S M and Samant S (2015). Design for Manufacture and Assembly (Dfma) Analysis of Burring Tool Assembly. *International Research Journal of Engineering and Technology*, 2395–56.

[12] Usman I M S and Wahab D A (2008). *DFM and DFA Approach on Designing Pressure Vessel*. (May 2015), 147–151.

[13] Ahmad M, Adeera N, Osman M H and Wahid M (2016). The Significant Improvement on The Design of Pedestrian Traffic Light Using Boothroyd Dewhurst Design for Assembly (DFA) Method: A Case Study. *Journal of Advanced Research Design ISSN (Online): 2462-1943*, ***25*(1)**, 1943–2462.

[14] Shukor A I A and Adam A (2018). Evaluation of design efficiency using boothroyd dewhurst method for PCB drilling machine products. *International Journal of Simulation: Systems, Science and Technology*, ***19*(5)**, 4.1-4.8.

[15] Teni M H M, Naroh A, Maliki K A A and Tukiran A (2019). *Design Development of a Coffee Maker using Design for Assembly Method*. **(2)**, 4041–4047.

# Study of wear performance of crosslinking UHMWPE acetabular liner for artificial hip joint made from CNC milling

**W D Lestari[1,*], A Nugroho[2], R Ismail[3], J Jamari[3] and A P Bayuseno[3]**

[1] Department of Mechanical Engineering, Faculty of Engineering, University of Pembangunan Nasional Veteran Jawa Timur, 60294, Indonesia
[2] PUTP (Pusat Unggulan Teknologi Plastik) Politeknik ATMI, 57102, Indonesia
[3]Department of Mechanical Engineering, Faculty of Engineering, Diponegoro University, 50275, Indonesia

*Corresponding author's email: wahyu.dwi.tm@upnjatim.ac.id

**Abstract.** Crosslinked ultra-high molecular weight polyethylene (UHMWPE) is a preferred material used as a bearing material in the artificial hip joint. The purpose of this study was to analyze the effect of dose cross-linking acetabular liner manufactured by CNC milling on wear depth. The wear test is carried out with a tribometer machine by loading the acetabular liner in a certain cycle. The test results found that specimens given a crosslink dose of 50 kGy with a load of 800 N resulted in a greater wear depth compared to specimens given a crosslink dose of 100 kGy with a load of 800 N. Specimen with crosslink dose 50 kGy yield greater wear depth than specimens given a crosslink dose of 100 kGy. Based on these results it can be concluded that the greater the crosslink dose given to the specimen will increase the wear resistance of the specimen so that the test results in a smaller wear depth.

**Keywords:** Crosslinking; UHMWPE; Wear Depth; Artificial Hip Joint.

## 1. Introduction

The joint is the link between the bones so that the bones can be moved. The hip joint is a very important joint in the lower limbs, because it carries the greatest load compared to other joints in the human body. Some of the permanent damage that can occur in the hip joint is the result of calcification, aging and accidents that require replacement with an artificial hip joint. Artificial hip joint replacement is a surgical method to replace the damaged part with a metal, ceramic, or polymer model so that the hip can move normally [1]. A total of 427,181 joint replacement procedures were performed in the United States of Amerika between 2012 and 2015 [2]. The most commonly used artificial pelvic pairs are stainless steel and UHMWPE pairs. UHMWPE with a molecular mass between 3500 and 7500 kDa has properties such as, lubricity, chemical stability, abrasion and impact resistance which are widely used in orthopedic applications [3]. However, it is known that the wear particles of the UHMWPE polymer in the artificial hip joint have been a major factor causing osteolysis and joint loss, thus limiting the lifetime of the implant and ultimately requiring surgical revision. Based on this, the wear and tear of UHMWPE has been studied extensively in industry and academia [4], where most of the focus is on the negative effects of wear debris on the body [5], and techniques to reduce the total wear volume, by means of crosslinking [6]. Crosslinking is the process in which carbon atoms of same or different polyethylene chains are combined together to form a threedimensional network structure [7]. One way of the cross-linking process is by immersing UHMWPE specimens in 1,7-octadiene, methylacethylene and ethylene, and exposing the sample to an electron beam at doses of 25, 50 and 100 kGy [10]. Therefore, cross-linking assistances in reducing elongation at break and yield index while improving creep resistance, impact strength, and resistance to fatigue and environmental stress cracking resistance (ESCR) [11]. Crosslinked ultra-high molecular weight polyethylene (UHMWPE) is extensively operated as a bearing material in artificial hip joint components [12].

Several researchers conducted research on crosslinked UHMWPE. Saiko [13] investigation the effect of lubrication on wear factor of highly crosslinked UHMWPE and unirradiated UHMWPE using multidirectional pin on disk tests. The outcomes presented that the wear factor of high cross-linked UHMWPE was 5.7 times lower than that of unirradiated UHMWPE. Garcia et al [2] studied the effect of oxygen plasma on crystallinity, hardness, surface chemistry and structural composition of UHMWPE at different treatment times. The results show that the cross-linking of the polymer can increase micro-hardness. Vidya et al [11] perform cross linking to reduce the wear rate of UHMWPE and compare its properties with conventional UHMWPE and its biocompatibility. The results show that an increase in cross-linking efficiency is followed by an increase in the wear resistance and mechanical properties of UHMWPE. Sirimamilla and Rimnac [14] investigated the resistance of crosslinked UHMWPE to crack initiation for clinically relevant notches under static and cyclic loading conditions. The results show that cross-linked UHMWPE is more resilient to crack initiation from the notch under static loading conditions compared to fatigue loading. The objective of this study was to investigate the effects of radiation dose on UHMWPE acetabular liner wear rate under cyclic/static loading condition. This study tested the wear depth of acetabular liner as a result of the optimization of the crosslinked UHMWPE cutting parameters with a load variation of 400 N and 800 N using a tribometer testing machine in dry condition (without lubrication).

## 2. Materials and methods

### *2.1. Materials / Sample preparation*

The specimen used in this study was the acetabular liner specimen and its partner, the femoral head. The acetabular liner specimen is made of UHMWPE polymer material, while the femoral head is made of SS 316L material. Both materials are approved by their nature to be used in hip joint replacement. The computer numerical control milling machine (YSM 1020 EV 20) with 3 axes used to process the acetabular liner under dry condition (without lubricant). The acetabular liner specimen produced had an outer radius of 18.5 mm and an inner radius of 14.1 mm which corresponded to dimensional tolerances. According to ASTM F2033-12 specifications for acetabular liners, the results of this milling process must have a dimension tolerance between +0.3 and -0.0 and a roughness value below 2 μm. The 4 acetabular liner sample (Figure 1), was then given a crosslink treatment with a dose of 50 KGy and 100 KGy. Before testing, the specimens are cleaned using 70% alcohol to remain sterile.

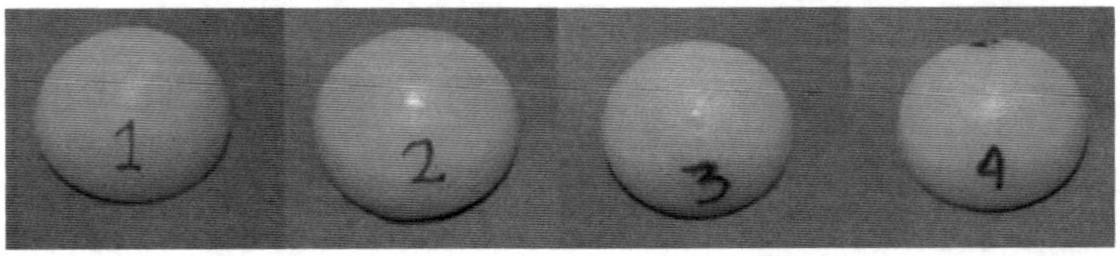


**Figure 1.** Acetabular liner specimen with crosslink 50 kGy (1 and 3), 100 kGy (2 and 4).

### *2.2. Wear test using hip joint simulator*

The joint prototype resulting from the fabrication process should be verified in a simulator that models biomechanical loads, motion, and environmental conditions in the body [15]. Acetabular liner wear testing results of the milling process in this study were carried out with tribometer test equipment that has been specially modified to carry out testing (Figure 2). The working principle of this machine is like a human hip joint where the femoral head is in contact with the human femur, and the cup is the human pelvis. This test machine has an acceptable maximum load of 3000 N. The resultant contact strength of the fixed hip joint is 2500 N, corresponding to about 70 kg for 3-4 times body weight [16]. The test design applied is illustrative and practical with respect to movement and physical loading between the femoral head and the acetabular liner.

**Figure 2.** Tribometer machine.

The two parameters that underlie this test are the number of cycles and the force applied. As with previous studies conducted by [17], [18], these acetabular liner test parameters are in accordance with ISO 14242 standards. ISO 14242 standards state that wear testing up to 5 million cycles. Wang et al [19] tested specimens for intervals of a total of 2 million cycles by physiological loading under diluted alpha serum lubricating conditions. Saikko and Shen [20] tested the UHMWPE acetabular liner for up to 5 million cycles. The number of cycles in this study is dissimilar from earlier studies. The focus of this research is to observe the effect of the machining process of acetabular liner products which are then given crosslink treatment on their wear value. The amount of force applied to the wear test is in accordance with the load received by the pelvis on the human body, which is 800N [21]. The acetabular liner sample was attached to the femoral head and packed in the specimen chamber as shown in Figure 3. The specimens were then tested for 30000 cycles. The wear test was carried out at room temperature with a temperature of 20°C with conditions without lubrication. The process of testing the UHMWPE crosslink acetabular liner wear is shown in the flow diagram in Figure 4. Measurement of roughness, dimensional accuracy, and mass is also done to find out the comparison between before and after wear testing.

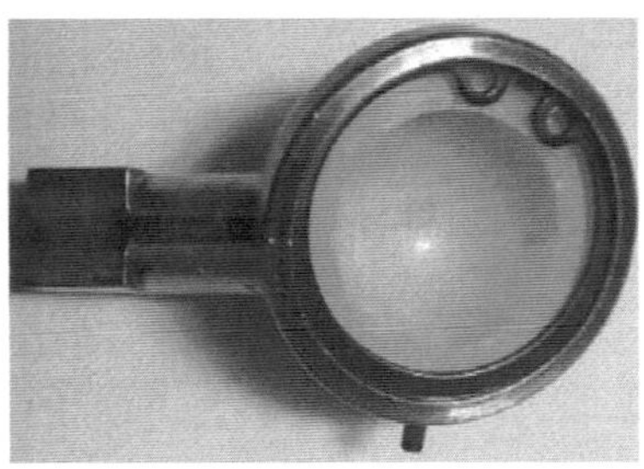

**Figure 3.** Setting acetabular liner on the tribometer machine chamber element.

### *2.3. Measurement of surface shape*

The measurement of dimensional accuracy after all machining steps and wear tests had been finished by a coordinate measurement machine (CMM). Measurement after the machining process is carried out to determine the difference in dimensions between the design and the product machining results. The measurement of dimensional accuracy after wear testing is to observe the difference in dimensions that occur due to wear. The measurement procedure using CMM follows the standard found in the tool. In this study, the measurement of dimensional accuracy using a CMM machine was carried out by measuring the acetabular liner specimens in a space that had length, width and height which were then translated into X, Y and Z Cartesian coordinate systems and converted into diameters. The wear volume of implants (hip) both in vivo and ex vivo in some experiments was also measured using CMM, which illustrates its potential in calculating the exact wear volume with suitable [22].

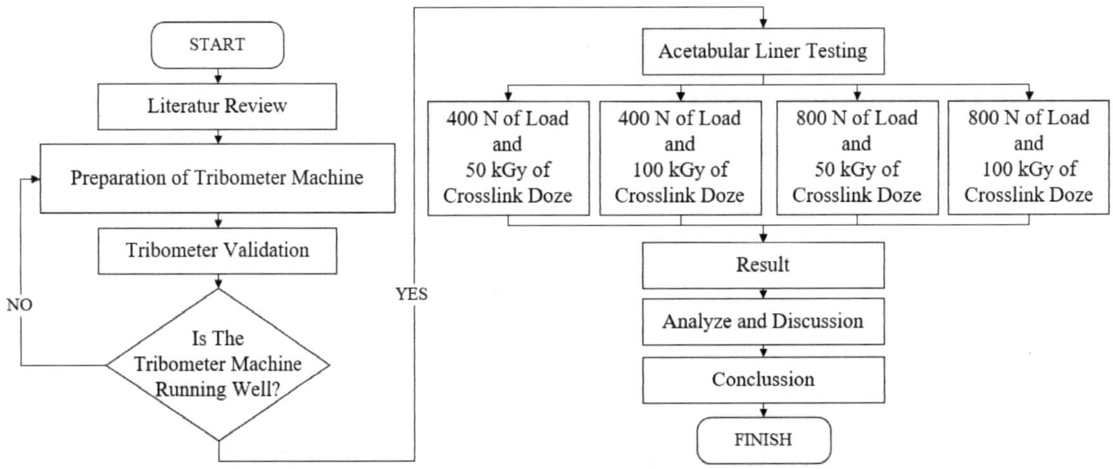


**Figure 4.** Flow chart of the UHMWPE acetabular liner testing procedure with a tribometer machine.

## 3. Experimental results

### *3.1. Measurement of wear mass*

The specimen was weighed to determine the change in the volume of the acetabular liner after testing. Figure 5 shows the graph of the test results of acetabular liner specimens with crosslink doses of 50 KGy and 100 KGy and with loadings of 400 N and 800 N. Based on the graph, it can be seen that the value of the wear depth on specimen 1 reaches 0.417 mm for 30,000 test cycles. For specimen 2, it shows a wear depth value of 0.359 mm, specimen 3 reaches a wear depth value of 0.377 mm and specimen 4 reaches a wear depth value of 0.266 mm. The following table 1 shows the initial weight and final weight of the acetabular liner after testing.

After the initial and final weighing after testing, it was found that specimen 1 and 4 had a difference in initial and final weight of 0.0005 grams, specimen 2 had a weight difference of 0.0004 grams and specimen 3 had a weight difference of 0.0007 grams. From these data, it can be seen that specimen 3 has a greater difference in initial and final weight after testing than specimens 1, 2 and 4. The change in weight for these 4 specimens proved that the wear occurred during the test which was simulated for a 5 year gait cycle. Figure 6 is a graphical comparison of the wear mass for specimens 1, 2, 3 and 4.

**Table 1.** The initial and final weight of the acetabular liner after testing.

| Wear Mass | | | |
|---|---|---|---|
| Specimens | Before testing (gr) | After testing (gr) | Deviation (gr) |
| Specimen 1 | 6.7535 | 6.753 | 0.0005 |
| Specimen 2 | 5.685 | 5.6846 | 0.0004 |
| Specimen 3 | 6.4644 | 6.4637 | 0.0007 |
| Specimen 4 | 6.5237 | 6.5232 | 0.0005 |

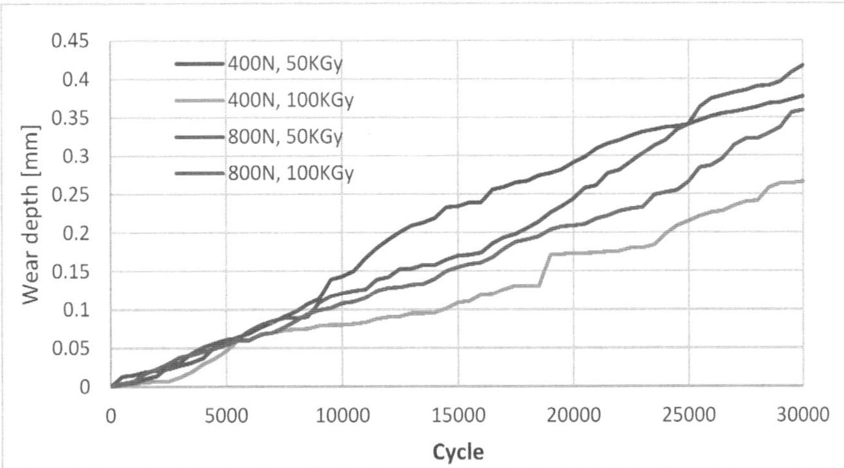


**Figure 5.** Comparison of the wear depth test results of acetabular liner specimens.

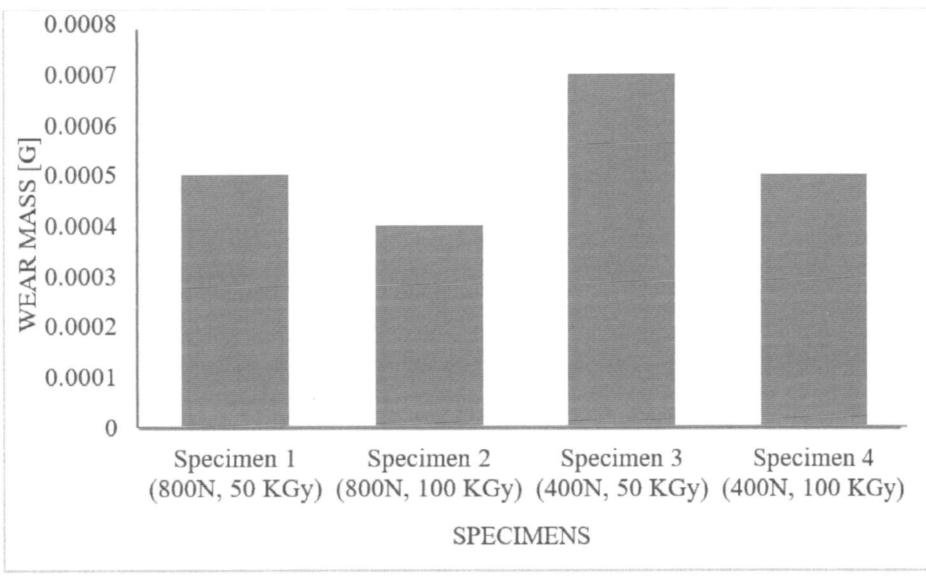


**Figure 6.** Comparison chart of wear mass specimens 1, 2, 3 and 4.

The mass measurement of the UHMWPE acetabular liner specimen was carried out to determine the difference in mass that occurred after wear testing. The results of mass measurement of acetabular liner specimens are shown in Table 4.8. Based on the table, it can be seen that the UHMWPE acetabular liner specimen with crosslink variation shows that specimen 3 has a greater difference between the beginning and the end after testing than specimens 1, 2, and 4. The weight change in all UHMWPE acetabular liner specimens proves that wear occurs during the test which is simulated for a 5 year gait cycle.

Based on Table 1, it can be seen that the mass difference proves the presence of particles and wear that occurred during the testing process. However, for the test of 30,000 cycles, the particles are still not clearly visible because the wear mass that occurs is still very small. Based on the table, it can also be seen that the crosslink treatment has an effect on the change in mass (wear mass), where the results show that if the greater the crosslink dose is given, the wear resistance is greater and the mass change is smaller.

### *3.2. Dimensional accuracy measurement of acetabular liner*

Understanding the level of laboratory wear is an important characteristic in preclinical authentication of prostheses. In this study, the wear test of the acetabular liner component made from UHMWPE on the hip joint simulator was carried out for 30,000 cycles to evaluate the wear behavior of the material. The elements are machined and finished following an organized standard implant specification. The measurement of joint wear and its evaluation is very important for the design and use of new materials. This is because it is imperative for researchers to continually create new joints and evaluate new materials to continuously improve implant design and performance [15]. To do this, a coordinate measurement machine (CMM) is needed to measure dimensional accuracy before and after wear testing.

The process of machining acetabular liner products made from UHMWPE in this study was carried out in a laboratory and obtained a quality product. Dimensions of UHMWPE acetabular liner products were measured using CMM and all data were recorded. Measurements were made of the outer and inner radii of the acetabular liner. The target dimensions of outer and inner radii of acetabular liner based on CAD design are 18.50 mm and 14.10 mm, respectively. According to ASTM F2033-12 (mm), the product must be within dimensional tolerances in the range of +0.3 mm and -0.0 mm. The dimensional accuracy for the four components of the acetabular liner made from UHMWPE which is subjected to wear testing is given in Table 2.

**Table 2.** Inner acetabular liner dimensional accuracy value before and after wear testing.

| Acetabular Liner | Dimensional Accuracy (mm) | |
|---|---|---|
| | Before | After |
| Speciment 1 | 14,384 | 14,304 |
| Speciment 2 | 14,469 | 14,392 |
| Speciment 3 | 14,234 | 14,154 |
| Speciment 4 | 14,466 | 14,386 |

Uddin et al [23] reported that the average linear and volumetric wear rates of acetabular components made from polyethylene as measured by CMM were found to be 0.12 mm / year and 37.18 $mm^3$ / year, respectively. The linear and volumetric wear rates of acetabular liner with XLPE material were also reported by Uddin [24] to be 0.024 mm / year and 4.5 $mm^3$ / year. In this study, the four acetabular liner specimens with crosslink treatment had almost the same difference in dimensional accuracy, which was around 0.08 mm during the 30,000 cycle test, where this result was smaller than the research conducted by Uddin et al [23].

## 4. Discussion

### *4.1. Validation*

Validation is needed to ensure whether the tool used is good enough before the actual test is carried out using a tribometer machine. The validation method used in this study is to calculate the value of the wear coefficient resulting from the test using the Archad equation with the radius of the femoral head 14 mm and the radius of the acetabular liner 14.1 mm. The load used for validation is the same as that used for testing, namely 800N with as many as 30000 cycles. The value of the coefficient of wear in this validation is $1.945 \times 10^{-4}$ $mm^3$ / Nm. This value is then compared with research from Dowson and Jobbins [25] by including the parameters in the formula generated from the study. Dowson and Jobbins [25] conducted a test and obtained two values for the wear coefficient of the two tests, namely $1.35 \times 10^{-7}$ $mm^3$ / Nm and $4.49 \times 10^{-7}$ $mm^3$ / Nm. The validation results can be seen in Figure 7. The wear depth formula from Dowson and Jobbins, [25] is:

$$p = \frac{0,001186}{\pi}\left(\frac{kBN}{R_1}\right)\left[1 + \sqrt{1 + 1686\pi\frac{R_1(R_2 - R_1)}{kBN}}\right] \quad (1)$$

where P is penetration or wear depth (mm), k is the wear coefficient ($mm^3\ N^{-1}\ m^{-1}$), B is body weight (N), N is the number of cycles, $R_1$ is the radius of the femoral head (mm), and $R_2$ is the radius of the acetabular liner (mm).

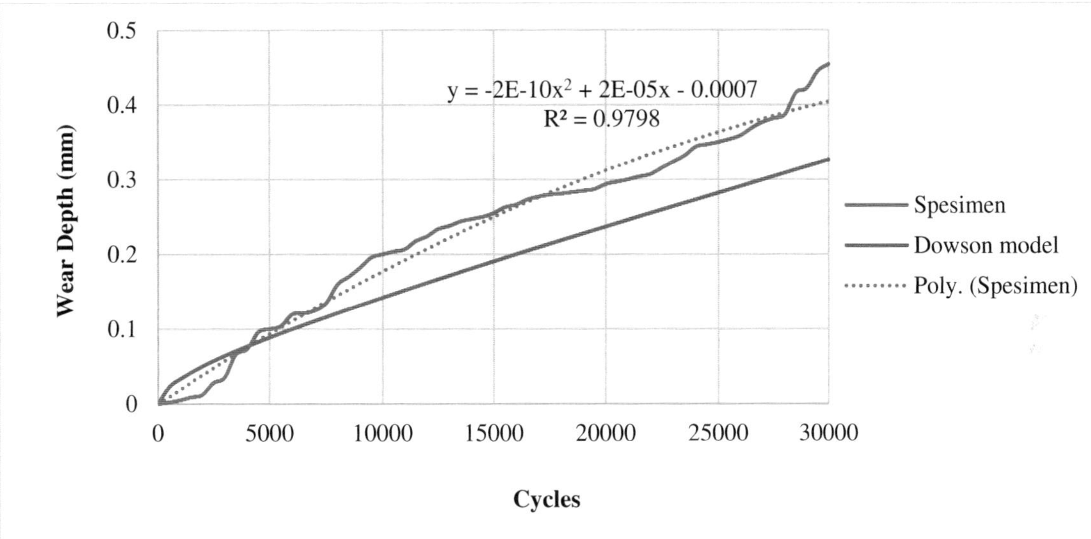


**Figure 7.** Testing validation with the Dowson model.

The calculation of the wear coefficient for the validation sample of this tribometer using a UHMWPE acetabular liner specimen with a crosslink variation of 50 kGy and a load of 400 N yields a value of 4.09 x $10^{-4}$ $mm^3$/ Nm. This value is then entered into the Dowson equation formula and input into the calculation. The calculation results obtained are then plotted on a graph comparing the test results with Dowson's [26] model, the results of which can be seen in Figure 7. The validation results show that the two graphs have almost the same trend. However, in the test results the wear value obtained in this study is greater than the results from the Dowson equation. Thus it can be concluded that the tools used to carry out the test are good enough so that they can produce the same wear trend as the research.

### *4.2. Wear behaviour*

The wear rate of a bearing material can be estimated through testing for various specifications and different activities with orientation to the actual body conditions. In this study, the wear test of the acetabular liner made from UHMWPE was carried out using a tribometer machine on the current cycle. The wear test of the acetabular liner specimens with UHMWPE was carried out by varying the load and dose crosslink for 30,000 cycles. There were 4 acetabular liner samples that were subjected to wear testing with these variations as shown in Table 3.

The data generated from the wear test of UHMWPE acetabular liner specimens are the number of cycles and the depth of wear. Then the data is processed and a graph of the relationship between the number of cycles and the depth of wear is made as shown in Figure 5. Based on the graph, it can be seen that the value of the wear depth on the crosslink specimen shows the value of the wear depth on specimen 1 which reaches 0.417 mm, specimen 2 reaches 0.359 mm, specimen 3 reaches 0.377 mm and specimen 4 reaches a wear depth value of 0.266 mm.

**Table 3.** Acetabular liner specimen variation.

| Specimens of Acetabular Liner | Load | Crosslink Dose |
|---|---|---|
| Specimen 1 | 800 N | 50 kGy |
| Specimen 2 | 800 N | 100 kGy |
| Specimen 3 | 400 N | 50 kGy |
| Specimen 4 | 400 N | 100 kGy |

The acetabular liner wear test with crosslink and load variations showed almost the same trend. Based on the graph in Figure 5, it can be seen that each variation of the crosslink dose and load given shows different results. Specimens given a crosslink dose of 50 kGy with a load of 800 N resulted in a greater wear depth compared to specimens given a crosslink dose of 100 kGy with a load of 800 N. Based on these results it can be concluded that the greater the crosslink dose given to the specimen will increase the wear resistance of the specimen so that the test results in a smaller wear depth. Based on the crosslink condition, the lowest wear coefficient value is owned by the specimen with the crosslink dose of 100 kGy with a load of 400 N, and the highest value is the specimen with the crosslink dose 50 kGy with a load of 400 N. The results of the calculation of the wear coefficient for the loading parameter of 800 N specimens with a crosslink dose of 50 kGy have a value of $1.46 \times 10^{-4}$ $mm^3$ / Nm and a specimen with a crosslink dose of 100 kGy has a value of $9.32 \times 10^{-5}$ $mm^3$ / Nm while for the loading parameter of 400 N the specimen with a crosslink dose of 50 kGy has a value of $4.09 \times 10^{-4}$ $mm^3$ / Nm and a specimen with a crosslink dose of 100 kGy has a value of $4.55 \times 10^{-5}$ $mm^3$ / Nm. The greater the crosslink specimen dose, the smaller the wear coefficient value.

The amount of polyethylene wear in the body generally ranges from 50-100 mg per year [27]. The value of polyethylene wear found in the simulator experiment is in the range of 20-35 mg per 1 million cycles [27][28][28]. In addition, the total wear of UHMWPE acetabular liner after testing for five million cycles was 7672 mg (EtO sterilized acetabular liner) and 6378 mg (gamma irradiated acetabular liner) [18]. Trommer and Maru [29] obtained a UHMWPE wear rate corresponding to a linear wear rate of 0.16 mm / year about 48 mg / 106 cycles. Shibo et al [29] resulting in a wear value of 31.73 mg / 1 million cycles and 15.20 mg / 1 million cycles with each loading 784 N and 392 N in distilled water lubrication. Bragdon et al [28] in his study resulted in mean annual wear from simulator tests similar to that obtained from metal on polymer (MoP) total pelvic reconstruction of 25 mg / 1 million cycles. The lowest weight loss of acetabular liner component made from UHMWPE after wear testing in this study was owned by a specimen with a crosslink dose of 100 kGy and a load of 800 N of 0.0004 g (0.4 mg / 30,000 cycles). The results of this experiment cannot be compared directly with previous studies that have used a total pelvic endoprosthesis simulator, because it uses a simplified wear mechanism.[17] [18]. However, the wear rate of UHMWPE acetabular liner from manufacturing using a CNC milling machine is in accordance with the in vivo wear rate requirements in the literature when plotted into a graph and calculated based on the one million cycle linear line equation [27]. Based on the results of the experiments conducted, it can be concluded that the variations in the load and the crosslink dose have an effect on the level of wear depth and the wear coefficient of the acetabular liner specimen.

## 5. Conclusion

Based on the experimental results obtained in this study, it is concluded that the trend of the acetabular liner test results obtained shows the same results, namely that the depth of wear continues to increase with the increasing number of cycles. Then for the results obtained from the two loading parameters and the dose cross-link given showed different results. This shows that the cross-linking treatment of the UHMWPE acetabular liner milling product has the effect of increasing resistance to wear depth in the normal gait cycle.

**Acknowledgements**
Thanks to PUTP (Pusat Unggulan Teknologi Plastik) Politeknik ATMI for facilitating the CMM machine to measure the accuracy dimensions in this research.

**References**
[1] Plumlee K G and Schwartz C J 2019 *Wear*, **426–427**, no. January, pp. 171–179.
[2] García A, Domínguez-Díaz M, Paniagua R, Hernández-Pérez J A, and Martínez H 2019 *Nucl. Instruments Methods Phys. Res. Sect. B Beam Interact. with Mater. Atoms*, **445**, no. February, pp. 8–12.
[3] Kurtz S M, *UHMWPE biomaterials handbook : ultra-high molecular weight polyethylene in total joint replacement and medical devices*. 2015.
[4] Bracco P, Bellare A, Bistolfi A, and Affatato S 2017 *Materials (Basel)* **10**, no. 7
[5] Dumbleton J H, Manley M T, and Edidin A A 2002 *J. Arthroplasty* **17**, no. 5.
[6] Oral E, Christensen S D, Malhi A S, Wannomae K K, and Muratoglu O K 2006 *J. Arthroplasty*, **21**, no. 4, pp. 580–591.
[7] Tamboli S M, Mhaske S T, and Kale D D 2004 *Indian J. Chem. Technol.* **11**, no. 6, pp. 853–864.
[8] Asano T, Akagi M, Clarke I C, Masuda S, Ishii T, and Nakamura T 2007 *J. Biomed. Mater. Res. - Part B Appl. Biomater.* **83**, no. 2, pp. 615–622.
[9] Muratoglu O K, Bragdon C R, O'Connor D O, Jasty M, and Harris W H 2001 *J. Arthroplasty*, **16**, no. 2, pp. 149–160.
[10] Bracco P, Brunella V, Luda M P, Zanetti M, and Costa L 2005 *Polymer (Guildf).* **46**, no. 24, pp. 10648–10657.
[11] Vidya V, Anoop C A, Jinan S, Anwar R, and Senthil Saravanan M S 2019 *Mater. Today Proc.*, **27** 2748–2751.
[12] Glyn-Jones S, Thomas G E R, Garfjeld-Roberts P, Gundle R, Taylor A, McLardy-Smith P, Murray D W 2015 *Clin. Orthop. Relat. Res* **473** no. 2, pp. 432–438.
[13] Saikko V 2019 *Wear* **430–431**, no. January, pp. 76–80.
[14] Sirimamilla A and Rimnac C M 2019 *J. Mech. Behav. Biomed. Mater.* **91** no. August 2018, pp. 366–372.
[15] Blunt L, Bills P, Jiang X, Hardaker C, and Chakrabarty G 2009 *Wear* **266** no. 3–4, pp. 424–431.
[16] Hua X, Wroblewski B M, Jin Z, and Wang L 2012 *Med. Eng. Phys.* **34** no. 3, pp. 318–325.
[17] Shibo W, Shirong G, Hongtao L, and Huang X 2010 *J. Biomimetics, Biomater. Tissue Eng.* **7**, pp. 7–25.
[18] Affatato S, Bordini B, Fagnano C, Taddei P, Tinti A, and Toni A 2002 *Biomaterials* **23** 1439–1446.
[19] Wang A, Lee R, Herrera L, and Korduba L 2013 *Wear* **301** no. 1–2, pp. 157–161.
[20] Saikko V and Shen M 2010 *Wear* **268** 617–621.
[21] Saikko V 2015 *Biotribology* **1–2** pp. 2–10.
[22] Spinelli M, Carmignato S, Affatato S, and Viceconti M 2009 *Wear* **267** 753–756.
[23] Uddin M S, Mak C Y E, and Callary S A 2016 *Wear* **364–365** 193–200.
[24] Uddin M S 2014 *Tribol. Trans* **57 (5)** 37–41.
[25] Dowson D and Jobbins B 1993 *Wear* **164** 162–164.
[26] Jin Z M, Dowson D, and Fisher J 1993 *Tribol. Ser* **25** 545–555.
[27] Kabo J M, Gebhard J S, Loren G, Amstutz H C 1993 *J. Bone Jt. Surg* **75-B (2)** 254–258.
[28] Bragdon C R, Connor D O, Lowenstein J D, Jasty M, and Syniuta W D 1996 *Proc Instn Mech Engrs* **210**.
[29] Trommer R M and Maru M M 2016 *Rev. Bras. Ortop* **52** 251–259.

# Investigation on linear and nonlinear dynamic equation for vehicle model in numerical simulation

**Li Maoqi**[1,2,*], **M I Ishak**[1,2] **and P M Heerwan**[1,2]

[1] Automotive Engineering Focus Group, Faculty of Mechanical and Automotive Engineering, Universiti Malaysia Pahang (UMP), 26600 Pekan, Pahang, Malaysia
[2] College of Engineering, Universiti Malaysia Pahang (UMP), 26300 Gambang, Pahang, Malaysia

*Corresponding author: limaoqi1995@163.com

**Abstract.** With the increase in vehicles in the world today, more accidents have occurred. Therefore, more safety systems are needed for conventional vehicles. However, physical testing is dangerous and costly compared with numerical simulation. So, a good numerical simulation is especially important. In general, the vehicle dynamics equations were used in the numerical simulation. There are two types of vehicle dynamics equations that are linear and nonlinear. This paper researched which one is more practical between linear and nonlinear equations in the simulations versus real-time experiments. In this paper, the equation of motion was used to simulate the state of vehicle motion during cornering conditions in the MatLab Simulink software. The results show that the value of the linear and non-linear yaw is close to the yaw from experimental at lower speed conditions. With the speed increase, the linear yaw value will decrease and farther and farther away from experimental yaw results. Although the linear equation is relatively simple to calculate, the results of the nonlinear equations are closer to the results of real experiments. In summary, the nonlinear equation is more applicable, while the linear equation is not applicable to simulate the motion of the vehicle.

**Keyword.** Numerical Simulation; Steering Performance; Passive Control; Vehicle Dynamics.

## 1. Introduction

Most of the traffic accidents that happened were due to the vehicle motion of cornering [1]. These accidents can be summaries as two factors which are a human factor and vehicle factor [2][3]. If the accident happened is because of the people wrong operating, then it can be solved most accidents by strengthen driver safety awareness, increase road signs, etc. If the vehicle factor makes the accident occur, more vehicle safety systems should be developed. Usually, the vehicle will have an understeering and oversteering phenomenon happen when the vehicle turns [4]. Understeer (US) and oversteer (OS) are important criteria for measuring vehicle handling steering balance [5][6]. It can be understood professionally as the true steering angle of the vehicle is greater or less than the steering angle of the wheels when the vehicle doing cornering, and the vehicle will easily lose control at this time [7][8]. The OS phenomenon occurs because when the vehicle is turning, the vehicle speed will be slightly lower than the normal driving speed. And the center of gravity will be shifted forward from the position [9]. The yaw of the vehicle increases at this time, so the lateral force of the vehicle will be greater than the tire and the ground of friction [10]. Therefore, the vehicle will skid. And the vehicle OS is mostly because of the fact that the driver fails to turn to the appropriate steering angle in time to make the front wheels of the car skid [11][12]. And the rear wheels have sufficient grip during this time, which makes the vehicle uncontrollably deviate to the outside of the road then leads to accidents occur. We cannot avoid many traffic accidents, but the safety performance of the vehicle itself can be developed as much as possible[13]. However, physical vehicle testing is both dangerous and very costly [14]. Therefore, it is very important to have a suitable numerical simulation system [15]. This paper is mainly to analyze which of the linear equation and the nonlinear equation is more suitable for analyzing the movement of the vehicle when the vehicle dynamics equation is used for numerical simulation. In order to accurately determine which equation is more suitable, the author used an experimental vehicle to conduct vehicle steering experiments before the simulation started. The process of the test is that the test vehicle drives around a roundabout to get the parameters of the vehicle in the process of turning. Then compare with the results of the simulation results. The equations of motion (EOM) use to simulate the trajectory of the vehicle [16]. And the EOM has two types which the linear vehicle dynamics (VD) equation and the nonlinear VD equation [17]. in order to study which of the linear and nonlinear equations is more suitable for the numerical simulation of vehicles, this paper uses the parameters from the testing car to do the numerical simulation for the linear and nonlinear equations respectively [18][19]. Although the linear equation is transformed from the nonlinear equation, the linear equation simplifies the calculation of most details. And it is still very different from the result obtained by the nonlinear equation, and a conclusion can be drawn from the analysis of the experimental results: When the vehicle speed is low, the result of the linear equation is very close to the true value. However, as the speed of the car gradually increases, the result obtained by the linear equation will be more and more offset from the true value.

## 2. Experimental vehicle

The Figure 1 is Malaysian national car and this car using for doing experimental, this test car from University Malaysia Pahang, under the Automotive Engineering Center. The car is a sedan type with 1.6-liter engine capacity with a manual transmission. In the research, this test car will be used for the experimental of the 2WS. This experimental use to validate the linear equation and non-linear that which one is more suitable for using in the simulation.

**Figure 1.** Testing car form automotive engineering center.

### *2.1. Place to doing the experimental*

This experiment was done at Universiti Malaysia Pahang campus Pekan. Due to the vehicle need to doing cornering to see the steering performance, the roundabout was selected as the experimental place. The testing roundabout located in the middle of the campus. There are three exits at this roundabout, and the diameter of it is about 55 meters. So, during the experiment, when there are some other vehicles drive in or some danger happens the testing vehicle will quickly exit from this roundabout.

## 3. Vehicle dynamics

The vehicle dynamics equation is used for the numerical simulation to simulate the movement of the vehicle. The results of the linear and nonlinear equations are used to compare with the results of the experiment. The tire characteristics are used to calculate the vehicle state of motion when the vehicle doing cornering. The vehicle's state of motion is affected by many factors, and the tire is the only vehicle in contact with the ground during movement. The steering of the vehicle will also be intuitively expressed on the tires. Therefore, the characteristics of tires are essential when simulating the movement of a vehicle.

### *3.1. Non-linear Equation*

Vehicle dynamics is a part of engineering and also refers to vehicle dynamics, mainly derived from classical mechanics. The dynamic equation of motion is derived from the application of Newton's law in the inertial reference frame. The dynamic equations are used in this article to simulate the longitudinal velocity, lateral velocity, and yaw rotation velocity of the model vehicle during its movement.

$$m\left(\frac{du}{dt}-v\gamma\right)=(X_{FR}+X_{FL})\cos\theta_F+(X_{RR}+X_{RL})\cos\theta_R-(Y_{FR}+Y_{FL})\sin\theta_F-(Y_{RR}+Y_{RL})\sin\theta_R \tag{1}$$

$$m\left(\frac{dv}{dt}+u\gamma\right)=(X_{FR}+X_{FL})\sin\theta_F+(X_{RR}+X_{RL})\sin\theta_R+(Y_{FR}+Y_{FL})\cos\theta_F+(Y_{RR}+Y_{RL})\cos\theta_R \tag{2}$$

$$\begin{aligned} I\frac{d\gamma}{dt}=\; & l_F[(X_{FR}+X_{FL})\sin\theta_F+(Y_{FR}+Y_{FL})\cos\theta_F] \\ & +l_R[(X_{RR}+X_{RL})\sin\theta_R+(Y_{RR}+Y_{RL})\cos\theta_R] \\ & +\frac{d_F}{2}[(X_{FR}+X_{FL})\cos\theta_R+(Y_{FR}+Y_{FL})\sin\theta_F] \\ & +\frac{d_R}{2}[(X_{RR}+X_{RL})\cos\theta_R+(Y_{RR}+Y_{RL})\sin\theta_R] \end{aligned} \tag{3}$$

### *3.2. Tire Characteristics*

The wheel is an important factor in the simulation. Therefore, during the modelling process, slip ratio, tire side slip angle, and weight distribution is all take into consideration in calculating the friction force and the side lateral force. The deformation of the tire tread rubber is also used to derive these shown below equations

When $\xi_s > 0$, then the longitudinal force and lateral force can be written as

$$\begin{aligned} F_x &= -K_s s\xi_s^2-6\mu F_z\cos\theta\left(\frac{1}{6}-\frac{1}{2}\xi_s^2+\frac{1}{3}\xi_s^3\right) \\ F_y &= -K_\beta(1+s)\tan\beta\,\xi_s^2-6\mu F_z\sin\theta\left(\frac{1}{6}-\frac{1}{2}\xi_s^2+\frac{1}{3}\xi_s^3\right) \end{aligned} \tag{4}$$

And when $\xi_s \le 0$, then

$$\begin{aligned} F_x &= -\mu F_z\cos\theta \\ F_y &= -\mu F_z\sin\theta \end{aligned} \tag{5}$$

Where:

$$\tan\theta=\frac{K_\beta\tan\beta\,(1+s)}{K_s s} \qquad \cos\theta=\frac{s}{\lambda}$$

$$sin\,\theta = \frac{K_\beta\,tan\,\beta\,(1+s)}{K_s\lambda}$$

During the simulation, the point at which the contact surface moves from the adhesive area to the slip area is defined as the following equations:

$$\xi_s = 1 - \frac{K_s}{3\mu F_z}\lambda \tag{6}$$

Where:

$$K_s = \frac{bl^2}{2}K_x \qquad K_\beta = \frac{bl^2}{2}K_y$$

$$\lambda = \sqrt{s^2 + \left(\frac{K_\beta}{K_s}\right)^2 (1+s)^2\,tan^2\,\beta}$$

The side-slip angle equation for each tire is shown below:

$$\beta_{FL} = tan^{-1}\left(\frac{v + l_F\gamma}{u + d_F\frac{\gamma}{2}}\right) - \theta_F \quad ; \quad \beta_{RL} = tan^{-1}\left(\frac{v - l_R\gamma}{u + d_R\frac{\gamma}{2}}\right) - \theta_R$$

$$\beta_{FR} = tan^{-1}\left(\frac{v + l_F\gamma}{u - d_F\frac{\gamma}{2}}\right) - \theta_F \quad ; \quad \beta_{RR} = tan^{-1}\left(\frac{v - l_R\gamma}{u - d_R\frac{\gamma}{2}}\right) - \theta_R \tag{7}$$

Tire slip ratio $s$ is used to measure and express the slipping behavior of the vehicle tire. The equation as below:

$$s = \frac{u - r\omega}{r\omega} \tag{8}$$

The coefficient of tire friction μ can be approximated by the following equation:

$$\mu = -1.10k \times (e^{35\rho} - e^{0.35\rho}) \tag{9}$$

*3.3. Linear equations*

The linear equation is assumed the sideslip angle is small. When the sideslip angle is small, it can be assumed that the direction of the vehicle almost coincides with the lateral direction. And the fundamental equations of motion for the linear is:

$$mV\frac{d\beta}{dt} + 2(K_f + K_r)\beta + \left\{mV + \frac{2}{V}(l_fK_f - l_rK_r)\right\}\gamma = 2K_f\theta \tag{10}$$

$$2(l_fK_f - l_rK_r)\beta + I\frac{d\gamma}{dt} + \frac{2(l_f^2K_f + l_r^2K_r)}{V}\gamma = 2l_fK_f\theta \tag{11}$$

## 4. Passive control system

This article aims to verify which of the nonlinear equations and linear equations in the vehicle dynamic equation is more suitable for numerical simulation of vehicles. So, this study uses a passive control system. At first, the values of lateral speed, longitudinal speed, and yaw angle when the vehicle is turning at different speeds are obtained by turning tests on the vehicle. Then the vehicle is simulated and

IOP Conf. Series: Materials Science and Engineering 1078 (2021) 012010 doi:10.1088/1757-899X/1078/1/012010

compared with the experimental results. The vehicle speed is controlled by changing the tire rotation speed to make it consistent with the experimental speed. In this research, the author wants to test the phenomenon that the vehicle during cornering when the vehicle speed is 25km/h, 30km/h, and 35km/h. However, since the vehicle cannot reach the exact integer value when driving, the average value of each speed when turning is finally taken, namely 26.367km/h, 30.688km/h, and 34.699km/h. In order to get more accurate simulation results to compare with the experimental, three significant figures are reserved.

## 5. Simulation procedure

This research was completed by using the MatLab software to do the simulation. First, complete the modelling in the MatLab Simulink by using the EOM. The modelling has two main parts which are linear VD and nonlinear VD, the independent variable is the tire rotational speed. Then, change the speed of the vehicle by changing the tire rotation speed so that the speed of the vehicle is the same as the speed reached by the vehicle in the experiment. When the simulation results are consistent with the experimental results, analyse the linear equation and the nonlinear equation by comparing the yaw angle values output from the experiment and the simulation, which one is more suitable for modelling the vehicle

## 6. Results and discussion

### *6.1. Experimental results*

The experimental results come from the test car, and the export of the results is using a software called DEWEsoft. The testing car itself is equipped with an additional computer, which is only used to receive the data sensed by various sensors on the vehicle. Then it is analyzed and processed by DEWEsoft software, and the changes of the data can be seen intuitively in the software also can export to an excel file. The 25km/h, 30km/h, and 35km/h appearing in the results below are not the final speeds of the vehicles in the test results. These three speeds are the target speeds, but because the driver cannot artificially control the speed exactly to each target speed when the vehicle is turning, these three speeds are only used as a reference, and the real vehicle speed will be used in the simulation and finally compared with the simulation results. From Figure 2 to Figure 7, it can be clearly seen that the data obtained by the experiment is uneven and it seems that the data is still very large. This is because the sensitivity of the sensor was adjusted to collect data every millisecond during the experiment. And although the driver tries to control the speed of the vehicle to keep the vehicle at a constant speed during the test, it still cannot maintain a perfectly constant speed. So, this has led to the fact that the result does not look ideal, but it is not the case.

Each of the figures from Figure 2 to Figure 7 has a dotted line and an equation. This line is the trend line of the set of data, and the equation is the expression formula of the trend line. Because the experimental results are uneven, the author uses the trend line of each set of data to analyse. It is not difficult to see from each equation that the independent variable X (time) has a very small effect on the value of the corresponding dependent variable Y (actual vehicle speed or yaw moment). Comparing Figure 3, Figure 5, and Figure 7, the yaw moment of the vehicle is gradually increasing when the vehicle turning with the vehicle speed increases. The amount of the increased yaw moment is basically the same. This proves that the results of these data can be used for comparison with the results of the simulation.

*25km/h*

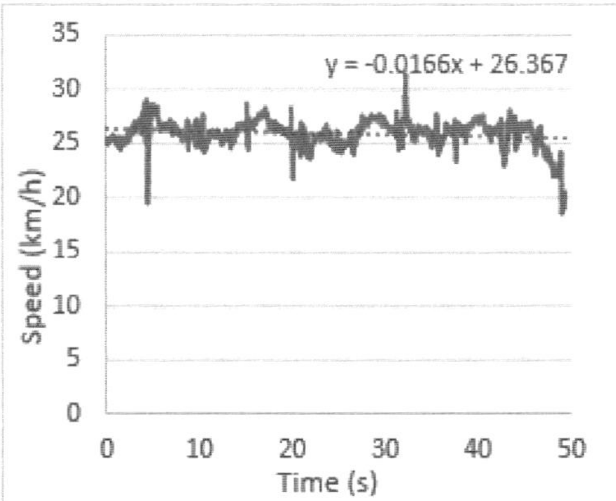


**Figure 2.** The actual vehicle speed for the experimental at 25km/h

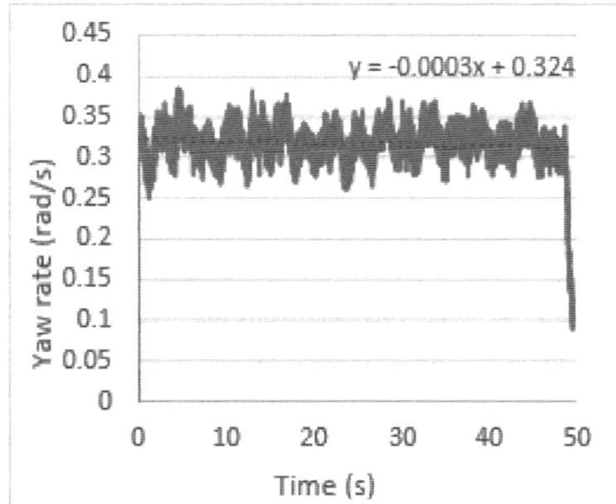


**Figure 3.** The yaw moment for the vehicle speed at 25km/h

*30km/h*

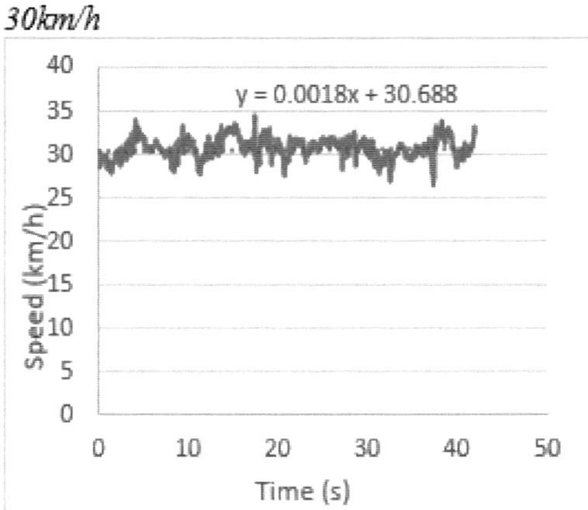


**Figure 4.** The actual vehicle speed for the experimental at 30km/h

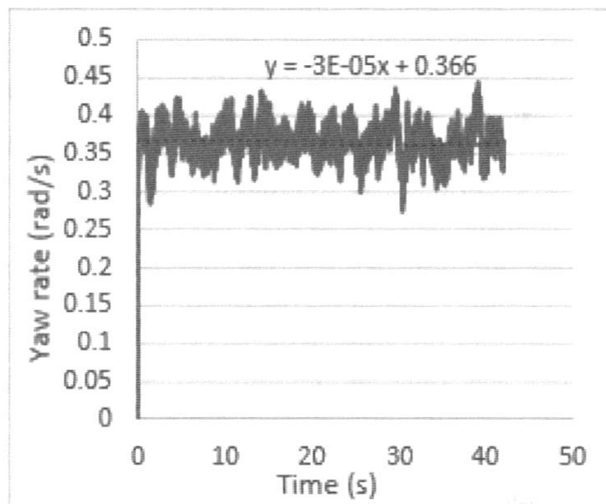


**Figure 5.** The yaw moment for the vehicle speed at 30km/

*35km/h*

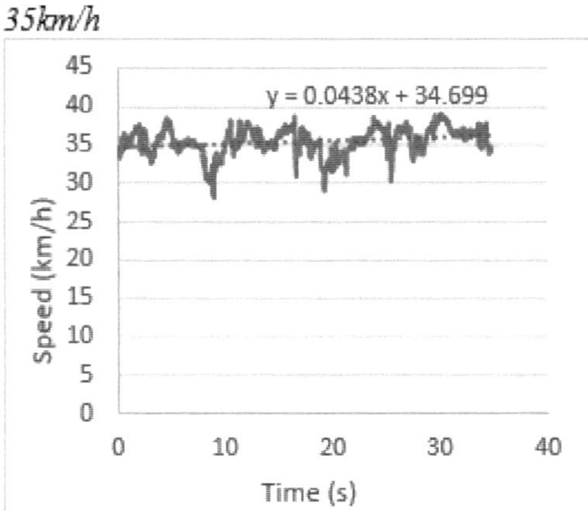


**Figure 6.** The actual vehicle speed for the experimental at 35km/h

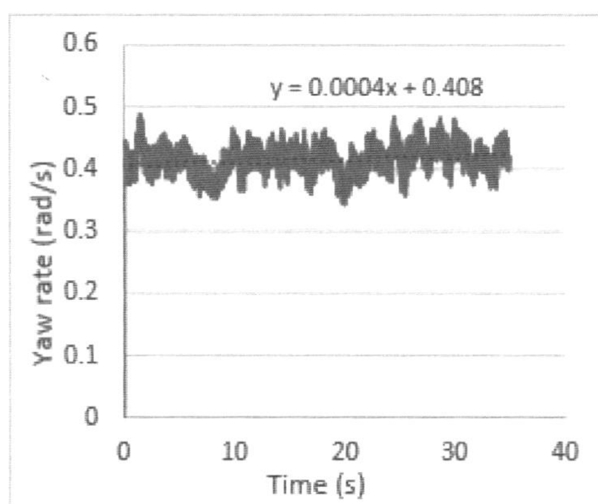


**Figure 7.** The yaw moment for the vehicle speed at 35km/h

### *6.2. Simulation results*

Figure 8 and Figure 9 are the results of the linear VD simulated by simulation, respectively corresponding to the vehicle speed and yaw moment derived from the linear equation. "LV1", "LV2", and "LV3" in Figure 8 and Figure 9 refer to linear speeds 1, 2, and 3. Corresponding to this are the target speeds of 25km/h, 30km/h, and 35km/h mentioned above. The straight-up part in the figure represents that the vehicle is accelerating until it reaches a constant speed. The declining part of each line represents that the vehicle starts to turn after reaching a constant speed and driving in a straight line, so the speed of the vehicle drops rapidly, but in the end, all reach a constant value. This also means that the vehicle can steer in a steady state at this moment. Comparing Figures 8 & 10, the simulated speed results are basically the same whether it is a linear or a nonlinear equation. However, in Figures 9 & 11, the yaw moment simulated by the linear equation decreases with the increase of speed, and the result of the nonlinear equation is that the yaw moment increases with the increase of speed. As the speed increases, the time required for the vehicle to reach the average speed gradually becomes longer in Figure 8, and "LV3" is obviously longer than the time required for the "LV1" to reach the average speed. And as the speed increases, the difference between the speed of the vehicle after turning and the speed of the vehicle before turning becomes larger at a different speeds condition. It can be speculated that when the speed is 40km/h. It takes longer for the vehicle to reach the average speed.

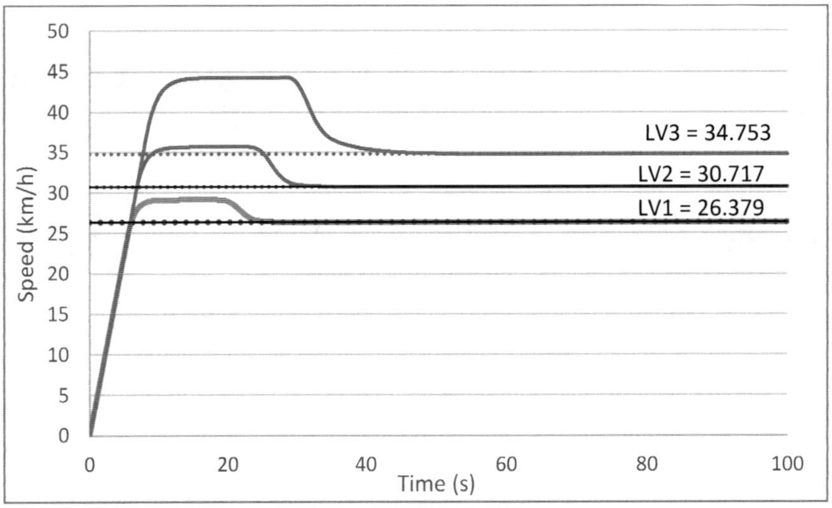


**Figure 8.** The linear VD simulation results for the vehicle speed

In Figure 9, "r" represents the yaw moment. And "Lr1", "Lr2" and "Lr3" are the yaw moment generated by the vehicle at the corresponding speed. Combining Figure 8 shows that the yaw moment of the vehicle is always zero before the vehicle turns, which also indirectly proves that the calculation of the equation is correct. When the vehicle is not turning, there is no deviation of the vehicle, so there is no yaw moment. Because zero yaw moment usually means that the vehicle is not moving or traveling in a straight line, and when the vehicle has a yaw moment and still maintains a constant value, then the vehicle is in a state of constant turning. After the vehicle starts to turn, the yaw moment increased extremely rapidly, and when it reaching the peak value, the speed at which the yaw moment increased slowly until it reaches a constant value. The Figure 9 shows that as the speed increases, the fluctuation before the yaw moment reaches a constant value will become larger and larger. This also means that when turning at high speeds, the vehicle shakes more severely. And as the speed increases, the shaking will increase. Therefore, it is not recommended that the driver turn the steering wheel drastically during high-speed driving because it is very easy to cause the vehicle to lose control.

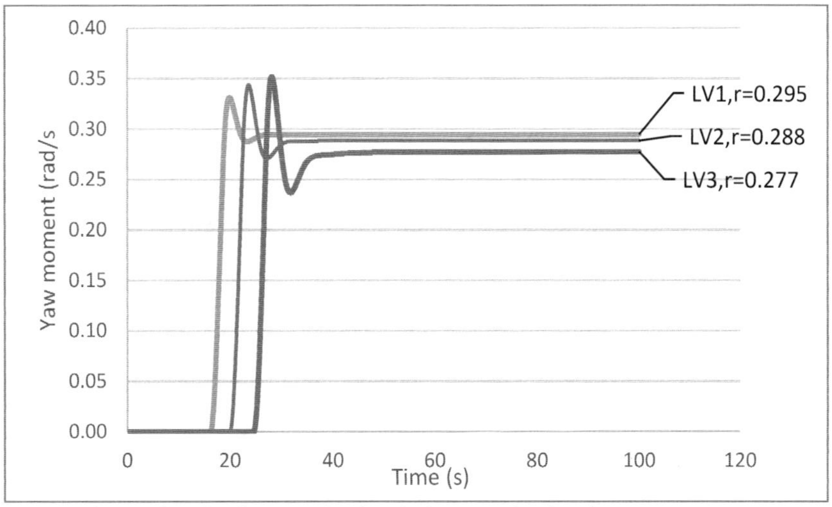


**Figure 9.** The linear VD simulation results for the yaw moment

Figures 10 and 11 show the simulation results of the nonlinear VD equation, Figure 10 shows the speed of the vehicle in the simulation results, and Figure 11 shows the yaw moment of the vehicle. Among them, "NLV1", "NLV2" and "NLV3" in the two figures correspond to the three speeds of the vehicle in the simulation. "NLr1", "NLr2" and "NLr3" are the yaw moment of the vehicle at this speed.

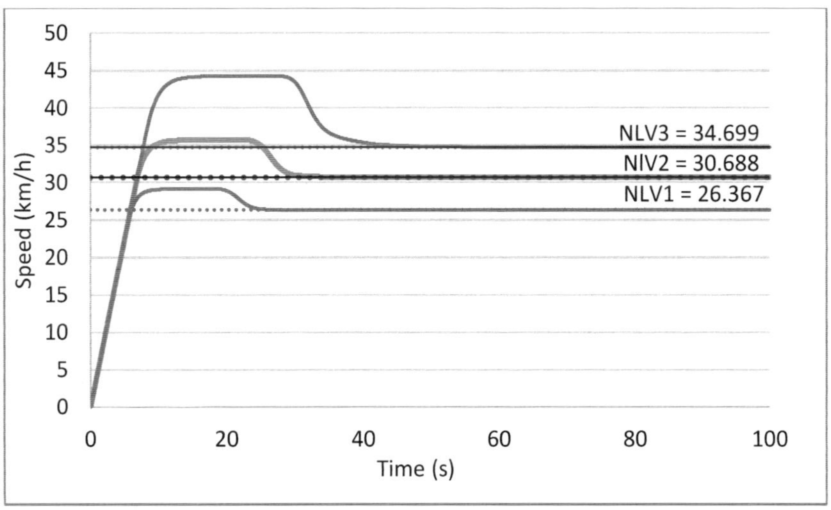


**Figure 10.** The non-linear VD simulation results for the vehicle speed

In Figure 12, the blue line represents the result of the test using the testing car. With the speed increases, the yaw moment increased gradually, and the ascent rate increases after 30km/h. The orange line represents the result of the linear equation simulation. As the speed increases, the yaw moment is slowly reduced. The yaw is lower than the experimental value at the speed of around 25km/h, and the yaw value differs as the speed increases. Is also getting bigger. However, compared with the non-linear result, when the speed is small, the result of the linear equation is closer to the result of the experiment. The gray line represents the results of the non-linear yaw moment simulation, and the yaw gradually increases with the increase of speed. It can be seen from the figure that when the speed is high, nonlinear equations are more suitable for vehicles to do the numerical simulation than linear equations.

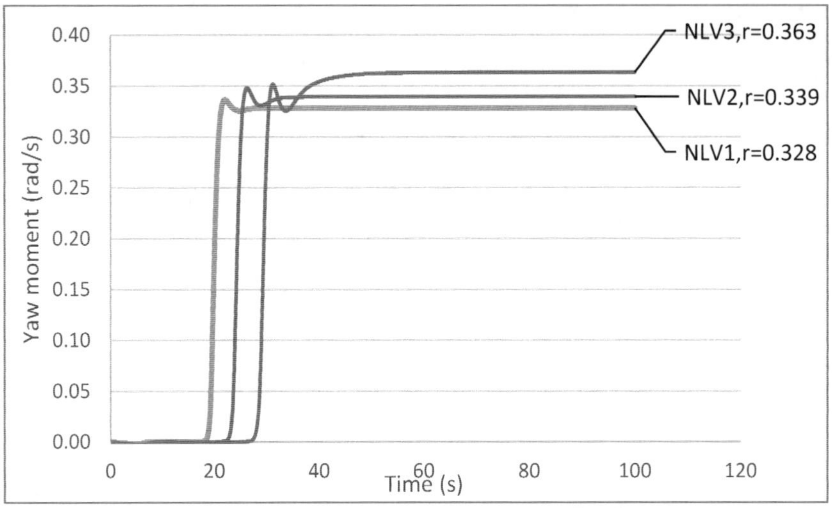


**Figure 11.** The non-linear VD simulation results for the yaw moment

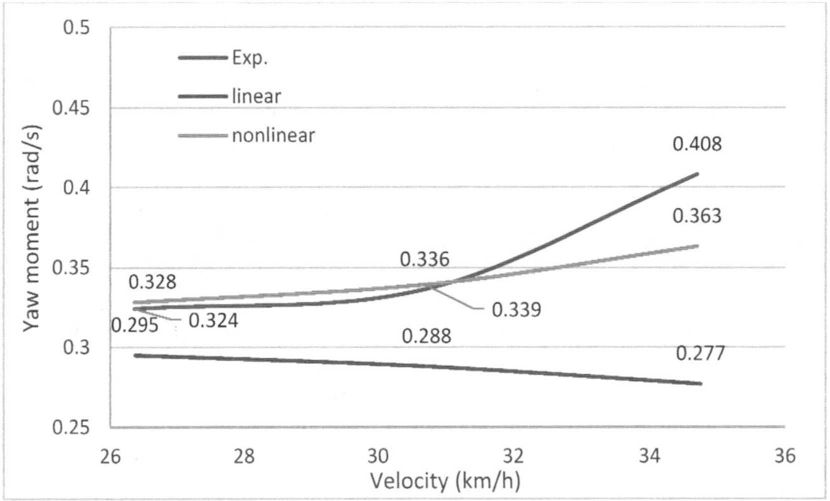


**Figure 12**. The change in yaw moment with increasing speed

## 7. Conclusion

This article mainly studies which equation is more suitable to do the numerical simulation between either linear equations or nonlinear equations. The whole study is divided into three stages. the first stage is to do the road testing using the testing car; the second stage is using the parameter of the testing car to build the simulation model and to do the simulation test; the third stage is to collect all the data and doing the analysis and compare the results between simulation and experimental. By comparing the results of the experiment and simulation, when the vehicle is running at a slower speed, the yaw moment simulated by the linear equation differs from the experimental yaw moment by 0.029 rad/s, while the yaw moment simulated by the nonlinear equation differs from the experimental result by 0.004 rad / s. As the speed increases, the results of the linear equation simulation become smaller and deviate more and more from the experimental results. The nonlinear simulation results are gradually rising with the speed increase and compared with the linear equation, it is closer to the experimental results, and when the speed is about 31km/h, the results of the nonlinear equation simulation are consistent with the experimental results. So, it can be said that the nonlinear equations are more suitable as equations for simulating vehicle dynamics.

**Acknowledgments**
This research was supported by the Ministry of Higher Education Malaysia through the Fundamental Research Grant Scheme FRGS/1/2019/TK08/UMP/02/5. Special thanks to Automotive Engineering Center, Universiti Malaysia Pahang (www.ump.edy.my) for providing test car and technical support and Dr. Muhammad Izhar Bin Ishak for this paper guidance.

**References**
[1] Chen W, Liang X, Wang Q, Zhao L and Wang X 2020 Extension coordinated control of four wheel independent drive electric vehicles by AFS and DYC *Control Eng. Pract.* **101** 104504
[2] Petrovic D, Mijailović R and Pešić D 2020 Traffic Accidents with Autonomous Vehicles: Type of Collisions, Manoeuvres and Errors of Conventional Vehicles' Drivers *Transp. Res. Procedia* **45** 161–8
[3] Han W and Zhao J 2020 Driver behaviour and traffic accident involvement among professional urban bus drivers in China *Transp. Res. Part F Traffic Psychol. Behav.* **74** 184–97
[4] Maoqi L, Ishak M I and Heerwan P M 2019 The effect of parallel steering of a four-wheel drive and four-wheel steer electric vehicle during spinning condition: A numerical simulation *IOP Conf. Ser. Mater. Sci. Eng.* **469**
[5] Ubaidillah, Setiawan B A, Aridharma A P, Lenggana B W and Caesar B P P 2018 Steering characteristic of an articulated bus under quasi steady maneuvering *AIP Conference Proceedings* vol 1931 p 30039
[6] Marzbani H, Vo D Q, Khazaei A, Fard M and N. Jazar R 2017 Transient and steady-state rotation center of vehicle dynamics *Procedia Comput. Sci.* **112** 1404–11
[7] Penmetsa P, Pulugurtha S S and Duddu V R 2018 Factors associated with crashes due to overcorrection or oversteering of vehicles *IATSS Res.* **42** 24–9
[8] Balkwill J 2018 Chapter 5 - Cornering *Performance Vehicle Dynamics* ed J Balkwill (Butterworth-Heinemann) pp 95–179
[9] Ni J, Hu J and Xiang C 2017 Control Engineering Practice Relaxed static stability based on tyre cornering stiffness estimation for all-wheel-drive electric vehicle *Control Eng. Pract.* **64** 102–10
[10] Tandy D F, Colborn J, Bae J C, Coleman C and Pascarella R 2015 The true definition and measurement of oversteer and understeer *SAE Int. J. Commer. Veh.* **8** 160–81
[11] Steindl A, Edelmann J and Plöchl M 2020 Limit cycles at oversteer vehicle *Nonlinear Dyn.* **99** 313–21
[12] Gim G and Kim J 2020 A Practical Concept of Vehicle Dynamics on Transient Steer Characteristic A Practical Concept of Vehicle Dynamics on Transient Steer Characteristic
[13] Hou G, Chen S and Chen F 2019 Framework of simulation-based vehicle safety performance assessment of highway system under hazardous driving conditions *Transp. Res. Part C Emerg. Technol.* **105** 23–36
[14] Henning K-U and Sawodny O 2016 Vehicle dynamics modelling and validation for online applications and controller synthesis *Mechatronics* **39** 113–26
[15] Symposium I and Symposium C 2016 ScienceDirect Independent Independent wheel control Independent wheel wheel steering based on
[16] Wikipedia contributors 2020 Equations of motion
[17] Abe M 2015 *Vehicle handling dynamics: theory and application* (Butterworth-Heinemann)
[18] Ishak M I, Ogino H and Yamamoto Y 2016 Numerical simulation analysis of an oversteer in-wheel small electric vehicle integrated with four-wheel drive and independent steering *Int. J. Veh. Technol.* **2016**
[19] Pan Y, Dai W, Xiong Y, Xiang S and Mikkola A 2020 Tree-topology-oriented modeling for the real-time simulation of sedan vehicle dynamics using independent coordinates and the rod-removal technique *Mech. Mach. Theory* **143** 103626

# Hydrogen fuel cells for sustainable energy: Development and progress in selected developed countries

**N A Baharuddin[1,*], W N A Wan Yusoff[1], A J Abd Aziz[1] and N N Mohd Tahir[1]**

[1]Fuel Cell Institute, Universiti Kebangsaan Malaysia, 43600 UKM Bangi, Selangor, Malaysia

*Corresponding author's e-mail: akidah@ukm.edu.my

**Abstract.** The sustainable development goals concept towards zero carbon emission, set forth by the Paris Agreement, is the foundation of decarbonisation implemented in most developed countries worldwide. One of the efforts in the decarbonisation of the environment is through hydrogen fuel cell technology. A fuel cell is an energy converter device that produces electricity via the electrochemical reaction, with water as the by-product. The application of fuel cells is strongly related to the economic aspect, including local and infrastructure costs, making it more relevant to be implemented in a developed country. This work presents a short review of the development and progress of hydrogen fuel cells in a developed country such as Japan, Germany, USA, Denmark, and China (in transition between developing to developed status); which championed hydrogen fuel cell technology in their region.

**Keywords:** Hydrogen; Sustainable Energy; Fuel Cell; Decarbonisation.

## 1. Introduction

The technology of fuel cells is one of the innovations established and developed worldwide in many countries. Compared to traditional power sources, fuel cell systems have been noted for their multiple benefits, such as high performance, decent durability, less time-consuming, and even environmentally friendly. As for that, as this technology is rapidly evolving all over the globe, particularly in developed countries, many countries are trying to substitute ordinary power sources with fuel cells from time to time. A fuel cell can be described as an electrochemical cell in which the chemical energy of conventional fuel is transformed directly and effectively into electrical energy [1]. The fuel cell also does not differ radically in its activity from the primary cells known to all of us as the Leclanche and Mercury-Oxide-Zinc dry cells [2]. In 1932, Francis Bacon invented a fuel cell derived from hydrogen and oxygen. This was the first ad to use a fuel cell [3]. In 1959, a team headed by Harry Ihrigh designed a 15kW fuel cell tractor for Allis-Chalmers, which was seen at state fairs around the USA. In 1838, the first fuel cells were invented by Sir William Grove [3]. Since the discovery of the hydrogen-oxygen fuel cell by Francis Bacon in 1932, fuel cells' first commercial use came more than a century later. Since the mid-1960s, alkaline fuel cells have been used in NASA space programmes to produce electricity for satellites and space capsules, known as the Bacon fuel cell after its inventor [4]. During that time, the hydrogen fuel cell is considered an exclusive-costly technology because it has been utilized in high-end applications such as NASA spaceship. Similar to other technology, the hydrogen fuel cell continues to develop as time passes by. One of the main developments and effort in achieving a low-carbon society is through the utilization of the fuels cell and hydrogen technology. Hydrogen is an energy transporter, but it can be contained in numerous ways and transported in various modes without exceptions. Together with a fuel cell, the hydrogen provides a stable and carbon-free route to produce electricity that is flexible and decentralized in many zero-emitting applications. In many other uses, fuel cells have to be used because of that. In the commercial, industrial and residential buildings and remote or inaccessible areas, fuel cells are used for primary and back-up control. Power cells, including forklifts, trucks, taxis, vessels and submarines, are also used in the vehicle.

Recently, to support decarbonisation effort, several countries are focusing on hydrogen fuel cell technology. Fuel cell technologies have been known for their advantages, such as high efficiency, good reliability, less time-consuming, and environmentally beneficial compared to conventional power sources. Many countries are trying to replace the familiar power sources with fuel cell by time to time as this technology is developing rapidly all over the globe, especially in developed countries. Among these countries, the developed country like Japan, Germany, USA, Denmark and China seems to lead the research and development in a hydrogen fuel cell in their region and the application of this technology. This paper aims to discuss the development, applications, achievements and challenges in bringing the hydrogen fuel cell technology into the local and global market in the as-mentioned developed countries briefly. The hydrogen fuel cell development or approach reviewed in this work will help others understand the overall aspects of utilizing hydrogen fuel cell technology.

## 2. Japan

Moonlight and Sunshine Project are among the initiatives initiated in the 1970s, proving the Japan involvement in hydrogen and fuel cell technology for almost 40 years, attaining the global leading expertise and experience in its sector. For the records, a as polymer electrolyte fuel cells (PEFCs) were developed for the first time in the 1950s by Generic Electric. Meanwhile, in space application advancement, a 1 kW fuel cell was integrated. The development of PEFC in Japan began in 1992 as part of the Sunshine initiative led by the Ministry of Economic Affairs, Trade and Industry (METI) [5]. In 1993, R&D activities began via the New Energy and Industrial Technology Development Organisation (NEDO) projects as part of the New Sunshine Initiative's sequel. Since 2000, more specific projects on real applications are being formulated and implemented within the Millennium Plan's scope, including a start-up production and a development project on the popularisation of fuel cells.

The development work of phosphoric acid fuel cells (PAFCs) already began in 1981 to design and build cells close to those already developed in the United States under the Moonlight Program of the Department of Industrial Science and Technology (now the National Institute of Advanced Industrial Science and Technology) [5]. Current Sunshine Scheme continues in making the innovation possible until 1997 while the current trend is more focussed on research for commercialization. Fundamental studies on molten carbonate fuel cells (MCFCs) were carried out from 1981 to 1986 and then expanded to 100 kW stack research from 1987 to 1994. The device studies were conducted between 1994 and 1999 and between 2000 and 2004 based on practical implementations. Besides PAFCs, research works for MCFC has begun under the Moonlight Initiative. Since 1988, its activities have been guided by the MCFC Research Association. The development currently ranges from analysis to practical execution. The manufacturing of PAFC has gained traction that can contribute to earlier commercial viability than other forms of PAFCs since 198s0. About 200 models of PAFC, including prototypes, are developed in which more than 20 units can be used for more than 40,000 hours. Dozens of units have been sold every year for field trials since the early 1990s.

### *2.1. Applications*

The fuel cell is a device that utilises the reaction process of hydrogen and oxygen in producing energy and heat and reversing the electrolysis of water. Fuel cells can be categorised through electrolytes employed, such as PEFC, PAFC, MCFC and solid oxide fuel cell (SOFC). In the early 1980s, the government and multiple other corporations have begun to develop fuel cells in Japan. Several decades have passed, Japan and other countries are still actively researching fuel cells. The study focussed can be narrowly divided into the following four groups which are PAFC, SOFC, MCFC and PEFC [6].

The bright future of Japan's hydrogen fuel cell is anticipated due to the support of the government as well as giant companies [5,7,8]. As this country is prone to the natural disaster such as earthquake, the needs to vary the energy source and power distribution not just for stationary used but also for mobile/portable application becomes Japan's main aim. This aim can reduce the dependence on the grid. With this spirit, Japan installed 265,000 residential used ENE-FARM units in 2018 with up to 5 kW capacity. Japan is currently planning to deploy about 5.3 million of these housing units by 2030 [10,11]. The exclusion of these fuel cells from the power grid moves Japan towards a future of self-sustaining energy. Japan has been planning to have more than 80,000 fuel cell vehicles and 1200 buses running on the roads by 2030 which is seen as initiatives to elevate the development of transportation of fuel cells and hydrogen thrived by the market growth and government initiatives. Fuel cell industrial use has also continued to progress, with the goal of 10,000 fuel cell forklifts operating in Japan by 2030 as well [10]. Japan's contribution to fuel cells has enhanced its drive towards a stable and clean energy future.

As for May 2018, Toyota has announced plans to accelerate the fuel cell vehicle (FCV) technology which aims for 30,000 increments of FCVs in the market by 2020 globally [11]. The increment in FCV units will require an excellent hydrogen trail across the country. Several organisations' collaboration is formed to develop a project that focuses on renewable energy - hydrogen supply chain involving Toyota City, Chia City, Toho Gas and Chubu Electric Power. This infrastructure is located at Toyota City and Chia City and sponsored the Aichi Low Carbon Hydrogen Supply Chain Promotion Organisation [11]. As fuel cell is shown high potential in the transportation sector, Toyota has upgraded the FCV from the light-duty passenger vehicle into fuel cell-powered bus (FCB) [12,13]. The 222 kW (111 kW/unit fuel cell stack) FCBs namely as Sora FCB was brought into the transportation market in March 2018 is powered by 10 units of 600 litres of hydrogen tanks. The Sora FCB is currently running in Tokyo Metropolitan Area and also will be used in Tokyo Olympic 2020 (postponed to 2021 due to pandemic Covid-19). The interesting fact about this Sora FCB is that the bus can also be used as 235 kWh portable power generator in case of emergency or disaster occurs [14].

**Figure 1.** Sora Fuel Cell Bus in Tokyo.

The continuous effort by Toyota results in a new partnership with Seven-Eleven Japan in June 2018 that agreed on the utilisation of hydrogen fuel cell technology at their distribution facilities [15,16]. Toyota has also announced the launch of a Hydrogen Mobility collaboration with the East Japan Railway Company (JR East) [17]. This collaboration aims to build hydrogen station on JR EAST facilities not just for the use of FCBs and FCVs, but also to develop hybrid battery- a hydrogen fuel cell-powered railway vehicle [18]. Besides, to support the transportation sector specifically fuel cell-powered forklift, Toyota has announced its intentions to generate and supply hydrogen to the Takahama's Aichi Prefecture facility using the Toshiba Energy Systems and Solutions (ESS) system [17].

Japan has been expanded the hydrogen fuel cell technology into the portable power market through the introduction of mobile charger namely as 500 JAQ Hybrid [19]. This Japan's Power Sourcing Equipment certified fuel cell mobile charger had been sell by Lightec across Japan. By referring to all the progress mentioned above, it is no doubt that Japan is leading in hydrogen fuel cell technology. From stationary to mobile and portable application, with power ranging from mW to MW, Japan made a significant remark involving hydrogen fuel cell technology in all of the related applications.

### *2.2. Future plan*

The opportunities for fuel cells in Japan are being deepened. For example, they were planned to use the 2020 Olympics as a forum to show the pledge that fuel cells are like simple hydrogen technologies. By 2050, they aim to turn hydrogen as a modern approach to green energy. CO2-free hydrogen is the aim that Japan is planning to accomplish with this simple hydrogen strategy. They would also substitute gas generation, replace gas stations, replace traditional fuel versatility and replace current suburban electricity networks with large FCVs.

## 3. Germany

In Europe, Germany is undeniably the number one fuel cell nation. The fuel cell company includes several thousand workers at universities, research institutes, and businesses. From the early days of the fuel cell, several of these organisations have already been involved. E.g., Siemens, which has been working since the 1950s on many different fuel cell technologies. The German Association of Hydrogen and Fuel Cells (DWV) is the national body for the commercial adoption of hydrogen and fuel cell technologies. DWV supports the introduction of hydrogen, especially in the oil and transport industries, as an energy carrier. It brings together specialists, corporations, research institutes, etc. to encourage them to share their expertise, contribute to the distribution of results, and foster public interest in the matter.

### *3.1. Background of fuel cell technology in Germany*

The hydrogen fuel cell in Germany is used in cogeneration applications based on fuel cells (also known as combined heating and control, CHP). SOFC and polymer exchange membrane fuel cell (PEMFC) are the most widely used type of fuel cell. The utilization of SOFC in Germany market for residential usage is known as a micro-CHP system and mini fuel (Figure 2). In 2018, the deployment pace recently increased significantly as a minimum of 3600 mini fuel cell units installed. And this number keeps rising and reached 5500 units and expected to continuously increase due to the support by the Ministry of Economic Affairs and Energy (BMWi) [20]. Prices are a significant barrier to the dissemination of these systems on the market. It is still projected that some CHPs based on fuel cells can be mounted to some 3 million buildings.

Germany has the most SOFC based-CHP cell device built under the Ene.field Project which mostly utilized synthetic gas [22]. The International Environment Agency (IEA) (2019) lists around 500 active passenger cars, 15 currently in operation for fuel cell bus (50 units to be launched) [20]. With 75 hydrogen filling stations in service and another 28 in planning or growth, Germany is officially ranked 4th in the world in the hydrogen fuel cell. Besides passenger vehicles (car and bus), Germany launched its first fuel cell train in 2018 using the existing non-electrified railway. So far, fuel cell technology usage has been very limited in the naval application, which only hydrogen fuel cell-powered tourist cruise available in Hamburg. In 2010, an alliance called as Clean Power Net consisting of playmakers from the research and business industries was created with regard to Uninterrupted Power Supply (UPS) systems to advance market adoption of UPS fuel cell applications. Approximately 200 fuel cell-powered UPS systems were installed from 2010 to 2016, e.g. at police and fire stations as back-up power supplies for the digital radio network. They were funded by the Federal Ministry of Transport and Digital Infrastructure (BMVI) [23].

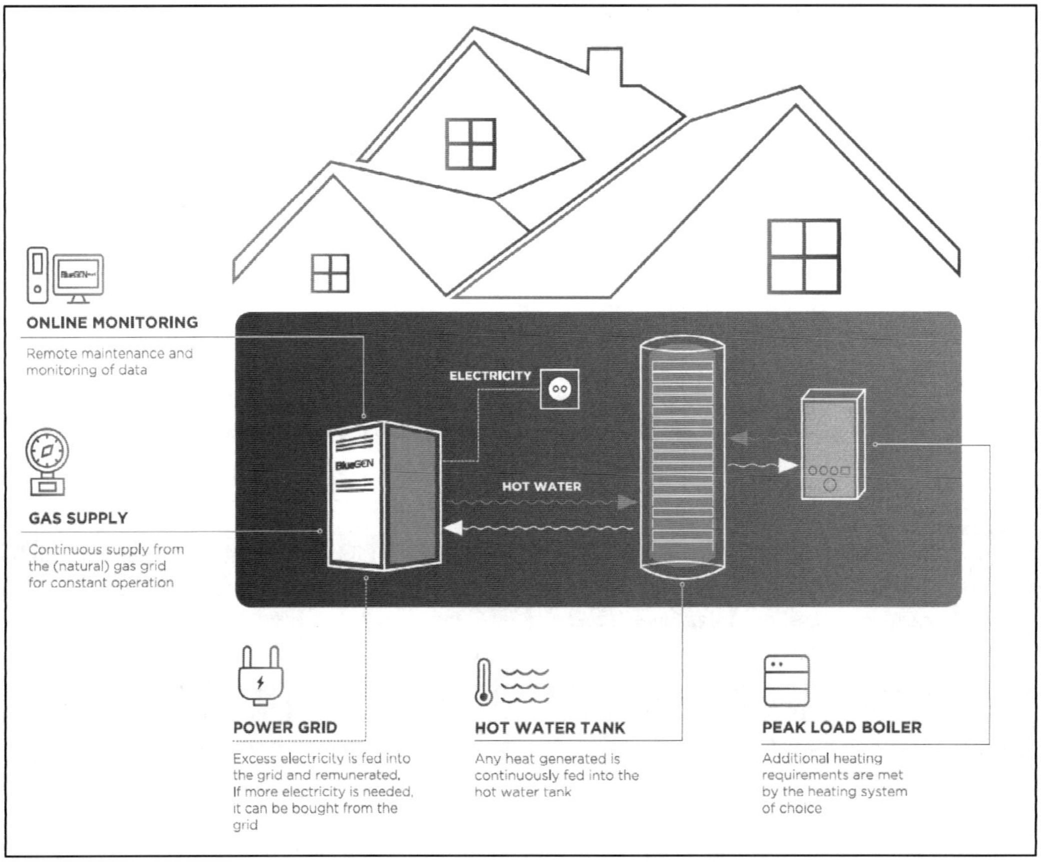


**Figure 2.** A compact micro CHP unit (BlueGEN) unit installed in residential and commercial buildings in addition to an existing heating system [21].

In addition, under the financing guideline for the 2018 BMVI market activation, 505 UPS fuel cell systems are financed for the wireless radio network. At the moment, these ventures are already ongoing. Several further fuel cell demonstration ventures are underway in different fields but only with a limited number of fuel cell systems including in sensor technology, water supply, traffic system and many more. The German Federal Government launched the first German National Hydrogen Programme at the end of 2019. The European Collaborative Undertaking for Fuel Cells and Hydrogen (FCH-JU) includes the European Commission, Hydrogen Europe's manufacturing stakeholders and Hydrogen Europe's research stakeholders. This public-private partnership focuses on the financing of fuel cell and hydrogen technology systems R&D and demonstration projects. The expenditure for the first cycle (2014-2020) is EUR 1.3 billion.

Joint R&D programs to promote clean energy supply for Europe would be available for financing, including converting and conserving renewable energy, based on a Directive agreed between the two countries. Hydrogen and fuel cells are specifically referred to as targeted innovations. The Promoter Jülich shall administer the support. As a division of the German Government for Hydrogen Innovation and Sustainable Mobility, the National Association for Hydrogen and Fuel Cell Technology (NOW) is one of the key stakeholders. Acting as a gateway between politics, academy and industry, NOW coordinates and facilitates operations across national and foreign networks. The National Development Program for Hydrogen and Fuel Cell Technology (NIP) and the Financial Recommendations for

Electromobility and Charging Facilities are some of today's most visible activities. In addition, the Mobility and Fuel Strategy (MKS) program has been set up to allow projects related to the usage and production of alternative fuels, including PtX. The Renewable Energy Export Program has expanded its activities on behalf of the Environment (BMU) to evaluate and facilitate hydrogen and fuel cell technology used worldwide. The NIP I was the global hydrogen technology platform from 2007 to 2016. In this period, EUR 1.4 billion has been invested in hydrogen and fuel cell technologies, generated in almost equal amounts by private and public sectors. The NIP II started in 2017 and is projected to continue until 2026 [24].

Although transport remains insight in industries, further funding will be geared towards the market introduction and integrated ventures in hydrogen regions. Regional initiatives are also underway in Germany, such as the Baden Württemberg Fuel Cell Project, a network to promote the development and expansion of renewable fuel cells and battery-based energy generation and storage technologies for stationary, mobile and portable applications. HySOLUTIONS in Hamburg is a public-private partnership that promotes fuel cells, hydrogen and power supply systems. To expand the hydrogen fuel cell market and to speed up the commercialisation progress, the North-Rhine Westphalia Fuel Cell and Hydrogen Network is linking related players in the hydrogen fuel cell field [20].

### *3.2. Applications*

The fuel cell and hydrogen market in Germany encompasses a wide variety of companies and events. This market covers all levels of device growth and nearly all forms of fuel cell types and implementations. Siemens and SiemensWestinghouse (submarines and stationary SOFC), DaimlerChrysler, Opel, Audi, BMW, VW, MAN, MTU CFC Solutions (stationary MCFC), Umicore (MEAs), Smart Fuel Cell (portable direct methanol fuel cell), Ballard (PEMFC), Proton Motor (PEMFC), Vaillant (CHP), Linde (gas and infrastructure) and ARAL (gas and infrastructure) are some of the best-known German firms with fuel cell operations. In CHP technology, waste heat is stored during service and can be used for space heating, domestic hot water preparation, hot water washing, and swimming pools or spas. CHP decreases the amount of energy lost by almost half, and the device can supply power with an efficiency of more than 90% while dramatically reducing emissions per kWh generated. CHP can be defined as the sequential or simultaneous generation of multiple forms of useful energy within a single integrated system. The CHP device's cumulative utility can be expressed as the ratio of the sum of the net power to the useful thermal energy output divided by the total energy of the fuel consumed. Energy efficiencies of up to 85 % to 90 % (combined electrical and thermal power) can be produced by CHP systems, which is much higher than the power of systems producing electricity and usable heat in separate processes. This change in energy efficiency will decrease costs in contrast to traditional ways of producing heat and power separately and will reduce greenhouse gas (GHG) emissions [25]. Countries in cooler climates, such as Germany or the United Kingdom, use more than 70% of their energy for room heating and 9% of their energy for water heating.

Worldwide development of FC-CHP systems is now taking place. Germany leads the way in Europe in showing and improving domestic FC-CHP programmes. CHP accounted for approximately 12.6 % of Germany's total electricity, with an installed capacity of 21 GW in 2005. New CHP laws in Germany are projected to double the output of electricity by 2020 [26]. Germany is well equipped for smaller-scale industrial and residential applications in both the biogas CHP and m-CHP. Germany ranks second in the implementation of FC-CHP systems with 350 systems were deployed between September 2008 and the end of 2012 under the CALLUX lighthouse initiative of the National Organisation for Hydrogen and Fuel Cell Technology [27]. Panasonic and Viessmann formed a European development agreement. Panasonic role is to create the fuel cell in Japan then transported to Germany for Viessmann integration. With a total efficiency of 90%, the machine produces 750 W of electricity and 1 kW of thermal energy. In Germany, it is anticipated that device sales under field tests will commence [28]. As for the portable application, SFC Smart Fuel Cell AG together with German Federal Army's Military Vehicle and Tank Technology Department WTD 41 announced their collaboration in developing portable fuel cell power supply namely SFC C20-II in 2005 [29].

### *3.3. Future plan*

The hydrogen fuel cell plan for this country lies under Germany's National Hydrogen Strategy [30]. There are several aims and goals for under this strategy, including establishing the hydrogen chain not just in Germany, but also across Europe. Besides that, this country also aims to focus on trading the blue hydrogen in the European market. The German government's goal is to use green hydrogen ($CO_2$-neutral), support a rapid market ramp-up, and establish corresponding value chains. As the hydrogen market growth, it also becomes their goal of pushing down the hydrogen price globally with promoting the decarbonisation via utilisation of hydrogen energy and fuel cell technology [30].

## 4. United States of America

U.S. Energy reported in 2019 that the total carbon emission originated from petroleum use at about 46%, 33% belongs to burning of natural gas and 21% from burning of coal where the transportation sector itself emits more CO2 compared to the industrial sector due to the massive dependence on petroleum fuels [31]. Most of this carbon output is partly related to light-duty vehicles (LDVs), including light-duty trucks and passenger car. To minimise the release of $CO_2$ and the other types of greenhouse gases originated from engine combustion, the government have developed and introduced their own regulatory and economic policies to reduce the emissions from LDVs [32]. In complying the newly designed legislative, the utilisation of hydrogen as future transportation is mainly attributed to its feature as a high potential for domestic production and fuel cell, which can operate with zero-emission and high efficiency.

The fuel cell can operate at two or three times more efficiently than a gasoline-powered internal combustion engine. Thus, fuel cell research and development's primary subject is more focused on hydrogen use in vehicles. Besides, a few vehicle manufacturers have started to make a light-duty hydrogen fuel cell electric vehicle in several regions equipped with hydrogen fueling stations such as Southern and Northern California. Not to mention that the test vehicles were already made available even though only in limited numbers charged to organisations with easy access to hydrogen fueling stations. For example, hydrogen-fueled vehicles are mostly automobiles and transit buses operating on the electric motor powered by a hydrogen fuel cell. Despite the excellent features of this technology, the major production and implementation of this application are still limited due to the high cost of fuel cells and the restriction on the number of available hydrogen fueling stations [33]. However, this problem will not become a constraint as fuel cell applications' usage expands over the years in this country.

### *4.1. Applications*

The fuel cell and A hydrogen fuel cell can generate energy through the mixing process of hydrogen and oxygen atoms. The electrochemical process will then occur that is almost similar to that of a battery system to generate electricity [34]. Besides, the products are water and a small quantity of heat as a result of the process. There are several different types of fuel cells suitable for a wide variety of applications. The smaller fuel cell version may operate laptop, computers, mobile phones and not limited to military applications. Meanwhile, a larger version of fuel cell could provide additional electricity or energy for as a reserved or backup power in a particular building and supply power to the areas that were not connected to the electrical power grid [35]. It was reported at sometime around the end of October 2019 that there have been approximately about 80 fuel cell power plants that already operated in the United States with the sum of 190 MW of electricity generation capacity. The largest power plant equipped with a capacity of 27 MW and utilized hydrogen, which was derived from the landfill gas to power the fuel cell located in Red Lion Energy Centre, Delaware [33].

Amongst the early step taken onto the futuristic green energy application, Government of the USA in the United States the Federal Transit Administration established the national Fuel Cell Bus (FCEB) program in 2005 to advance commercialization FCEBs that were enacted in Public Law 109-59 [36]. The program has provided nearly $90 million since 2006 to promote the development and testing of cleaner, greener sources of fuel for the transit industry. As of August 2013, there are 18 active FCEBs in demonstrations at six locations where 14 of the 18 busses are in California [4]. Since then, the demonstration of FCEB progress across the country was administered by the US Department of Energy (DOE) and US Department of Transportation (DOT, through the Federal Transit Administration [FTA]) [37]. A prominent aspect of the DOE/DOT FCEBs ventures is the FCEB fuel economic program was aimed at 8 MPDGE (miles per diesel gallon equivalent, or 3.4 km/l). Both DOE and DOT are making collaboration in 8 MPDGE goals in 2012, relying on the pre-2012 FCEB vehicle research and performance reviews and the stakeholders' input as a common transit bus operated on diesel achieved about 4MPDGE, while the 8 MPDGE targets for the FCEBs program gave off to a fuel economy to the ratio of 2.

As of February 2020, reported in Fuel Cells Bulletin, the heavy-duty transit buses named The Xcelsior Charge H2$^{TM}$ were distributed by New Flyer of America Inc, a subsidiary of global bus manufacturer NFI Group [38]. The buses include a fuel cell of 85 kW from Ballard Power Systems and deliver a range of more than 265 miles (425 km). The buses were also fitted with the ability to take advantage of regular regen incidents, high-speed service, and to stretch range if the bus was out of fuel [39]. Figure 3 are the models of Xcelsior Charge H2 supplied by New Flyer. The launch of new fuel cell buses incorporated into the OCTA (Orange County Transportation Authority) fleet to start operating in regions. The current hydrogen refuelling platform in the Santa Ana Bus Depot of OCTA is the largest public transport fast-fill facility in the US capable of refuelling up to 50 fuel cell buses per day. The station can fuel up the transit buses with an average of 28 kg of hydrogen per bus in just 6–10 min while offering back-to-back bus refuelling for up to 30 buses and several lanes for continuous refueling [40].

**Figure 3.** Xcelsior Charge H2$^{TM}$ model [41].

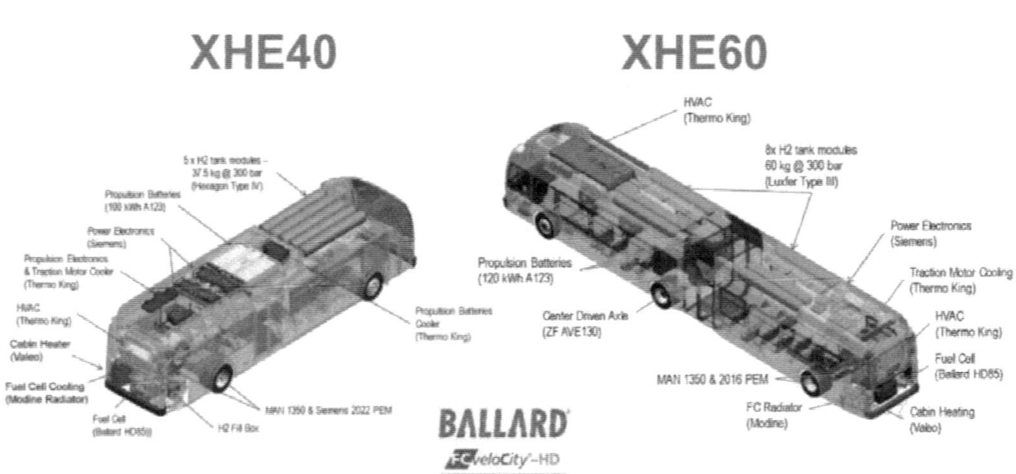


**Figure 4.** Parts of Xcelsior Charge H2$^{TM}$ models adapted from [39].

In July 2020 as reported in Fuel Cells Works news, Intelligent Energy announced that a manufacturing company named Zepher in the US specialized in aerospace and defense had purchased its 800 W PEM Fuel Cell Power Modules (FCPMs). This FCPMs will then be designed into something of a custom-built electric vertical take-off and landing (eVTOL) unmanned aerial vehicle (UAV, or drone) being developed for the US Army [42]. Intelligent Energy has also established a distribution agreement with Parry Labs in the US as confirmed in Intelligent Energy news in July 2020 [43]. Meanwhile, Zepher intends to make the UAV technology accessible during the first 12 months of the contract. To provide more detail on the project is funded in support of a US Army Special Operations Command requirement. Intelligent Energy technologies have been chosen due to their ability in delivering longer flight time considerably, work silently and last but not least require little maintenance. If the first integration has been completed, Zepher will have a concept approach ready for initial development. The architecture will incorporate two 800 W FCPMs [FCB, October 2018, p6] with IE's Power Path Module (PPM) connecting them to a 1.6 kW output. The goal of the flying time of this technology is 14 hours [44].

For the past decade, the implementation and development of stationary fuel cell development have been growing in a strong pace, currently being led by the United States and South Korean market with almost 95% of the installed capacity. There are still significant variations in the methods employed at the national level within the United States, mostly with the bulk of the capability built-in two states, California and Connecticut. To facilitate large-scale stationary fuel cell operations, fuel cell producers are also primarily dependent on public support, whether by technology drive or demand-pull activity [45]. Bloom Energy is another prominent player in the advanced distributed energy generation platform that operates for 24 hours without a break serving the highest efficiency of any power solution globally. Bloom Energy had supplied their product in various sectors where most of the fuel cell technology used SOFC. In July 2018, Bloom Energy based in California, made a headline as their 1.8 MW onsite SOFC technologies power up the SUNY Downstate Medical Centre in the New York City borough of Brooklyn [46]. In January 2019, Bloom Energy, based on California, offered help to Partners Healthcare in Massachusetts with SOFC-based power systems. The Eaton Blackout Tracker announced that it experienced more than 100 electrical power outages in 2017, the last 55 hours in total, which affected more than 630 000 people across Massachusetts. Thus, Partners have agreed to extend to improve the reliability at particular facilities by installing 4.1 MW Bloom Energy's OFC-based power system at a location across Massachusetts [47].

### *4.2. Future plan*

Although the horrific coronavirus contagion continues to impact many businesses, individuals and cities adversely, the new hydrogen economy is expected to shed light, as many rallies behind a green revival, with hydrogen and fuel cells playing a significant role in this endeavour. 2020 portrayed the first year of growth and achievements for the US hydrogen economy. Ever since a new year is on the horizon, H2 Perspective is focusing on several of the highlights of the US hydrogen and fuel cell sector since the hasty year, and how those activities would positively affect the country's advancement [48]. A report made by The Navigant regarding the promising development of the stationary fuel cell industry in the US over the last few years has been propelled by the foreign governments' interest in scalable and efficient energy sources as well as global demand that could rise more than 50 GW. Soon, approximately 2.5 million telephone towers will be powered by a fuel cell-based reserved power system globally by the end of 2020 [45].

## 5. Denmark

Denmark is a developed country which spends more on hydrogen than any other country in the world. The Danish energy source is strongly dependent on renewable energy technologies. Nearly half (47%) of Danish power output in 2013 comes from renewables, primarily from wind farms [49]. From 1996 to 2019, Denmark was self-sufficient in net energy because of oil and gas output. Independence has only ended in recent years due to reduced production for supply and Denmark is forecast to be a major exporter of petroleum by 2020 and gas by 2035 [50]. Consequently, Denmark has signed an aggressive 2020 political energy deal to this, for example, to double the output and power supply of wind power. The Danish government has further endorsed an ambitious goal of making the energy sector renewable towards fossil fuel-free and fossil-fuel-free economies by 2035 [51]. Denmark is currently the highest exporter for a renewable energy source in the Europe region through its transmission with Netherland, Sweden, Norway, and Germany. However, during the last decade, the trend of electricity import in Denmark is decreasing. To maintain as the biggest supplier in renewable energy in Europe, Denmark has to venture into new technology [52]. Hydrogen is expected to play an essential role in this future scenario [53].

Denmark has been one of the world leaders in hydrogen and fuel cell developments and becomes the world's first country to establish an effective chain of hydrogen refuelling stations. Denmark is increasing compliance in fuel cell technology, from R&D to manufacturing and hydrogen development to deployment of infrastructure. In Denmark, the government agreed on an ambitious energy strategy for 2012-2020, which specified by 2050, the national infrastructure would be completely carbon-free, including transport. Therefore, hydrogen and fuel cells are given priority as a possible key technology in Denmark, facilitating the required energy system balance [49]. The Danish hydrogen and fuel cell strategy are divided into hydrogen technology development and fuel cell technology development. For hydrogen technology, it involves electronics, hydrogen infrastructure, and hydrogen storage while developing fuel cell technology will be focused on research, development, and testing. Two different applications of hydrogen have had the broadest social impact. The first is strength. In 2012, the Danish government signed an ambitious energy agreement, with the key concerns relating to the transition from fossil electricity production to wind-based renewable energy. This approach inspired entrepreneurs to participate in the hydrogen fuel cell technology's business development projects in the electricity sector to store electricity, offset the charges, and be considered a great potential, at least in principle, concerning residential heat and power solutions. Residential heating and power solution have growth potential of approximately 20% of Danish homes, a proportion not related to district heating [54].

As concerning mobility, a comparatively small $1.8 million has been allocated to hydrogen infrastructure. Despite these challenges, Denmark is the most expensive country in Europe to purchase and own a vehicle[55]. Hydrogen cars are also excluded from entry tax and vehicle excise duties, making them more economically efficient for customers. However, no tax or price incentives are in place to use hydrogen petrol technologies. For Denmark, the future short-term growth potential is mainly restricted to the production and implementation of maintenance and hydrogen petrol stations. According to the internal research carried out by H2Logic, a leading hydrogen fuelling station manufacturer and HydrogenLink, a network of hydrogen road production, about half the Danish automotive industry would in 2050 be made of hydrogen fuel cell cars corresponding to around 1.4 million automobiles [56].

The hydrogen fuel cell landscape in Denmark consists of about 20 local firms and a few supporting organisations and networks. They are mostly small and medium-sized companies (SME) engaging in public/private collaboration research and development work with Danish universities, but only 3 Danish companies in the top 200 worldwide in the fuel cell business. The Danish hydrogen and fuel cells collaboration promotes cooperation with a partnership with a formal organisation and enables threefold cooperation and promotes the production and marketing of technology for fuel cells. In addition to the alliance, there is a range of smaller networks with broader targets. One of them is Cemtec, a network promoting industry for businesses based in Denmark's regional region and the hydrogen transportation network HydrogenLink, which work together with other Scandinavian countries and automotive manufacturers internationally to extend their hydrogen vehicle networks.

*5.1. Applications*

Public transport and freight transport account for around 25% of carbon emission in Denmark. The introduction of zero-emission electric fuel cell buses will contribute significantly to Danish climate policy, which is promising 70% below carbon emission by 2030 and will achieve complete carbon neutrality by 2050 [57]. Danish municipalities play an important role in developing a potential public transit infrastructure, but trade actors need to work together to create zero-emission buses. Several major players in the hydrogen fuel cell chain have founded the H2BusEurope consortium committed to supporting the fuel cell electric bus (FCEB) infrastructure to advance de-development.

For more than ten years, FCEB has already been in the commercial venture in places around the globe. In March 2020, the first three FCEBs were deployed in Denmark [58]. The three busses represent a very significant step or public transport with zero-emission in Denmark, as the busses fill the gap between demonstrations and the broader usage of FCEB. In the city of Aalborg, Denmark, the first three fuel cell buses were operational. The three coaches are part of the ambitious EU-funded 3Emotion project which is financed by the FCH-JU program. This project is one of Ballard, and 17 other investors' several projects are taking part to support fuel cell in Denmark. Keeping fuel cells technology would minimise reliance on fossil fuels and protect the expansion of fuel cells. This 12-meter A330 bus can carry up to 78 passengers. Van Hool builds the vehicle, while Ballard built the fuel cell system for this bus. Siemens provides the electric drive with up to 160kW of output for the A330 bus. This bus's hydrogen tanks can carry up to 38 kilograms hydrogen, which should be 350 kilometres long enough until the next hydrogen fuelling station. FCEB is operated as a traditional bus with a range of up to 400 kilometres depending on the hydrogen tank size, rendering it the most versatile zero-emissions option.

Denmark is a pioneer in introducing clean energy sources such as wind turbines and by electrolysis surplus. Thus, in Denmark wind power can be transformed into hydrogen as an alternative zero-emission replacement for fossilised –fuel combustion engine vehicles. Hydrogen production will also meet much of the balancing needs for clean energy integration in Denmark by 2050. Under the H2BuusEurope programme, 1,000 FCEBs are deployed by 2023 with 200 buses for Denmark. The hydrogen fuel cell bus from H2BusEurope is anticipated as the cheapest proper zero-emission choice that possible, with the target single-decker bus price below EURO375,000, hydrogen costs between EURO5 and EUO7 per kilogram, and the bus operation price below EURO0.30 per kilometre. This amount could be achieved by the link of FCEBs across Europe countries and cities including London, Aberdeen, Cologne, Hamburg, Oslo, Aalborg and others. The more prominent usage of the fuel cell will reduce the cost of the technologies itself.

In the Copenhagen area, the Danish partnership aims to merge the demand and supply of renewable fuel with a vision to become one of the world's largest electricity and sustainable energy facilities. The electrolyser will be installed in three phases, with a 1.3-GWelectrolyser system provided by renewable energy from offshore winds. COWI and BCG advisory groups serve as consultants for the initiative funded by the Copenhagen municipality. Airport Copenhagen, shipping giant A.P. Moller Maersk, DSV Panalpina, DFDC shipping firm, SAS airline operator and Orsted have developed an alliance to establish an advanced facility for the manufacture of hydrogen and e-fuel by 2023 [59]. The plan will provide green hydrogen to support the fuel cell buses operating from the Movia transit agency, heavy-duty lorries from the DSV Panalpina, aircraft form A.P. SAS, air freight from Moller-Maersk and clean jet fuel from Copenhagen airport. This project would mark a significant supply of renewable energy that could be generated through an offshore wind farm at the Orsted Ronne Banke site in Bornholm Island.

A 10 MW electrolyser is the first phase that could be operational in 2023, that will deliver green hydrogen for the direct use of refuelling buses and trucks. The second phase will involve a 250 MW electrolyser system, which will be operational by 2027. This phase is targeted to operate when the first offshore Bornholm wind power facilities are ready. This facility will integrate green hydrogen production with sustainable carbon capture from points in the greater Copenhagen region to generate marine renewable methanol and aviation e-kerosene. For the third phase, which could become operational by 2030 when Bornholm offshore wind potential is fully developed and operate. In this stage, it will raise the electrolysis capability's ability to 1.3 GW and produce more carbon dioxide; enough to provide sustained fuels for all transportation involve to more than 250,000 tons per year [60].

Denmark is one of the world leaders within hydrogen and fuel cell technologies and was the first country in the world to establish a nationwide network of hydrogen refuelling stations. Denmark is also fully engaged in fuel cell technology; from R&D to product manufacturing and hydrogen production to infrastructure deployment. There is an 11-hydrogen station built across Denmark (Figure 5). There are 6 hydrogen stations with an onsite electrolyser, 3 hydrogen station operate with the hydrogen is supplies from the Strandmollen electrolyser plant, 1 hydrogen station under construction, and another under planned. Every station has a pressure of 70MPa, which is consistent with new hydrogen fuel cell vehicle, and it is derived from the electrolyser of renewable electricity [61]. An internationally uniform refuelling dock is used at the 700-bar. SAE J2601 refuelling standard is applied in the whole hydrogen fuelling station across Denmark. This standard is the same uniformly in hydrogen refuelling at less than three minutes for each fuel cell vehicle refuelling [62].

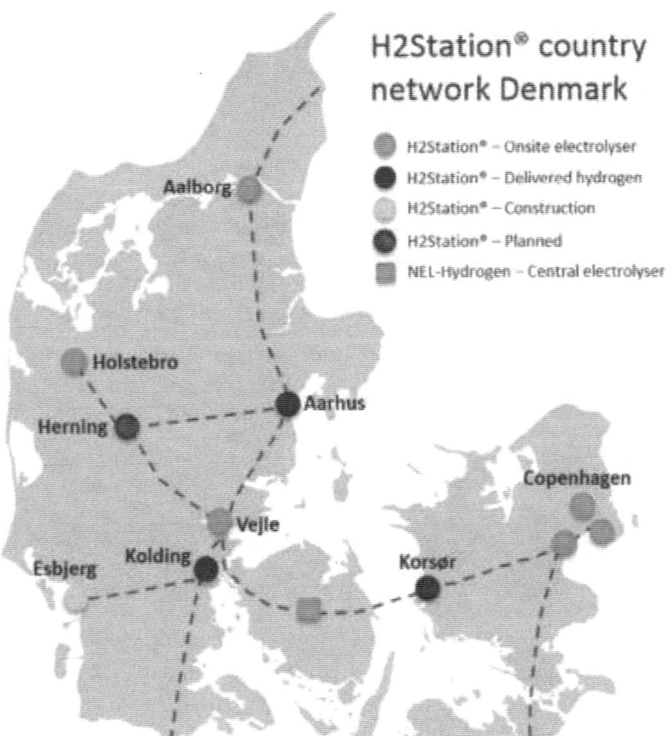


**Figure 5.** Hydrogen Fuelling Station in Denmark [61].

### *5.2. Future plan*

For Denmark, when the facilities complete in 2030, it will provide an annual amount of 250,000 tons of fuel to power fuel cell buses, cars and vehicle. It is a target to minimizing annual carbon emission up to 850,000 tons. The link of hydrogen fueling station will connect through Europe and enable more fuel cell vehicles to be used across Denmark and Europe. Infrastructure for building the Scandinavian hydrogen route will form about half of the Danish car market by 2050 of hydrogen fuel cell vehicles, which around 1.4 million cars. Denmark's possible short-term growth opportunity is mainly constrained to create and deploy support for fuel cell and expand fueling station for hydrogen.

## 6. China

In an era of revolution, many developed countries worked to take part in current evolving technologies of fuel cell energy. This country included The Great Wall country or well-heard China. As one of the developed countries, China is currently focusing on enhancing hydrogen fuel cells' utilisation in their country. Hydrogen is the central pillar in the contribution of a low-carbon to meet the Paris Agreement 2015 commitment, hence, zero-emission energy system [63][64]. According to Yinfei et al., stated that three main reasons why using hydrogen energy are crucial. (1) Hydrogen is one of the key proponents to lead to a low-carbon or zero-emission energy system. (2) Hydrogen is theoretically to be considered as energy storage. (3) Hydrogen is commonly used in a variety of industries (International Energy Agency) [63]. Due to these alternatives, China announced to utilise hydrogen fuel fully. Besides, China was reported to have a significant amount of hydrogen energy; thus, these reasons lead to the alternative made [65].

### *6.1. Applications*

China has been focused mainly on the primary factor in developing fuel cell technology which is research and development. Research and development industry (R&D) play a vital role in developing a product, business, and others to achieve the said goals or purposes. China is said to have done many kinds of research in developing the application of fuel cell technology. China is said to have started its research 45 years ago and continued developing its research throughout the 1990s' reaching over a dozen Chinese universities and institutes, involving in the study of various cell fuel types. However, China started to fall behind in developing fuel cells when its research had faced several problems. Its research started to become smaller, irregular, and incomparable to the level of investments exemplified by other industrial countries. Besides that, China had spent thinly in all research institutes, including fuel cell-related topics that had caused the nation's overall fuel cell development to become weak. However, in 1996, China started to rise again in the development of fuel cell technology. The Global Environmental Facility (GEF) had provided great international assistance in developing fuel cell vehicles which included a kind of study or research about the fuel cell vehicle. The contribution of GEF also benefited in the way of capturing the China government's attention and support to increase the research and development of the fuel cell vehicle in China.

There are several applications and developed technologies of hydrogen fuel cells were invented by China [53,63,73,65–72]. But, the most dominant application is in transportation (MAXUS, Feichi Bus and Zhongtong Bus). China's government has announced a range of policies and initiatives to promote fuel cell vehicles (FCVs) [66]. China took serious in developing the fuel cells vehicle in the industry because it gives advantages to China in many ways such as enhancing energy security, greater industrial competitiveness, improving urban air quality and reducing greenhouse gas emissions. The development of this sector in China can be seen when the government used Olympics 2008 and World Expo 2010 to show their capabilities to the world. During the Beijing Olympics 2008, the government used 20 fuel cell cars and 3 fuel cell buses to transport special guests, press and officials from the Olympics organising committee. According to Zhang and Cooke (2010), the cars used were designed by Shanghai's Volkswagen Passat while the buses were developed by Tsinghua University and Beiqi Foton Motor Co [74]. Two years after the Olympics, the World Expo 2010 was held in Shanghai. According to Gupta, Basile and Veziroglu (2016), the government used 196 fuel cell vehicles, including 90 fuel cell cars, 6 fuel cells buses and 100 fuel cells tourist vehicles. This vehicle's function is to transport visitor in and around the exposition site [75]. 2 permanent and 2 mobile hydrogen refuelling stations were constructed and operated at the Expo.

National Development and Reformation of China reported that 20% of energy consumption in China comes from transportation. This report shows the transportation sector requires much energy consumption compared to another industry [73,76]. Hydrogen-based fuel vehicles are essential due to their aptitude to reduce the greenhouse gas effect plus air pollution. This fuel cells vehicle is also a great choice due to its ability as a life-cycle carbon-free option for the transportation system [69]. Refer to the news from China, the city of Changshu planned to open a hydrogen station for the transportation of hydrogen-based vehicles in 2022. Most of China's developers and investors got an eye-catching on this developed hydrogen-based fuel cells vehicles. This reason comes with the better service from these new hydrogen vehicles because these hydrogen fuel cells vehicles have improved service life, cell power density also millage. Those explanations lead to these vehicles' hydrogen fuel cells as an alternative to safer and conventional future technology [65].

Besides the fuel cell application in the transportation sector that is already synonymous in China, this country is also planning for its first hydrogen-based ENE-FARM in northern Hebei province's Guyuan county [77]. This ENE-FARM project aims to power up the Winter Olympic 2022 and already 95 % finish. This project combines wind farm (200 MW) with 10 MW electrolyser system [77]. In addition, the hydrogen yields from electrolysis are said to replace the usage of gas and oil, e.g. in the transportation sector. This ENE-FARM project is funded by the Hebei Construction & Investment Group. The decarbonisation of energy also supported by one of the largest energy company in China, Sinopec. With the goals to implement hydrogen fuel cell vehicle during Winter Olympic 2022, Sinopec built its hydrogen production unit in Yanshan, Beijing [77].

On the other hand, there are minor drawbacks for the beneficiary. As a result of recent renewable energy shifts, these hydrogen fuel cells are currently in the in-house deployment process. Work is still under the jurisdiction of the government. The cons that were identified were the high cost of development, and the infrastructure's fragility also missed the quality. China's technology lags behind other developing countries, giving it a high cost of production. Other cons of this alternative concept of turning to hydrogen are commercial storage and the hydrogen filling network. This storage technology could give up a high cost of hydrogen if the handling method is inaccurate. The concern is that hydrogen can only be transported using the gas form, but that costs five times the storage and delivery of liquid hydrogen. Finally, there is a lack of regulation and a reference standard for handling hydrogen or a method in the industry that may contribute to the acceptance of the product and its spread. Plus, from an environmental point of view, the problems often emerge from the geographical variations in hydrogen supplies. However, the difficulties of getting society to embrace the latest invented technology fully may take some time as well as lack of infrastructures and policies [63,65,78].

To infer the above point, the development of hydrogen fuel cells energy contributes significantly to China's growth. It provides a green technology with less pollution and reduces the dependence of non-renewable energy. In contrast, it also has several drawbacks that require more research and development from the government. The weakness should be improved on the infrastructure if the hydrogenation stations, reduce the production cost also establish some reasonable policies and standard of procedures that can be used as the guideline. Overcoming the challenges of the drawback may be improved the smoothness of utilising hydrogen energy in China because China has a great potential to develop this hydrogen technologies due to its privilege as a developed country.

Table 1 summarised the major application that utilises hydrogen fuel cell in Japan, Germany, USA, Denmark and China. All countries have been focused on the transportation and energy production sector as these two sectors produce large carbon. Success in decarbonisation in these two sectors will result in a drastic reduction in carbon emission worldwide.

**Table 1.** Summary of the hydrogen fuel cell in major sectors (transportation and energy farm).

| Country | Applications (including R&D phase) — Transportation | Applications (including R&D phase) — ENE-FARM & Stationary Power Generator |
|---|---|---|
| Japan | • Toyota Mirai<br>• 222 kW SORA fuel cell bus (FCB)<br>• > 800,00 fuel cell vehicles (FCVs), 1,200 busses & 10,000 forklifts on the roads by 2030<br>• Toyota-JR East hydrogen mobility alliance | • Latest: Osaka ENE-FARM project<br>• 5.3 million of 5 kW residential units by 2030 |
| Germany | • ~500 active fuel cell passenger vehicles, 15 buses with an additional 50 accepted for launch<br>• Hamburg tourist ship | • Ene.field project<br>• 3,600 microCHP units deployed in 2018<br>• 5,500 microCHP units in 2019 sponsored by BMWi<br>• 505 fuel cell UPS systems for the digital radio network of vital organisations funded by BMWi |
| USA | • National Fuel Cell Bus (FCEB) program in 2005<br>• 18 active FCEBs (14 of the 18 busses are in the state of California) in 2013<br>• 85 kW Xcelsior Charge H2TMheavy-duty transit bus in Feb 2020<br>• 800 W unmanned aerial vehicle (UAV) or drone for military in July 2020 | • 190 MW fuel cell power plants in 2019<br>• 27 MW plant at Red Lion Energy Centre, Delaware<br>• California and Connecticut stationary fuel cell projects<br>• 1.8 MW onsite SOFC technologies power up the SUNY Downstate Medical Centre in July 2018<br>• 4.1 MW SOFC-based power systems (collaboration of Bloom Energy with Partners Healthcare in Massachusett) in January 2019 |
| Denmark | • 160 kW fuel cell electric bus (FCEB) with capacity of 78 passengers 350 km/full tank | • 1st phase: 10 MW electrolyser system by 2023<br>• 2nd phase: 250 MW electrolyser system by 2027.<br>• 3rd phase: 1.3 GW electrolyser system by 2030 |
| China | • carbon-free option for the transportation system<br>• hydrogen station for the hydrogen-based vehicles located at Changshu in 2022<br>• MAXUS, Feichi Bus and Zhongtong Bus (Fuel cell vehicles | • 10 MW electrolyser system in northern Hebei province's Guyuan county by Hebei Construction & Investment Group<br>• Sinopec hydrogen production in Yanshan, Beijing<br>*both for 2022 Winter Olympic |

## 7. Conclusion

The fuel cell industry has evolved quickly, and a variety of research has been undertaken to bring the industry to light. As technology improves, deregulation and public assistance are projected to develop. Furthermore, non-renewable fuels such as oil, coal and natural gas are reduced through fuel cell utilisation. The fuel cell industry has faced obstacles as it meets a time of stagnation and completes the transition from R&D to commercialisation. Overall, the conditions have survived incredibly complicated. While some fuel cell businesses are still far from sustainable, future growth prospects are fascinating. In recent years, the popularity of some application segments means that a step has been made to integrate specific technology into a common reference design for a particular type of fuel cell. This situation has led to fuel cells being rapidly being developed as flexible energy technologies capable of serving various market segments, be they APUs or power devices such as unmanned aerial vehicles.

Conclusively, the development of fuel cell technology in the selected developed countries has grown rapidly and efficiently. This successful development happened with the contribution of a lot of parties. There was a lot of research institute had been built to conduct fuel cell research. A lot of research had been done by the experts, and many more are still ongoing. Fuel cell industry in those countries has gained a lot of support from the researchers, developers including the government itself to make sure that all the research conducted will ensure a worth return to the industry's development. It will give advantages and merits economically and helps to lift the country's reputation.

**Acknowledgments**

This work was supported by Universiti Kebangsaan Malaysia under research grant number DIP-2019-011. The author would like to express the sincere gratitude to members of Group 1, Group 5 and Group 6 of the Fuel Cell Technology Class for their contribution to this works.

## References

[1] Andújar J M and Segura F 2009 Fuel cells: History and updating. A walk along two centuries *Renew. Sustain. Energy Rev.* **13** 2309–22
[2] Douglas D L and Liebhafsky H A 1960 Fuel cells: History, operation, and applications *Phys. Today* **13** 26–30
[3] Brandon N 2004 Fuel Cells *Encycl. energy* **2** 749–58
[4] Burke K 2012 *Current perspective on hydrogen and fuel cells* vol 4 (Elsevier Ltd.)
[5] Kato Y, Koyama M, Fukushima Y, Nakagaki T, Energy F, Based S and Beyond F T 2016 *Energy Technology Roadmaps of Japan* (Springer Nature)
[6] JETRO 2006 *Japan ' s Fuel Cell Industry*
[7] Futamura S, Muramoto A, Tachikawa Y, Matsuda J, Lyth S M, Shiratori Y, Taniguchi S and Sasaki K 2019 SOFC anodes impregnated with noble metal catalyst nanoparticles for high fuel utilization *Int. J. Hydrogen Energy* **44** 8502–18
[8] Nagashima M 2018 *Japan's Hydrogen strategy and its economic and geopolitical implications*
[9] Popov S and Baldynov O 2018 The Hydrogen Energy Infrastructure Development in Japan *E3S Web Conf.* **69** 1–11
[10] Chaube A, Chapman A, Shigetomi Y, Huff K and Stubbins J 2020 The role of hydrogen in achieving long term Japanese energy system goals *Energies* **13** 1–17
[11] Toyota Motor Corporation 2018 Toyota moves to expand mass-production of fuel cell stacks and hydrogen tanks towards ten-fold increase post-2020 *Toyota Newsroom*
[12] Lucas L 2017 Toyota unveils Sora fuel cell bus, Fine-Comfort Ride saloon *Fuel Cells Bull.* **2017** 2
[13] Lucas L 2018 Toyota begins sales of Sora fuel cell bus *Fuel Cells Bull.* **2018** 2–3
[14] Association F C & H E 2020 Japan ' s Hydrogen Olympics *FCHEA* 1–4
[15] Samantha O 2017 Seven-Eleven Japan , Toyota Partner on Hydrogen Test *cspdailynews*
[16] HLDGS S 2020 Expansion of the Small Fuel Cell ( FC ) Truck Demonstration Experiment

Area *7andi.com*
[17] Lucas L 2018 Toyota tech in fuel cell buses for Caetanobus, Japan rail partnership *Fuel Cells Bull.* **2018** 2
[18] Corporation T M 2020 JR East, Hitachi and Toyota to Develop Hybrid (Fuel Cell) Railway Vehicles Powered by Hydrogen *Fuel Cells Bull.* 6
[19] Myfcpower 2018 myFC ships JAQ Hybrid order to Lightec in Japan *Fuel Cells Bull.* **2018** 8–9
[20] Jensterle M, Narita J, Piria R, Samadi S, Prantner M, Crone K, Siegemund S, Kan S, Matsumoto T and Shibata Y 2019 *The role of clean hydrogen in the future energy systems of Japan and Germany* vol 6
[21] BlueGen HOW BLUEGEN WORKS *Power, Solid*
[22] Gandiglio M, Ferrero D, Lanzini A and Santarelli M 2020 Fuel cell cogeneration for building sector: European status *REHVA J.* 21–5
[23] Nina P 2019 FIVE FEDERAL STATES TO RELY ON FUEL CELLS *https://www.now-gmbh.de/* 60–2
[24] Commision E 2003 *Hydrogen Energy and Fuel Cells*
[25] Tom K 2009 CHP / DHC Country Scorecard: Germany *IEA* 1–12
[26] EPA 2007 Combined Heat and Power (CHP) Partnership *Agency, United States Environ. Prot.*
[27] Office A M 2016 *Combined heat and power technology fact sheet series - Fuel Cells*
[28] Anon 2013 Panasonic, Viessmann for European home cogen *Fuel Cells Bull.* **2013** 1
[29] Shepard J 2005 SFC to Develop Portable Fuel Cells for German Federal Army
[30] Amelang S 2020 Germany ' s National Hydrogen Strategy
[31] EIA 2019 *Energy and the environment explained: Where greenhouse gases come from*
[32] Borlaug B, Holden J, Wood E, Lee B, Fink J, Agnew S and Lustbader J 2020 Estimating region-specific fuel economy in the United States from real-world driving cycles *Transp. Res. Part D Transp. Environ.* **86** 102448
[33] EIA 2020 Hydrogen explained: Use of hydrogen *U.S. Energy Inf. Adm.*
[34] Alaswad A, Baroutaji A, Achour H, Carton J, Al Makky A and Olabi A G 2016 Developments in fuel cell technologies in the transport sector *Int. J. Hydrogen Energy* **41** 16499–508
[35] Mekhilef S, Saidur R and Safari A 2012 Comparative study of different fuel cell technologies *Renew. Sustain. Energy Rev.* **16** 981–9
[36] Lee D Y, Elgowainy A and Vijayagopal R 2019 Well-to-wheel environmental implications of fuel economy targets for hydrogen fuel cell electric buses in the United States *Energy Policy* **128** 565–83
[37] Alvarez-Meaza I, Zarrabeitia-Bilbao E, Rio-Belver R M and Garechana-Anacabe G 2020 Fuel-cell electric vehicles: Plotting a scientific and technological knowledge map *Sustain.* **12**
[38] FuelCellBulletin 2020 *Orange County unveils fleet of fuel cell buses, largest H2 station in US* vol 2020
[39] California Transit Association 2019 *Real life Experience with Fuel Cell Electric Buses*
[40] OCTA Hydrogen Fuel Cell Electric Buses *Orange Cty. Transp. Auth.*
[41] Anon 2020 Xcelsior CHARGE H2™ Fuel cell-electric and zero-emission
[42] FuelCellsWorks 2020 *Intelligent Energy UAV Fuel Cells Power Fixed Wing eVTOL Endurance Drone for the US Army* (Loughborough)
[43] Energy I 2020 Intelligent Energy and Parry Labs LLC Announce New Distribution Agreement *Inteliigent Energy*
[44] Brief I N 2020 Intelligent Energy fuel cells power endurance drone for US Army *Fuel Cells Bull.* **2020** 5–6
[45] Rajalakshmi N, Balaji R and Ramakrishnan S 2021 *Recent developments in hydrogen fuel cells: Strengths and weaknesses* (INC)
[46] FuelCellBulletin 2018 Bloom SOFC system powers SUNY medical centre in Brooklyn *Fuel Cells Bull.* **2018** 7
[47] FuelCellBulletin 2019 *Bloom onsite SOFC power for Partners HealthCare in MA* vol 2019

[48] Burgess M, Cockerill R and Sampson J 2020 Hydrogen fuel cell developments in the US *H2 View*
[49] Andreasen K P and Sovacool B K 2014 Mapping and interpreting critical hydrogen stakeholders in Denmark *Int. J. Hydrogen Energy* **39** 7634–7
[50] Sovacool B K 2013 Energy policymaking in Denmark: Implications for global energy security and sustainability *Energy Policy* **61** 829–39
[51] Jillian Ambrose 2020 Denmark to end new oil and gas exploration in North Sea *theguardian* 1
[52] Aziz M and Huda M 2019 ScienceDirect ScienceDirect ScienceDirect Utilization of electric vehicles for frequency regulation in Danish The and Cooling Utilization electric vehicles for frequency regulation in Danish electrical grid electrical grid Assessing the Muhammad feasibili *Energy Procedia* **158** 3020–5
[53] Ren J, Andreasen K P and Sovacool B K 2014 Viability of hydrogen pathways that enhance energy security: A comparison of China and Denmark *Int. J. Hydrogen Energy* **39** 15320–9
[54] Https://www.iea.org/ 2020 Danish Energy Agreement for 2012-2020 *IEA/IRENA Renewables Policies Database* 1
[55] Ã U K and Kuhfeld H 2007 The diverse structures of passenger car taxation in Europe and the EU Commissions proposal for reform **14** 306–16
[56] Workshop H L 2010 Danish Partnership for Hydrogen and Fuel Cells Public R , D & D budgets for Hydrogen and Fuel Cells in 2008 **78**
[57] Buses Z and Denmark I 2020 Fuel Cell Electric Buses Zero – Emission Fuel Cell Electric Buses for Denmark .
[58] Nina Hjorth 2020 Fuel cell electric buses: Hitting the road in Denmark and around the world *https://blog.ballard.com/*
[59] East S 2020 ABB control solution to
[60] Reuters Staff 2019 Offshore wind developer Orsted secures money for renewable hydrogen project *https://www.reuters.com/* 10
[61] https://insideevs.com/ 2016 With 9 Hydrogen Fuel Stations, Denmark Is 1st Country With Basic National Network *https://insideevs.com/*
[62] Gmbh O 2011 Shell opens hydrogen station with Linde tech in Germany 2011
[63] Li Y and Kimura S 2021 Economic competitiveness and environmental implications of hydrogen energy and fuel cell electric vehicles in ASEAN countries: The current and future scenarios *Energy Policy* **148** 111980
[64] Zhong X Z, Wang G C, Wang Y, Zhang X Q and Ye W C 2010 Monomeric indole alkaloids from the aerial parts of Catharanthus roseus *Yaoxue Xuebao* **45** 471–4
[65] Pan A, Liu J, Liu Z, Yang Y, Yang X and Zhang M 2020 Application of Hydrogen Energy and Review of Current Conditions *IOP Conf. Ser. Earth Environ. Sci.* **526**
[66] He X, Wang F, Wallington T J, Shen W, Melaina M W, Kim H C, De Kleine R, Lin T, Zhang S, Keoleian G A, Lu X and Wu Y 2020 Well-to-wheels emissions, costs, and feedstock potentials for light-duty hydrogen fuel cell vehicles in China in 2017 and 2030 *Renew. Sustain. Energy Rev.*
[67] Liu W, Zhang Z, Chen J, Jiang D, Wu F, Fan J and Li Y 2020 Feasibility evaluation of large-scale underground hydrogen storage in bedded salt rocks of China: A case study in Jiangsu province *Energy* **198** 117348
[68] Zhou J, Guo Y, Huang Z and Wang C 2019 A review and prospects of gas mixture containing hydrogen as vehicle fuel in China *Int. J. Hydrogen Energy* **44** 29776–84
[69] Zhang G, Zhang J and Xie T 2020 A solution to renewable hydrogen economy for fuel cell buses – A case study for Zhangjiakou in North China *Int. J. Hydrogen Energy* **45** 14603–13
[70] Xu X, Xu H, Zheng J, Chen L and Wang J 2020 A high-efficiency liquid hydrogen storage system cooled by a fuel-cell-driven refrigerator for hydrogen combustion heat recovery *Energy Convers. Manag.* **226** 113496
[71] Sun Z Y, Liu F S, Liu X H, Sun B G and Sun D W 2012 Research and development of

hydrogen fuelled engines in China *Int. J. Hydrogen Energy* **37** 664–81
[72] Tian M W, Yuen H C, Yan S R and Huang W L 2019 The multiple selections of fostering applications of hydrogen energy by integrating economic and industrial evaluation of different regions *Int. J. Hydrogen Energy* **44** 29390–8
[73] Chang X, Ma T and Wu R 2019 Impact of urban development on residents' public transportation travel energy consumption in China: An analysis of hydrogen fuel cell vehicles alternatives *Int. J. Hydrogen Energy* **44** 16015–27
[74] Zhang F and Cooke P 2010 Hydrogen and fuel cell development in China: A review *Eur. Plan. Stud.*
[75] Ram B. Gupta, Basile A and Veziroglu T N 2013 *Compendium of hydrogen energy Volume 2: hydrogen storage, distribution and infrastructure*
[76] Li W, Long R, Chen H, Chen F, Zheng X, He Z and Zhang L 2020 Willingness to pay for hydrogen fuel cell electric vehicles in China: A choice experiment analysis *Int. J. Hydrogen Energy* **45** 34346–53
[77] Yihe X 2020 ENERGY China set to fire up wind farm-linked hydrogen scheme with Winter Olympics in its sights
[78] Ren X, Dong L, Xu D and Hu B 2020 Challenges towards hydrogen economy in China *Int. J. Hydrogen Energy* **5**

# Investigation of low frequency fibre Bragg grating accelerometer based on thermoplastic cantilever beam

**Z M Hafizi[1,2] and E Vorathin[2,*]**

[1]Automotive Engineering Centre (AEC), Universiti Malaysia Pahang (UMP), 26600 Pekan, Pahang, Malaysia
[2]Advanced Structural Integrity and Vibration Research (ASIVR), Faculty of Mechanical & Automotive Engineering Technology, Universiti Malaysia Pahang (UMP), 26600 Pekan, Pahang, Malaysia

*Corresponding author's email: vora.91.11@gmail.com

**Abstract.** Vibration measurement technique is very important in structural integrity monitoring. Various fibre Bragg grating (FBG) based accelerometers have been developed for vibration measurement. However, most of the researchers focused on high frequency monitoring and only few reported works are based on low frequency measurement. Therefore, this paper presented a low frequency FBG accelerometer based on Polyphenylene Ether (PPE) thermoplastic cantilever beam. The proposed FBG accelerometer was attached to a shaker and vibration signals were given with variations in frequency and acceleration. As a result, the FBG accelerometer has a sensitivity of 110 pm/g and natural frequency of 9 Hz. The proposed accelerometer capable to detect low frequency of 2 Hz at 0.04 g which is suitable for utilisation in seismic monitoring of earthquake.

**Keywords:** Fibre Bragg Grating (FBG); Accelerometer; Thermoplastic Cantilever Beam.

## 1. Introduction

Structural integrity based on vibration measurement technique has been a critical concern in various engineering applications such as oil and gas, railway and earthquake [1, 3]. Therefore, accelerometer has been an essential tool to monitor the vibration signals. Over the years, various sensors have been utilised to develop the accelerometer such as capacitive, piezoelectric fibre optic sensor [2]. Of all sensors, fibre optic sensor based on fibre Bragg grating (FBG) shows great advantages in terms of electrical isolation, immune to electromagnetic and radio frequency interference [4, 5].

Au et al. [3] bonded an FBG on an isosceles triangular cantilever beam. The cantilever beam has the dimension of 32 mm×0.25 mm for length×thickness. The tip of the cantilever beam was attached with a mass of 29.8 grams. As a result, a sensitivity of 21.6 pm/g was obtained across the range of 25 Hz to 150 Hz. Liu et al. [6] developed a copper alloy diaphragm based FBG accelerometer with the diameter of 24 mm. The centre of the diaphragm was attached with a mass of 20 grams. The proposed accelerometer obtained the sensitivity of 36.6 pm/g across the range of 10 Hz to 200 Hz. Compliant cylinder based FBG accelerometer proposed by Zhang et al. [7] obtained a sensitivity of 42.7 pm/g across the range of 30 Hz to 300 Hz. Combination of two cantilever beams with diaphragm FBG accelerometer proposed by Weng et al. [2] obtained the sensitivity of 100 pm/g across the range of 10 Hz to 120 Hz.

However, the aforementioned works reported on a minimum detectable frequency of only 10 Hz. Only few researchers [8-10] have reported on low frequency FBG accelerometer. Low frequency detection is very crucial especially in earthquake monitoring which has the usual frequency range from 0.01 Hz to 50 Hz [1]. Zhang et al. [8] proposed a double-semicircle cantilever beam FBG accelerometer with minimum detectable frequency of 5 Hz at the sensitivity of 1296 pm/g. Bending of spring plates method proposed by Liu et al. [10] capable to capture low frequency of 0.7 Hz. However, both of the proposed methods utilised complex structure. The used of distributed Bragg reflector fibre laser proposed by Zhang et al. [9] capable to detect low frequency of 1 Hz. However, the used of fibre laser is relatively expensive as compared to conventional FBG.

In this paper, a low frequency FBG accelerometer based on Polyphenylene Ether (PPE) thermoplastic cantilever beam has been proposed. The proposed accelerometer capable to capture low frequency of 2 Hz at 0.04 g. The sensitivity of the accelerometer was obtained at 110 pm/g with natural frequency of 9 Hz. Therefore, this paper has successfully presented a high sensitivity FBG accelerometer for low frequency measurement.

## 2. Design

The proposed FBG accelerometer consists of only three parts which are C-shaped beam for the support base, a "dog bone" shaped PPE cantilever beam and a mass. Figure 1 shows the design and interrogation system of the proposed accelerometer.

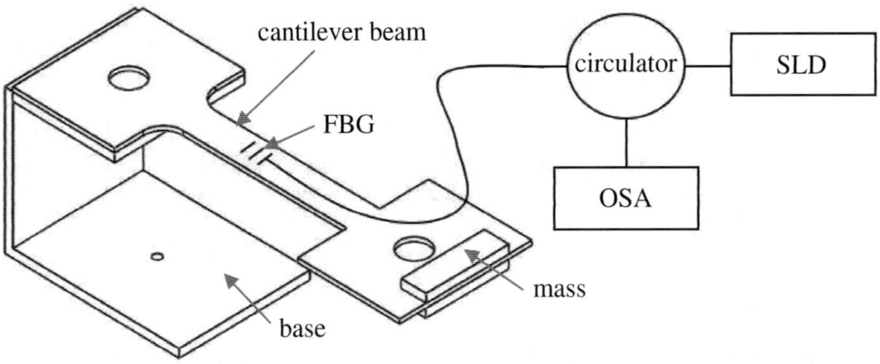


**Figure 1.** Design and interrogation system of the FBG accelerometer.

The base of the accelerometer was 3D printed made from Polylactic Acid (PLA) material with a height of 30 mm. The PPE cantilever beam has a total length of 95 mm and width at the mid-section of 8 mm. The tip of the cantilever beam was attached with two masses that have the total weight of 3.6 grams. A standard silica glass uniform grating FBG was bonded at the middle of the cantilever beam. A Superluminescent Diode (SLD) light was used to illuminate the FBG via the optical circulator. The reflected spectrum from the FBG was read by the optical spectrum analyser (OSA). The OSA has the wavelength resolution and sample rate of 1 pm and 5000 Hz. At the room temperature of 26.2 °C, the centre wavelength of the FBG was recorded at 1536.5129 nm. Under constant temperature, the centre wavelength change ($\Delta\lambda_B$) of the FBG can be expressed as [8]:

$$\Delta\lambda_B = \lambda_B(1 - p_e)\varepsilon \tag{1}$$

where $\lambda_B$ is the initial centre wavelength, $p_e$ is the effective elastic-optic coefficient and $\varepsilon$ is the axial strain that act on the fibre.

## 3. Experimental procedures and results

Three experimentations have been carried out in order to assess the characteristics of the proposed accelerometer. The experimental setup was as shown in Figure 2. The FBG accelerometer was attached to a vibration shaker manufactured by the Data Physics Corporation model V4. A DASYLab software with generator function module was used to excite the vibration of the shaker with sinusoidal wave driven at an amplitude of 5 V. A conventional electrical accelerometer manufactured by the PCB Piezotronics was used to monitor the acceleration of the shaker. The accelerometer has a sensitivity of 10 mV/g and measurement range of -5 g to 5 g where 1 g = 9.81 m/s$^2$. The sampling rate of the electrical accelerometer was set to 5000 Hz which is the same as the OSA.

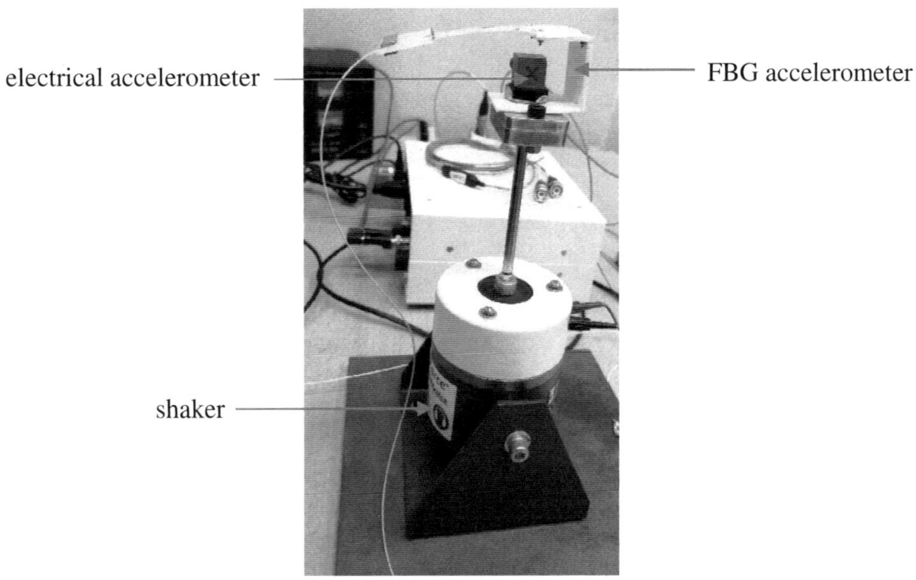


**Figure 2.** Experimental setup.

*3.1 Natural frequency*
In order to evaluate the natural frequency of the FBG accelerometer, the acceleration of the shaker was fixed at 0.1 g and the frequency was varied from 5 Hz to 15 Hz. Figure 3 shows the frequency response of the FBG accelerometer.

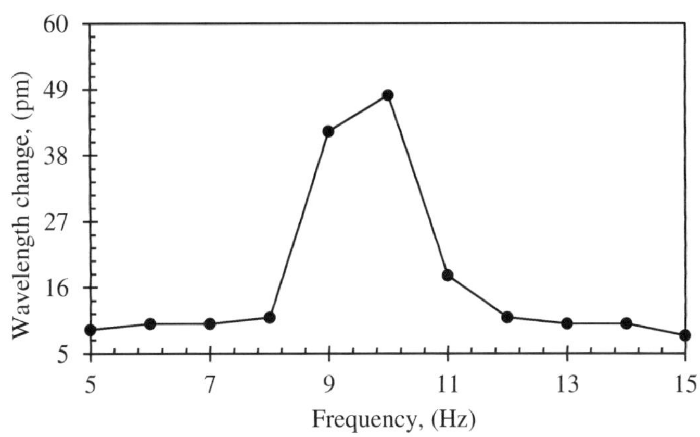


**Figure 3.** Frequency response of the FBG accelerometer.

By monitoring the wavelength change, a sudden increased in wavelength change was observed at the frequency of more than 8 Hz. The wavelength change at the frequency of 9 Hz and 10 Hz was recorded at 42 pm and 48 pm, respectively. The wavelength change at both the frequencies was four times much higher than the wavelength change from 5 Hz to 8 Hz. This is because the accelerometer will vibrate in a much larger amplitude under resonant frequency. Therefore, resulted in much larger wavelength change. From the result, the operating frequency of the proposed accelerometer was below 8 Hz. Meanwhile, the natural frequency was determined at about 9 Hz.

*3.2 Sensitivity*
The sensitivity of the proposed accelerometer can be determined at the flat response range from 5 Hz to 8 Hz. Here, the frequency was fixed at 5 Hz and the acceleration was varied from 0.02 g to 0.1 g with a step of 0.02 g. Figure 4 shows the sensitivity of the FBG accelerometer.

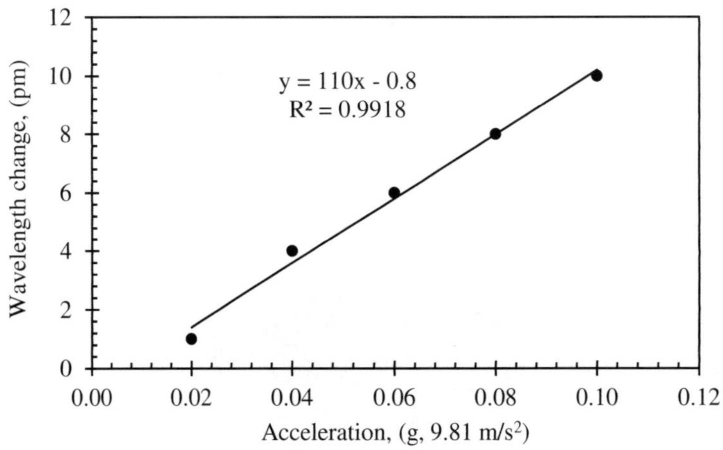


**Figure 4.** Sensitivity of the FBG accelerometer.

At the initial of 0.02 g, the FBG accelerometer recorded a wavelength change of 1 pm. The wavelength change increased to 10 pm at the maximum of 0.1 g. By determining the slope of wavelength change against acceleration, the sensitivity of the accelerometer was obtained at 110 pm/g. The obtained sensitivity at 110 pm/g was much higher as compared to the previous reported work as in [3, 6, 7].

*3.3 Vibration signals*
The vibration signals captured by the proposed FBG accelerometer was compared with the electrical accelerometer. Here, the minimum frequency was determined by varying the frequency and acceleration at 1 Hz and 0.01 g until the FBG accelerometer capable to capture the signal. It was observed that the proposed accelerometer capable to capture the lowest frequency signal of 2 Hz at the acceleration of 0.04 g as shown in Figure 5 (a). The signals captured by the FBG accelerometer was an unfiltered signal. On the other hand, the signals captured by the electrical accelerometer was Butterworth filtered with filter order 2 and limit frequency of 10 Hz. The frequency captured at 2 Hz was much lower as compared to the work by Zhang et al [8]. The capability of the proposed accelerometer in capturing low frequency vibration making it suitable for seismic measurement of earthquake. Figure 5 (b) shows the vibration signals at 5 Hz and 0.1 g. The frequency ($f$) captured by the signals can be calculated as [11]:

$$f = \frac{1}{T} \tag{2}$$

where $T$ is the period of time in second for one complete oscillation. From Figure 5 (a) and Figure 5 (b), clearly seen that the frequency signals captured by the FBG accelerometer at 2 Hz and 5.07 Hz was closed to the electrical accelerometer at 1.93 Hz and 5.09 Hz.

IOP Conf. Series: Materials Science and Engineering 1078 (2021) 012012 doi:10.1088/1757-899X/1078/1/012012

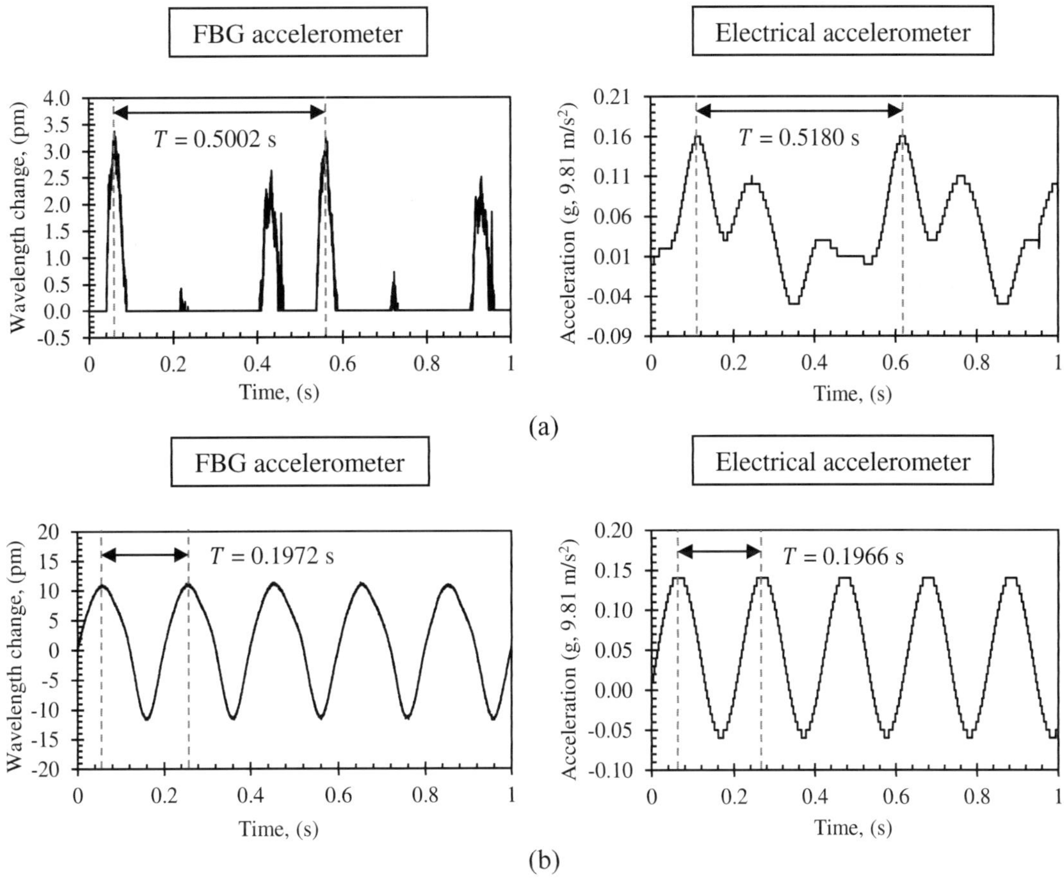


**Figure 5.** Vibration signals at (a) 2 Hz, 0.04 g and (b) 5 Hz, 0.1 g.

## 4. Conclusion

This paper has successfully presented a low frequency FBG accelerometer based on PPE thermoplastic cantilever beam. A series of experimentations have been carried out to determine the characteristics of the proposed accelerometer. From natural frequency test, the proposed accelerometer has a first mode natural frequency at 9 Hz. Therefore, the operating range of the FBG accelerometer was below 8 Hz with the sensitivity at 110 pm/g. Comparison of the vibration signals captured by the FBG accelerometers was similar to the electrical accelerometer. The lowest frequency capable to be captured by the FBG accelerometer was at 2 Hz and 0.04 g.

## Acknowledgments

The authors would like to thank the Faculty of Mechanical & Automotive Engineering Technology, Universiti Malaysia Pahang (http://www.ump.edu.my/) for providing the laboratory facilities. The authors would also like to acknowledge the Institute of Postgraduate Studies (IPS), Universiti Malaysia Pahang for funding through the Doctoral Research Scheme (DRS) scholarship. Finally, special thanks to the UMP Research and Innovation Department for providing the Postgraduate Research Grants Scheme (PGRS) under grant no. PGRS1903148; and internal research grant under grant no. RDU180387.

## References

[1] Pérez-Alonzo A and Sandoval-Romero G E 2019 Experimental study of a fiber Bragg grating accelerometer for seismic measurement *Vibroengineering Procedia* **22** 53-6

[2] Weng Y, Qiao X, Guo T, Hu M, Feng Z, Wang R and Zhang J 2011 A robust and compact fiber Bragg grating vibration sensor for seismic measurement *IEEE sensors journal* **12** 800-4

[3] Au H Y, Khijwania S K and Tam H Y 2008 Fiber Bragg grating based accelerometer *19th International Conference on Optical Fibre Sensors* **7004** 70042S

[4] Udos W, Lee Y S, Lim K S, Ong Z C, Zaini M K A and Ahmad H 2020 Signal enhancement of FBG-based cantilever accelerometer by resonance suppression using magnetic damper *Sensors and Actuators A: Physical* **304** 111895

[5] Yan B and Liang L 2019 A novel fiber Bragg grating accelerometer based on parallel double flexible hinges *IEEE Sensors Journal* **20** 4713-8

[6] Liu Q P, Qiao X G, Zhao J L, Jia Z A, Gao H and Shao M 2012 Novel fiber Bragg grating accelerometer based on diaphragm *IEEE Sensors Journal* **12** 3000-4

[7] Zhang Y, Qiao X, Liu Q, Yu D, Gao H, Shao M and Wang X 2015 Study on a fiber Bragg grating accelerometer based on compliant cylinder *Optical Fiber Technology* **26** 229-33

[8] Zhang X, Rong Q, Sun H, Yang S, Yuan L and Hu M 2014 Low-frequency fiber Bragg grating accelerometer based on a double-semicircle cantilever *Optical Fiber Technology* **20** 190-3

[9] Zhang B, Cheng L, Liang Y, Jin L, Guo T and Guan B O 2017 Low-frequency vibration measurement by a dual-frequency DBR fiber laser *Photonic Sensors* **7** 206-10

[10] Liu F, Dai Y, Karanja J M and Yang M 2017 A low frequency FBG accelerometer with symmetrical bended spring plates *Sensors* **17** 206

[11] Van Baak D A and Herold G 2014 Response of a lock-in amplifier to noise *American Journal of Physics* **82** 785-97

# Design and simulation of solar roof-top projects for an energy self-reliant university campus

**Ajithgopi[1,2], K Sudhakar[3,4,*] and Ngui Wai Keng[5]**

[1] Renewable Energy and Environmental Engineering Research Group, Universiti of Malaysia Pahang, Malaysia
[2] Agency for New and Renewable Energy Research and Technology (ANERT), Thiruvananthapuram, India
[3] Faculty of Mechanical and Automotive Engineering Technology, Universiti Malaysia Pahang, Pekan 26600, Malaysia
[4]Automotive Engineering Centre, Universiti Malaysia Pahang, Pekan 26600, Pahang Malaysia
[5]College of Engineering, Universiti Malaysia Pahang, 26300 Gambang, Kuantan, Pahang, Malaysia

*Corresponding author's email: sudhakar@ump.edu.my

**Abstract.** A University campus becoming self-reliant in terms of electricity generation is always important. The power requirement of a University campus is mainly for its academic blocks for different departments. The laboratories, libraries, and hostel facilities also require continuous power needs. Conventionally, the electricity needs are met utilizing the power received from the local utility and the University will be paying for it based on the tariff fixed by the local regulators and agreed by the utility. It is proposed to install Solar Roof-top PV power plants in shade-free rooftops of the buildings inside a University campus to offset the University's electricity needs and to make the University self-reliant in terms of electricity generation and consumption. This integrated and comprehensive solar project is intended to reduce the carbon footprint of the University and aiming for a carbon-neutral campus. Detailed feasibility studies have been conducted for installing solar roof-top power plants in the selected building roof-tops of the University. Based on the feasibility study, design and simulation have been carried out for a total 1.6MWp capacity of solar roof-top projects to completely offset the total electricity needs of the University based on the current electricity consumption patterns. The design and simulation are accomplished utilizing PVsyst software and the economic and environmental analysis has been carried out utilizing RETscreen software for arriving at the results.

**Keywords:** Solar PV; Self-Reliance; Solar Roof-Top; Simulation; Carbon-Neutral; Green Campus.

## 1. Introduction

Renewable energy sources are becoming a prominent energy source globally, especially in the last decade [1]. Solar energy is the most prominent renewable energy source and is replacing fossil fuel generations at a great pace worldwide [2]. The cost of solar power has reduced considerably in the last decade bringing more and more National Governments are embracing renewable and affordable resources with aggressive policy measures to cope up with the Paris Agreement regulations regarding climate change. A lot of countries have set targets for achieving 100% renewable energy-based generation [3].

Solar Photovoltaic power has reached a cumulative capacity of 627GW in 2019 with China leading in the case of total installations followed by the United States, Japan, Germany, and India. Solar PV is becoming the most sought renewable energy source in the world with a tremendous investment made all over the world. China leading the world as the biggest market in the world in terms of total capacity additions of 31.1GW followed by United State with 13.3GW and India in third place with 9.9GW [4]. International Renewable Energy Agency (IRENA) predicts that the total installation cost of solar PV projects would continue to decline in the next three decades. This would make solar PV highly competitive in many markets, with the average cost falling in the range of USD 340 to 834 per kilowatt (kW) by 2030 and USD 165 to 481/kW by 2050, compared to the average of USD 1 210/kW in 2018. Solar PV power plants are being installed worldwide, mainly in two forms:

1. Solar Roof-tops for residential purposes, industrial purposes, and for commercial purposes[5].
2. MW scale solar farms mainly feeding bulk power to the utility [6]

Among the above categories, solar roof-tops are decentralized power generating systems that can be installed with limited time and easiness. Academic institutions are generally holding large shade-free roof-tops usually eligible for installing solar PV plants. Power can usually be evacuated to the utility grid terminal in the building itself reducing the cost of power evacuation infrastructure and also reducing transmission losses [7].

This paper discusses an integrated and comprehensive approach for installing solar roof-top PV plants to make a University self-reliant in terms of electricity generation. This aims to make the University carbon-free in terms of electricity generation and intended to become a carbon-neutral campus [3]. Karyavattom University Campus under the University of the South Indian State Kerala has been selected for the proposed study.

## 2. Methodology

### *2.1 Site details*

A site visit has been conducted to assess the technical feasibility of the site to get precise inputs for the technical design of the proposed PV Power Plant. The details are:

**Table 1.** Site visit details.

| Site Name | Kariavattom Campus of the University of Kerala |
|---|---|
| Roof-top | Roof-top of the Dept. of Geology |
| Ownership | University of Kerala |
| District | Thiruvananthapuram |
| State | Kerala |
| Latitude | 080 33'' 54.143'' |
| Longitude | 760 53'' 14.707" |
| Roof -top Area available | 10634 square meters |

**The methodology involves:**
1. Feasibility study at the University by assessing shade-free roof-tops for installing the roof-top PV plants[8].
2. Design and Simulation of the project utilizing PV Syst Software [8]
3. Economic and Environmental Analysis of the project utilizing RETScreen software [9].

**Figure 1.** Google earth map of the proposed university campus.

*2.2 Feasibility study conducted at the university*
The feasibility study is an important step for any solar PV project to succeed [8]. Design, simulation, and economic analysis of a solar photovoltaic system were conducted by Shukla et al. [7]. Primary data was collected by visiting the Kerala University Camus. The various electrical loads operating in these facilities present a peak power demand of around 450 kVA and off-peak demand of around 150 kVA. Thus, the average daily electricity consumption on campus is around 6 MWh. Electrical load in the campus increases by about 10% annually, there is a need to have sufficient power facilities to meet up the future demand as well. Since the campus is having various buildings with sufficient availability of shade-free rooftops, solar power generation is an excellent option for power system management and to meet the peak load demand. At present the entire electric power requirement is supplied by the KSEB through an 11 kV substation located within the campus and from a 100kWp solar power plant is installed on the roof top of Golden Jubilee Building in 2015.

Following buildings were surveyed as part of the feasibility study for installing Roof-top PV power plants:

**Table 2.** List of buildings selected for solar roof-top PV plants.

| Sl. No. | University Building | Area (sq.m) | Estimated Power Generation (kWp) (Assuming 8 sq.m for 1kWp) |
|---|---|---|---|
| 1 | Dept. of Biochemistry | 758.54 | 100 |
| 2 | Dept. of Chemistry | 1031.55 | 130 |
| 3 | Dept. of Computer science(New) | 238.78 | 30 |
| 4 | Dept. of Computer science(Old) | 606.45 | 80 |
| 5 | Golden Jubilee building | 1883.78 | 240 |
| 6 | Dept. of Future studies | 1218.75 | 150 |
| 7 | SICC building | 706.92 | 90 |
| 11 | Dept. of Environmental Science (New) | 421.28 | 50 |
| 13 | Dept. of Environmental science(Old) | 1322.8 | 170 |
| 14 | Dept. of Malayalam and Tamil | 1129.33 | 140 |
| 15 | Dept. of Commerce | 1315.65 | 170 |
| 16 | Library Building | 400.24 | 50 |
| 17 | ONV Memorial Building | 1600.46 | 200 |
| | TOTAL projected capacity | | 1600kWp |

*2.3 PVSyst software*

PVsyst is a PC software package for the design and simulation of PV systems. The software is developed by the University of Geneva, Switzerland, and is widely used by the solar industry, developers, and academia. It deals with grid-connected, stand-alone, pumping, etc, and includes PV system component databases [10].

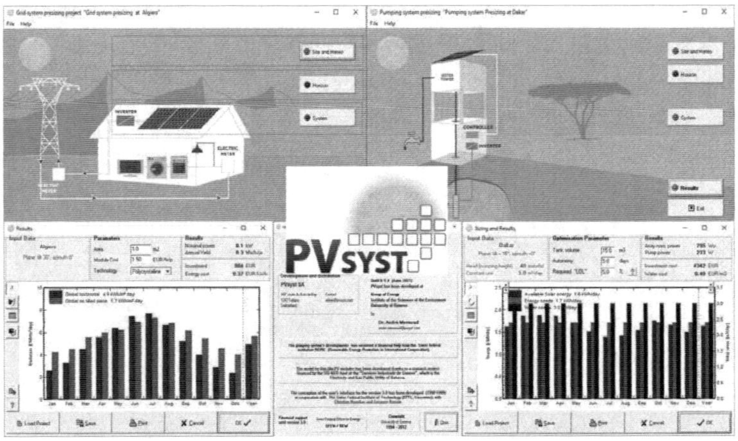


**Figure 2.** PV Syst software – screenshot.

*2.4 RETScreen software*

RETscreen Expert integrates several databases to assist the user, including a global database of climatic conditions obtained from NASA satellite data; benchmark database; cost database; project database; hydrology database, and product database. The economic and environmental analysis is to done using this software [10].

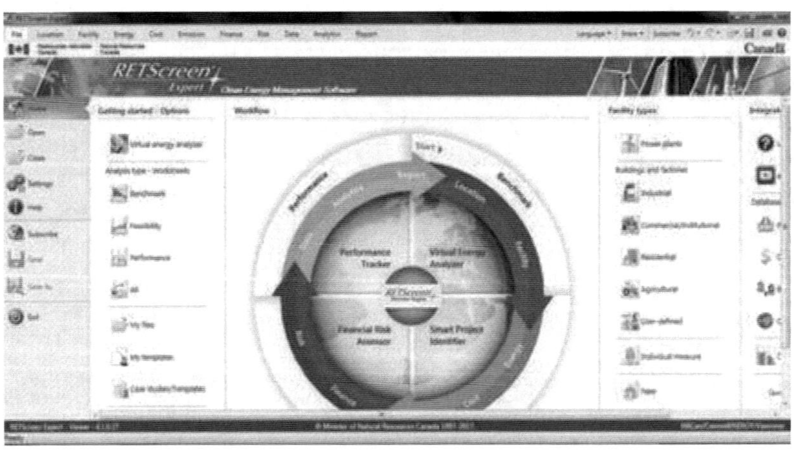


**Figure 3.** RETScreen software – screenshot.

## 3. Design and simulation of the project and results

*3.1 Design of the project utilizing PV Syst software*

PV Syst simulation for the project with a total capacity of 1.6MWp is carried out utilizing the freely available solar radiation data of NASA –SSE. The summary of the PV Syst simulation results for the project is given in the following Table 3 & 4 and in Figures 4 & 5. Solar Resource available at the site as a result of the PV syst simulation is given in Table 4.

Summary of the solar generation expected, specific production and the expected PR along with projections for Performance Ratio (PR) and the normalized productions per installed kWp are given as Figure 4. A loss diagram in PVsyst reflects the losses happening to the PV power plant. Loss Diagram obtained as a simulation result of the 1.6MWp Solar Plant at the Kerala University Campus is given as in Figure 5.

**Table 3.** Summary of the results of PVSyst simulation.

| | | |
|---|---|---|
| Proposed Installed Capacity | Unit | 1600kWp |
| Project Type | | Rooftop |
| Estimated Annual Energy production | Units (kWh/yr) | 2553000 (2.553 million units per year) |
| Specific Production | kWh/kWp/Year | 1596 |
| Performance Ratio (PR) | % | 76.7 |
| Solar resource available (GHI) | Wh/m$^2$ | 2039.6 |

**Table 4**. Solar resource available at the site as resulted from the PVsyt simulation.

| Months | GlobHor kWh/m$^2$ | T Amb $^0$C | Globlnc kWh/m$^2$ | GlobEff kWh/m$^2$ | EArray MWh | E_Grid MWh | EffArrR % | EffSysR % |
|---|---|---|---|---|---|---|---|---|
| January | 179.8 | 27.04 | 198.2 | 183.1 | 245.1 | 241.5 | 12.28 | 12.10 |
| February | 180.9 | 27.08 | 193.3 | 178.9 | 237.6 | 234.1 | 12.20 | 12.02 |
| March | 211.7 | 27.22 | 216.5 | 200.4 | 265.2 | 261.1 | 12.16 | 11.98 |
| April | 187.2 | 27.25 | 183.6 | 169.2 | 227.1 | 223.5 | 12.28 | 12.09 |
| May | 172.7 | 27.58 | 164.1 | 150.6 | 205.3 | 202.1 | 12.42 | 12.23 |
| June | 144.9 | 27.05 | 136.2 | 124.7 | 172.2 | 169.5 | 12.56 | 12.36 |
| July | 152.2 | 26.40 | 143.9 | 131.9 | 182.5 | 179.6 | 12.59 | 12.39 |
| August | 163.1 | 26.20 | 158.5 | 145.6 | 199.6 | 196.4 | 12.51 | 12.31 |
| September | 172.2 | 26.31 | 173.4 | 159.8 | 217.2 | 213.9 | 12.44 | 12.25 |
| October | 162.4 | 26.34 | 169.9 | 156.6 | 212.2 | 209.0 | 12.40 | 12.21 |
| November | 148.2 | 26.73 | 161.1 | 148.4 | 201.5 | 198.4 | 12.43 | 12.23 |
| December | 164.3 | 27.08 | 182.5 | 168.4 | 226.7 | 223.4 | 12.34 | 12.15 |
| Year | 2039.6 | 26.86 | 2081.2 | 1917.6 | 2592.3 | 2552.6 | 12.37 | 12.16 |

Where:
GlobHor: Horizontal global irradiation
T Amb: Ambient Temperature
Globlnc : Global incident in coil plane
GlobEff: Effective Global corresponding for IAM and shadings
EArray: Effective energy at the output of the array
E_Grid : Energy injected into grid
EffArrR: Effective Eout array/rough area
EffSysR : Effective Eout system/rough area

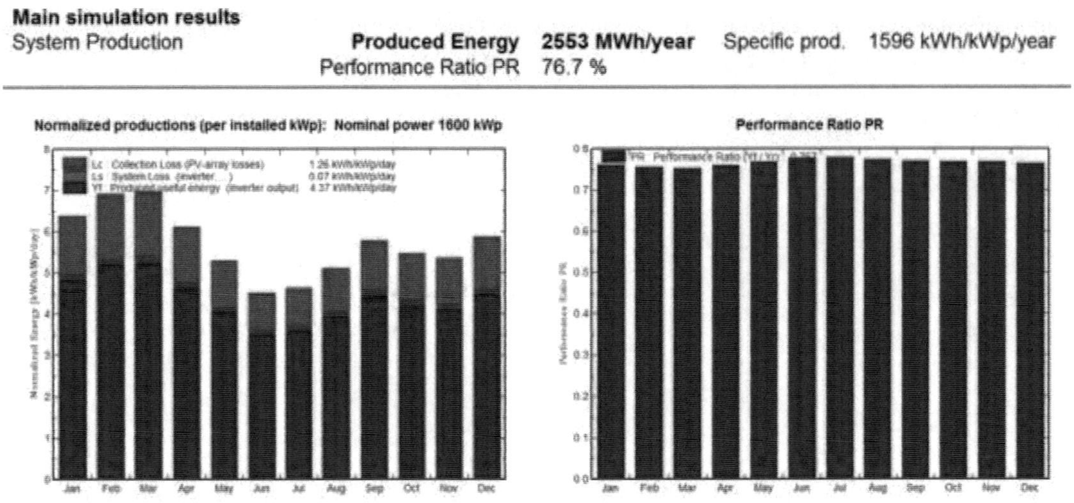


**Figure 4.** Summary of the Normalized productions (per installed (kWp)).

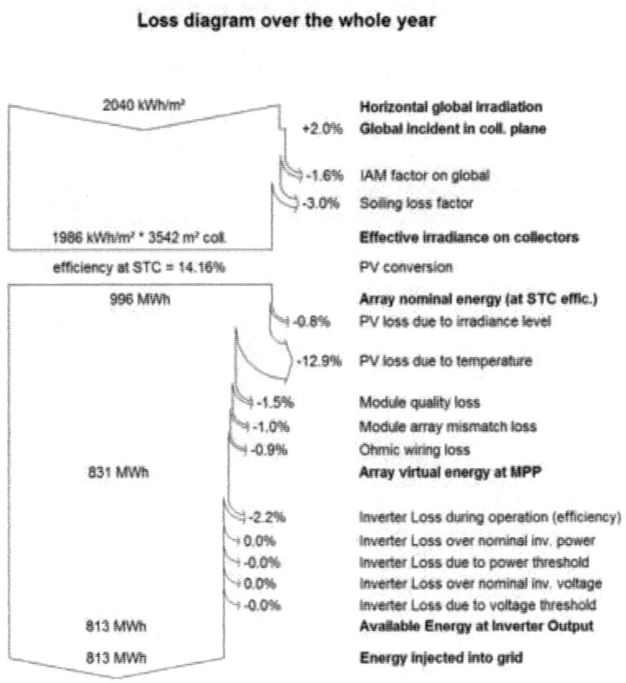


**Figure 5.** Loss diagram of the PV Plant as simulated for 1.6MWp solar plant.

*3.2 Data Analysis to size and define the project*

The project size is determined based on the consumption at the university campus and the near future of the addition of any loads. KSEB Ltd. (local public utility) has conducted an energy audit at the site and the report shows that the campus had an energy consumption of 1941497kWh from July 2017 to June 2018. The observation as per the energy audit report of KSEB Ltd is given as in Table 5.

It is observed that the energy consumption of the University from July 2019 to June 2020 was above 22MWh and expected an increase of 10% only in near future. By simulating a 1.6MWp solar power array, using a reputed design and simulation tool PVSyst it is found that we can completely offset this requirement. The area is already available for setting up the roof-top projects as per the feasibility study conducted as given in table 1. From the simulation report, it is clear that the roof-top projects with a total capacity of 1.6MWp can generate 2553 MWh/yr as per the PV syst simulation and this can completely offset the yearly consumption of the University campus in a futuristic perspective. The additional power generated from the PV plants is fed to the 11kV Grid which is going to the utility substation.

**Table 5.** Energy consumption as part of the energy audit report of KSEBL Ltd.

| Sl. No. | Month | Energy Consumption in the Campus (kWh) |
|---|---|---|
| 1 | July 2017 | 173035 |
| 2 | August 2017 | 140375 |
| 3 | September 2017 | 148000 |
| 4 | October 2017 | 151300 |
| 5 | November 2017 | 153200 |
| 6 | December 2017 | 174100 |
| 7 | January 2018 | 145542 |
| 8 | February 2018 | 145542 |
| 9 | March 2018 | 157493 |
| 10 | April 2018 | 185790 |
| 11 | May 2018 | 181250 |
| 12 | June 2018 | 185830 |
| | TOTAL | 1941497 |

## 4. Environmental and economic analysis

Environmental and economic analysis is needed to understand the viability and bankability of the project. Emission reduction studies can bring out figures related to GHG reduction etc. [3]. The design analysis is done using the industry famous software tool RETscreen. The analysis is done in two parts to find out the economic and environmental feasibility of the proposed design [11]. Emmanuel et al; has conducted an economic and environmental analysis of a 10kWp Roof-top PV plant [12]. Solar PV roof-tops are environmentally friendly and are good for mitigating climate change [13]. Fthenakis and Kim investigated the important environmental impacts on solar energy projects and observed its importance in addressing climate change mitigation activities [14]. Gallardo and Karlsson analyzed the PV roof-top plant installed at Galve University in Sweden [15].

### *4.1. Environmental analysis*

Environmental analysis is to find out the annual reduction in GHG emission and thus to determine whether the project is environmentally feasible or not. The analysis gives out the GHG emission reduction summary from where the Gross Annual GHG Emission Reduction can be found out.

The GHG emission reduction as explained above can be achieved by installing Solar Roof-top PV Plants with a total capacity of 1.6MWp. The annual GHG emission reduction is equivalent to various parameters as defined by the RETScreen Software. The analogy as derived by the software is given below as equivalent to the environmental benefits of installing Solar Roof-top PV plants.

**Table 6.** Summary of GHG emission reduction.

| Base case electricity system (Baseline)<br>Country region | Fuel type | GHG emission factor (excl. T&D) tCO2/ MWh | T&D losses % | GHG emission factor tCO2/ MWh |
|---|---|---|---|---|
| India | All types | 0.856 | 3.0 | 0.882 |
| Electricity exported to grid | MWh | 2,680 | T&D losses | 3.0 |
| GHG emission | | | | |
| Base case | $tCO_2$ | 2364.3 | | |
| Proposed case | $tCO_2$ | 70.9 | | |
| Gross annual GHG emission reduction | $tCO_2$ | 2293.4 | | |

**Table 7.** Comparison of GHG emission reduction.

| The amount of carbon dioxide Mitigation is equivalent to 2,293.4 tonnes | |
|---|---|
| 420 | Cars & light trucks not used |
| 9,85,406 | Liters of gasoline not consumed |
| 5,334 | Barrels of crude oil not consumed |
| 2,293.4 | People reducing energy use by 20% |
| 521.2 | Acres of forest absorbing carbon |
| 211 | Hectares of forest absorbing carbon |
| 790.8 | Tonnes of waste recycled |

### *4.2 Economic analysis*

The economic analysis has been carried out using RETscreen Software which is developed by the Government of Canada. RETScreen is an efficient software tool used by both industry and academicians. RETScreen is used to analyze the performance of a Solar PV plant of 5MWp capacity in Tamilnadu in India [11]. As per the Govt of India benchmark cost for grid-connected solar power plant, the total initial cost of 1.6MWp solar power plant is 6 crores. The software is used to find out the financial parameters like Payback Period, Internal Rate of Return (IRR), Net Present Value (NPV, Annual Life Cycle Savings, etc. of the solar power plant. The financial parameters found out through RETScreen software are given below in Table 5.

From the RETScreen simulation, it is found that the project has a Simple Payback Period of about 3 years at an electricity export rate of Rs8/- (INR). The project's IRR (equity) value of 33 and the positive NPV of Rs43,08,35,795/- (INR), indicate an economically viable project.

**Table 8.** Summary of financial parameters obtained for the 1.6MWp project.

| Particulars | Unit | Values/ Rs |
|---|---|---|
| Electricity Export Rate | | Rs 8/-(INR) |
| Pre-tax IRR–equity | % | 33.9 |
| Pre-tax MIRR–equity | | 8.8 |
| Pre-tax IRR- assets | % | 33.9 |
| Pre-tax MIRR- assets | | 8.8 |
| Simple payback | Yr | 2.9 |
| Equity payback | Yr | 2.9 |
| Net Present Value (NPV) | INR | 43,08,35,795 |
| Annual Life cycle savings | INR/Yr | 1,72,33,432 |
| Benefit-Cost (B-C) ratio | - | 8.2 |
| Debt service coverage | | |
| GHG reduction cost | INR/t$CO_2$ | -7,514 |
| Energy production cost | INR/kWh | 1.57 |

## 5. Conclusion

Design and Simulation of the proposed Solar Roof-top Projects with a total capacity of 1.6MWp was accomplished in this paper based on feasibility studies conducted at the University Campus and also utilizing industry-standard software like PV syst and RETScreen. This project will offset the electricity needs of the University Campus and thus helps the campus to become self-reliant in terms of electricity generation for meeting its power needs. The extra power will be fed to the utility grid. The economic analysis brings attractive figures for investment. The project has got simple payback of 2.9 years, a Pre-tax IRR equity of 33.9%, and a Benefit-Cost (B-C) Ratio of 8.2. This project can bring the GHG emission reduction of 2293.6 tonnes of carbon dioxide which is equivalent to 790.8 tonnes of waste recycled per annum, 211 hectares of forest absorbing carbon, 5334 barrels of crude oil not being consumed, 985406 liters of gasoline not consumed, and 420 light trucks and cars not being used. Thus, the proposed solar roof-top project is both environmentally friendly and economically feasible. A University becomes carbon-free in terms of electricity generation when it generates all its electricity needs from renewable energy sources. The University's effort to become self-sufficient in terms of energy is also full filled utilizing PV and its emerging technologies.

## Acknowledgments

The authors acknowledge the University for permitting conducting the feasibility study of the project. Also, the authors acknowledge the Agency for New and Renewable Energy Research and Technology (ANERT), the Kerala State Renewable Nodal Agency in India for the technical help during the visit. The authors are grateful for the financial support provided by the Universiti Malaysia Pahang (www.ump.edu.my) through the Postgraduate Research Grant Scheme, PGRS 1903172.

## References

[1] Members R E N 2020 *Renewables 2020 global status report 2020* ed Paris: REN21 Secretariat (Paris: Paris: REN21 Secretariat)

[2] Anon *Pvps, Iea. (2014). Task 13: Analytical Monitoring of Grid-Connected Photovoltaic Systems.*

[3] Spiegel R J, Greenberg D L, Kern E C and House D E 2000 Emissions reduction data for grid-connected photovoltaic power systems *Sol. Energy* **68** 475–85

[4] REN21 2020 *Renewables 2020 global status report 2020* (Paris: REN21 Secretariat)

[5] International I and Energy R 2018 *FUTURE OF Deployment , investment , technology , grid*

*integration and socio-economic aspects*
[6] Beylot A, Payet J Ô, Puech C, Adra N, Jacquin P, Blanc I and Beloin-Saint-Pierre D 2014 Environmental impacts of large-scale grid-connected ground-mounted PV installations *Renew. Energy* **61** 2–6
[7] Shukla A K, Sudhakar K and Baredar P 2016 Design, simulation and economic analysis of standalone roof top solar PV system in India *Sol. Energy* **136** 437–49
[8] Kalita P, Das S, Das D, Borgohain P, Dewan A and Banik R K 2019 Feasibility study of installation of MW level grid connected solar photovoltaic power plant for northeastern region of India Feasibility study of installation of MW level grid connected solar photovoltaic power plant for northeastern region of India *Sādhanā*
[9] Gürtürk M 2019 Economic feasibility of solar power plants based on PV module with levelized cost analysis *Energy* **171** 866–78
[10] Wijeratne W M P U, Yang R J, Too E and Wakefield R 2019 Design and development of distributed solar PV systems: Do the current tools work? *Sustain. Cities Soc.* **45** 553–78
[11] Sundaram S and Babu J S C 2015 Performance evaluation and validation of 5 MW grid connected solar photovoltaic plant in South India *Energy Convers. Manag.* **100** 429–39
[12] Emmanuel M, Akinyele D and Rayudu R 2017 Techno-economic analysis of a 10 kWp utility interactive photovoltaic system at Maungaraki school , Wellington , New Zealand *Energy* **120** 573–83
[13] Pazheri F R and Baby T 2014 Rooftop Solar Projects Enhance the Environmental Friendly Power Dispatch in Kerala Saudi Aramco Chair in Electrical Power 233–6
[14] Fthenakis V M and Kim H C 2011 Photovoltaics: Life-cycle analyses *Sol. Energy* **85** 1609–28
[15] Gallardo-Saavedra S and Karlsson B 2018 Simulation, validation and analysis of shading effects on a PV system *Sol. Energy* **170** 828–39

IOP Conf. Series: Materials Science and Engineering 1078 (2021) 012014 doi:10.1088/1757-899X/1078/1/012014

# Cotton waste research follows the effect of pre-treatment and the observation of physical appearance

**A S F Mahamude[1,*], W S W Harun[1], K Kadirgama[2], D Ramasamy[1] and K Farhana[3]**

[1]College of Engineering, Universiti Malaysia Pahang, 26300 Gambang, Pahang, Malaysia
[2]Faculty of Mechanical & Automotive Engineering Technology, Universiti Malaysia Pahang,
26600 Pekan, Pahang, Malaysia
[3]Department of Apparel Engineering, Bangladesh University of Textiles, Dhaka 1208, Bangladesh

*Corresponding author's email: mahamude.ump@gmail.com

**Abstract.** This paper aims to evaluate the waste cotton quality and its commercialization. The recycled cotton demand seems to be increased day by day: although the lower quality of the resulting items is the biggest barrier to cotton recycling. A thorough analysis of recycled cotton processes is therefore necessary from separation/shredding to completion. To that end, textile wastes made of pre-consumer cotton were systematically collected from the spinning mill. After pretreatment of cotton waste cellulose by caustic soda, alkali, and detergent in 100°C temperature for 30 minutes have produced white clean cotton. Following the next process by sodium hydroxide and isopropyl alcohol must produce Carboxymethyl Cellulose (CMC). CMC has big use in Textile printing, finishing, and warp sizing even in Ice-cream and food baking industry. Recent success stories will be addressed with the steady cotton processing to altering the properties of cell wall cotton and cotton fiber. The prospects of conducting experiments of waste cotton and checking the properties of treated cotton have a great value in the production of commercial products and the results relived that the value addition of waste cotton is a very easy method.

**Keywords:** Waste cotton; Treated Cotton; Nano-cellulose; Microcrystalline cellulose; Recycling.

## 1. Introduction

Environment protection and waste recycling have become two of the most important challenges facing the scientific and industrial community. The different textile processes create various waste materials in different stages such as fibre, sliver and yarn. Textile production and consumption are continually increasing. Generally, waste recycling requires accounting technical and economic considerations. Environment studies have to address these two interests and suggest solutions to persuade manufacturers to treat waste [1]. In our region, the textile industry has huge potential. The largest foreign currency earner has been raised. It spread throughout the area. In recent years, it has made a significant contribution to our GDP. This sector generates approximately 4.2 million workers. Waste management in this field has also become very important in the textile sector in Bangladesh. The recycling of textiles also has an immense effect on our economy [2]. The demand for cotton increases day by day in the world, the traditional approach used to assess the quantity of cotton litter, which includes high volume [3]. In the textile industry, cotton is the main raw material. It accounts for around 40% of the world's total refined fibers. Many quality parameters and origins can characterize cotton. Length, maturity, strength, and elongation parameters of base cotton fibers depend on the genetic structure of the variety, the area of cultivation, i.e. the soil and the soil/environment, irrigating, or non-irrigated conditions, the harvest, and ginning. Saw ginning can do more harm to fiber than rollers [4]. Mechanical cotton processing in yarn mills results in cotton waste [5]. Waste cotton is a highly mechanical textile fiber that can be re-used in different ways. To measure the waste cotton quality a world-class testing machine is designed to calculate the mechanical properties which are used in research [6]. For analyzing cotton quality, a set of machines developed from the very beginning of the industrial period. Standardized cotton testing instruments [7] and AFIS measurements are increasingly being used to acquire quality data from cotton breeders and fiber processors as a primary and routine tool [8]. HVI machine was a great creation of science. The HVI test is a primary and regular measurement that gives cotton researchers fiber properties [9]. Afterward, there was a requirement of trash or waste cotton analysis, thereby the precondition HVI machine has developed [10]. The demand for cotton increases day by day in the world, the traditional approach used to assess the quantity of cotton litter, which includes high volume [3]. After pretreatment, there is a change of cotton fiber in regards to its physical, chemical, and mechanical structure. For example, after liquid ammonia treatment cellulose converts to cellulose-III with a much more crystalline region [11]. In this research study, to explore the quality analysis parameters of raw cotton, wastage cotton, and treated cotton.

## 2. Methodology

The cotton-growing region in the world has been shown in Figure 1. Only the topmost 15 countries report evaluating in this report. Many factories in the world are manufacturing a lot of goods using waste cotton as their principal raw material [12]. In a bath containing industrial scoring and bleaching agents, the cotton is first treated. In the process of scoring and bleaching cationization can easily be done in a short time when alkaline and high-temperature conditions are used. Optimal conditions of pre-treatment are as follows: griges are treated with a bathtub of 4 grams of cotton and a hydrogen detergent of 6 g/L of 30% at 90 °C for 60 min, and 30 g/L of GTA and 3 g/L of sodium hydroxide are applied for further 15-minute treatment in the bath. Bathroom condition: Tests and evaluations of fiber performances, including whiteness, water absorption, time of diffusion and capillary effect. Analysis of X-ray diffraction, the morphology of the surface and thermal analysis of cotton that has been pre-treated, are also investigated [14].

This chapter is intended to clarify the techniques and designs used in this thesis for experimental testing. The flow diagram is the project process from start to finish as shown in Figure 2. The purpose of the flow chart is to graphically show the viewers the sequence of activities in a process to make them understandable. The whole analysis methodology phase flow chart was presented, if we see below flow diagram, cotton collection from the spinning mill which cotton already going to trash after that collected sample was tested by USTER HVI-1000 same time we compare spinnable cotton fiber with waste cotton fiber in same USTER HVI-1000 AFIS model machine. There are two types of waste normally we got from spinning mill one called Dropping-1 and another called Dropping-2. Later there was a treatment by a different chemical such as Dicholoro Mythene (DCM), Sodium Hydroxide (NaOH), Peroxide($H_2O_2$), and detergent the relevant output was Mono Crystalline cellulose.

### *2.1 Raw cotton and waste cotton source*

Raw cotton and waste cotton fiber collected from different spinning as per 5% Uster Statistics. The wastage from the Carding frame is called Dopping-01, Wastage from the blow room is called Dopping-02 and the good quality raw cotton collected from the imported fiber of the same spinning mill named Multazim Spinning Mills Limited. This cotton origin from Australia Raw Cotton(E-2160) Dt:18.09.18 Supplier: Shandong Jing, L/C No-0167618230007.

*2.2 Raw and waste cotton analysis*

In practice, the SCI could be used as the priority to pick bales and second priority, followed by micronaire, to monitor the selection of fibres. Since the SCI has six associated properties, the SCI and micronaire can be controlled to regulate the properties of all cotton. The Equation for Regression The following is used for the estimation of the SCI [15]:

$$SCI = -414.67 + 2.9\times strength - 9.32\times micronaire + 49.17\times UHML + 4.74\times UI + 0.65\times Rd + 0.36\times(+b) \quad (1)$$

Where UHML is the upper half mean length in inches; UI is the uniformity index; Rd is the reflectance degree, and (+b) is the yellowness of cotton fibre. Uniformity of cotton (the distribution of cotton fiber longitude, short fiber, is an important factor in the textile industry. In the processing of cotton of a given variety in standard machinery, there is a statistically significant relation between short fiber quality (SFC) and uniformity (UI) [16].

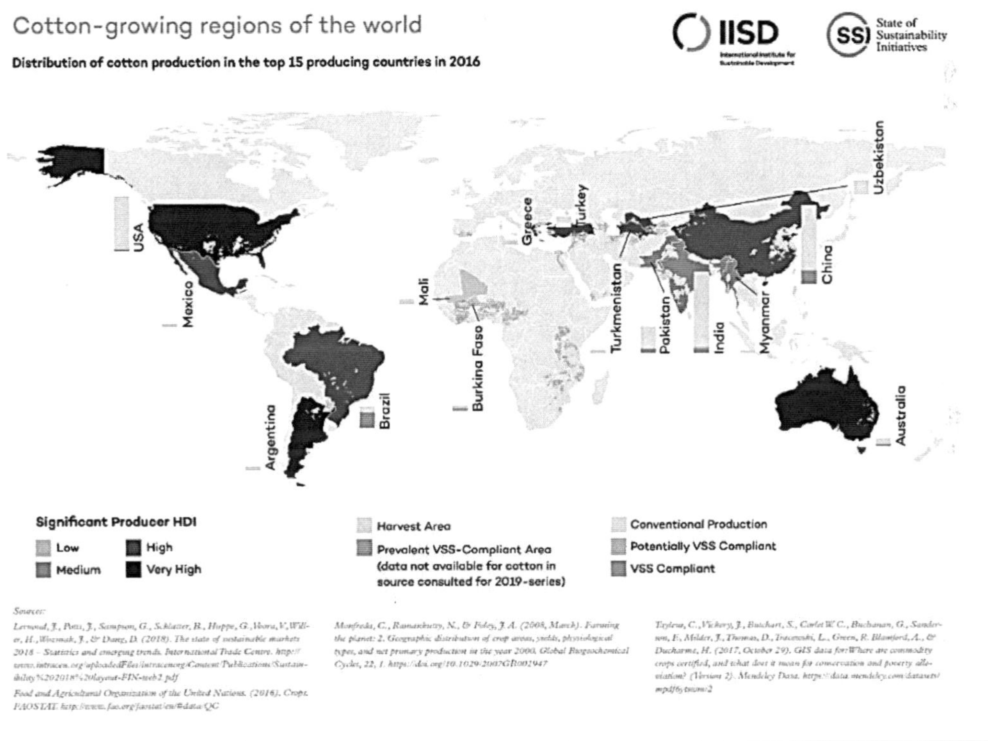


**Figure 1**. Map of cotton -growing countries in the world [13].

**Table 1**. Chemicals and sources.

| Chemical | Source | Purity/Grade |
|---|---|---|
| Dichloromethane | Merck, Germany | 99.8% |
| NaOH | Murali, India | 98% |
| Sodium chloride | Fisher Chemicals, Belgium | Analytical grade |
| Detergent | Rheopol BMW, Uk | 90% |
| Hydrogen Peroxide | HP chemicals, Bangladesh | 50% |

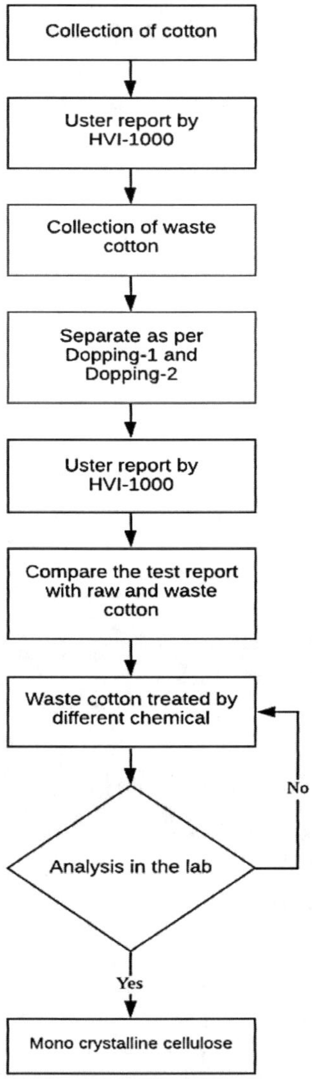


**Figure 2**. Process flow of preparing mono crystalline cellulose.

In order to research different manifestations and cellulose morphologies, we attempted to create a general characterization of SEM techniques. On the one hand the analysis of SEM and which simultaneously performs ensemble measures on multiple sections. On the other hand, scarce particles of a specific kind are detected, classified and fully defined by careful sample screening even in case of low concentration. [17]. After preparing CMC, Mono Crystalline Cellulose was analyzed in Carrif laboratory of University of Malaysia Pahang. The resultant images and data are spelled out below the data analysis and discussion part.

## 3. Result and discussion

### *3.1 Short fiber comparison*

A cotton quality parameter is important for short fibers in a cotton sample. The efficiency of yarn production and yarn quality is impaired by shorter cotton fibers. [18]. But due to only the availability of HVI, we have used the HVI USTER statistics regression model. The waste ratio from USTER can be regarded as a short fiber quality indicator for the fiber statistics with raw and waste fiber. Short fiber average ratio% was obtained that dropping-1, Dropping-02 and raw cotton were 17.5%, 24.8%, 8.7% consequently. CV% of dropping-1 and dropping-02 almost the same 2 and 1.8 but CV% of raw cotton was 1.3. But the standard deviation of dropping-1 and dropping-o2 was 2.8 and 0.8 whereas the standard deviation of raw cotton 1.6. In this Figure 3, it has shown that Sample-01 raw cotton short fiber percentage almost 8 but at the same time when this raw sample waste produces dropping-01 (Carding waste) short fiber percentage nineteen and dropping-02 (Blow room waste) short fiber percentage twenty-five percentages.

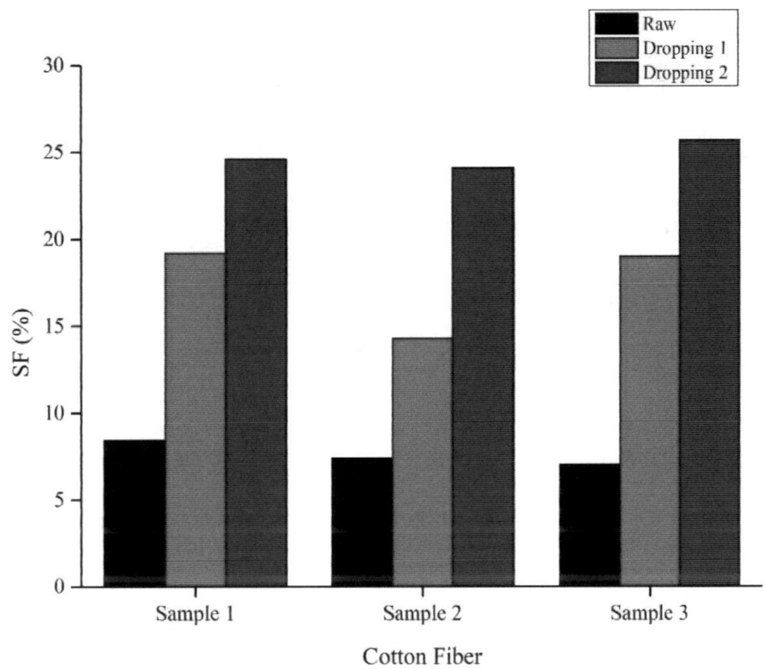


**Figure 3**. SF% comparison ratio of raw cotton and waste cotton (Dropping-01, Dropping-02).

*3.2 Upper half mean length*

The textile industry is interested in fiber quality as it is directly concerned with the production, efficiency, and quality of yarns. Fiber length is one of the most important marketing and processing properties of cotton fibers. The fiber length has a few parameters affecting production quality, the quantity of waste, fly generation, and cleaning [19]. There was a clear picture of raw cotton and waste cotton Upper Half Mean Length. Raw cotton upper UHML is on an average 30-33mm and waste cotton varies from 25-27mm. The optimal minimum upper half mean [20] fiber length in foreign markets is 28 mm. UHML is also linked to characteristics of strength and fineness, they allow increased spinning speeds. This is equivalent to increased output and lower costs per unit.

*3.3 Elongation*

The higher elongation fibers also developed higher yarn tensile and fewer yarn defects. Consequently, due to their less break-prone nature, the higher detent in fibers seems to improve their efficiency in spinning and other processing phases. Higher elongation would therefore ultimately increase textile products' efficiency. By the reference of journal and paper, elongation% is a must need part of cotton spinning. In Figure 5, we observed that raw cotton elongation percentage is more or less, but waste cotton fiber elongation percentage is varying from 5% to 6%.

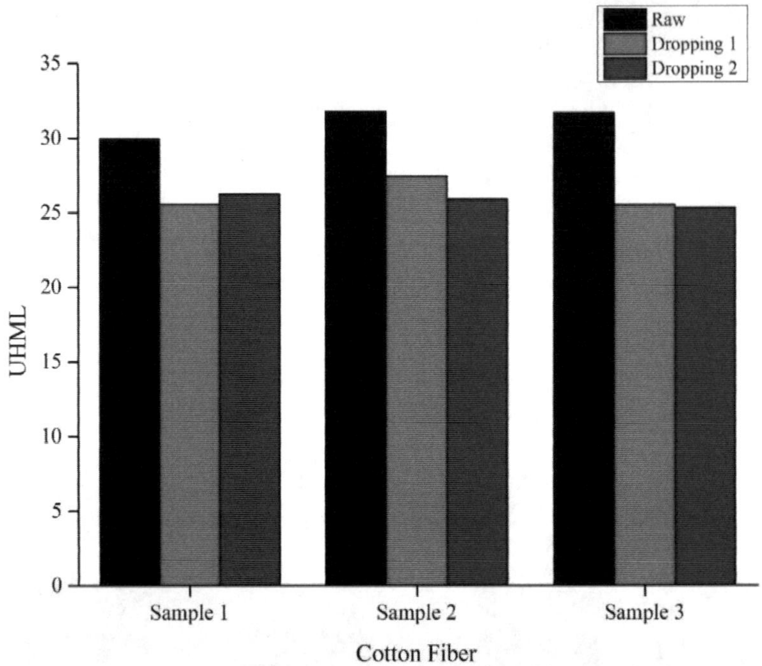


**Figure 4**. Upper half mean length(UHML) comparison ratio of raw cotton and waste cotton (Dropping-01, Dropping-02).

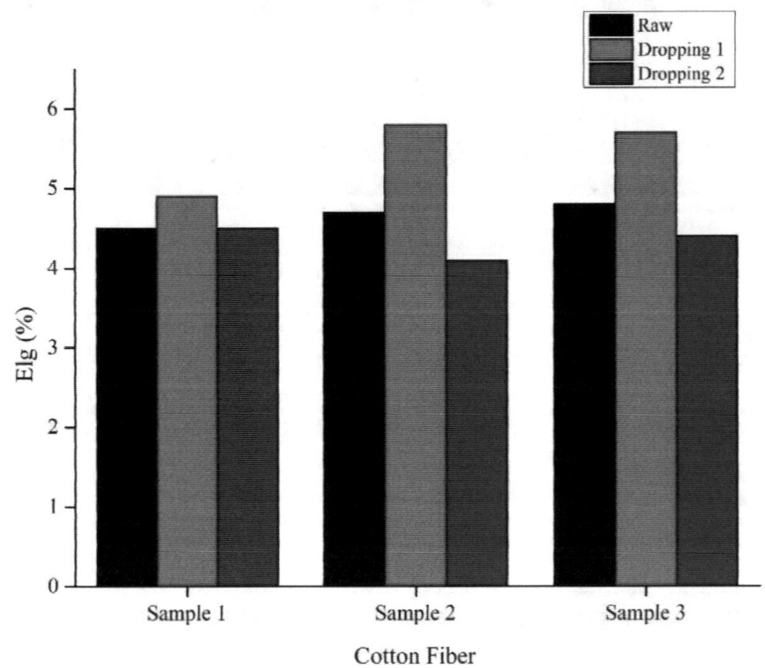


**Figure 5**. Elongation% (UHML) ratio of raw cotton and waste cotton (Dropping-01, Dropping-02).

*3.4 Spinning Consistency Index*

The spinning consistency index (SCI) serves to measure the overall strength of the cotton fiber as well as its spin ability. Most HVI measurements are used for the regression equation. In Figure 6 we have seen 11 operating points of cotton spinning. Till drawing frame and carding spinning consistency index may be floating on a different scale for example SCI was 115 to 150. Later after the drawing frame jumping 150 to 190. The fiber characteristics affect yarn characteristics. The fiber of cotton still is a very common textile industry fiber with physical properties varying depending on the growing area. Properties including fiber length, fineness, strength, and maturity influence the tensile tension of the yarn, equality, imperfection, and hairiness. SCI was the most important parameter to maintain good quality yarn [21]. If we compare Figure 6 with Figure 7, it has been shown that Figure 6 just shown the SCI value of different stages in the meantime I was measured the value of SCI of raw cotton and waste cotton. In Figure 6, sample-01 the raw cotton value of SCI is almost 140 whereas the waste cotton SCI value was 85 and 60. But is a matter of sorrow that below 110 it is very difficult to prepare or produce yarn. Sample-02 and sample-03 were showing almost the same results. In Figure 6, blow room and carding highest SCI value is 150 but in Figure 6, only raw cotton was spinnable but waste cotton was not spinnable due to low SCI value.

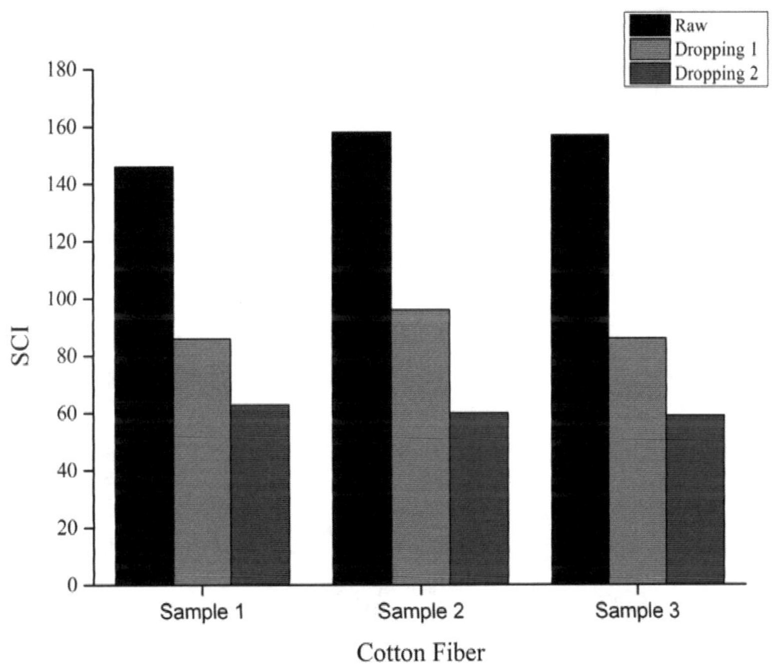


**Figure 6**. SCI value of raw cotton and waste cotton (Dropping-01, Dropping-02).

*3.5 Micronaire value of cotton (MIC).*

Buyers of cotton and textiles prefer a range of 3,8 to 4,5 Micronaire and a fiber length of 1 1/8 inch or longer. Thicker or thinner fibers cause problems in both spinning and uniform dying of yarn. The air permeability calculation of compressed cotton fibers is Micronaire (MIC). It is also used as a fiber fineness and maturity indication [22]. In Figure 7, we have seen that raw cotton MIC value in between 4 to 5, Dropping -01 value between 5 to 6 and at last Dropping-02 value in more than 6.

*3.6 Moisture Content of Cotton*

In determining cotton moisture before ginning, according to information from the USDA, the optimal moisture content for cotton is between 6.5% and 8%. When yarn dyed with this percent MC is of a higher quality than cotton is if the moisture content was substantially lower. In this Figure 8, we have seen that raw cotton moisture in the same level most 6.2%-6.3% on the other hand waste cotton Mst% value is almost the same in samples 1 and 2 but in sample 3 the waste cotton moisture content is almost 7.

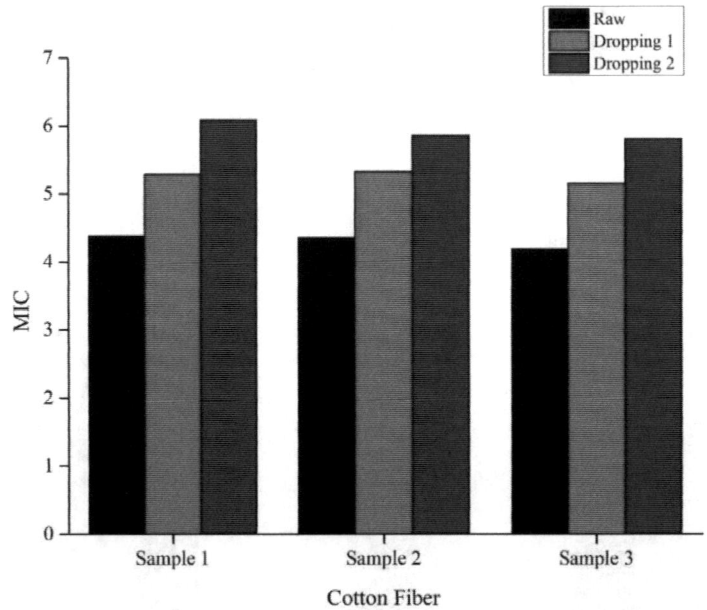


**Figure 7**. MIC value of raw cotton and waste cotton (Dropping-01, Dropping-02).

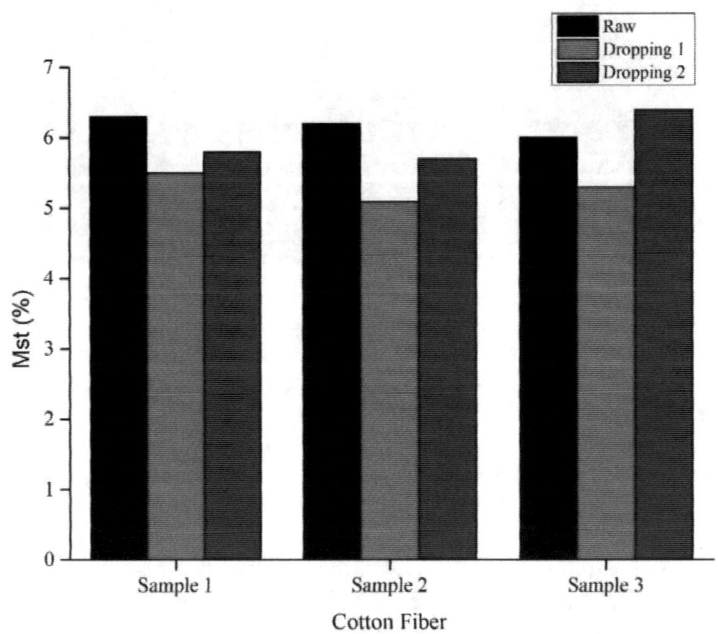


**Figure 8**. Moisture (Mst%) value of raw cotton and waste cotton (Dropping-01, Dropping-02).

IOP Conf. Series: Materials Science and Engineering 1078 (2021) 012014 doi:10.1088/1757-899X/1078/1/012014

*3.7 Summary of waste and raw cotton analysis*

By Uster ANOVA analysis we have analyze Spinning consistency index (SCI), Microniar value (MIC), Upper half mean length (UHML), Moisture(MS), Maturity(MAT), Uniformity Index (UI), Short fiber percentage (SF), strength (STR), elongation (EL). If we see all 3 tables, the average value of the SCI index of raw cotton is 143 at the same time average of dropping-02 and dropping-3 are 61 and 89. Moisture% is more or less the same in regards to raw and waste cotton. The average value of MIC value in raw cotton 4.62 same times in waste cotton was higher. The maturity content of all samples is almost the same varying from 0.89 to 0.92. The most important part of this table is UHML, the average UHML of dropping-02 was 25.83mm, dropping-01 was 26.18mm, these waste cotton were not used in spinning. The standard deviation of dropping-02 and dropping-01 also too high 1.8 and 1.6 on the other hand raw cotton or workable cotton UHML was 30.69mm and the standard deviation very satisfactory level 1.1. Same if we see uniformity index raw cotton UI% was higher than waste cotton. Short fiber is a curse for the spinning process if have observed in any kind of analysis that higher short fiber lower quality of cotton yarn. In analyzing average dropping-02 the short fiber content 24.8% and dropping-01 was 17.5% whereas raw cotton short fiber content was only 8.7% and the standard deviation also remarkable level only 0.9 which shows very good quality cotton. Strength% of all 3 characters of cotton were almost equal, the average value of dropping-02 was 31.5 g/tex, dropping-01 was 29.7 g/tex as well as raw cotton was 33.3g/tex which is higher than the waste cotton. Subsequently, elongation% of dropping-02 was 4.3% then dropping-01 was 5.5% and raw cotton have 4.3%. All three tables showed that the average, standard deviation, CV%, Q99%+/-, Min, and Maximum value of all parameters. The quality of the yarn is dependent on the quality and processing of the raw material. This is a reality, particularly in the current growing competition on the cotton market, leading to the use of high-performance processing equipment in spinning mills for the correct yarn quality. In spinning mills, on-line quality control, automated and integrated data processing systems and off-line quality control systems, laboratory testing systems, raw material, and yarn are also responsible for results [23]. Thereby, as a summary, we have seen the comparison of quality of raw material and waste cotton which is not useable in the spinning process.

**Table 2**. Resultant data of dropping-02 waste cotton.

**Matam Fibre** **USTE®** ***HVI***

System Testing - Individual Tests

Lot ID (Dropping-02)0 Trutzschler Line/Australia

Operator HVI SW Version 3.2.2.15

Print Date 19/3/2019 Serial Number 1007159

Print Time 11.40 am Test Method 4

Short/Weak Reference Upland Long/Strong ref Upland

n 3

| | **SCI** | **Ms [%]** | **Mic** | **Mat** | **UHML [mm]** | **UI [%]** | **SF [%]** | **Str [g/tex]** | **Elg [%]** |
|---|---|---|---|---|---|---|---|---|---|
| Average | 61 | 6 | 5.92 | 0.92 | 25.83 | 72.8 | 24.8 | 31.5 | 4.3 |
| Std. Dev | 2 | 0.3 | 0.15 | 0.00 | 0.46 | 1.3 | 0.8 | 1.1 | 0.2 |
| CV% | 3.4 | 5.8 | 2.5 | 0.3 | 1.8 | 1.8 | 3.3 | 3.6 | 4.4 |
| Q99%+/- | 12 | 2.0 | 0.85 | 0.01 | 2.63 | 7.5 | 4.7 | 6.5 | 1.1 |
| Min | 59 | 5.7 | 5.81 | 0.91 | 25.33 | 71.6 | 24.1 | 30.3 | 4.1 |
| Max | 63 | 6.4 | 6.09 | 0.92 | 26.24 | 74.2 | 25.7 | 32.6 | 4.5 |

**Table 3**. Resultant values of dropping-01 waste cotton.

**Matam Fibre** **USTE®** ***HVI***

System Testing - Individual Tests
Lot ID (Dropping-01) Trutzschler Line/Australia
Operator HVI SW Version 3.2.2.15
Print Date 19/3/2019 Serial Number 1007159
Print Time 11.33 am Test Method 4
Short/Weak Reference Upland Long/Strong ref Upland
n 3

| | **SCI** | **Ms [%]** | **Mic** | **Mat** | **UHML [mm]** | **UI [%]** | **SF [%]** | **Str [g/tex]** | **Elg [%]** | **Rd** |
|---|---|---|---|---|---|---|---|---|---|---|
| Average | 89 | 5.3 | 5.26 | 0.89 | 26.18 | 77.5 | 17.5 | 29.7 | 5.5 | 69.8 |
| Std. Dev | 5 | 0.2 | 0.09 | 0.01 | 1.10 | 1.6 | 2.8 | 1.4 | 0.5 | 0.4 |
| CV% | 6.1 | 4.0 | 1.7 | 0.6 | 4.2 | 2.0 | 16.0 | 4.6 | 9.2 | 0.6 |
| Q99%+/- | 31 | 1.2 | 0.50 | 0.03 | 6.28 | 9.0 | 16.1 | 7.7 | 2.9 | 2.3 |
| Min | 86 | 5.1 | 5.16 | 0.89 | 25.52 | 76.2 | 14.3 | 28.4 | 4.9 | 69.5 |
| Max | 96 | 5.5 | 5.33 | 0.90 | 27.45 | 79.2 | 19.2 | 31.1 | 5.8 | 70.3 |

**Table 4**. Resultant data of raw waste cotton.

**MULTAZIM SPG. MILLS LTD. UNIT-3**
**USTE®**

System Testing - Individual Tests
Australia Raw Cotton (E-2160) Dt:18.09.18; Supplier: Shandong Jinig, L/C No-016761823000
Operator HVI SW Version 3.4.1.71

Serial Number 1512345

Test Method 4

Short/Weak Reference Upland 34119 Long/Strong ref Upland 34119
n 3

| | **SCI** | **Ms [%]** | **Mic** | **Mat** | **UHML [mm]** | **UI [%]** | **SF [%]** | **Str [g/tex]** | **Elg [%]** | **Rd** |
|---|---|---|---|---|---|---|---|---|---|---|
| Average | 143 | 5.9 | 4.64 | 0.88 | 30.69 | 82.2 | 8.7 | 33.3 | 4.9 | 80.4 |
| Std. Dev | 9 | 0.3 | 0.24 | 0.01 | 0.87 | 1.1 | 0.9 | 1.6 | 0.4 | 1.9 |
| CV% | 6.2 | 5.5 | 5.1 | 0.8 | 2.8 | 1.3 | 10.8 | 4.9 | 7.4 | 2.3 |
| Q99%+/- | 1 | 0.0 | 0.01 | 0.00 | 0.05 | 0.1 | 0.1 | 0.1 | 0.0 | 0.1 |
| Min | 102 | 4.8 | 4.00 | 0.86 | 27.31 | 74.5 | 4.7 | 27.9 | 3.8 | 70.4 |
| Max | 175 | 7.1 | 6.60 | 0.93 | 38.75 | 86.4 | 12.3 | 40.7 | 6.5 | 83.8 |

### *3.8 Treated cotton image analysis*

The fiber's morphological structures show that waste cotton and after treatment waste cotton by different chemical chemicals. Figure 9 indicates that the treated cotton fiber is relatively smooth, while the Figure 10 illustrates that its raw surface is rough. SEM images describe that waste cotton was not so much organized not sufficient crystalline region but the size of fiber was almost the same in waste and treated cotton the diameter varying 11μm-15μm. The treated cotton having more air permeability area then the waste cotton. Waste containing materials is more pilled with blow room and textiles. Card waste materials have a lower degree of plundering and a higher content of short fiber but a shorter length of the fiber. Small fibers are possibly too small to render a card waste pill. Treated cotton is more organized than waste one and less short fiber. By the morphological test, we have seen that treated cotton diameter was more volatile.

It has been compared with the below images [24] than we see below images were described micro value to nano value of cotton. For further study of using these micro cotton waste fibers to nano cellulose which might be used commercially. Natural fiber is used because it is inexpensive, rich, sustainable, and biodegradable [25]. They also can be used to improve adhesives, mobile device parts, biomaterials, spumes, aerogels, and textiles [26-28]. So, at the end of the day, this waste cotton has great value if we recycled this cotton to make microcrystalline cotton and for further study to make nanocrystalline fiber. When it was waste it was very cheap but when after chemical treatment it was made a raw material of valuable nano cellulose preparation.

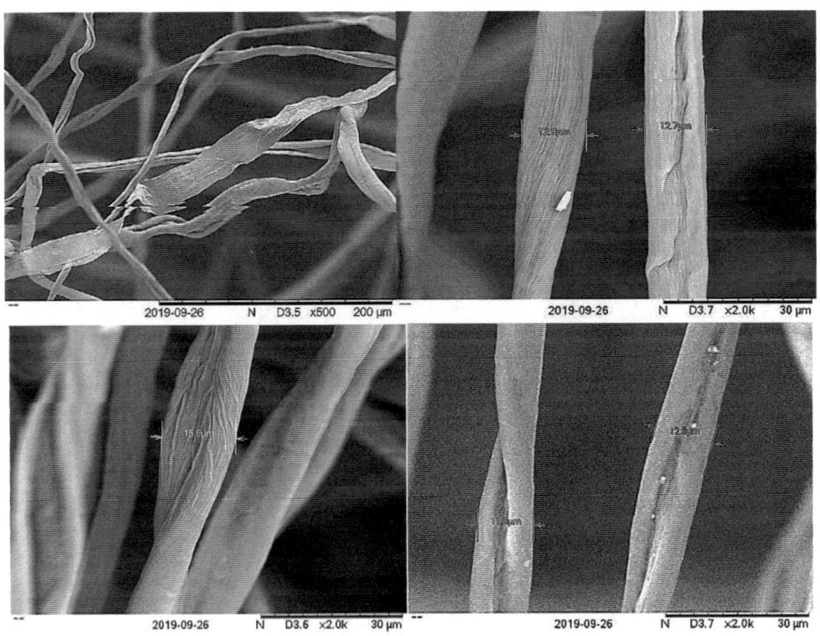


**Figure 9**. SEM images of waste cotton and treated cotton.

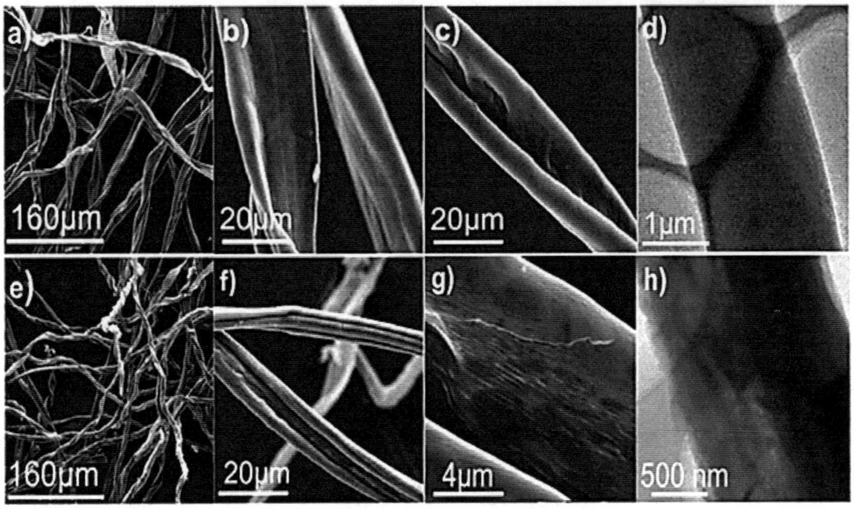


**Figure 10**. SEM images (a–c) and TEM image (d) of the commercial cotton; SEM images (e–g) and TEM image (h) of the NCCFs [24].

## 4. Conclusion

Initially, a method is built for more product developments to characterize waste and cotton treated. The Spinning Consistency Index is a measure for the prediction of cotton fiber overall quality and spinnability. Below 110 of Spinning Consistency Index cannot produce better yarn. So below this index is going to cotton waste. This process helps us to produce sustainably recycled and durable product which can also make nano thread, coated yarn, packaging material, food packaging material and so on. The crystallinity area was increased following the pre-treatment of waste cotton, a significant improvement in cotton strength. The most significant advancement is the properties of waste cotton after treatment, raising the volatile area and becoming a commercial commodity. We can thus generate a monocrystalline area and Nano-cellulose in better condition.

## Acknowledgments

The authors are grateful to University Malaysia Pahang for the Grant RDU191801-3 to aid financial support and lab works to do this study.

## References

[1] Halimi M T, Hassen M B, Sakli F J R, Conservation and Recycling 2008 Cotton waste recycling: Quantitative and qualitative assessment **52** 785-91

[2] Tabassum F, Bari Q, Rahman M, Mahmud S M M and Raj A GARMENTS WASTE RECYCLING IN DHAKA: A CASE STUDY OF MIRPUR AREA

[3] Bazrafshan E, Mohammadi L, Ansari-Moghaddam A and Mahvi A H 2015 Heavy metals removal from aqueous environments by electrocoagulation process- a systematic review *J Environ Health Sci Eng* **13** 74

[4] Frydrych I and Thibodeaux D J C t f t s c W D I C A C 2010 Fiber quality evaluation-current and future trends/intrinsic value of fiber quality in cotton 251-96

[5] Rajput D, Bhagade S, Raut S, Ralegaonkar R, Mandavgane S A J C and Materials B 2012 Reuse of cotton and recycle paper mill waste as building material **34** 470-5

[6] Shatkin J A, Wegner T H, Bilek E T, Cowie J J T J, Volume 13, Number 5, and 9-16. p 2014 Market projections of cellulose nanomaterial-enabled products-Part 1: Applications **13** 9-16

[7] Lipsitch M, Cohen T, Cooper B, Robins J M, Ma S, James L, Gopalakrishna G, Chew S K, Tan C C and Samore M H J S 2003 Transmission dynamics and control of severe acute respiratory syndrome **300** 1966-70

[8] Liu Y, Todd Campbell B, Delhom C and Martin V J T R J 2015 Variation and relationship of quality and near infrared spectral characteristics of cotton fibers collected from multi-location field performance trials **85** 1474-85
[9] Liu Y, Todd Campbell B and Delhom C J T R J 2019 Study to relate mini-spun yarn tenacity with cotton fiber strength **89** 4491-501
[10] Liu Y and Delhom C J T R J 2018 The relationship between instrumental leaf grade and Shirley Analyzer trash content in cotton lint **88** 1091-8
[11] Wakida T, Lee M, Park S-J and Saito M J S i G 2002 Effect of hot mercerization on liquid ammonia treated cottons **58** 185-7
[12] Ütebay B, Çelik P and Çay A J J o C P 2019 Effects of cotton textile waste properties on recycled fibre quality
[13] IISD 2016 Cotton-growing regions of the world. Distribution of cotton production in the top 15 producing countries in 2016
[14] Ma W, Shen K, Xiang N and Zhang S J M 2017 Combinative scouring, bleaching, and cationization pretreatment of greige knitted cotton fabrics for facilely achieving salt-free reactive dyeing **22** 2235
[15] Majumdar A, Majumdar P K and Sarkar B J A R J 2004 Selecting cotton bales by spinning consistency index and micronaire using artificial neural networks **4** 1-8
[16] Hequet E, Abidi N and Gannaway J R 2007 Relationship between HVI, AFIS, and yarn tensile properties. In: *World cotton research conference-4. Sept,* pp 10-4
[17] Krishnamachari P, Hashaikeh R and Tiner M J M 2011 Modified cellulose morphologies and its composites; SEM and TEM analysis **42** 751-61
[18] Thibodeaux D, Senter H, Knowlton J, McAlister D and Cui X 2008 A comparison of methods for measuring the short fiber content of cotton
[19] Braden C A and Smith C W J C s 2004 Fiber length development in near- long staple upland cotton **44** 1553-9
[20] Takabi B and Shokouhmand H 2015 Effects of Al 2 O 3–Cu/water hybrid nanofluid on heat transfer and flow characteristics in turbulent regime *International Journal of Modern Physics C* **26** 1550047
[21] Günaydin G K, Soydan A S, Palamutçu S J F and Europe T i E 2018 Evaluation of cotton fibre properties in compact yarn spinning processes and investigation of fibre and yarn properties
[22] Montalvo Jr J G 2005 Relationships between micronaire, fineness, and maturity. I. Fundamentals
[23] El Mogahzy Y 1998 Cotton blending: How the EFS system can help in producing optimum yarn quality. In: *Proc. 10th Annual Engineered Fiber Selection System Conf,* pp 8-10
[24] Fan L-Z, Chen T-T, Song W-L, Li X and Zhang S J S r 2015 High nitrogen-containing cotton derived 3D porous carbon frameworks for high-performance supercapacitors **5** 1-11
[25] Eichhorn S J, Dufresne A, Aranguren M, Marcovich N, Capadona J, Rowan S J, Weder C, Thielemans W, Roman M and Renneckar S J J o m s 2010 Current international research into cellulose nanofibres and nanocomposites **45** 1-33
[26] Beck-Candanedo S, Roman M and Gray D G J B 2005 Effect of reaction conditions on the properties and behavior of wood cellulose nanocrystal suspensions **6** 1048-54
[27] Eichhorn S, Baillie C, Zafeiropoulos N, Mwaikambo L, Ansell M, Dufresne A, Entwistle K, Herrera-Franco P, Escamilla G and Groom L J J o m S 2001 Current international research into cellulosic fibres and composites **36** 2107-31
[28] Ummartyotin S, Juntaro J, Sain M, Manuspiya H J I C and Products 2012 Development of transparent bacterial cellulose nanocomposite film as substrate for flexible organic light emitting diode (OLED) display **35** 92-7

# Formation of colours on SS304L stainless steel induced by laser colouring

**R Linggamm[1], M M Quazi[1,*], M Ishak[1], M H Aiman[1], A Q Zafiuddin[1] and Abdullah Qaban[2]**

[1]Joining, Welding and Laser Processing Lab (JWL), Faculty of Mechanical and Automotive Engineering Technology, University Malaysia Pahang, 26600 Pekan, Pahang, Malaysia
[2]Department of Mechanical Engineering and Aeronautics, City, University of London, London, UK

*Corresponding author's e-mail: moinuddin@ump.edu.my

**Abstract**. In this research, the laser colour marking process was carried out to determine the formation of colour on surface of SS304L stainless steel. The pulse laser nanosecond (1064nm) was used to irradiate the SS304, resulting in the formation of various colour permeating on the sample surface. The surface characterization and surface roughness were analysed by using the 3D optical microscope. The effect of laser processing parameters such as hatching distance, pulse width, and defocusing distance on the evolution of surface colours were investigated. The results show that a thin oxide film was essentially formed that resulted in the formation of four different colours. Each colour exhibited a different surface texture and surface roughness implying the effect of parameters significantly affects the surface characteristics. Grey colour formed registered the highest roughness ($R_z$) of about 3.5 microns followed by a decreasing roughness trend for blue, purple and green respectively.

**Keywords**: Laser colouring; SS304; Surface texture; Roughness.

## 1. Introduction

Laser based manufacturing processes such as laser surface modification laser [1] , laser hardening [2], laser welding [3] are now gaining widespread recognition due to robust and rapid production time. Laser Laser-induced colour marking of metals has been well known for the past decades but these days there has been a widespread growth in the new era of technological development. This process is rapidly replacing traditional marking methods such as painting, etching and printing [4]. However, this process adds colour to an object that is being marked merely "stains" of the surface. It has become a foremost industrial process today in many applications, such as the labelling of a serial number, date, barcode, quick response (QR) code, company logo, trade marking, and other product information. There are several types of laser colour marking processes which are etching, marking, and lithography. However, Laser colour marking boasts high quality, difficulty to forge, wear resistance and is a noncontact marking process that does not causes any contact pressure nor deformation whilst is sustainable [5].

The process of laser colour marking ensues when the oxide layer is created on the ferrous metal such as steel and high-grade steel through the localized heating. In general, the oxide layer that was created on the surface is black, but it can also be tailoured to obtain a variety of colours such as red, blue, and green etc [6]. Hence the colour produce on the surface of metal primarily depends on the temperature of the heated layers and the penetration of heat to the depth of the metal surface. The phenomena of colour production are also dependent on the chromium due changes in molecular structure and electrons on the metal surface. The laser acts as a heat source allowing forming of a thin transparent or semi-transparent oxide film. If a beam of parallel monochromatic light falls, at an angle, onto a transparent or partly transparent layer, this beam is repeatedly reflected from the surfaces limited by the oxide layer (from the surface of both oxide and metal) [5]. By increasing the thickness of oxide on the surface and as a consequence by changing the interference effect (each wavelength has a different degree of attenuation or amplification), a colour effect can be observed and controlled.

The most common laser employed in the laser colour marking is a pulsed wave mode nanosecond, picosecond, or femtosecond laser. This is due to the difference in the behaviour of their pulse width and pulse frequency. Therefore, through the use of femtosecond and picosecond lasers, the method of laser-induced periodic surface structure (LIPSS) are used to induced colour on the metal surface [7]–[9]. There is also paper demonstrate that the influence of the pulse frequency, mark depth width, and mark contrast on the colour marking process are depends on the interaction process of the laser beam and the material which was influence dramatically by the pulse frequency on the quality of the mark [10].

When we glance at the formation of colour on a metal surface that contributes by nanosecond laser, most of the researchers have analysed the parameter of the laser such as laser power, frequency, hatching distance, defocusing, and scanning speed on this colour mark process, where they summarized this factors may alter the colouring effect on the metal surface [6], [11]. Most of the study carried out by [10], [14], [15], it shows that the laser parameter directly affects the stainless-steel surface colour. However, the colour marking implementation and development of the metal surface is an ongoing process, since the formation of colour on the surface still not uniform throughout the laser-scanned area, where the laser process colour mark stability depends on many factors such as temperature, laser parameter and the quality of the metal surface [9]. Thus, a new experiment has been conducted to study the nanosecond laser mark process on the stainless steel of 304L material. The variation of parameters such as laser power, frequency hatch distance, an defocusing resulting from the distribution of numerous colours on the steel surface shall be made. Subsequently, the surface roughness parameters have been investigated corresponding to the colours generated by parameter variation.

## 2. Experimental Setup

### *2.1. Material and material preparation*

SS304L, austenitic stainless steel was chosen for the study, which is widely used in many industries. Therefore, for this study, the rectangular specimens of (1000 mm in length and width with a thickness of 2 mm) were cut from a large bar of SS304L stainless steel. Thereafter, the top surface of the metal surface was polished to minimize the initial surface roughness before the laser treatment. Prior to the laser irradiation, the specimen was cleaned in acetone. The grease and contaminants of the sample removed by using the ultrasonic machine that contained acetone.

Besides that, the laser beam is required to be focused on the sample surface through an F-theta lens in the scanning area of 10 x10 mm, where the scanning mode is shown in Figure 1. Thus, the line scans were completed from right to left by the laser beam and the galvanometer scan mirror will change the scanning path as shown in Figure 1. However, by moving the sample on the platform, the area of the laser scanning should not change where it always needed to be located directly below the laser beam before each scan begins. This will ensure the pulse energy density of the laser will not change remarkably.

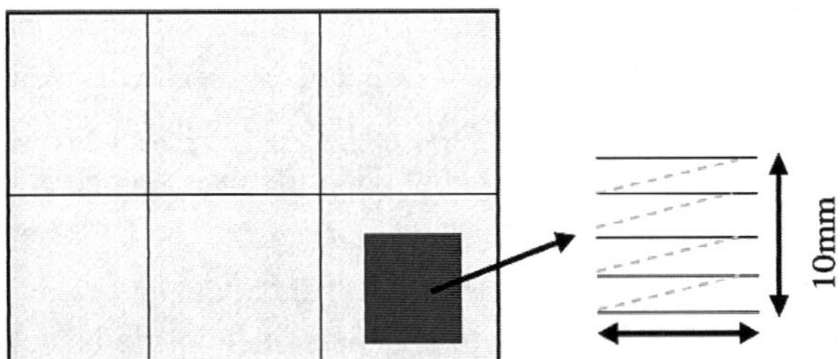


**Figure 1.** The schematic diagram of laser scanning on the sample.

*2.2. Laser color marking setup*
The colour marking process was conducted by using IPG-Fibber laser (Hanser Laser company). Then, this laser beam is a red light with a central wavelength of 1064 nm, and beam focusing distance is 245 mm. While the laser beam emitted diameter by this device was 0.5 mm. The beam expander of this device can be expanded the magnification until 5 mm which then will be transmitted to the scanning galvanometer. Lastly, the beam will be a focus on the target spot, which is nearly the gaussian spot with a diameter of approximately 40 μm. Apart from that, the maximum power output of this laser was 30 W and the pulse repetition frequency can vary within the range of 10-300 kHz. In comparison with the laser (wavelength of 532 nm), the optical path of the laser used in this study (wavelength is 1064 nm) which can be visible, where it is convenient to adjust and user-friendly. Hence, the 1064 nm laser has a larger heat-affected zone on the stainless steel and it can act as a good absorption rate of laser energy. In addition, these laser systems consist of an operating panel, a laser control system, an optical system, a cooling system, and a working platform of an industrial computer with a compact configuration. It is also easy to use a network with a reliable output. Lastly, the processing parameter of this laser is set by the special control software (Ezcad) from the supporting pc terminal, while the laser optical path structure is represented by Figure 2.

The selection and identification of the sample surface was been analysed by using a 3D optical microscope. The surface characterization and surface roughness of the sample was analysed by using LEXT OLS5000. This laser is a confocal scanning microscope, where it is the better device for 3D measuring for the analyses of non-contact and non-destructive imagery [12]. Therefore, it has no sample size or content limitations which enables the observation under normal environmental conditions. Hence the apparent reproduction of the object of the optical image is formed by a mirror or lens system which produce by reflected, refracted, or diffracted light waves. However, for determining the surface roughness of each sample on the colour formations, the optical and height image from this 3D laser microscope was recorded to evaluate the effect of each sample roughness of colour formation on the SS304L substrate. The finding of the surface roughness on the formation of colour on SS304L is important in our research to see the correlation change of colour on the surface roughness.

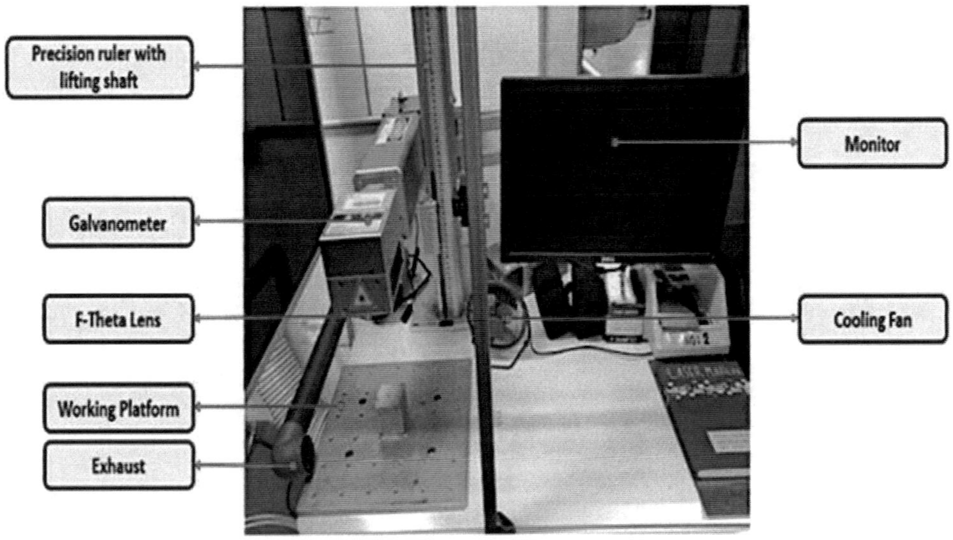


**Figure 2.** Laser machine use in laser colour marking.

### *2.3. Surface Roughness*

In this work, the roughness of the sample is calculated by using the OLS5000 laser confocal microscope, where it has precisely measured the shape and surface roughness at the level of submicron [12]. However, the quantitative roughness value is observed based on 2D line roughness and 3D surface profile with four-parameter of roughness recorded which is, average line roughness (Ra), a mean depth of line roughness (Rz), arithmetical mean height (Sa), and maximum height (Sz). From these four recorded values, the investigation of the value of Ra and Rz will be observed and analyzed due to the information of peak and valley of the surface roughness. With the value, the observation of the color change in each surface could be correlated with different roughness surfaces.

## 3. Result and Discussion

### *3.1. Surface color*

The obtain colour in our experiment has been analysed by using a matrix method to investigate the influence of laser pulse width, hatch distance, and different defocusing distance of the lasering process. The other parameter of the lasering remained unchanged during our experiment. Hence the influence of the three parameters on the colour plate has been shown in Figure 3. By referring to the experiment that were carried out, the defocusing distance, pulse width, and hatch distance was found to be affecting the colour marking of a stainless-steel surface. However, when the defocusing distance is on a lower side, the colour made by laser in the sample surface as shown was mainly grey and dark brown while the traces of material removal can be observed on the sample surface. This material removal was mainly due to the high laser energy absorbed by the material. Rich colour appears when the defocusing distance is in the range of 4mm, 5mm, and 6mm as showed in Figure 3. Moreover, the same figure shows the multiple colours formed on the sample with a defocusing distance of 5mm and 6mm respectively. If the defocusing distance is fixed, then the colour appears on the sample are in a similar sequence as shown in Figure 4 a and b. As an example, in Figure 3, the colour appears in the dark in both columns and rows as marked in yellow line where the colour appears are dark brown and brown. Meanwhile, it has been observed that the small pulse width and defocusing distance have tendency to produce dark surfaces that cannot reflect the light in that particular colour. Additionally, rich colour does not appear on the sample surface at the upper left corner of the sample. By simply increasing the pulse width and defocusing distance the rich colours start to appear in the sample surface in the central area. As the two-parameter increase, the shiny and transparent oxide layer is generated on the sample surface but when these two-parameter are adjusted to the smaller value the colour becomes dark as showed Figure 3 in the bottom right corner of the sample.

Table 1 show the range of machine parameter and Table 2 shows the laser repetition frequency and releasing time scanning speed and power, we set as 225kHz, 5micro second, 350mm/s, and 15W. The hatch distance range was from 1 to 10 micron with an incremental of 1 for each lasering process. The pulse width range is from 10ns to 170ns with an increment of a minimum of 10 -25ns for each colour marking and the defocusing distance was 6mm for every colour mark. From the colour mark sample, we have chosen four representative colours to further investigate the surface roughness of the colour mark on the sample surface.

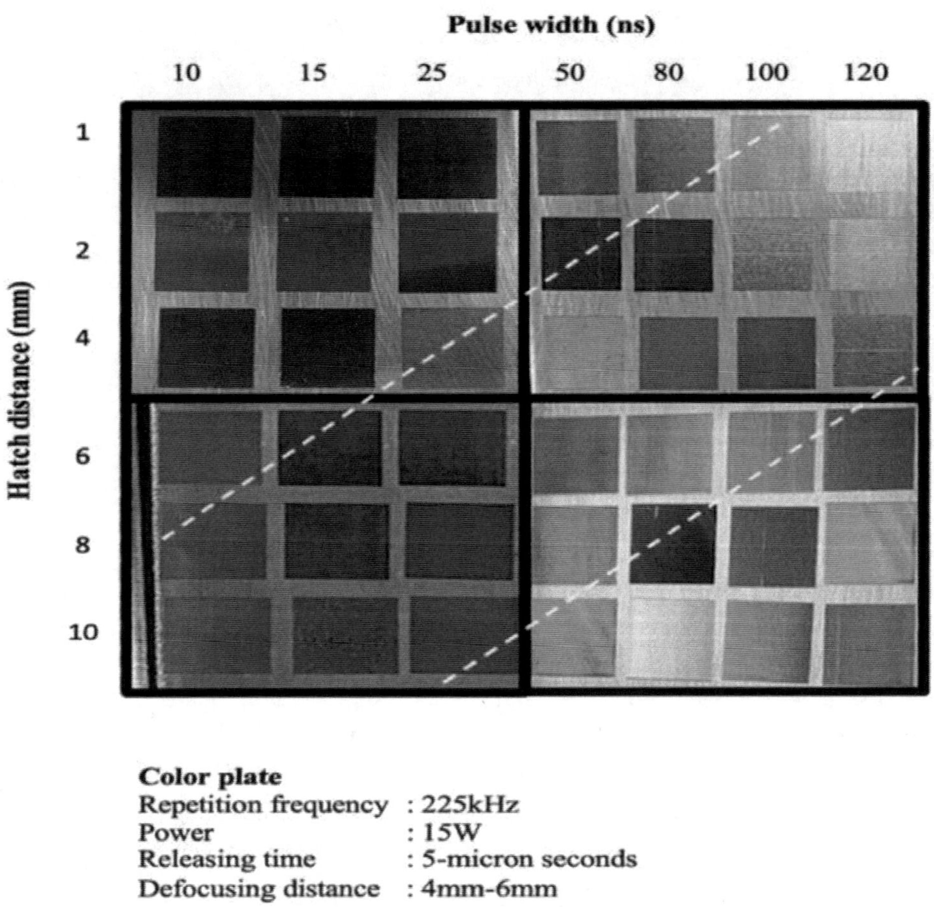


**Figure 3.** Influence of pulse width and hatch distance on the colour.

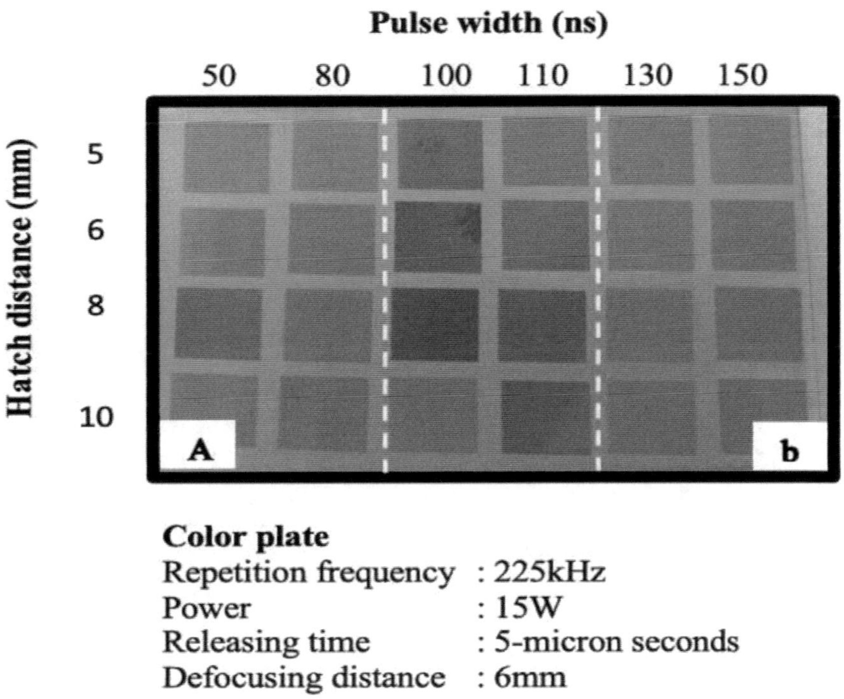


**Figure 4.** Color obtain in the fixed defocusing distance.

**Table 1.** Range of parameter of the machine.

| **IPG fiber laser machine** | |
|---|---|
| **Parameter** | **Range of parameter** |
| Power | 30W |
| Frequency | 300KHZ |
| Scanning speed | 10000 mm/s |
| Pulse width | 300ns |
| Focal plane | - |
| Hatching distance | - |
| Workpiece dimension | 150mm x150mm |

**Table 2.** Laser parameter for coloring process.

| Laser parameter | Symbol | Unit | Value | Increment |
|---|---|---|---|---|
| **Repetition frequency** | F | (kHz) | 225 | - |
| **Average power** | P | (W) | 15 | - |
| **Releasing time** | Qt | (Microsecond) | 5 | - |
| **Defocusing distance** | D | (mm) | -6mm to +6mm | 1 |
| **Hatch distance** | H | Micron | 01-Oct | 2 |
| **Scanning speed** | V | (mm/s) | 350 | - |
| **Pulse width** | Pw | (ns) | 170 | 10 |

*3.2. Surface characterization*

The analysis of surface roughness on the colour making of the sample surface has been an investigation on their roughness in corresponding to the formation of colour. The four representative colours were picked for this study, which is green, purple, blue, and grey colour as shown in Figure 5. Therefore, the relevant parameter for producing this colour mark has been shown in Table 2.

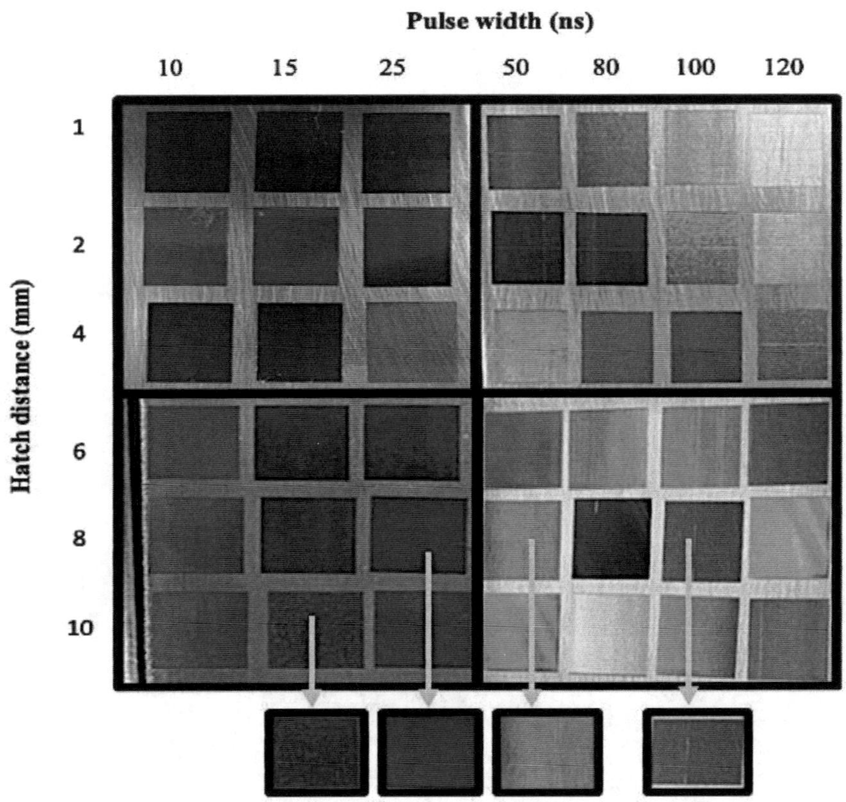


**Figure 5.** The four representative colours chosen for surface characterization.

However, in Figure 4 a and b the same colours have been appearing several times at different locations. It shows that the colour can be obtained by the combination of the different parameter such as with the smaller pulse width and hatching distance. Although it can produce the same colour by different combination parameters, but the brightness and saturation of the colour are different. This is why we have chosen the four representative colours to further investigate their surface roughness on the colour mark.

*3.3. Surface roughness*

Table 3 shows the parameter used to colour marking the four distinct colours by varying the pulse width, hatch distance, and defocusing distance while the other parameter was fixed, which are scanning speed, power, and repetition frequency. By adjusting these three parameters as stated, the formation of colour on the sample surface was different, while a small change of such value could produce same dark colours as shown in Figure 3. However, with the large value of pulse width and defocusing distance, the colour produce was brighter while the colour was reflecting the lights such as the blue, green, and purple colour. Apart from the effect of these three parameters on the formation of the colour, the roughness of the sample surface is an important aspect of the colour marking where different roughness produces dissimilar colour. Hence, the result of the 2-dimensional image and 3-dimensional image of the four representative colours has been capture by using an 3D optical microscope. Then, each colour has been comparing with the hatching distance where it is the distance of scanning track of laser, the pulse width is the power produce by laser during lasering process and the defocusing distance is the focus point of the laser beam and the workpieces surface where this comparison value has been shown in Figure 6.

These figures have summarized the adjustable parameter, which is the pulse width, defocusing distance, and hatch distance in marking of these colours and the fixed parameter of these colours has been presented in Table 3. The formation of the four-color comes at different defocusing distance. The hatching distance also a major part but it only differs slightly compare to the defocusing distance. Therefore, the defocusing distance needed to precisely be adjusted as compare with the adjustment of another parameter to form the colour.

The roughness test of 4 different colours is compared and showed in Figure 7. In this comparison the surface roughness value for four representative colours has been concluded, where it shows that the grey colour produces on the roughest surface with the $R_a$ value of 0.433 and $R_z$ value of 3.63. Hence, the rougher the surface the colour tends to be darker. This is due to an uneven roughness surface of the sample without polishing the entire surface. For the next three representative colours, there seems to be a slight change in the value of Rz, which is the maximum peak. In contrast with the different roughness distribution at the metal surface laser treatment where it differently affects the surface during the colour marking process. Moreover, from the macrostructure showed in Figure 6. It can be seen that the smoother the surface, the heat is distributed evenly to form a thin film layer. Hence, the colour can appear in solid nature. But when the surface is rougher the hatch distance is observed in the form of scanned tracks. Likewise, when the surface is rougher it is hard to interact with the material with laser treatment where it can lead to appearing dark colour. From the glance we can conclude that many of the researcher has mentioned the important of polishing the metal surface to even surface roughness before the laser treatment [14]. However, further studies should be conducted to investigate the effect of surface roughness in different grit paper and polish surface.

**Table 3.** The processing parameters corresponding to different colours.

| Color | Scanning speed(mm/s) | Defocusing distance(mm) | Hatch distance(mm) | Repetition frequency (kHz) | Pulse width(ns) | Average power(w) |
|---|---|---|---|---|---|---|
| **Purple** | 350 | 4mm | 10mm | 225kHz | 15 | 15W |
| **Green** | 350 | 5mm | 8mm | 225kHz | 25 | 15W |
| **Gray** | 350 | 6mm | 8mm | 225kHz | 50 | 15W |
| **Blue** | 350 | 4mm | 8mm | 225kHz | 100 | 15W |

IOP Conf. Series: Materials Science and Engineering 1078 (2021) 012015 doi:10.1088/1757-899X/1078/1/012015

| Sample overview | HD | DD | PW | 2D | 3D |
|---|---|---|---|---|---|
|  Purple | 10mm | +4mm | 15ns | 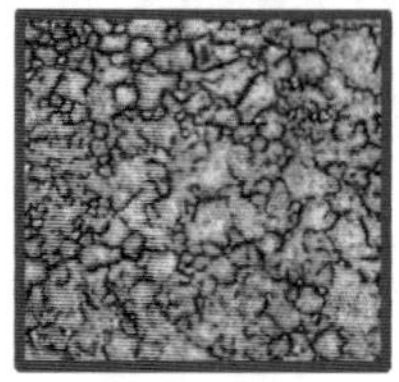 |  |
|  Green | 8mm | -5mm | 25ns | 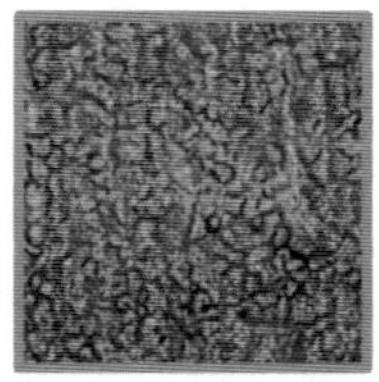 |  |
|  Blue | 8mm | +4mm | 50ns | 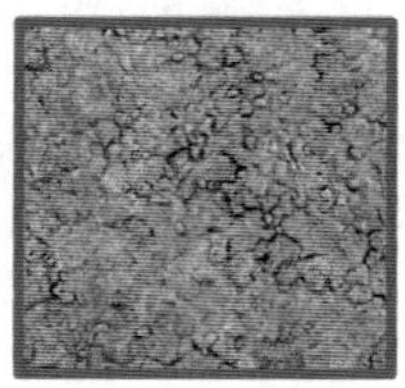 |  |
|  Grey | 8mm | 6mm | 100ns |  | 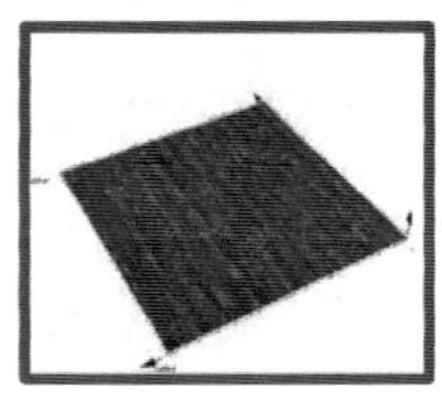 |

**Figure 6.** Overview of 2d and 3d sample by 3d optical microscope of four-color marking on different hatch distance, defocusing distance and pulse width.

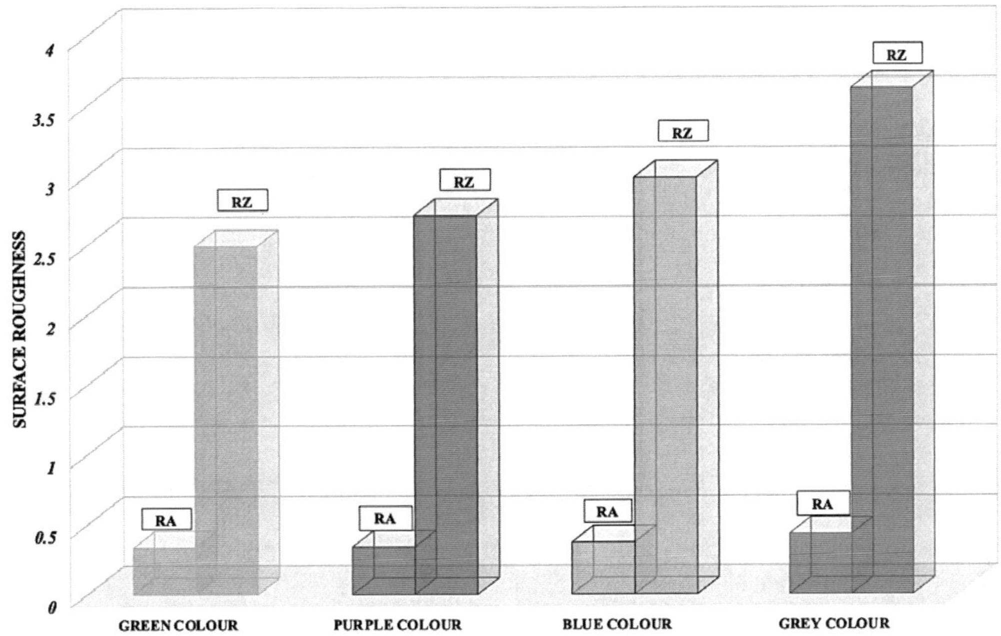


**Figure 7.** Roughness test of four colours with its RA and RZ value.

## 4. Conclusion

In the conclusion of this study, we have provided some intuition into the proper selection of parameters such as laser pulse width and defocusing distance during the laser colour marking process on the SS304L metal surface. Following results can be concluded from this study.

1. The sample surface produces different colours with the changes of pulse width and focus distance due to the changes in the thermal interaction of laser with the metal surface and due the absorption effect. However, it is indicated that an accurate selection of laser treatment parameters, such as smaller Pulse width and defocusing distance will result in the formation of colour are mostly in dark for example brown, black and dark brown colour.
2. By increasing the pulse width, hatch distance, and defocusing distance samples could form a brighter colour. With the increase of these parameters the rich colour from where the heat is distributed evenly through the entire laser treatment space to a solid colour.
3. From the roughness result, it can conclude that the formation of colour at each surface is different. This is due to a raw surface with uneven surface roughness distribution that can affect the laser treatment process. The rougher the surface the colour produced will be darker. In particular, the effect of pulse width and defocusing distance on colouring stability was greater than the hatch distance.

**Acknowledgments**

The author would like to thank the technical staff of University Malaysia Pahang for all the work by providing laboratory facilities within which the experiments were conducted. Hence the work was supported by University Malaysia Pahang research grant funding RDU1903119 and RDU191109.

## References

[1] A. Q. Zaifuddin, M. H. Aiman, M. M. Quazi, M. Ishak, and T. Ariga, "Effect of Laser Surface Modification (LSM) on laser energy absorption for laser brazing," *IOP Conf. Ser. Mater. Sci. Eng.*, vol. 788, no. 1, 2020, doi: 10.1088/1757-899X/788/1/012013.

[2] M. N. M. Salleh, M. Ishak, M. H. Aiman, Q. Zaifuddin, and M. M. Quazi, "The effect of laser surface hardening on the surface hardness of mild steel," *IOP Conf. Ser. Mater. Sci. Eng.*, vol. 788, no. 1, 2020, doi: 10.1088/1757-899X/788/1/012014.

[3] M. M. Quazi *et al.*, "A comprehensive assessment of laser welding of biomedical devices and implant materials: recent research, development and applications," *Crit. Rev. Solid State Mater. Sci.*, vol. 0, no. 0, pp. 1–43, 2020, doi: 10.1080/10408436.2019.1708701.

[4] E. H. Amara, F. Haïd, and A. Noukaz, "Experimental investigations on fiber laser color marking of steels," *Appl. Surf. Sci.*, vol. 351, pp. 1–12, 2015, doi: 10.1016/j.apsusc.2015.05.095.

[5] H. Liu, W. Lin, and M. Hong, "Surface coloring by laser irradiation of solid substrates," *APL Photonics*, vol. 4, no. 5, 2019, doi: 10.1063/1.5089778.

[6] Z. L. Li *et al.*, "Analysis of oxide formation induced by UV laser coloration of stainless steel," *Appl. Surf. Sci.*, vol. 256, no. 5, pp. 1582–1588, 2009, doi: 10.1016/j.apsusc.2009.09.025.

[7] J.-M. Guay *et al.*, "Enhanced plasmonic coloring of silver and formation of large laser-induced periodic surface structures using multi-burst picosecond pulses," no. September, 2016, doi: 10.1002/adom.201800189.

[8] A. Y. Vorobyev and C. Guo, "Multifunctional surfaces produced by femtosecond laser pulses," *J. Appl. Phys.*, vol. 117, no. 3, 2015, doi: 10.1063/1.4905616.

[9] A. Y. Vorobyev and C. Guo, "Spectral and polarization responses of femtosecond laser-induced periodic surface structures on metals," *J. Appl. Phys.*, vol. 103, no. 4, pp. 16–19, 2008, doi: 10.1063/1.2842403.

[10] A. J. Antończak, D. Kocoń, M. Nowak, P. Kozioł, and K. M. Abramski, "Laser-induced colour marking - Sensitivity scaling for a stainless steel," *Appl. Surf. Sci.*, vol. 264, pp. 229–236, 2013, doi: 10.1016/j.apsusc.2012.09.178.

[11] X. Ma, J. Guo, X. Si, and J. Zhao, "The Influence of Laser Parameters on the Coloring Effect and Microstructure of Stainless Steel Surface," *IOP Conf. Ser. Mater. Sci. Eng.*, vol. 538, no. 1, 2019, doi: 10.1088/1757-899X/538/1/012024.

[12] W. Kaplonek and K. Nadolny, "Advanced 3D laser microscopy for measurements and analysis of vitrified bonded abrasive tools," *J. Eng. Sci. Technol.*, vol. 7, no. 6, pp. 661–678, 2012.

[13] E. S. Gadelmawla, M. M. Koura, T. M. A. Maksoud, I. M. Elewa, and H. H. Soliman, "Roughness parameters," *J. Mater. Process. Technol.*, vol. 123, no. 1, pp. 133–145, 2002, doi: 10.1016/S0924-0136(02)00060-2.

[14] X. Ma *et al.*, "Effect of nanosecond pulsed laser parameters on the color making of 304 stainless steel," *Opt. Laser Technol.*, vol. 126, no. October 2019, p. 106104, 2020, doi: 10.1016/j.optlastec.2020.106104.

[15] E. M. Garcell, S. C. Singh, H. Li, B. Wang, S. A. Jalil, and C. Guo, "Comparative study of femtosecond laser-induced structural colorization in water and air," *Nanoscale Adv.*, 2020, doi: 10.1039/c9na00804g.

# Enhancement of tribological behaviour and thermophysical properties of engine oil lubricant by Graphene/Co-Cr nanoparticle additives for preparation of stable nanolubricant

**M Sandhya[1], D Ramasamy[1,2,*], K Kadirgama[1], W S W Harun[1], M Samykano[3] and A Ameer[3]**

[1]Faculty of Mechanical and Automotive Engineering Technology, Universiti Malaysia Pahang, 26600 Pekan, Pahang, Malaysia
[2]Automotive Engineering Centre, Universiti Malaysia Pahang, 26600, Pekan, Malaysia
[3]College of Engineering, Universiti Malaysia Pahang, Lebuhraya Tun Razak, 26300 Gambang, Kuantan, Pahang, Malaysia

*Corresponding author's email: deva@ump.edu.my

**Abstract.** In the current study the tribological properties of Aluminum alloy (Al 25) with moment in time was examined by adding nanoparticles as additives. Nanoparticles exhibit high thermal conductivity, stable viscosity and can be dispersed well in oil. The purpose of this research is to determine the effect of $Al_2O_3$, $TiO_2$, Graphene (G), Co-Cr, Ni nanoparticles on lubricity using linear wear testers under reciprocating sliding motion. SEM and EDX characterization techniques are used to study the composition and morphology on worn surfaces. The findings showed that the tribological properties of the substrate lubricant were significantly increased with the addition of nanoparticles. Friction and Wear are reduced with Graphene, $Al_2O_3$, $TiO_2$, Ni & Co-Cr nanoparticles as additives for the same load value and can be used as multifunctional additives. Co-Cr at 0.03wt% has provided best performance among the other nanoparticles and can be used as Nano lubricant in the future.

**Keywords:** Nanoparticle, Graphene, Nanolubricant, SEM-EDX, Engine Oil.

## 1. Introduction

Lubrication is among the foremost issues in manufacturing, transportation, and aerospace [1-3]. Automotive engines are completely dependent on lubrication [4]. Engine oil plays a vital role to keep running the engine with sustainability. Performance of engine oil is directly linked by the formulation (Additive content). More than a decade several chemical substances are used as additives in engine oil to succeed in decreasing tribological tests like friction, wear and tear, corrosion, deterioration, and oxidation [5]. Also, massive hard work has been performed in order to attain best thermal performance through multi-grade oils, phosphates, Organic poly sulphides, dithiopho-sphates, dithio carbamates, molybdenum di-sulphide, zinc di-alkyl dithiophosphates (ZDDP), etc. persist in being famous as anti-wear additives [6]. Meanwhile from previous decade nano particles are being thoroughly inspected as possible performance improver for conventionally used anti-friction and anti-wear additives [7]. Allotropes of Carbon grew attention because of its exclusive morphology as well as properties [8]. Nevertheless, the selected nanoparticles are equitably novel in the lubrication engineering, astoundingly great thermo-physical and tribological properties have made it a striking research subject [6].

Lubrication is administration of friction and wear by the presentation of a rubbing lessening film between moving surfaces included. The oil utilized be a fluid, solid, or plastic substance. In spite of the fact that this is a legitimate definition, it neglects to understand all that lubrication really accomplishes. Various materials can be utilized in order to grease up the surface. Oil along with grease are the greatest and widely recognized [9]. The nano lubricants earned enormous attention because previous experiment results proved effective in efficiency improvement inferable from diminish frictional improvement power misfortunes straightforwardly as result brilliant working and adjusting as per the car motors necessities. The basics to have chosen this kind of nano-added substances is sufficiently reacted by this examination. The Nanomaterials have captivating highlights as high thermal conductivity and outstanding mechanical properties [10-12]. Nanomaterials offer superior chemical stability, also they can serve adequately in various tribological applications. Graphene have bunch of possibilities, for example, anti-friction/wear, going about as strong oils on sliding worn interfaces, which lead to a more drawn-out life expectancy for motor parts and improving motor toughness. Likewise, nanomaterials give powerless powers of the van der Waals and static communications powers between the 2D layers, that encourage inside the solidness of Nano lubricants [6].

Current research paper aims to investigate the effect of $Al_2O_3$, $TiO_2$, Graphene (G), Co-Cr, Ni nanoparticles on lubricity under reacting to sliding movement utilizing Linear Wear Tester. Two types of contact settings are used to measure the friction wear rate. Afterward linear wear test, the morphology and composition of worn surfaces are characterized using SEM and EDX in order to deliver an understanding on the Nanomaterial's tribological mechanisms [9]. The key reason for resistance and wear of the worn surfaces are vitality dissemination in automobile engines. 17-19% of power produced by engine is wasted due to frictional losses. Now the whole world is looking forward to increasing the efficiency of engine many ways. Among of the ways is to diminish the friction and wear in engine by improving the engine oil. Some of the countries are dependent on importing oil at an extraordinary scale, which can prompt an imbalanced exchange and nation security encounters [13, 14]. If we look at the economic loss worldwide due to frictional losses it will be around US$120 billion per year. From this information we can clearly come to a point that we need to focus on strategies to utilize our natural resources more efficiently [15]. The objectives of this research are to study the structure, characteristics & uses of G, $Al_2O_3$, $TiO_2$, Ni & Co-Cr in Lubricants and to mix G, $Al_2O_3$, $TiO_2$, Ni & Co-Cr and base engine oil (10W40) in proper concentration and later to evaluate tribology characteristics of newly prepared nano-lubricant and compare the data to show the tribological improvement and lubricants performance [16-19].

## 2. Materials and methods

The area of this examination is to show the mechanism liable for the tribological properties (anti friction & anti wear) of the piston ring and cylinder body interfaces utilizing nanomaterial enriched nano lubricants compared to the usual engine oil (10w-40) [15, 20, 21]. Moreover, the configuration and morphology of the tribo-science films shaped on the well-used surfaces are studied by tests of SEM, EDS, 3D surface profiler [6].

### *2.1 Graphene preparation & formulation of nano lubricants*

The Graphene is obtained in the form of GO (Graphene Oxide) (8.64 mg/ml). Vacuum Oven was used to make it in Dry form. In this study the commercial lubricant 10W-40 was used as reference oil or base oil. This 10W-40 engine oil chosen because it is one of the widely used engine oil worldwide [22-24]. The technical data of these specific lubricants has been presented in Table 1. The G, $Al_2O_3$, $TiO_2$, Ni & Co-Cr nano lubricants were prepared in 3 concentrations (0.01, 0.03, 0.05 wt.%) to obtain a balanced dispersion and to avoid sedimentation in the base oil. The composition of G, $Al_2O_3$, $TiO_2$, Ni & Co-Cr nano lubricant is presented in below Table 2. The mixing of nano lubricant samples was

IOP Conf. Series: Materials Science and Engineering 1078 (2021) 012016 doi:10.1088/1757-899X/1078/1/012016

done by a mechanical stirring device for 4 h and 2 hours of ultrasonic mixing for a homogeneous stable suspension [15].

**Table 1.** Physical Data of sample base lubricant 10w-40.

| Parameters | Method | Unit | Typical Value |
|---|---|---|---|
| Appearance | - | - | B&C |
| Density 15°C | ASTM D 4052 | g/cm C $^2$ | 0.875 |
| Viscosity @100°C | ASTM D 445 | cSt | 14.2 |
| Viscosity Index | ASTM D 2270 | - | 150 |
| Flash Point Flash Point CI. O. C | ASTM D 92 | °C | >200 |
| TBN | ASTM D 2896 | mg KOH/g | 10.0 |
| Pour Point | ASTM D 97 | °C | < -30 |
| Foam @24°C | ASTM D 892 | cc/s | Traces/0 |

**Table 2.** The Composition of Gr Nanolubricant sample.

| Lubricant oil type | Concentration (wt.%) | Reference oil(10W-40) (ml) | Additive Solution (gm) |
|---|---|---|---|
| Base Oil | 0 | 50 ml | 0 % additives |
| G Nano lubricant | 0.01, 0.03, 0.05 | 50 ml | 0.0004, 0.0013, 0.0021 |
| Co-Cr Nanolubricant | 0.005, 0.01, 0.03, 0.05 | 50 ml | 0.0571, 0,1714, 0.2859 |
| $Al_2O_3$ Nano lubricant | 0.01, 0.03, 0.05 | 50 ml | 0.0226, 0.0677, 0.1129 |
| $TiO_2$ Nano lubricant | 0.01, 0.03, 0.05 | 50 ml | 0.0242, 0.0725, 0.1209 |
| Ni Nano lubricant | 0.01, 0.03, 0.05 | 50 ml | 0.0509, 0.1526, 0.2549 |

*2.2 Sedimentation*

A stable solution of the mixture is necessary for the proper functioning of the nano lubricant to be verified. Achieving sufficient nano lubricant dispersion consistency is an essential problem for the implementation of nano lubricants in any particular application. Graphene has superior properties in its pristine form, but it is enormously susceptible to agglomeration. As a result, in the presence of reducing agents, the pristine Graphene, $Al_2O_3$, $TiO_2$, Ni & Co-Cr layers appear to coagulate due to heavy π-π interaction and van der Waals tempting forces between the planes of Graphene, $Al_2O_3$, $TiO_2$, Ni & Co-Cr. As Graphene, $Al_2O_3$, $TiO_2$, Ni & Co-Cr are propagated in oil-based lubricants, agglomeration is further differentiated [25]. Researchers have found that graphite-derived graphene platelets contain functional groups of hydroxyl and carboxyl on the sheet edges that make them soluble in water. As a consequence, the propensity to coagulate in oil is greater [26]. Few research studies indicate Co-Cr as anticorrosive material and $Al_2O_3$ being easily available & low cost where as $TiO_2$ & Ni being used for its special characterization. Visual analysis has been carried to conduct sedimentation test for 1 month and further. Below Fig 1. Shows the Prepared Nano lubricant Samples for visual stability.

*2.3 Thermal conductivity & viscosity measurements*

The thermal conductivity of the prepared Graphene nano lubricants was calculated using a KD2 PRO thermal conductivity probe (Decagon Devices Inc., USA) [27]. It is the recommended way in the effort to calculate the thermal conductivity of nanofluids used by previous researchers [7, 27-29]. To test the viscosity of the prepared Graphene nano lubricant, a Brookfield LVDV-III Ultra Rheometer (Viscometer) was used.

*2.4 Linear wear test & worn surface analysis*

The piston skirt-liner tribometer was used to measure the Graphene, $Al_2O_3$, $TiO_2$, Ni & Co-Cr Nano lubricants' friction and wear conduct. Under lubricated sliding conditions, the wear test was performed. Wear checking involves creating a linear motion to a cylinder-piston ring pair working in realistic condition in the same way. The Tribo-tester acts within the engine like a touch piston-ring. Before and after the experiment, all materials are washed in acetone and dried in hot air. Varying loads, velocities and temperatures for all Graphene, $Al_2O_3$, $TiO_2$, Ni & Co-Cr nano lubricant sample concentrations as shown in Table 3 [30]. Scanning Electron Microscopy (SEM) by Hitachi TM3000 with Energy Dispersive X-ray Spectroscopy was used to test the morphology and element distribution of the worn surface on the disk (EDS) [30].

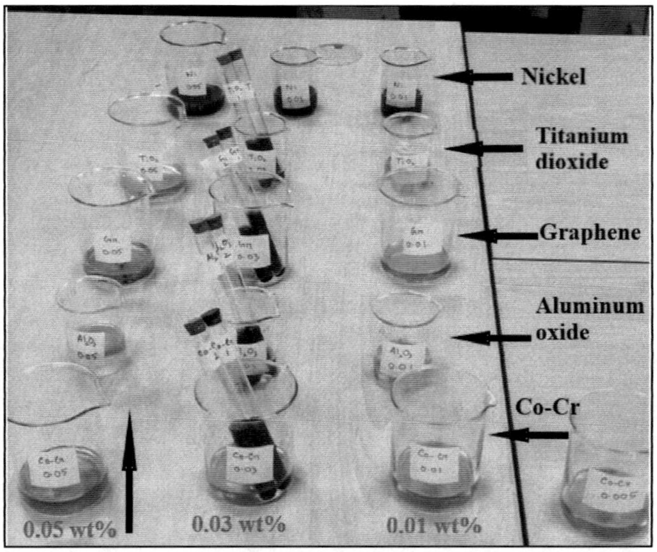


**Figure 1:** Prepared nano lubricant samples.

**Table 3.** Experimental conditions for tribometer.

| Load (N) | Speed (rpm) | Temp. (°C) | Concentration (wt%) | Nanoparticles |
|---|---|---|---|---|
| 38.5 | 400 | 25 | 0.01, 0.03, 0.05 | Gr, $Al_2O_3$, $TiO_2$, Ni & Co-Cr |
| 98.5 | 200 | 25 | 0.01, 0.03, 0.05 | Gr, $Al_2O_3$, $TiO_2$, Ni & Co-Cr |

## 3. Results and discussions

### *3.1 Thermal conductivity and viscosity*

Experimental analysis can declare that one of the chosen nanoparticle Co-Cr is most suitable among Graphene, $Al_2O_3$, $TiO_2$, Ni & Co-Cr nanoparticles for using in engine oil because the viscosity readings are most stable. Fig 2. show the thermal conductivity results. High thermal conductivity means more heat dissipation from engine as engine oil's one of the most important purpose is to keep the engine temperature in control and Co-Cr shows the highest thermal conductivity. Viscosity is preferred to be stable in different condition. One of the main reasons to care about engine oils viscosity is to keep the lubrication in proper manner inside engine. As far we see from the results the most stable viscosity is shown by Co-Cr.

Figure 2 (b) & (c) shows that Co-Cr has the best value for thermal conductivity with 0.3429 W/m-k & 0.3030 W/m-k at 70°c for 0.05 &0.03 wt% respectively. Whereas Graphene is showing the second highest at 0.03% but the most changed according to temperature is Nickel. For 0.03wt% the highest thermal conductivity value obtained for Co-Cr followed by Nickel with 0.2139 W/m-k. Fig 2 (a) shows the thermal conductivity as highest for Co-Cr and followed by $Al_2O_3$ with 0.2234 W/m-k and Graphene(G) shows the most gradual increment at 0.01 & 0.05 wt%.

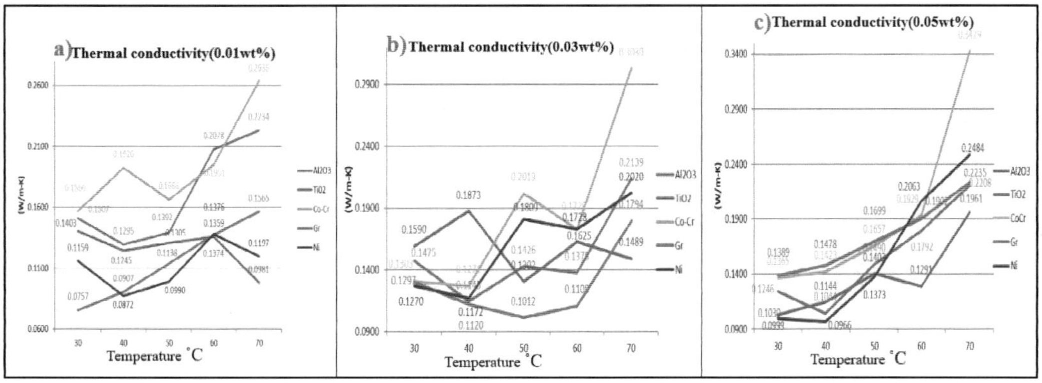


**Figure 2.** Thermal Conductivity for different nano lubricants a) 0.01wt%, b) 0.03wt%, c) 0.05wt%.

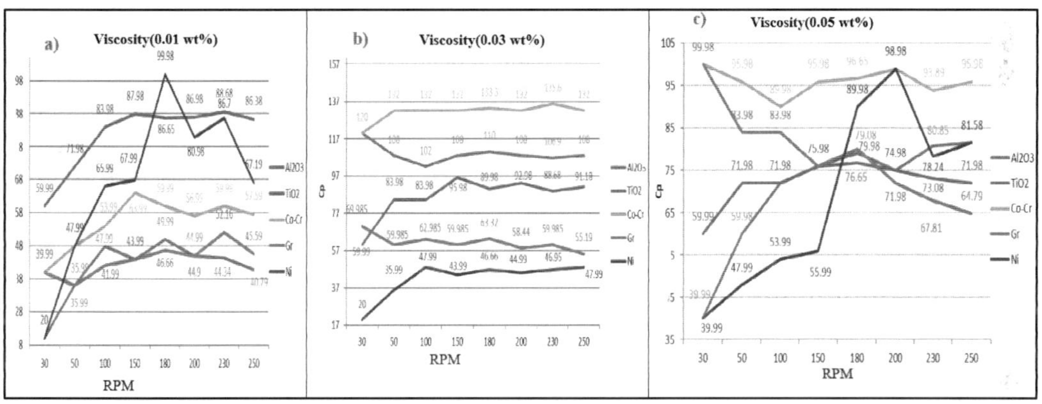


**Figure 3.** Viscosity measurements for different nano lubricants a) 0.01wt% b) 0.03wt% c) 0.05 wt%.

The Figure 3 indicates that the most stable viscosity reading is attained by Co-Cr at 0.03 &0.05 wt% at varied speed. viscosity at 0.03wt% shows very critical to choose stable reading but still we can see that CoCr shows the most stable viscosity. In Figure 3 (a) shows the stability almost similar for Co-Cr and Graphene.

*3.2 SEM and EDX analysis (linear wear test nano lubricant)*

For Linear wear test Co-Cr 0.03 wt% nano lubricant has been chosen as it has shown the highest thermal conductivity and most stable viscosity reading among all other materials in different concentration (from Figure 2 & 3). Below Figure 4. shows the nano lubricant SEM images of Co-Cr at 0.03 wt% with linear wear test (200 rpm and 98.5N load). SEM images of Linear wear test (400 rpm and 38.5N load) of Co-Cr 0.03wt% enriched nano-lubricant is shown in Figure 5. It can be clearly observed from SEM images that there are worn surface which contains the layer (of Co-Cr) after Linear wear test. The green highlighted data from EDS result shows the percentage of Cobalt and Chromium found in the layer after linear wear test. Actually, this layer is going to reduce the friction making the surface of components slide. In EDS result Carbon found due to wear and friction, Aluminium found because the friction plate was Aluminium alloy and oxygen found due to decomposition. Aluminium plate was photographed shown in Figure 6.

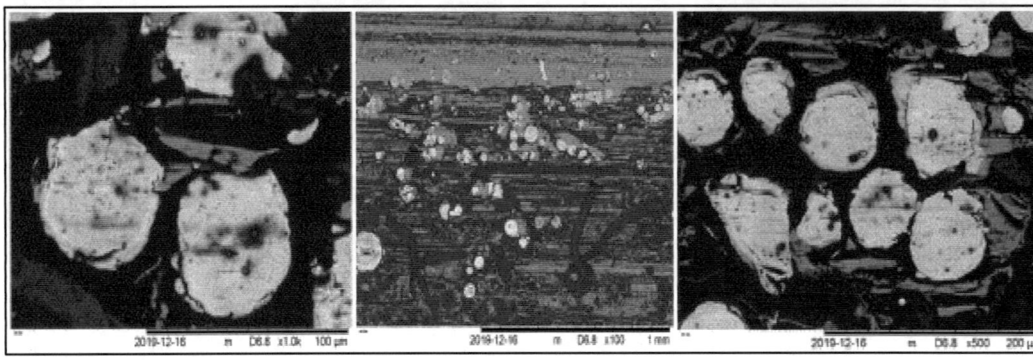

**Figure 4.** SEM of Linear wear test (200 rpm and 98.5N load) of Co-Cr 0.03wt% nano lubricant.

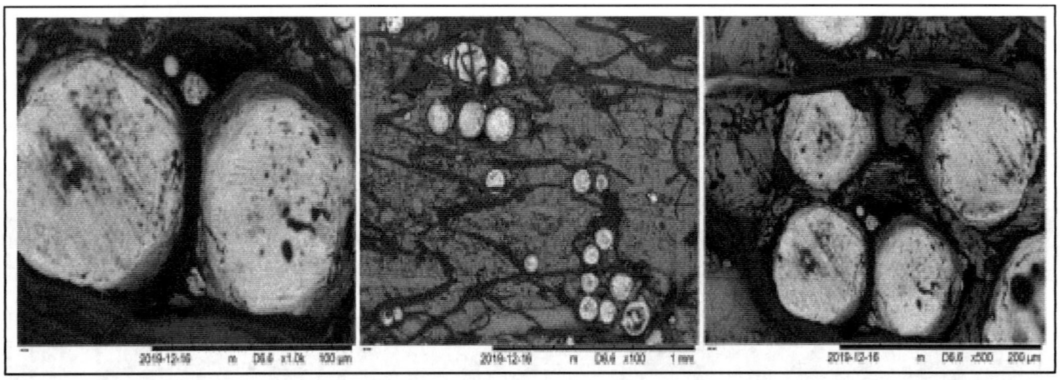

**Figure 5.** SEM of Linear wear test (400 rpm and 38.5N load) of Co-Cr 0.03wt% nano lubricant.

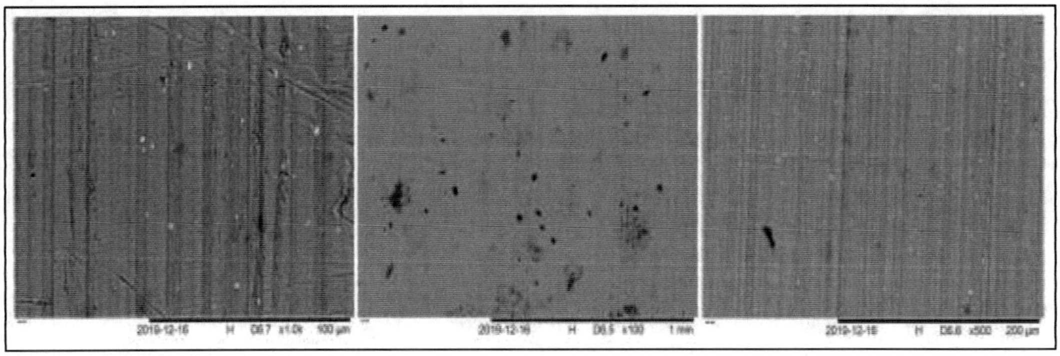

**Figure 6.** Scanning Electron photograph of plane Aluminium plate.

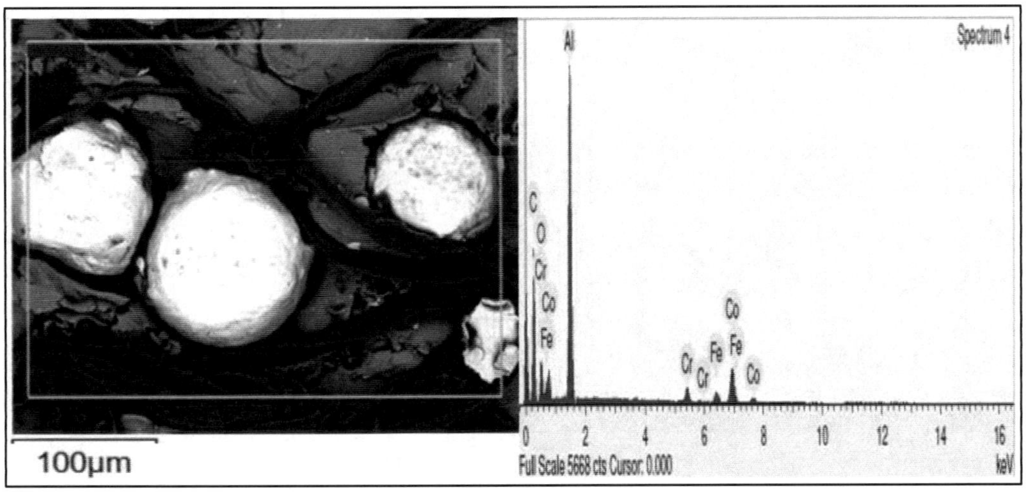


**Figure 7.** EDX results of Linear wear test (200 rpm and 98.5N load) of Co-Cr 0.03wt% enriched nano-lubricant.

Linear wear test was done at the 200 rpm for 98.5N load. Scanning electron microscopy was done to understand the surface modifications. The Energy Dispersive X ray testing was performed with elemental analysis. as shown in Figure 7 and the observed elements are listed in the Table 4. Chromium and cobalt were found with the highest weight % as Chromium was 6.610 % and Cobalt was 36.212% respectively.

**Table 4.** Composition of elements by XRD analysis of Co-Cr 0.03wt% (200 rpm and 98.5N load).

| Element | Weight % | Weight % σ | Atomic % |
|---|---|---|---|
| Carbon | 50.662 | 0.351 | 80.859 |
| Oxygen | 2.480 | 0.251 | 2.971 |
| Aluminum | 1.547 | 0.049 | 1.099 |
| **Chromium** | **6.610** | **0.126** | **2.437** |
| Iron | 2.490 | 0.123 | 0.855 |
| **Cobalt** | **36.212** | **0.318** | **11.779** |

The Figure 8 showing the linear wear test observed under 400 rpm and 38.5 N load conditions. The EDX analysis found that as the rpm increases, and load decreased the amount of Chromium and Cobalt. Table 5 showed Chromium reduced to 4.868 wt% and Cobalt to 24.774 wt% the total percentage reduction observed as by 26% chromium and 31% cobalt.

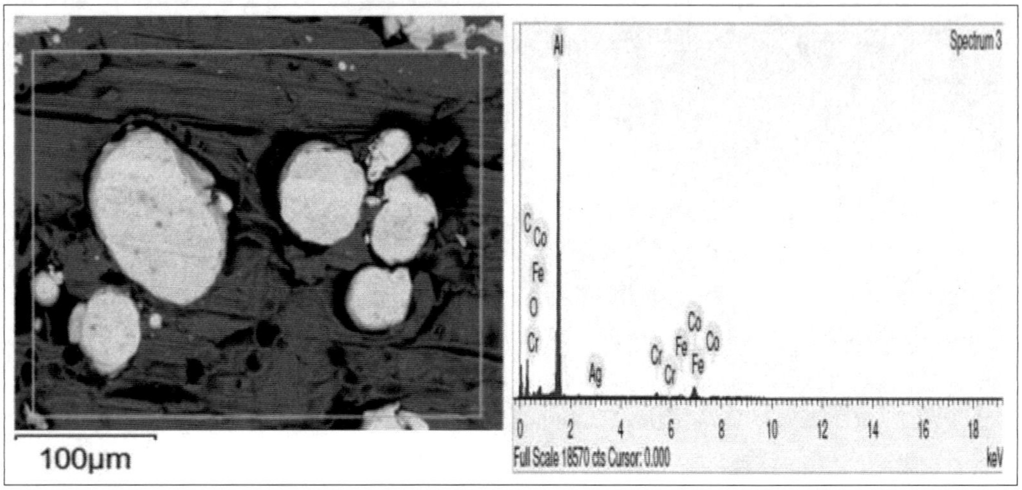


**Figure 8.** EDX results of Linear wear test (400 rpm and 38.5N load) of Co-Cr 0.03wt% enriched nano-lubricant.

**Table 5.** Composition of elements by EDX analysis of Co-Cr 0.03wt% (400 rpm and 38.5N load).

| Element | Weight % | Weight % σ | Atomic % |
|---|---|---|---|
| Carbon | 53.400 | 0.333 | 79.158 |
| Oxygen | 7.037 | 0.273 | 7.832 |
| Aluminum | 2.217 | 0.050 | 1.463 |
| **Chromium** | **4.868** | **0.103** | **1.667** |
| Iron | 6.407 | 0.147 | 2.043 |
| **Cobalt** | **24.774** | **0.265** | **7.485** |
| Zinc | 1.297 | 0.113 | 0.353 |

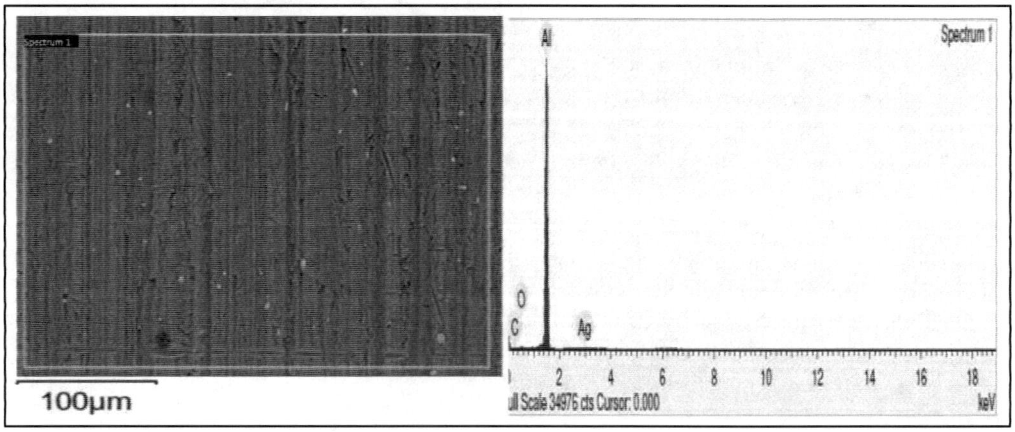


**Figure 9.** EDX results of Linear wear test (400 rpm and 38.5N load) of plane aluminium plate.

The aluminium plate was observed under the SEM as shown in Fig. 9. The clear and smooth surface was observed. The energy dispersive x ray analysis found very few numbers of elements observed as listed in Table 6. The total percentage of aluminium was found 77.175 %. Result confirmed that the aluminium plate composed of pure aluminium element.

**Table 6.** Composition of elements by EDX analysis of plane Aluminium plate.

| Element | Weight % | Weight % σ | Atomic % |
|---|---|---|---|
| Carbon | 16.125 | 0.695 | 29.662 |
| Oxygen | 4.905 | 0.260 | 6.774 |
| **Aluminum** | **77.175** | **0.681** | **63.196** |
| Silver | 1.796 | 0.143 | 0.368 |

## 4. Conclusion

The tribological and thermophysical properties of engine oil were improved by the addition of nanoparticles G, $Al_2O_3$, $TiO_2$, Ni & Co-Cr. As additives in the engine oil, G, $Al_2O_3$, $TiO_2$, Ni & Co-Cr nanoparticles were used to display good friction reduction, anti-wear activity, improved thermal conductivity and consistent viscosity. This effect may be due to the rolling between the contact surfaces of spheres such as nanoparticles, thereby reducing friction. The thermal conductivity increased with an increasing concentration of nanoparticles with the stability of oil viscosity. The anti-wear mechanism can be due to the accumulation on the worn surface of G, $Al_2O_3$, $TiO_2$, Ni & Co-Cr nanoparticles, which in turn reduced the resistance to shearing and thus improved the tribological characteristics. On rubbing surfaces, nanoparticles may be known as nano-bearings. The nanoparticles G, $Al_2O_3$, $TiO_2$, Ni & Co-Cr are deposited under mixed and boundary lubrication only. Experimental studies have confirmed that the deposition on the fluid layers of G, $Al_2O_3$, $TiO_2$, Ni & Co-Cr nanoparticles improves the tribological properties of the base lubrication oil, showing decreased wear and friction. Aluminum alloy wear (Al 25) was observed to increase with an increased load value over time. As a multifunctional additive, G, $Al_2O_3$, $TiO_2$, Ni & Co-Cr nanoparticles can therefore be used. As Co-Cr 0.03wt% has been chosen as the most successful among our other experimented nanoparticles enriched nano lubricant so we can say that in future after considering the health & nature hazard it can be marketed for the usage of lubricant.

**Acknowledgment**

The authors would like to thank University Malaysia Pahang (UMP) for Grant RDU190323, FRGS/1/2017/TK03/UMP/02/25 and UMP Flagship RDU192204 for financial assistance and facilities provided.

## References

[1] Fusaro, R.L., Lubrication of Space Systems (c). 1995.
[2] Rudnick, L.R., Lubricant additives: chemistry and applications. 2009: CRC press.
[3] Leger, L.J. and K. Dufrane, Space station lubrication considerations. 1987.
[4] Wong, V.W. and S.C. Tung, Overview of automotive engine friction and reduction trends–Effects of surface, material, and lubricant-additive technologies. Friction, 2016. 4(1): p. 1-28.
[5] Wu, Y., W. Tsui, and T. Liu, Experimental analysis of tribological properties of lubricating oils with nanoparticle additives. Wear, 2007. 262(7-8): p. 819-825.
[6] Rasheed, A., et al., Heat transfer and tribological performance of graphene nanolubricant in an internal combustion engine. Tribology International, 2016. 103: p. 504-515.
[7] Hamdan, S. and W. Chong. Simulation of lubricant Stribeck curve for SAE grade engine lubricants. in Proceedings of Asia International Conference on Tribology 2018. 2018. Malaysian Tribology Society.
[8] Nasir, S., et al., Carbon-based nanomaterials/allotropes: A glimpse of their synthesis, properties and some applications. Materials, 2018. 11(2): p. 295.
[9] Su, Y., L. Gong, and D. Chen, An investigation on tribological properties and lubrication mechanism of graphite nanoparticles as vegetable based oil additive. Journal of Nanomaterials, 2015. 2015.

[10] Ma, P.C., B.Z. Tang, and J.-K. Kim, Effect of CNT decoration with silver nanoparticles on electrical conductivity of CNT-polymer composites. Carbon, 2008. 46(11): p. 1497-1505.
[11] Pil Jang, S. and S.U. Choi, Effects of various parameters on nanofluid thermal conductivity. 2007.
[12] Zhou, W.X., et al., Thermal conductivity of amorphous materials. Advanced Functional Materials, 2020. 30(8): p. 1903829.
[13] Jianbin, L. and X. Zhou, Superlubricitive engineering—Future industry nearly getting rid of wear and frictional energy consumption. Friction, 2020. 8(4): p. 643-665.
[14] Khan, K. and S. Rehman, Wonders of Nanotechnology. Nanomaterials and their Fascinating Attributes, 2016. 1: p. 3.
[15] Ali, M.K.A., et al., Novel approach of the graphene nanolubricant for energy saving via anti-friction/wear in automobile engines. Tribology International, 2018. 124: p. 209-229.
[16] Syahir, A., et al., Tribological improvement using ionic liquids as additives in synthetic and bio-based lubricants for steel–steel contacts. Tribology Transactions, 2020. 63(2): p. 235-250.
[17] Jia, X., et al., Monodisperse Cu nanoparticles@ MoS2 nanosheets as a lubricant additive for improved tribological properties. Applied Surface Science, 2019. 494: p. 430-439.
[18] Singh, A., P. Chauhan, and T. Mamatha, A review on tribological performance of lubricants with nanoparticles additives. Materials today: proceedings, 2020. 25: p. 586-591.
[19] Zhang, R., et al., Well dispersive TiO2 nanoparticles as additives for improving the tribological performance of polyalphaolefin gel lubricant. Industrial & Engineering Chemistry Research, 2018. 57(31): p. 10379-10390.
[20] Ali, M.K.A., et al., Fuel economy in gasoline engines using Al2O3/TiO2 nanomaterials as nanolubricant additives. Applied energy, 2018. 211: p. 461-478.
[21] Elagouz, A., et al., Frictional performance evaluation of sliding surfaces lubricated by zinc-oxide nano-additives. Surface Engineering, 2020. 36(2): p. 144-157.
[22] Esfe, M.H., et al., Using artificial neural network for investigating of concurrent effects of multi-walled carbon nanotubes and alumina nanoparticles on the viscosity of 10W-40 engine oil. Physica A: Statistical Mechanics and its Applications, 2018. 510: p. 610-624.
[23] Krulish, J., H. Lowther, and B. Miller, An update on synthesized engine oil technology. 1977, SAE Technical Paper.
[24] Kumbár, V. and A. Sabaliauskas, Low-temperature behaviour of the engine oil. Acta Universitatis Agriculturae et Silviculturae Mendelianae Brunensis, 2013. 61(6): p. 1763-1767.
[25] Singh, R.P., Nanocomposites: Recent Trends, Developments and Applications. Advances in Nanostructured Composites: Volume 1: Carbon Nanotube and Graphene Composites, 2019: p. 16.
[26] Paul, G., et al., Nanolubricants dispersed with graphene and its derivatives: an assessment and review of the tribological performance. Nanoscale, 2019. 11(8): p. 3458-3483.
[27] Ramachandran, K., et al., Thermophysical properties measurement of nano cellulose in ethylene glycol/water. Applied Thermal Engineering, 2017. 123: p. 1158-1165.
[28] Nguyen, C., et al., Temperature and particle-size dependent viscosity data for water-based nanofluids–hysteresis phenomenon. International journal of heat and fluid flow, 2007. 28(6): p. 1492-1506.
[29] Lee, S.W., et al., Investigation of viscosity and thermal conductivity of SiC nanofluids for heat transfer applications. International Journal of Heat and Mass Transfer, 2011. 54(1-3): p. 433-438.
[30] Awang, N., et al., Study on friction and wear of Cellulose Nanocrystal (CNC) nanoparticle as lubricating additive in engine oil. International Journal of Heat and Mass Transfer, 2019. 131: p. 1196-1204.

# A review on the prospect of wind power as an alternative source of energy in Malaysia

**S N Ashwindran[1], A A Azizuddin[1,*], A N Oumer[2] and M Z Sulaiman[1]**

[1] Faculty of Mechanical & Automotive Engineering Technology Universiti Malaysia Pahang, Pekan, Malaysia.
[2] College of Engineering,Universiti Malaysia Pahang, Pekan, Malaysia.

*Corresponding author's email: azizuddin@ump.edu.my

**Abstract**. Fossil fuels has dictated the energy sector for several decades, rising as the primary and prominent source of energy production for various sector of industries namely electricity (power), transportation, manufacturing and others. However, extensive usage of fossil fuels majorly contributed to pollution issues and degradation of the environment. Hence, governments and research agencies are providing supports and resources to facilitate the growth of renewable energy sector (RES). This manuscript provides a brief view of the challenges of wind renewable energy sector in Malaysia relative to geographical wind condition, government policies and challenges in initiation of wind technologies and global perspective of green energy in terms of RER. It is found that data presented by previous researchers on wind speed data are considered unreliable to be utilized for WT power density assessment and design. Notably, wind speed in Malaysia is seasonal based and highly inconsistent. The wind flow pattern and intensity experience by Malaysia are merely an aftereffect of weather change in neighbouring regions. Nevertheless, it is found that selected regions of east-coast peninsular Malaysia and selected region of Sabah and Sarawak is potential for harvesting wind power due to substantial and adequate wind pattern behavior during monsoon season. Therefore, it is evident that geographical wind condition of Malaysia is appropriate for wind power devices as a form of RER in green energy sector of Malaysia. However, it is also found that fundamental design modification is required on conventional design of WT in terms of blade morphology and configuration in order to adapt to unsteady monsoon wind speed pattern.

**Keywords:** Wind Engineering; Renewable Energy; Wind Speed Characteristic in Malaysia.

## 1. Introduction

Renewable energy (RE) is an alternative approach to depleting non-renewable energy resources. In response to increasing awareness to environmental issue such as rise of greenhouse gases (GHG), and high carbon lifecycle assessment (LCA) [1]. Governments and research agencies are providing supports and resources to facilitate the growth of renewable energy sector (RES) [2]. There are different types of RE namely on solar, wind, tidal or hydro motion, nuclear and biomass. Today wind turbines (WT) are the prominent form of renewable energy for direct energy harvesting. Joselin Herbert et al. [3] stated that wind power is an indirect form of solar energy, where the differential heat radiated by the sun on earth surface influence the wind potential. Considering in terms of yearly growth per installed energy harvesting technology, wind energy is considered as the fastest growing in comparison to other renewable energy resources (RER) [4,5]. Technologies on RER are sufficient and well developed. However, more extensive research and development is being conducted to improve the efficiency of the respective RE system [6]. Wind turbine technology has grown significantly as a prominent source of RE harvesting due to its zero emission, reliability and low cost [7–9]. Cognet et al. [10] stated that, conventional WT performs efficiently only at specific working condition (angle of attack and aerodynamic characteristic of blade) and environment (wind speed) in which makes wind energy conversion poor in general. Moreover, lack of development and research on wind power devices leads to energy and economic loses. Hence, different approach is required on convention WT blade design in order to improve the operating regime and power extraction performance. In comparison to between the two types of WT configuration, HAWT is widely preferred in wind power generation than VAWT [11,12]. However, there are disadvantages to HAWT type domains such as unstable physical structure, requires high material criteria for blade fabrication and low wind energy capturing area [13]. The wind energy capturing performance of VAWT is lower in comparison to HAWT [14,15]. However, research shows that VAWT framework indicated more substantial potential than HAWT for low wind speed condition regions and offshore application [16–18]. VAWT has received limited attention among researcher over the past decades due to the aerodynamic complexities relative to HAWT [19]. Lately, VAWT received attention among researcher on the aerodynamic performance properties and design mechanism [15].

Generally conventional WTs are design to operate at high wind speed ranging from 10-15 m/s [20]. This constrains the WT to harvest adequate power at low wind speed condition [21]. Sustainable amount of research is being conducted in effort of improving the conventional WT or in other words design reconfiguration such as blade configuration and morphology, rotor configuration, pitch angle variation and others [22]. Research shows that, design configuration adjustment and optimization has improve the efficiency in power extraction [23–26]. Currently, researcher is actively working on blade morphology and configuration for optimal wind interaction. Ikeda et al. [27] stated that conventional WTs work within the narrow zone center at different TSR at turbulent wind environment. Hence, limits the aerodynamic performance of the turbine. Li et al. [28] mentioned that WTs operates below atmospheric boundary layer, which is comparatively higher than inflow turbulence of aviation flight. The author also stated that interference pose by geographical obstacles in low wind speed region will further increase the turbulent intensity. Hence such environment creates unreliable wind flow condition for the blade. Bianchini et al. [16] stated that design optimization needs to be implement on conventional VAWT in order to improve the aerodynamic efficiency. Design improvement and optimization of WT is conducted at various level and aspects namely from blade morphology to rotor control system. The primary governing factor that influences the performance of the WT is the blade shape. Although, numerous concept and method emerged over the years, current research trend in WT blade modelling is smart blade [29] and biomimicry or bioinspiration [30]. Research shows that WT blade modelling via bioinspiration technique has improved the power extraction capabilities. Moreover, understanding the aerodynamic behavior of bio-elements helps to comprehend the complexity in flow behaviour exhibited by WT. Nonetheless, the geometrical properties of the blade morphology also play an important role in blade performance. The blade shape influences the flow field properties and the behaviour of flow response. Traditional WT design struggles with fundamental design error which consequently impacted the operational capabilities. Compounded structural design issues especially blade structure constitutes to the inefficiency of the WT [31]. Despite over 30 years of experience, manufacturers and engineers are adherently working to improve the efficiency of WT [31]. In spite of other improvements namely material structure, gearbox modification, turbine and control system enhancement in WT technology, less attention is paid to blade morphology design.

Conventional WTs can be classified into two configurations in terms of rotor orientation that is vertical axis wind turbine (VAWT) and horizontal axis wind turbine (HAWT) [32–34]. HAWT is commonly used in large megawatt scale wind power application or known as large wind turbine (LWT). Where else VAWT is used for domestic and small wind power application (SWPA) or mini wind turbine (MWT). The propulsion of WT is governed by aerodynamic forces that is drag and lift

[29]. Meanwhile, VAWT can be further classified into two types namely drag and lift type or commonly known as Savonius and Darrieus respectively [35–38]. Apart from generating power wind power devices is used for various applications such as wind power heating, seawater distillation, ship navigation and etc [39]. Moreover, WT are also classified based wind rotor performance with respect to power extraction coefficient ($C_p$) and tip speed ratio (TSR) and namely Savonius, horizontal axis three blade, vertical axis Darrieus [40], multi-blade [41], Dutch-four blade [42], high speed two blade [43,44] and single blade [45]. The minimal cut-in speed required for WTs to initiate rotation is from 3.5 m/s - 4 m/s [46], that is dependent on the configuration of blade type scale of turbine and propulsion configuration of wind turbine (lift or drag type). However, regions of South East Asia onshore sites have low wind speed potential, in which conventional wind power turbines can't produce sufficient power [47]. Hence design improvement and adaption are being implemented in the motive of obtaining a more efficient and reliable wind power device for low wind speed regions. The design adaption and improvement are mainly focused on propulsion aerodynamic load, blade configuration, blade morphology and etc. Researcher has proposed several innovative approaches to design optimization such surrogate modelling, parametric based optimization, specialized modified algorithm-based optimization, design adaptation and implementation and others. Based on the all the novel approach proposed by engineers and researchers on specialized WT, it might take years for the proposed design or mechanism to be considered feasible for commercial use [42]. Case in point, Musgrove rotor WT developed by Professor Musgrove is commercially available ranging from 25 kW to 1000 Mw [48]. However, Musgrove rotor is not prominent as other WT configuration counterparts. The aim of the manuscript is to provide a brief view of the challenges of wind renewable energy sector in Malaysia relative to geographical wind condition, government policies and challenges in initiation of wind technologies and global perspective of green energy in terms of RER.

## 2. Usage of fossil fuels

The alternative approach of discouraging the usage of fossil fuels as the source of power generation is on the investment and development of RE resources. The politics and policies revolving around fossil fuel has encourage greed among individuals on the control of power generation via fossil fuels [49]. Chances on harvesting electricity from thermal based fuel cheaper has decreased due to the extended learning process over the years on improving the harvesting process. The cost on fossil fuel has increase due to the rise of consumption and shortage of resources. Moreover, the cost of power plant has increased due to the rise of living standard in terms of minimizing pollution and enhanced safety standards [50]. Consequently, the rise in prices on non-renewable energy supply has impacted the development of economy. At the initial stage the initiation cost on renewables will be costly but as rise in knowledge on harvesting and enhancing technologies on RER, the energy cost will subsequently decrease. As in July 2016 in northern Europe, off-shore wind prices were reported at 70 EUR/MWh, the prices drop to 60 EUR/MWh in September and reach below 50 EUR/MWh by November [50]. Meanwhile in Malaysia, high consumption of fossil fuel led to high amount of carbon dioxide ($CO_2$) emission. Since 1980, there is continuous rise of carbon dioxide emission due to non-renewable fuel usage [51]. United States Energy Information Administration (EIA) reported that in 2017, Malaysia emitted 229 million metric tons of $CO_2$ due to high amount non-renewable fuel consumption [52]. Case in point, relative to Kyoto protocol in 2002, Prime minister of Malaysia made a promise in reducing the $CO_2$ by 40% by 2020 in United nations climate change conference (UNCCC) [53]. However, Malaysia government has not taken any serious steps in reducing the $CO_2$ index. This can be seen by the rise of $CO_2$ emission as shown in Figure 1 from 2002 till 2017.

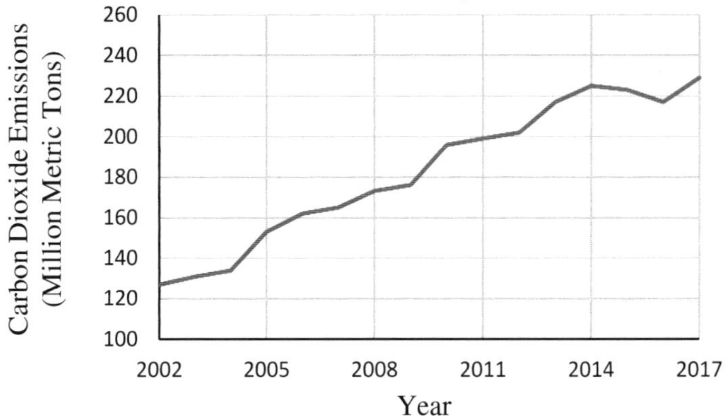


**Figure 1.** Carbon Dioxide Emissions by year in Malaysia. Source: United states.EIA 2020 Energy analysis of Malaysia [54].

**Table 1**. Countries examining "100% renewable energy mechanism (REM)"

| Countries | Analysis |
|---|---|
| Denmark [4] | The author stated that on global scale renewable energy constitutes less than 15 % of primary energy supply. The major key players are hydropower and fuel. The author concluded that with appropriate energy strategy it's possible for Denmark energy system to adapted 100% REM with the combination of biomass, wind power and photovoltaics RER. |
| Ireland [57] | Since the research were conducted in technical standpoint not economical perspective, the author stated that its technically feasible for Ireland to achieve 100% REM due to Ireland energy mix system. The author also mentions that there are several methods in achieving 100% REM apart from the presented methodology. |
| New Zeeland [58] | Researcher concluded that 32% of fossil fuel utilization for thermal generation can be replaced by 100% REM consisting renewable energy generation mix (REGM) of 53-60% hydro, 22-25% wind power, 0.2-0.3 bio-gas, 0.8-0.9 wood thermal and 12-14% geothermal. |
| Porto Santo island [59] | The authors developed a model for energy planning and optimization. The analysis was conducted at two different scenario which is peak shaving and 100% renewable power system. Results indicated that wind solar energy mix for peak shaving would be more effective than its counterpart. Where else, 100% renewable power system could be more cost effective. |
| Portugal [60] | The theoretical analysis conducted by the author indicated that Portugal could attain 100% REM energy system in 10 years with the assistance of suitable energy system strategy. The author utilized $H_2$RES software for energy plan system modelling. RER meteorological data such as wind speed and solar radiation were compared against the result obtained from $H_2$RES model. |
| Australia [61] | Proper and restructured planning on energy supply and demand makes it feasible for 100% REM to penetrate Australian National Energy Market (ANEM). Specific attention needed to be given to 100% REM with respect to wind and solar RER for balanced energy supply and demand. This is due to absence of wind and solar during winter evenings which demands high energy supply. |
| Croatia [62] | The author concluded that it requires for Croatia to carry out comprehensive planning on all economy sectors with regards to the implementation of 100% REM. The energy calculation was carried in EnergyPLAN software. Results illustrated that 2000 MW WTs could contribute to 10% total electricity demand granting that the new power plans operate at minimal load of 400 MW. |

Unfortunately, at current status quo Malaysia is strongly dependent on fossil fuel by 90% for electrical power generation. Although solar energy supply is considered as the main contributor Malaysia's RE sector, the contribution only reached up to 2%. RE sector in Malaysia is targeted to reach 25% by the year 2025 [55]. Although other counties are rapidly evolving into RER, even as introducing the "100% renewable energy mechanism", where energy supply are entirely from RER. Several countries are involved in "100% renewable energy mechanism (REM)" where its being analyzed relative to their energy supply and demand [56]. Refer to Table 1 for details on counties that are examining on the concept of "100% renewable energy mechanism". Needless to say, 100% REM requires strategic energy planning in terms of technically and in economical perspective to maintain balance in energy demand and supply. Strategic energy planning would assist high penetration of RE to Malaysia energy supply chain. Relative to Malaysia, restructuring RE infrastructure and energy supply chain would displace the dependency of fossil fuel as main domain of electricity supply. However, Malaysian government is still struggling and reluctant in establishing a proper RE power supply. Therefore, Malaysia should carry out extensive and comprehensive research on RE power supply to cater for a sustainable and cleaner future.

## 3. Energy policies in Malaysia

Energy policies in Malaysia are regulated and overseen by Energy department of Economic Planning Unit (EPU) under the jurisdiction of the Prime Minister. Prior to 1979, there were several other energy policies, hence an overall energy policy was introduced known as National Energy Policies (NEP) 1979. The policy comprise of comprehensive guidelines on nation's energy objectives and strategies [63]. There are three main objectives to the 1979 energy policy, that is supply, utilization and environmental [64]. Later in 2000, Eight Malaysia Plan energy mix fuel policy were revised Four Fuel Policy to Five Fuel Policy by including RE [65]. In Five Fuel Policy, the target was set to 5% as for RE's contribution to energy mix. It is estimated that over the period of 5 year, it could save Malaysia RM 5 billion [66]. Electrification plans based on RE for 59,960 housing units was proposed in 9$^{th}$ Malaysian Plan (2006-2010) for rural areas specially Sabah and Sarawak region [67]. Besides, in 10$^{th}$ Malaysian Plan, the government suggested to increase the amount of electricity generation of RE from 1 % to 5.5 % of total energy by year 2010 [68]. Small Renewable Energy Power (SREP) program was initiated in May 2001 to motivate the utilization of small scale RER [69]. Concurrently, Special Committee on Renewable Energy (SCORE) under the supervision of Ministry of Energy, Green Technology and Water (KeTTA) were initiated to assist and coordinate SREP strategies in order to boost the development of RER [70]. The initial plan of SREP was to provide additional 500 MW via the installation of biomass, mini hydrokinetic turbine, biogas, solar photovoltaics and municipal solid waste from 2001 to 2005. Regrettably, wind energy was not included as one of the RER in SREP program. However, by the end of 2005, only 12 MW was able to be achieved. Nonetheless, the program was extended for another five years and reduced the target to 350 MW, yet only 61.7 MW was achieved [71]. Relentless, in 2010 National Renewable Energy Policy (NREP) were formulated, in which the principal objective of the policy is to attain 11 % from the total energy production via RER by 2020 [72]. Regardless of the initiatives done by the government, SREP program composed of complicated rules, regulation and legislation relative to project developers involved in SREP. Needless to say, this would be troublesome for companies to go through complicated procedures [71]. Government of Malaysia formed sustainable energy development authority (SEDA) to advise and assistance ministries and government entities on matters pertaining to RE policies law and to promote sustainable energy.

In 2011, Renewable Energy Act 2011 (Act 725) was formed, where the role of Act 725 is to stimulate the growth of RES, where the Act is composed of comprehensive implementation procedures and information pertaining to Feed in Tariffs mechanism [53],[54]. Although several countries managed to develop sturdy and stable FiT mechanism on wind energy, Malaysia do still struggle on developing a robust FiT framework for wind power technologies. Unreliable political policies such as unattractive energy tariff rates, hike in land prices and lake of financial support hinders the development of wind power sector [75]. Despite of the issues, based on the efforts and support given by several local parties it is evident that wind power harvesting is feasible in Malaysia. For an instance, realizing the prospect of wind energy during monsoon season, in 2005 Malaysia government installed 150 kW WT at Pulau Terumbu Layang Layang, Sabah [76]. Other than that, in 2007 Tenaga Nasional Berhad (TNB) with Kuala Terengganu state government collaborated in Pulau Perhentian project solar-wind-diesel project [77]. The hybrid system is comprised of two 100 kW WT generator (WTG) [78], 100 kW solar-photovoltaic and 100 kW diesel generator [79]. The hybrid system is backed up with 480 kWh battery to store the energy [80]. Moreover, eight 5-10 kW MWT were installed in rural areas of Sabah and Sarawak by Ministry of rural and regional development [81,82].

## 4. Wind assessment in Malaysia

In order to assess wind power characteristic at particular region, comprehensive investigation is required relative to the location's topology. Therefore, information on wind speed dominance namely period (season), location and direction is crucial in wind energy potential assessment [83]. In Southeast Asia, Malaysia is located at $2^0$ 30'N $112^0$ 30'E near to equator on Sunda shelf. Consequently, the wind flow pattern and behavior is governed by monsoon seasons namely Southwest (SW) (June-September) and Northeast (NE) (November-March). Masseran & Razali, (2016) [83] investigated the behavior of wind speed direction in Peninsular Malaysia during monsoon seasons via finite von Mises-Fisher density function numerical model (FVMF). The author stated that monsoon seasons occurs due to the influence of change in pressure and temperature. Since the pressure above landmass is low in which causes to warm up the air faster in comparison to air over ocean. Due to the difference in gradient, this causes the wind over ocean to blow to shore landmass. Studies showed that during SW monsoon season air temperature at Asian continent is high from due to low pressure and in Australian continent has low air temperature and high-pressure environment. Consequently, this phenomenon causes the wind to advect and convect to Penisular Malaysia and South China Sea (SCS). During SW monsoon season, wind speed can reach to approximately 7.71 m/s. Meanwhile, Northwest monsoon season influence the wind speed to reach high speed margin due to the geographical region and weather phenomena. Wind speed during NE monsoon seasons can reach up to 15.4 m/s [83–85]. This indicates that wind speed during NE is higher than SW [86]. Moreover, Ocean and land breeze can achieve 5-7.5 m/s on sunny days. Nonetheless, valley breeze can exceed 9 m/s, while mountain breeze can rise to 11 m/s. Monthly average wind speed of Mersing passes 3 m/s. Annual wind speed of Mersing and Kuala Terengganu is 3.29 m/s and 2.67 m/s respectively [87].

Researchers in Malaysia carried out wind speed assessment at Kota Kinabalu, Melaka, Labuan, Alor Star, Kuala Terengganu, Tawau, Mersing, Cameron Highlands, and Kuching. The investigation was carried out to collect wind speed data from local metrological station using an anemometer at the height of 10 m [88]. The researcher found out that Mersing and Kuala Terengganu has the wind speed potential in comparison to other geographical locations. Industrias Metalurgicas Pescarmona S.A (IMPSA) and TNB collaborated in 2009 to investigate the wind speed potential in Malaysia to generate electricity at utility scale. IMPSA research indicated that Langkawi, Mersing, Kuala Terengganu and Kota Kinabalu are potential locations. The wind roses data were recorded using 80 m high mast. IMPSA stated that Malaysia has the credibility to produce 500 - 200 MW of power from wind [49]. Wind speed data from 18 m high were collected at Megabang Telipot indicated that maximum monthly arithmetic mean (AM) wind speed is 6.54 m/s and minimum monthly AM is 2.04 m/s. High AM wind speed occurs during the NE monsoon. Refer to Figure 2 on monthly AM wind speed at Megabang Telipot and Perhentian Island.

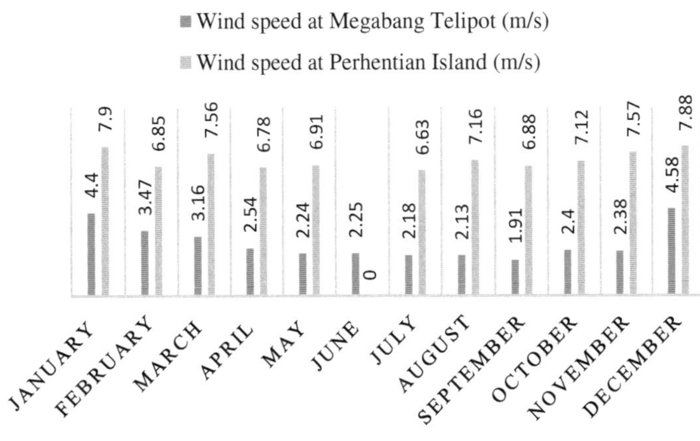


**Figure 2.** Monthly AM wind speed at Megabang Telipot and Perhentian Island. Source: Ho (2016) [49].

Between monsoon periods, at particular period selected area experience strong wind flow. Figure 3 illustrates real time Metrological terminal air report (METAR) wind speed on 18-April-2020. As manifested in the Figure 3, selected regions indicated wind speed of 6 knots, that is approximately 3.08 m/s. Strong and gustier windstorms take place at particular period between the two monsoon seasons (April-October) or otherwise known as inter-monsoon. Hence the windstorm exhibits potential treat, therefore being classified as metrological disaster subgroup [89]. Typhoons frequently occur across the Pacific West and move to westwards to Philippines. Consequently, Sabah and Sarawak experiences wind more than 10.3 m/s during the month from April to November [90]. Approximately, 6-7 tropical cyclones move across Philippines and SCS on yearly basis [91]. Refer to Table 2 for overview of wind speed potential and energy in Malaysia.

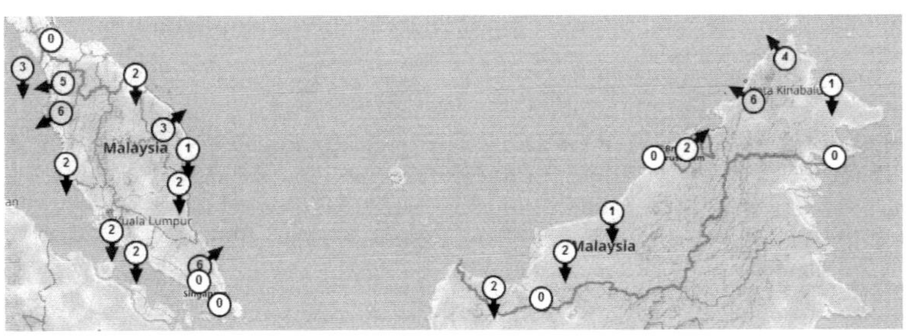


**Figure 3.** METAR wind speed mapping of Malaysia on 18-April-2020.

**Table 2.** Overview on wind speed characteristic and power in Malaysia.

| Authors | Overview | Wind speed characteristic |
|---|---|---|
| Siti Khadijah et al. [92] | Geographically Malaysia is situated at equatorial region in which the weather is influence by monsoon. Thus, making Malaysia a strategical venue for wind energy harvesting at selected location and time period. East coast (EC) regions namely Kelantan and Kuala Terengganu indicated great potential with high wind speed potential during monsoon season. Moreover, the author also added that wind energy offers great theoretical | East Coast state's average wind speed is 15.4 m/s during monsoon season. |

| | | |
|---|---|---|
| | potential as an alternative energy resource. | |
| Daut et al.[93] | On 9th March 2011, data on solar radiation and wind speed were collected over a period of 24 hours via weather station at Universiti Malaysia Perlis at 21.7 m height. Electrocorder were utilized to measure the output every per minute. | Over the period of 24 hours, it is found that the maximum and average wind speed is 7.61 m/s and 3.03 m/s respectively. |
| Daut et al. [94] | Since the 70's wind energy technology has evolved rapidly. Due to the advance in wind engineering, WTs are made available at reasonable cost. Where numerous new blade profile has been introduced with higher efficiency then the conventional blade, advanced control systems and turbine electronics has become easily accessible at low cost. | Maximum, arithmetic mean and minimum of daily wind speed (m/s) data of Perlis on month April 2011, are 3.2274, 1.7133 and 0.1992 respectively. |
| Masri et al. [95] | Two low speed wind speed were installed at USM Nibong Tebal. The turbines were placed at different heights in order to analyse the power extraction process relative to the wind speed at given height. At height 10 m, average monthly energy produced by 2 kW WT is approximately 163 Wh. The author utilized Davis instrument health weather link for data acquisition. | The maximum wind speed recorded at 10 m was 6 m/s is between 2 pm to 6 pm. However, at 30 m height the maximum wind speed is 18 m/s. |
| Masseran et al. [96] | In three-years period, 18.2 % of the time the wind speed exceeds 4 m/s at Mersing Station. Consequently, only selected period in a year, substantial amount of energy is harvested. | Although maximum wind speed (m/s) at Kuantan and Alor Setar is 11.8 and 11.5 respectively. However, the average wind speed did not exceed 3 m/s. Since the minimum cut in speed for WT power generation is 3 m/s. Therefore, it's not suitable to electricity generation. Nonetheless, wind speed lies between 6 -12 m/s, thus making it possible to harvest energy at selected period in a year. |
| Albani et al. [97] | Northern part of Kudat has potential for small scale wind farm based on wind speed data collected via Wind Atlas analysis and application program (WAsP). Based on the AM wind speed value, MWT installation has potential for energy harvesting due to cut-in speed of 4.0 m/s. | Based on interpolated wind speed data collected at Kudat, it is found that the peak wind speed value is 5.4 m/s and minimum value of 4.3 m/s. |
| Lawan et al. [98] | Due to Malaysia geographical properties, it is considered as light wind speed zone. Therefore, WTs need to be installed in the appropriate or at potential location with minimum cut in speed. Mini scale wind turbine would be suitable to electrical power generation. | Wind speed in Malaysia are inconsistent caused by variation in seasons and wind pattern. Wind speed potential in Malaysia ranges from 2.0-8.93 m/s. |
| Ahmadian et al. [99] | Wind energy density investigation were conducted on Butterworth to asses wind harvesting feasibility. Five commercially available WTs were utilized for the development of turbine power curve. The analysis conducted using HOMER energy simulator will assist the engineers in the selection of WT to be installed in Butterworth. | Studies shows that, the average annual wind speed of Butterworth is 2.87 m/s and frequency of occurrence more than 3 m/s is 20.2 %. Hence, WTs with cut in speed of 3 m/s would be suitable for this site. |

| | | |
|---|---|---|
| Akorede et al. [100] | In 2009 the Prime Minister launched the green technology policy which resulted in launching of FiT in 2011 to further encourage RES. However, wind power sector is yet to be enrolled and scrutinized. | At 70 m height, on month January Mersing recorded an average monthly wind speed of 6.95 m/s. However, at 10 m hub tower height, maximum average monthly wind speed resulted in 5.4 m/s. |
| Ahmadian et al. [101] | HOMER energy simulator was utilized to simulate the suggested WTs for the study in terms of energy production and capacity factor relative to wind speed data in Kuala Terengganu. Analysis relative to the performance WTs, it is found that LWT is not suitable for Kuala Terengganu site. | Kuala Terengganu were considered as potential site for wind power production due to its 50 % probabilistic occurrence of wind speed more than 3 m/s. |
| Ibrahim et al. [102] | Malaysia is considered as mesoscale wind where the wind flow is influenced by local wind movement. Since Malaysia is maritime country, wind flow pattern is governed by sea and land breeze. Mountain and valley breeze also have an effect on wind flow. | Southwest wind speed is below 7.5 m/s and steady wind speed of 5 – 10 m/s north-eastly wind. East coast states of Peninsular Malaysia can reach up to 15 m/s. Average valley breeze win speed is $V_m$ = 9 m/s and mountain breeze can reach to $V_m$ = 11 m/s. |
| Daryabor et al. [103] | Due to monsoon current circulation properties of SCS during SW, Penisular Malaysia experience strong western boundary current and eddies. Thus, making SCS an anti-cyclonic region. The eddies are generated due to temperature difference and baroclinic instabilities. Eddies formation are also caused by wind stress due to wind stratification resulted by wind shear. | Generally, SW wind speed is below 7 m/s. |
| Belhamadia et al. [104] | Wind speed data over the period of 2011-2012 from National Aeronautics and Space Administration (NASA) surface metrology database was verified against data from MMD for data reliability. Wind speed data of seven cities indicated that, highest annual average wind speed record was Mersing flowed by Kota Bharu, Miri and Kuala Terengganu. Wind turbines with cut in speed of 2.5 m/s are able to operate for 5084 hours in Mersing. | Annual AM wind speed in Mersing is 2.74 m/s with peak value of 3.97 m/s during mid of NE monsoon. |
| Deros et al., [105] | Data collected from RADAR-SAT SAR were validated against QUICKSCAT wind vector data. | Standard-2 RADAR-SAT SAR data indicated that Langkawi experiences an average yearly wind speed ranging from 3-5 m/s. |
| Malaysia Meteorological Department [106] | MMD yearly report stated that, in 2017 Malaysia experienced on average normal weather and climate condition. Nonetheless, Kedah experienced extreme weather resulting in drastic flooding. This is due to tropical depression which resulted in heavy rainfall and strong wind. | On 31 December 2017, Kota Bharu, Kelantan recorded the highest daily average wind speed of 7.1 m/s. |
| Ahmad Zaman et al. [107] | Malaysia territory is divided into regions; Peninsular Malaysia and Borneo facing SCS. Meanwhile, west cost of Peninsular Malaysia is face by straits of Malacca. Since the conditions of the oceans surrounding Malaysia is less turbulent, steadier wind with less wind shear. Thus, making it more prospective to offshore wind power energy. | From month April to September, Sabah and Sarawak experiences high wind speed reaching up to 10 m/s due to the occurrence of typhoon at Philippines. Wind speed data shows below 7 m/s during SW monsoon and high wind speed of 15 m/s during NE monsoon. |
| Kubota & | Studies were carried out to investigate the wind | The AM wind speed of the six towns is |

| | | |
|---|---|---|
| Ahmad [108] | speed potential of major towns in Malaysia. The major towns were Penang, Kuala Terengganu, Kuala Lumpur, Johor Bahru, Temerloh and Kota Bharu. The local climate is influenced by the land-sea breeze due to the continuous solar radiation throughout the year. Penang has higher sea breeze effect then Kuala Lumpur due to location of the island at south coast facing straits of Malacca. Unlikely, Kuala Lumpur is situated far from the costal line. | from 1-3 m/s from 1988-2002. The AM wind velocities between Kota Bharu and Kuala Terengganu town is 2.7 m/s during NE monsoon. Termerloh indicated the weakest wind speed strength due to its inland location. Termerloh AM velocities is from 0.2-0.3 m/s. |

## 5. Classification of wind turbine

The description on the intensity of the wind speed with regards to a region is indecisive and complicated. Since wind speed intensity is a qualitative description, which varies according to benchmarks defined by local standards and geographical location. Moreover, this predicament has pondered manufactures on WT classification for product marketing purpose. Hence, for the purpose of WT design classification manufactures adapted an international technical standard, documented and proposed by International Electrotechnical Commission (IEC) which is IEC 61400: 2019 [109–111]. Table 3 reports the parametric description of WT classes complying to IEC 61400-1:2019. As illustrated in Table 3, WT are classified in to three types namely type I (up to 10 m/s), type II (up to 8.5 m/s) and type III (up to 7.5 m/s) based on $V_m$. Meanwhile, $V_{ref}$ is used as reference for 10 min average wind speed with 50-year period. As for $V_{ref,,T}$, Selamat et al. (2013) [112] stated that the 50-year return period wind speed is 64.6 knots which is approximately 33.23 m/s. The prefixes A+, A, B and C is used to represent the turbulent intensity (*I*) of the wind. Since, actual or experimental data on turbulent parameters are unavailable, therefore value of *I* is defined according to class A which is *I* = 0.16. In regards to literature survey on wind speed assessment in Malaysia, the adapted wind speed range is from 5.5 m/s to 7.5 m/s. Hence, for this study the proposed design is defined as IIIA WT with regards to IEC 61400: 2019 standard. Moreover, IIIA class WT is considered as low wind speed WT.

**Table 3. Parametric description of WT classes.**

| Wind turbine classes | Description | I | II | III |
|---|---|---|---|---|
| $V_m$ (Max) | Annual average wind speed. | 10 m/s | 8.5 m/s | 7.5 m/s |
| $V_{ref}$ | 10 min average wind speed with 50-year return period. | 50 m/s | 42.5 m/s | 37.5 m/s |
| $V_{ref,T}$ | 10 min average wind speed with 50-year return period in typhoon condition | 57 m/s | 57 m/s | 57 m/s |
| A+ | Turbulent intensity | 0.18 | 0.18 | 0.18 |
| A | Turbulent intensity | 0.16 | 0.16 | 0.16 |
| B | Turbulent intensity | 0.14 | 0.14 | 0.14 |
| C | Turbulent intensity | 0.12 | 0.12 | 0.12 |

## 6. Discussion
Although Malaysian researchers presented an attractive form of investigation on wind speed potential assessment. However, the methodology presented by the authors are highly doubtful and questionable. This is due to the lack of crucial information on the methodology adapted for wind speed assessment. Conventionally, surface wind gust is measured by mast at standard height of 10 m above ground [113]. However, several authors did not present the height of the measurement were taken or did not comply to the standards. For an instance, several authors relied on wind speed data collected by Sopian et al. [88]. Meanwhile, Sopian et al. [88] gather the data from 10 MMD stations through Malaysia. Unfortunately, the height at which the data were collected varies by station. Although, Sopian et al. [88] corrected the data to standard height of 10 m via numerical mode. Factors such as location where the reading via MMD's anemometer were collected at different environment and condition namely near open sea, at airport and flat area. Data gather from MMD facilities are near or from airports that is considered bias due to influence of airplanes on anemometer reading and airports are located at low wind speed regions [49,80,87,114,115]. Masseran et al. [96] stated that location that faces the sea and with low surface roughness would be an ideal location for 10 m height wind speed data logging. Moreover, researcher also did not present information on instrumentations utilized for measurement such as hardware specification, sensor calibration parameters and techniques. Uncalibrated or inappropriate calibration of anemometer may establish bias data collection [116]. Nonetheless, this does not imply that the data provided by MMD is inadequate because the role of MMD is to gather wind speed data relative to weather forecasting and meteorological study not for WT wind density assessment. Since wind speed assessment techniques for WT power performance need to comply international standard IEC 61400-12-2 [117]. Hence, it is found that data collected by previous researchers on wind speed data are considered unreliable to be utilized for WT power density assessment and design. Notably, wind speed in Malaysia is seasonal based and highly inconsistent. The wind flow pattern and intensity experience by Malaysia are merely an aftereffect of weather change in neighbouring regions. Nevertheless, it is found that selected regions of east coast peninsular Malaysia and selected region of Sabah and Sarawak is potential for harvesting wind power due to substantial and adequate wind pattern behavior during monsoon season. Therefore, it is evident that geographical wind condition of Malaysia is appropriate for wind power devices as a form of RER in green energy sector of Malaysia. However, it is also found that fundamental design modification is required on conventional design of WT in terms of blade morphology and configuration in order to adapt to unsteady monsoon wind speed pattern regions.

## Acknowledgments
This research work was conducted under Grant number FRGS/1/2019/TK10/UMP/02/4. We would like to thank Ministry of Higher Education, Malaysia for making this research possible.

## References
[1] Cheng M-Y, Wu Y-F, Wu Y-W and Ndure S 2019 Fuzzy Bayesian schedule risk network for offshore wind turbine installation *Ocean Eng.* **188** 106238
[2] Arvizu D, Bruckner T, Christensen J, Devernay J-M, Faaij A, Fischedick M, Goldstein B, Hansen G, Huckerby J, Jager-Waldau A, Kadner S, Kammen D, Krey V, Kumar A, Lewis A, Lucon O, Matschoss P, Maurice L, Mitchell C, Moomaw W, Moreira J, Nadai A, Nilsson L J, Nyboer J, Rahman A, Sathaye J, Sawin J, Schaeffer R, Schei T, Schlomer S, Sims R, von Stechow C, Verbruggen A, Urama K, Wiser R, Yamba F and Zwickel T 2011 Summary for Policy Makers: Intergovernmental Panel on Climate Change Special Report Renewable Energy Sources (SRREN) (United States)
[3] Joselin Herbert G M, Iniyan S, Sreevalsan E and Rajapandian S 2007 A review of wind energy technologies *Renew. Sustain. Energy Rev.* **11** 1117–45
[4] Lund H 2007 Renewable energy strategies for sustainable development *Energy* **32** 912–9
[5] Ackermann T and Söder L 2002 An overview of wind energy-status 2002 *Renew. Sustain.*

*Energy Rev.* **6** 67–127
[6] Xia L, Ma Z, Kokogiannakis G, Wang Z and Wang S 2018 A model-based design optimization strategy for ground source heat pump systems with integrated photovoltaic thermal collectors *Appl. Energy* **214** 178–90
[7] Azad A K, Rasul M G, Islam R and Shishir I R 2015 Analysis of Wind Energy Prospect for Power Generation by Three Weibull Distribution Methods *Energy Procedia* **75** 722–7
[8] Suresh A and Rajakumar S 2019 Design of small horizontal axis wind turbine for low wind speed rural applications *Mater. Today Proc.*
[9] Wizelius T 2012 Design and Implementation of a Wind Power Project *Comprehensive Renewable Energy* ed A B T-C R E Sayigh (Oxford: Elsevier) pp 391–430
[10] Cognet V, Courrech du Pont S, Dobrev I, Massouh F and Thiria B 2017 Bioinspired turbine blades offer new perspectives for wind energy *Proc. R. Soc. A Math. Phys. Eng. Sci.* **473** 20160726
[11] Nayar C V, Islam S M, Dehbonei H, Tan K and Sharma H 2011 Power Electronics for Renewable Energy Sources *Alternative Energy in Power Electronics* ed M H B T-A E in P E Rashid (Boston: Elsevier) pp 1–79
[12] Qin N, Howell R, Durrani N, Hamada K and Smith T 2011 Unsteady Flow Simulation and Dynamic Stall Behaviour of Vertical Axis Wind Turbine Blades *Wind Eng.* **35** 511–27
[13] Peng Z, Chen B, Qin F G F, Jiang R, He Q, Tao S, Xiao H, Chen Y and Yang M 2019 Numerical simulation of aerodynamic performance of an airfoil combined lift and drag *Energy Procedia* **158** 6282–7
[14] Chong W T, Muzammil W K, Wong K H, Wang C T, Gwani M, Chu Y J and Poh S C 2017 Cross axis wind turbine: Pushing the limit of wind turbine technology with complementary design *Appl. Energy* **207** 78–95
[15] Eriksson S, Bernhoff H and Leijon M 2008 Evaluation of different turbine concepts for wind power *Renew. Sustain. Energy Rev.* **12** 1419–34
[16] Bianchini A, Balduzzi F, Bachant P, Ferrara G and Ferrari L 2017 Effectiveness of two-dimensional CFD simulations for Darrieus VAWTs: a combined numerical and experimental assessment *Energy Convers. Manag.* **136** 318–28
[17] Shahizare B, Nik-Ghazali N, Chong W T, Tabatabaeikia S, Izadyar N and Esmaeilzadeh A 2016 Novel investigation of the different Omni-direction-guide-vane angles effects on the urban vertical axis wind turbine output power via three-dimensional numerical simulation *Energy Convers. Manag.* **117** 206–17
[18] Wang Y, Sun X, Dong X, Zhu B, Huang D and Zheng Z 2016 Numerical investigation on aerodynamic performance of a novel vertical axis wind turbine with adaptive blades *Energy Convers. Manag.* **108** 275–86
[19] Joselin Herbert G M, Iniyan S, Sreevalsan E and Rajapandian S 2007 A review of wind energy technologies *Renew. Sustain. Energy Rev.* **11** 1117–45
[20] Akour S N, Al-Heymari M, Ahmed T and Khalil K A 2018 Experimental and theoretical investigation of micro wind turbine for low wind speed regions *Renew. Energy* **116** 215–23
[21] Riyanto, Pambudi N A, Febriyanto R, Wibowo K M, Setyawan N D, Wardani N S, Saw L H and Rudiyanto B 2019 The Performance of Shrouded Wind Turbine at Low Wind Speed Condition *Energy Procedia* **158** 260–5
[22] Pambudi N A, Pristiandaru D L, Basori, Wijayanto D S, Sriwardani N, Rahman N, Bugis H, Wahyudi B D, Sudibyo C, Karno M W and Subagsono 2017 Experimental Investigation of Wind Turbine Using Nozzle-lens at Low Wind Speed Condition *Energy Procedia* **105** 1063–9
[23] Butbul J, MacPhee D and Beyene A 2015 The impact of inertial forces on morphing wind turbine blade in vertical axis configuration *Energy Convers. Manag.* **91** 54–62
[24] Kosasih B and Saleh Hudin H 2016 Influence of inflow turbulence intensity on the performance of bare and diffuser-augmented micro wind turbine model *Renew. Energy* **87** 154–67

[25] Sotoudeh F, Kamali R and Mousavi S M 2019 Field tests and numerical modeling of INVELOX wind turbine application in low wind speed region *Energy* **181** 745–59
[26] Vaz J R P and Wood D H 2016 Aerodynamic optimization of the blades of diffuser-augmented wind turbines *Energy Convers. Manag.* **123** 35–45
[27] Ikeda T, Tanaka H, Yoshimura R, Noda R, Fujii T and Liu H 2018 A robust biomimetic blade design for micro wind turbines *Renew. Energy* **125** 155–65
[28] Li X, Zhang L, Song J, Bian F and Yang K 2019 Airfoil design for large horizontal axis wind turbines in low wind speed regions *Renew. Energy* **145** 2345–57
[29] Schubel P J and Crossley R J 2014 Wind turbine blade design *Wind Turbine Technol. Princ. Des.* 1–34
[30] Seidel C, Jayaram S, Kunkel L and Mackowski A 2017 Structural Analysis of Biologically Inspired Small Wind Turbine Blades *Int. J. Mech. Mater. Eng.* **12** 19
[31] Papaelias M and Márquez F P G 2020 Wind turbine inspection and condition monitoring *Non-Destructive Testing and Condition Monitoring Techniques for Renewable Energy Industrial Assets* ed M Papaelias, F P G Márquez and A B T-N-D T and C M T for R E I A Karyotakis (Boston: Elsevier) pp 19–29
[32] Kaldellis J K and Zafirakis D 2011 The wind energy (r)evolution: A short review of a long history *Renew. Energy* **36** 1887–901
[33] Wenehenubun F, Saputra A and Sutanto H 2015 An Experimental Study on the Performance of Savonius Wind Turbines Related With The Number Of Blades *Energy Procedia* **68** 297–304
[34] Zhao D, Han N, Goh E, Cater J and Reinecke A 2019 Chapter 1 - General introduction to wind turbines ed D Zhao, N Han, E Goh, J Cater and A B T-W T and A E H Reinecke (Academic Press) pp 1–20
[35] Ahire R, Gawli S, Geedh3 K, Gade A and Patil S S 2017 Review on Vertical Axis Wind Turbine *Int. J. Adv. Sci. Res. Eng.* **03** 5–9
[36] Moh Saad M M and Asmuin N 2014 Comparison of Horizontal Axis Wind Turbines and Vertical Axis Wind Turbines *IOSR J. Eng. www.iosrjen.org ISSN* **04** 2250–3021
[37] Ramadan A, Yousef K, Said M and Mohamed M H 2018 Shape optimization and experimental validation of a drag vertical axis wind turbine *Energy* **151** 839–53
[38] Yang W, Hou Y, Jia H, Liu B and Xiao R 2019 Lift-type and drag-type hydro turbine with vertical axis for power generation from water pipelines *Energy* **188** 116070
[39] Zheng C wei, Li C yin and Xu J jun 2019 Micro-scale classification of offshore wind energy resource ——A case study of the New Zealand *J. Clean. Prod.* **226** 133–41
[40] Möllerström E, Gipe P, Beurskens J and Ottermo F 2019 A historical review of vertical axis wind turbines rated 100 kW and above *Renew. Sustain. Energy Rev.* **105** 1–13
[41] Yoon S-H, Lim H-C and Kim D-K 2013 Study of several design parameters on multi-blade vertical axis wind turbine *Int. J. Precis. Eng. Manuf.* **14** 831–7
[42] Schubel P J and Crossley R J 2012 Wind Turbine Blade Design *Energies* **5** 3425–49
[43] Clark R N 2014 Wind Turbine Components and Descriptions *Small Wind* ed R N B T-S W Clark (Boston: Elsevier) pp 39–67
[44] Yoshida S 2016 In-plane self-excitation of two-bladed horizontal axis wind turbine rotors due to blade elasticity and gravity *Wind Eng.* **40** 343–59
[45] Bangga G, Hutomo G, Wiranegara R and Sasongko H 2017 Numerical study on a single bladed vertical axis wind turbine under dynamic stall *J. Mech. Sci. Technol.* **31** 261–7
[46] Misaran M S, Rahman M M, Muzammil W K and Ismail M A 2017 Assessment of C-Type Darrieus Wind Turbine Under Low Wind Speed Condition *IOP Conference Series: Materials Science and Engineering* vol 217 p 012019
[47] Gitano-Briggs H 2012 Low Speed Wind Turbine Design *Advances in Wind Power* (InTech)
[48] Mathew S 2006 *Wind Energy* (Berlin, Heidelberg: Springer Berlin Heidelberg)
[49] Ho L-W 2016 Wind energy in Malaysia: Past, present and future *Renew. Sustain. Energy Rev.* **53** 279–95

[50] Kåberger T 2018 Progress of renewable electricity replacing fossil fuels *Glob. Energy Interconnect.* **1** 48–52
[51] Administration U S E I 2013 United States Energy Information Administration (USEIA)
[52] Administration U S E I 2017 United States Energy Information Administration
[53] Anon 2012 *IEA International CCS Regulatory Network (Network): 4th CCS Regulatory Network Meeting Workshop Report 2012* (International Energy Agency (IEA) Paris, France.)
[54] United states.EIA 2020 Energy analysis of Malaysia
[55] Abdullah W S W, Osman M, Kadir M Z A A and Verayiah R 2019 The potential and status of renewable energy development in Malaysia *Energies* **12** 2437
[56] Cosi B and Kraja G 2012 A 100 % renewable energy system in the year 2050 : The case of Macedonia **48** 80–7
[57] Connolly D, Lund H, Mathiesen B V. and Leahy M 2011 The first step towards a 100% renewable energy-system for Ireland *Appl. Energy* **88** 502–7
[58] Mason I G, Page S C and Williamson A G 2010 A 100% renewable electricity generation system for New Zealand utilising hydro, wind, geothermal and biomass resources *Energy Policy* **38** 3973–84
[59] Duić N and Da Graça Carvalho M 2004 Increasing renewable energy sources in island energy supply: Case study Porto Santo *Renew. Sustain. Energy Rev.* **8** 383–99
[60] Krajačić G, Duić N and Carvalho M da G 2011 How to achieve a 100% RES electricity supply for Portugal? *Appl. Energy* **88** 508–17
[61] Elliston B, Diesendorf M and MacGill I 2012 Simulations of scenarios with 100% renewable electricity in the Australian National Electricity Market *Energy Policy* **45** 606–13
[62] Krajačić G, Duić N, Zmijarević Z, Mathiesen B V, Vučinić A A and Da Graa Carvalho M 2011 Planning for a 100% independent energy system based on smart energy storage for integration of renewables and CO2 emissions reduction *Appl. Therm. Eng.* **31** 2073–83
[63] Jalal T S and Bodger P 2009 National Energy Policies and the electricity sector in Malaysia *2009 3rd International Conference on Energy and Environment (ICEE)* (Malacca: IEEE) pp 385–92
[64] Lim C H and Elias S 2006 Current and Future Trends of Renewable Energy In Malaysia *Int'l J. Sustain. Trop. Des. Res. Pract.* **1** 32–9
[65] Economic Planning Unit 2001 *Eighth Malaysia Plan, 2001-2005* (Economic Planning Unit, Prime Minister's Department)
[66] Rahman Mohamed A and Lee K T 2006 Energy for sustainable development in Malaysia: Energy policy and alternative energy *Energy Policy* **34** 2388–97
[67] Economic Planning Unit 2006 *The Ninth Malaysia Plan 2006–2010*
[68] Economic Planning Unit 2010 *Tenth Malaysia Plan 2011-2015* (Economic Planning Unit, Prime Minister's Department)
[69] Shamsuddin A H 2012 Development of renewable energy in Malaysia strategic initiatives for carbon reduction in the power generation sector *Procedia Eng.* **49** 384–91
[70] Mekhilef S, Barimani M, Safari A and Salam Z 2014 Malaysia's renewable energy policies and programs with green aspects *Renew. Sustain. Energy Rev.* **40** 497–504
[71] Sovacool B K and Drupady I M 2011 Examining the small renewable energy power (SREP) program in Malaysia *Energy Policy* **39** 7244–56
[72] Kardooni R, Yusoff S B, Kari F B and Moeenizadeh L 2018 Public opinion on renewable energy technologies and climate change in Peninsular Malaysia *Renew. Energy* **116** 659–68
[73] Hashim H and Ho W S 2011 Renewable energy policies and initiatives for a sustainable energy future in Malaysia *Renew. Sustain. Energy Rev.* **15** 4780–7
[74] Khor C S and Lalchand G 2014 A review on sustainable power generation in Malaysia to 2030: Historical perspective, current assessment, and future strategies *Renew. Sustain. Energy Rev.* **29** 952–60
[75] Irena and Iea-etsap 2016 *Wind Power Technology Brief*

[76] Farriz M B, Azmi A N, Said N A M, Ahmad A and Baharin K A 2010 A study on the wind as a potential of renewable energy sources in Malaysia *ECTI-CON2010: The 2010 ECTI International Confernce on Electrical Engineering/Electronics, Computer, Telecommunications and Information Technology* pp 651–5
[77] Ibrahim H N 2011 Towards Renewable Energy *Tenaga Link Malaysia* **1** 1–20
[78] Ong H C, Mahlia T M I and Masjuki H H 2011 A review on energy scenario and sustainable energy in Malaysia *Renew. Sustain. Energy Rev.* **15** 639–47
[79] Karim O A, Zuhairuse M D, Nor Atikah H, Siti Nurhidayah A M, Mohd Azhar A R and Khairul Nizam A M 2009 The Development of Hybrid Integrated Renewable Energy System ( Wind and Solar ) for Sustainable Living at Perhentian *Eur. J. Soc. Sci.* **9** 557–63
[80] Albani A and Ibrahim M Z 2017 Wind energy potential and power law indexes assessment for selected near-coastal sites in Malaysia *Energies* **10**
[81] Borhanazad H, Mekhilef S, Saidur R and Boroumandjazi G 2013 Potential application of renewable energy for rural electrification in Malaysia *Renew. Energy* **59** 210–9
[82] Liang T S 2016 *Rural Electrification in East Malaysia - Achieving optimal power generation system and sustainability of rural electrification projects* (KTH School of Industrial Engineering and Management)
[83] Masseran N and Razali A M 2016 Modeling the wind direction behaviors during the monsoon seasons in Peninsular Malaysia *Renew. Sustain. Energy Rev.* **56** 1419–30
[84] Satari S Z, Zubairi Y Z, Hussin A G and Hassan S F 2015 Some Statistical Characteristic of Malaysian Wind Direction Recorded at Maximum Wind Speed: 1999-2008 *Sains Malaysiana* **44** 1521–30
[85] Anon 2020 Asia wind energy association: Malaysia wind market overview
[86] Saberi Z, Fudholi A and Sopian K 2019 Potential Evaluation of Wind Energy in Kuala Terengganu, Malaysia through Weibull Distribution Method *IOP Conf. Ser. Earth Environ. Sci.* **268**
[87] Wen T W, Palanichamy C and Ramasamy G 2019 Small wind turbines as partial solution for energy sustainability of Malaysia *Int. J. Energy Econ. Policy* **9** 257–66
[88] Sopian K, Othman M Y H and Wirsat A 1995 The wind energy potential of Malaysia *Renew. Energy* **6** 1005–16
[89] Sabah S H A, Ramli N I, Majid T A and Zaini S S 2020 Development of site-specific wind hazard map for peninsular Malaysia via spatial modeling *Energies* **13**
[90] Islam M R, Saidur R and Rahim N A 2011 Assessment of wind energy potentiality at Kudat and Labuan, Malaysia using Weibull distribution function *Energy* **36** 985–92
[91] Pacheco B M, Rosaria N E, Aquino R E R and Garciano L E O 2007 Historical review of wind speed maps in the Philippines for various purposes: toward further development and use as typhoon hazard maps *Proceedings of the 4th Civil Engineering Conference in the Asian Region (4CECAR)*
[92] Siti Khadijah N, Azami Z, Ahmad Mahir R, Mohd Said Z, Kamarulzaman I and Kamaruzzaman S 2009 Analyzing the East Coast Malaysia Wind Speed Data *Int. J. Energy Environ.* **3**
[93] Daut I, Irwanto M, Irwan Y M, Gomesh N, Rosnazri and Ahmad N S 2011 Potential of solar radiation and wind speed for photovoltaic and wind power hybrid generation in Perlis, Northern Malaysia *2011 5th Int. Power Eng. Optim. Conf. PEOCO 2011 - Progr. Abstr.* 148–53
[94] Daut I, Razliana A R N, Irwan Y M and Farhana Z 2012 A study on the wind as renewable energy in perlis, northern Malaysia *Energy Procedia* **18** 1428–33
[95] Masri S, Mohamad N and Hariri M H M 2012 Wind speed analysis at Nibong Tebal *2012 Int. Conf. Power Eng. Renew. Energy, ICPERE 2012* 1–4
[96] Masseran N, Razali A M, Ibrahim K and Wan Zin W Z 2012 Evaluating the wind speed persistence for several wind stations in Peninsular Malaysia *Energy* **37** 649–56

[97] Albani A, Ibrahim M Z, Yong K H and Terengganu K 2013 Wind Energy Investigation In Northern Part Of Kudat , Malaysia *Int. J. Eng. Appl. Sci.* **2** 14–22
[98] Lawan S, Abidin W, Chai W, Baharun a and Masri T 2013 Reviewing Wind Speed and Energy Distribution in Malaysia *Eur. Acad. Res.* **I** 2104–22
[99] Ahmadian J, Ghorbanian M J, Shams S, Goodarzvand F and Selvaraj J 2013 Butterworth, Malaysia wind energy potential assessment *CEAT 2013 - 2013 IEEE Conf. Clean Energy Technol.* 13–8
[100] Akorede M F, Mohd Rashid M I, Sulaiman M H, Mohamed N B and Ab Ghani S B 2013 Appraising the viability of wind energy conversion system in the Peninsular Malaysia *Energy Convers. Manag.* **76** 801–10
[101] Ahmadian J, Ghorbanian M J, Shams S, Goodarzvand F and Selvaraj J 2013 Kuala Terengganu, Malaysia wind energy assessment *CEAT 2013 - 2013 IEEE Conf. Clean Energy Technol.* 214–9
[102] Ibrahim M Z, Yong K H, Ismail M and Albani A 2014 Wind speed modeling for Malaysia *Int. J. Renew. Energy Res.* **4** 912–23
[103] Daryabor F, Tangang F and Juneng L 2014 Simulation of southwest monsoon current circulation and temperature in the east coast of peninsular Malaysia *Sains Malaysiana* **43** 389–98
[104] Belhamadia A, Mansor M and Younis M A 2014 A study on wind and solar energy potentials in Malaysia *Int. J. Renew. Energy Res.* **4** 1042–8
[105] Deros S N M, Asmat A and Mansor S 2015 Low wind speed direction extraction using wavelet transform technique *Int. Conf. Sp. Sci. Commun. Iconsp.* **2015-Septe** 358–61
[106] Meteorological Malaysia Department 2017 *Meteorological Malaysia Department: Annual Report 2017*
[107] Ahmad Zaman A A, Hashim F E and Yaakob O 2019 Satellite-Based Offshore Wind Energy Resource Mapping in Malaysia *J. Mar. Sci. Appl.* **18** 114–21
[108] Kubota T and Ahmad S 2006 Wind Environment Evaluation of Neighborhood Areas in Major Towns of Malaysia *J. Asian Archit. Build. Eng.* **5** 199–206
[109] Katsigiannis Y A, Stavrakakis G S and Pharconides C 2013 Effect of Wind Turbine Classes on the Electricity Production of Wind Farms in Cyprus Island *Conference Papers in Energy* vol 2013, ed C Sourkounis, A Poullikkas, P Demokritou and Y Al-Assaf (Hindawi Publishing Corporation) p 750958
[110] Zhang M H 2015 *Wind Resource Assessment and Micro-siting: Science and Engineering* (John Wiley & Sons)
[111] International Electrotechnical Commission 2019 *International Standard:IEC 61400-1*
[112] Selamat I M, Liew M S, Nasir M and John K V 2013 Extreme Value Analysis and Joint Density of Metocean Loads for Malaysian Water *Malaysian J. Civ. Eng.* **52** 40–52
[113] Suomi I, Vihma T, Gryning S-E and Fortelius C 2013 Wind-gust parametrizations at heights relevant for wind energy: a study based on mast observations *Q. J. R. Meteorol. Soc.* **139** 1298–310
[114] Kadir E A, Miskon M T, Rashid N A and Yunus M Y 2018 A Study of Vertical Wind Turbine for Application in Low Wind Speed Condition in UiTM Terengganu, Malaysia *J. Power Energy Eng.* **06** 38–48
[115] Uti M N, Din A H M and Omar A H 2017 Reliability of wind speed data from satellite altimeter to support wind turbine energy *ISPRS - Int. Arch. Photogramm. Remote Sens. Spat. Inf. Sci.* **XLII-4/W5** 215–24
[116] Demurtas G, Pedersen T F and Zahle F 2016 Calibration of a spinner anemometer for wind speed measurements *Wind Energy* **19** 2003–21
[117] International Electrotechnical Commission 2005 *Iec 61400-12-1* vol 2005

# The effect of temperature on low temperature vacuum drying with induced nucleate boiling for stingless bees honey

**N F Halim[1] , M F Basrawi[1,*] , N A M Azman[2] , S A Sulaiman[3] and S B A Razak[4]**

[1]Meliponini Engineering Laboratory (MePEL), Faculty of Mechanical and Automotive Engineering Technology, Universiti Malaysia Pahang, Pekan, Pahang, Malaysia
[2]Faculty of Chemical & Natural Resources Engineering, Universiti Malaysia Pahang, Kuantan, Malaysia
[3]Department of Mechanical Engineering, Universiti Teknologi PETRONAS, Bandar Seri Iskandar, 32610, Perak, Malaysia
[4]Apis and Meliponine Special Interest Group, School of Fishery and Food Sciences, Universiti Malaysia Terengganu

*E-mail: mfirdausb@ump.edu.my

**Abstract.** Low Temperature Vacuum Drying (LTVD) with induced nucleate boiling is a new method that has been invented to reduce the water content of honey at a lower temperature without damaging the nutrient content of honey. The objective of this study is to investigate the effect of temperature on the dewatering rate by this LTVD method. The honey sample was placed in a pressure vessel and the air from the vessel was removed to create a vacuum condition. This experiment was conducted by using three different temperature which was 40 °C, 45 °C and 50 °C. The honey sample was dewatered for 5 minutes and each condition was repeated three times. The water content before and after the experiment was measured by a digital refractometer. Final water content decrease from 26.5 % to 25.1 % from temperature 40 °C to 50 °C. The dewatering rate increase from 0.14 %/min to 0.28 %/min when the temperature increase from 40 °C to 50 °C. The highest dewatering rate was obtained at 50 °C. However, considering the quality conservation of honey, it is suggested to apply a temperature of 45 °C to get a high dewatering rate while considering the quality of honey. It could be concluded that the temperature effect the final water content and dewatering rate of honey.

**Keywords:** LTVD; Nucleate Boiling; Dewatering; Stingless Bees Honey.

## 1. Introduction

Honey is a primary product of bees which is generally known as a natural remedy that obtains from flower nectar or honeydew. Stingless bees (Meliponini) is one of the bee species that can produce honey and play an important role in flower pollination. Stingless bees produce three main products such as honey, bee bread and propolis. Each product has its advantages that beneficial for its users. Stingless bees honey (SBH) is known for its high nutritious and therapeutic value. SBH also has stronger antioxidant properties compared to other honey types [1]. In recent years, SBH has been introduced to the public in Malaysia as a nutritious superfood rich with health benefits. The specialty of SBH is it has a unique taste which is a little sour and has a unique aroma that makes it different from other honey. Nowadays, there was a lot of improvement in the development of meliponiculture (stingless bees keeping) technology, this includes bee bread dryer, propolis extractor, honey extractor and also honey dryer.

Water content is one of the parameters that determine the quality of honey. Honey with high water content tends to ferment easily. SBH has been reported contain higher water content, acidity, ash and HMF but low total sugar compared with honey from honey bees (Apis Mellifera) [2]. Honey spoilage is the main problem in storing SBH as it has high water content. To overcome this issue a drying process is required to reduce water content while preserving honey's nutritional content. Department of Standards Malaysia (2017) suggested that the water content of SBH should be kept below 22%. Since honey is known as heat sensitive material, the process should consume low temperature to drying honey. High dewatering temperatures could damage the nutrition content of honey. As reported by Tosi et al., heating led to a reduction in honey quality as the diastase activity decreases and hydroxymethylfurfural (HMF) content increases [3].

In order to lower the temperature for the drying process, the process should consume low pressure or vacuum condition to lower the boiling point of the material. This technique is well recognized as low temperature vacuum drying (LTVD). LTVD is commonly used to reduce the water content from heat sensitive materials such as food, pharmaceuticals and biological products [4]. There are various types of LTVD available including microwave vacuum drying, freeze-drying, vacuum evaporation, vacuum drying and spray drying. Microwave vacuum drying was able to enhance the reduction of water content honey, but the radiation emitted from the microwave can damage the nutrient content of honey [5]. While freeze-drying works by freezing the material and reduce the pressure surrounding it to sublimate the frozen water in the material [4]. However, this method is costly as it consumes a large amount of energy to freeze and dehydrated material [6]. Vacuum evaporation invented by Shinya et al., was able to dry the honey, but this method is very costly to operate [7]. Meanwhile, vacuum drying developed by Tabouret (1977) manages to reduce about 4 % of the water content of honey within 90 minutes at 68 °C [8]. However, the operating temperature was considered too high for drying honey. There is also spray-drying combine with LTVD, but this method can only producing a dry powder from a liquid and operation temperature was exceed the limit for drying honey [9]. Therefore, an LTVD with induced nucleate boiling has been developed to meet a requirement for drying honey without jeopardizing its nutritional content [10].

Nucleate boiling is a phase in which bubbles nucleate, grow and depart from the heated surface. This type of boiling is the most desirable for the chemical process due to the high heat rates. The performance of LTVD with induced nucleate boiling could be affected by temperature. As reported by Kretavičius et al., heating honey below 50 °C does not affect the enzyme activity in honey [11]. Therefore, to prevent degradation of the honey quality, it should be heated at temperatures below 50 °C. From the literature review, there is no study reported on the effect of temperature at range 40 °C, 45 °C and 50 °C on LTVD with induced nucleate boiling dewatering honey. Therefore, the objective of this study is to investigate the effect of temperature on the dewatering rate of SBH.

## 2. Materials and methods

SBH sample used in this experiment was obtained from a farm at Padang Terap, Kedah, Malaysia and stored in bottles. Figure 1 shows an experimental setup that has been developed to dewatering 200 g of honey. 200 g of honey sample was placed in a pressure vessel that was connected with a vacuum pump to create a vacuum condition. A pressure gauge was used to monitor the pressure inside the vessel. A stainless steel tube was used as a heating element that transfers heat to dry the honey sample inside the vessel. For heating the honey sample, water inside the water tank was heated until obtaining the required temperature and the water pump was used to transfer hot water from the water tank to stainless steel inside the vessel then flow inside the water tank again. Temperature inlet, outlet and honey were monitored by using thermocouples. Three different temperatures were conducted which were 40 °C, 45 °C and 50 °C to investigate the effect of temperature on the dewatering rate of honey. The pressure was set at -96 kPa for every experiment. The initial and final water content of honey was measured by using the digital refractometer ATAGO. The honey sample was dewatered for 5 minutes and the process was repeated 3 times for each temperature. The dewatering rate was determined by using Equation 1:

$$\eta_D = \frac{\Delta W_h}{t} \tag{1}$$

where $\eta_D$ is dewatering rate [%/min], $\Delta W_h$ is the changes in water content [%] and t is the time take for the dewatering process [min].

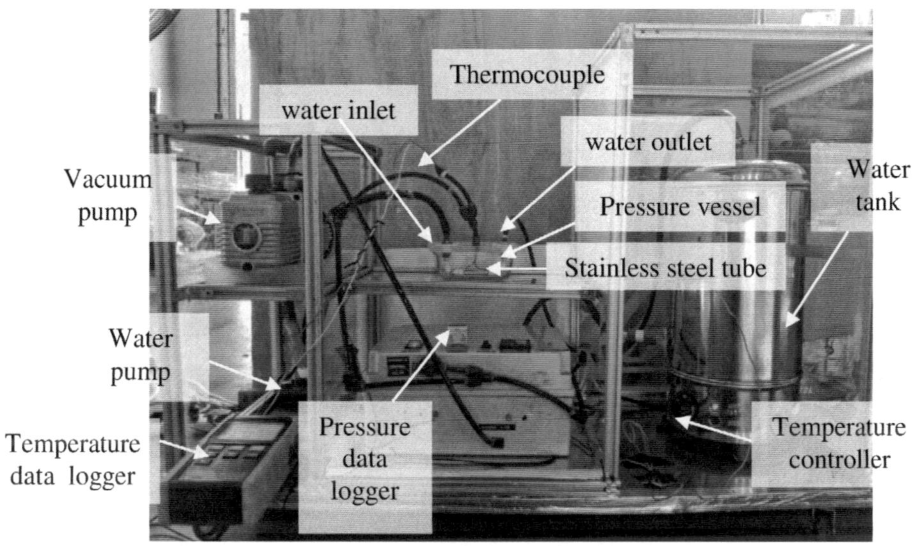


**Figure 1.** Experiment setup

## 3. Result and discussion

Figure 2 shows the graph of the average water content of SBH for each temperature after the dewatering process for 5 minutes. While Figure 3 shows the dewatering rate for each dewatering temperature. It shows the highest reduction of water content was at 50 °C which was reduced from 26.5 % to 25.1 %. As the result at temperature 50 °C could obtain the highest dewatering rate than others which was 0.28 %/min. However, it is not suitable for honey to be heated at 50 °C because it could overshoot and exceed the temperature limit for drying honey that can damage the nutrient content of honey. The lowest reduction of water content and dewatering rate was at temperature 40 °C which was only reduced by 0.7 % from initial water content and the dewatering rate was 0.14 %/min. While dewatering at temperature 45 °C was able to reduce from 26.5 % to 25.8 % water content after 5 minutes and the dewatering rate was around 0.16 %/min. The same trend also was observed by Ramli et al., higher dewatering temperature resulted in a much faster water removal rate [12]. The water content decreases faster at higher temperatures due to the increases in temperature difference between the sample and the heating element [13].

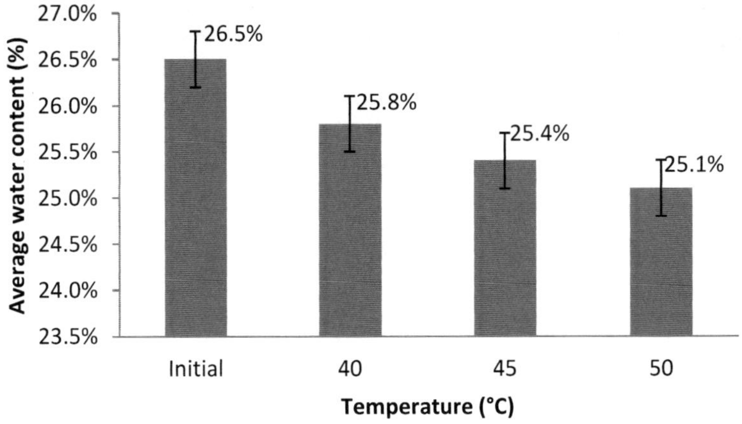


**Figure 2.**The average water content of SBH after the dewatering process

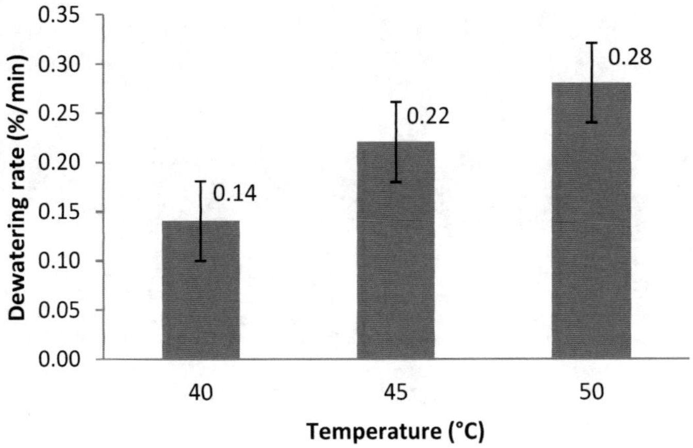


**Figure 3.** Dewatering rate for each dewatering temperature

From the result, the dewatering rate at 45 °C in this experiment was slightly higher from the previous study by Ramli et al., [12] which also uses LTVD with induce nucleate boiling method to reduce the water content of honey below 20 %. The dewatering rate at 45 °C from the previous study was around 0.2 %/min while in this experiment was 0.22 %/min. This different result due to the initial water content of honey which was the initial water content used in the previous study was around 26 % while in this experiment was around 26.5 % which was also slightly higher than the previous study. Dewatering rate higher at higher water content due to the evaporation of the remaining free water content [14]. At lower water content higher energy was needed to evaporate the bound water make the dewatering rate decrease.

Figure 4 shows the dewatering rate from the present study and other methods. The dehumidifier invented by Paysen shows the lowest dewatering rate which is approximately 0.00033 %/min. This method can reduce 0.6 % water content of 57 tonnes of honey in 30 hours [15]. Tabouret reduces the water content of honey with low pressure vacuum and the dewatering rate was about 0.044 %/min at 68 °C [8]. However, the dewatering temperature used is considered too high for dewatering honey.

IOP Conf. Series: Materials Science and Engineering 1078 (2021) 012018 doi:10.1088/1757-899X/1078/1/012018

Meanwhile LTVD with induced nucleate boiling by Ramli et al., able to reduce the water content at the rate of 0.2 %/min at 45 °C which is 5 times faster than the dewatering rate achieved by Tabouret [12]. Meanwhile, the dewatering rate from the present study with the same dewatering temperatures was around 0.22 %/min which is showed an increment due to initial water content. Microwave vacuum drying shows the fastest drying method compare to others as the dewatering rate of this method is around 2.56 %/min [5]. However, this method cannot be considered as the best method because the radiation emitted from the microwave has the potential to damage the nutrient and quality content of the honey [16]. Thus, LTVD with induced nucleate is the better method for dewatering honey without causing damage to the nutrition content of honey compared to other methods.

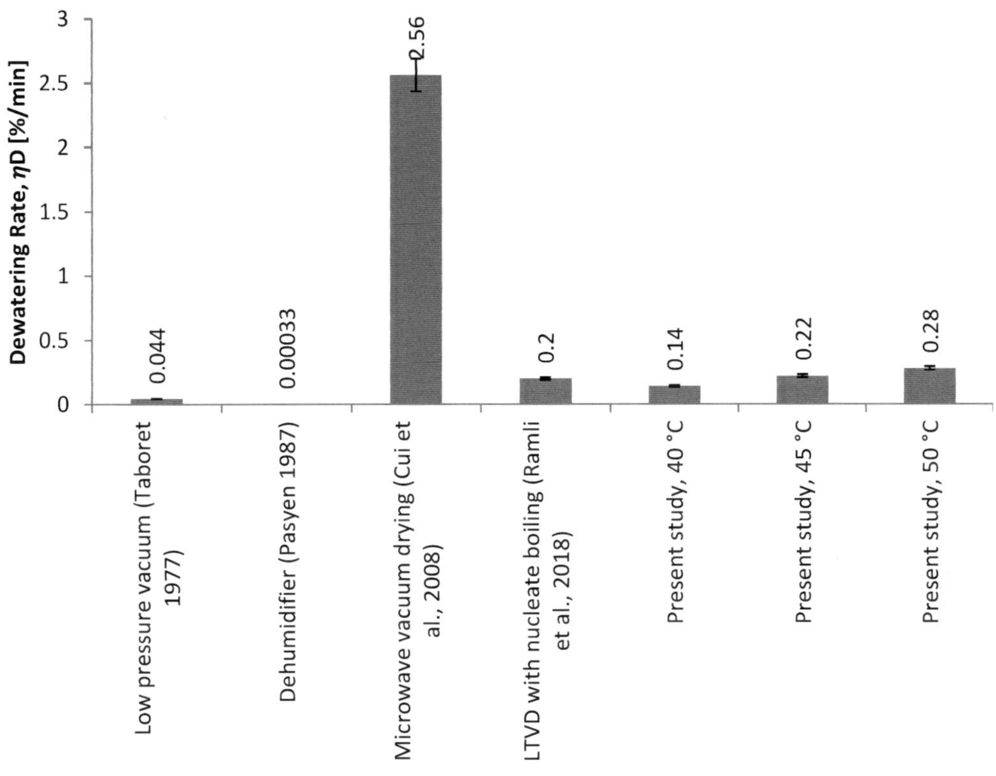


**Figure 4** Comparison of the dewatering rate of the present study with other methods

## 4. Conclusion

The performance of LTVD with induced nucleate boiling increases as the dewatering temperature increase. From the range of 40 °C to 50 °C, the final water content of honey was decreased. Thus the dewatering rate increase as the water content decrease. The highest dewatering rate was at temperature 50 °C, but this temperature is not suitable for heating honey. Dewatering honey at 45 °C was highly recommended to minimize the deterioration of honey quality during the dewatering process.


## Acknowledgments

Authors acknowledge the financial support by the Ministry of Education Malaysia for FRGS/1 /2018/TK03/UMP/02/25 (RDU190192).

**References**

[1] Ramli N Z, Chin K Y, Zarkasi K A and Ahmad F 2019 The beneficial effects of stingless bee honey from Heterotrigona itama against metabolic changes in rats fed with high-carbohydrate and high-fat diet *Int. J. Environ. Res. Public Health* **16** 1–17

[2] Chuttong B, Chanbang Y, Sringarm K and Burgett M 2016 Physicochemical profiles of stingless bee ( Apidae : Meliponini ) honey from South East Asia ( Thailand ) *FOOD Chem.* **192** 149–55

[3] Tosi E A, Ré E, Lucero H and Bulacio L 2004 Effect of honey high-temperature short-time heating on parameters related to quality, crystallisation phenomena and fungal inhibition *LWT - Food Sci. Technol.* **37** 669–78

[4] Parikh D 2015 Vacuum Drying : Basics and Application

[5] Cui Z W, Sun L J, Chen W and Sun D W 2008 Preparation of dry honey by microwave-vacuum drying *J. Food Eng.* **84** 582–90

[6] Bae S H, Nam J H, Song C S and Kim C J 2010 A numerical model for freeze drying processes with infrared radiation heating *Numer. Heat Transf. Part A Appl.* **58** 333–55

[7] Aozuka R 1983 United States Patent (19)

[8] T.Tabouret 1977 Vacuum drying of honey *Apiacta*

[9] Cuevas-Glory L F, Pino J A, Sosa-Moguel O, Sauri-Duch E and Bringas-Lantigua M 2017 Optimization of the Spray-Drying Process for Developing Stingless Bee Honey Powder *Int. J. Food Eng.* **13**

[10] Ramli A S, Basrawi F, Mohamad D, Daing N, Yusof H, Ibrahim T K, Mustafa Z and Sulaiman S A 2017 A new dewatering technique for stingless bees honey **03014**

[11] Kretavičius J, Račys J, Čeksteryte V and Kurtinaitiene B 2010 Inactivation of glucose oxidase during heat-treatment de-crystallization of honey *Zemdirbyste* **97** 115–22

[12] Ramli A S, Basrawi F, Yusof M H Bin, Oumer A N, Johari N A, Muhamad A, Mamat M R, Habib K and Ibrahim T K 2019 Experimental analysis on a novel low-temperature vacuum drying with induced nucleation technique for dewatering stingless bees honey *Dry. Technol.* **37** 149–55

[13] Devahastin S, Suvarnakuta P, Soponronnarit S and Mujumdar A S 2004 A comparative study of low-pressure superheated steam and vacuum drying of a heat-sensitive material *Dry. Technol.* **22** 1845–67

[14] Vaxelaire J and Puiggali J R 2002 Analysis of the drying of residual sludge: From the experiment to the simulation of a belt dryer *Dry. Technol.* **20** 989–1008

[15] Paysen J 1987 Wet honey--a method for drying honey on a commercial scale *Am. bee J.*

[16] Kowalski S 2013 Changes of antioxidant activity and formation of 5-hydroxymethylfurfural in honey during thermal and microwave processing **141** 1378–82

# Suction effect on an unsteady Casson hybrid nanofluid film past a stretching sheet with heat transfer analysis

**Nur Ilyana Kamis[1,*], Md Faisal Md Basir[1], Sharidan Shafie[1], Taufiq Khairi Ahmad Khairuddin[1] and Lim Yeou Jiann[1]**

[1]Department of Mathematical Sciences, Faculty of Science, Universiti Teknologi Malaysia, 81310 UTM Johor Bahru, Malaysia

* Corresponding author's email: mfaisalmbasir@utm.my

**Abstract.** This paper studies the heat transfer in the blood fluid-based copper and alumina nanoparticles over an unsteady permeable stretching sheet. The model is governed by the governing equations consist of a series of ODEs that are reduced from PDEs by implementing the similarity transformations subjected to mixed boundary conditions. The results of the transformed equations has been obtained by using the Keller-box method in MATLAB software. This paper focuses on the characteristics of thin-film flow and heat transfer through the governing parameters; unsteadiness parameter, nanoparticles volume fraction, Casson parameter, and intensity of suction. From this study, it is observed that the behavior of both fields for nanofluid is lower than hybrid nanofluid under the suction effect. It is noticed that enhance the physical parameters increase the velocity field of the fluid. Further, increase the physical parameter also deteriorate the temperature field except for nanoparticles volume fraction.

**Keywords:** Thin-film; Nanoparticles; Stretching Sheet; Keller-box Method.

## 1. Introduction

The development of technologies nowadays attracted researchers to explore and analyze the issues on heat transfer. The passage of particles from different temperatures is regarded as the transfer of heat. As far as the thermodynamic system is concerned, heat transfer is the flow of heat across the device boundary due to the difference in temperature between the system and the environment. The heat transfer in the system can also be performed at different points inside the system because of temperature differences. Numerous researchers studied and explored the issues of heat transfer in theoretical and numerical works included the following literature: Anitha et. al. [1], Bai et. al. [2], Hashimoto, Kurazono and Yamauchi [3], Hussein et. al. [4], Izadi et. al. [5] and Waini, Ishak and Pop [6].

The presence of nanoparticles in the fluid additionally influenced the heat transfer of fluid. The mixes of nanoparticles known as hybrid nanoparticles will be improved the heat transfer contrasted with regular fluids [7]. Hashimoto et al. [4] and Ali [8] discussed the presence of hybrid nanofluid that would affect the heat transfer. The perfect combination of various nanoparticle properties has demonstrated an outstanding heat transfer coefficient enhancement with a low-pressure decrease limit. The host fluid with the hybrid nanoparticles studied by Yazid et al [9] increased the heat flux of the fluid. The numerical studied of natural or free convection heat embedded with hybrid nanoparticles in a nanofluid also has been explored by Bai [5]. Besides that, the increase of the material industrial productions attracted the researcher to explore and study the similar phenomena in the non-Newtonian fluids flow since the fluids are widely used in that industry. Firstly, the numerical behavior of Casson fluid due to a stretching sheet has been done by Abas et al. [10]. The empirical and theoretical analysis of the Casson fluid with the inclusion of nanoparticles used in the industry has been cross-examined by several scientists, including [11-14]. Rawi et. al.[12] claimed that the enhancement of the temperature profile in Casson fluid is due to an increase in the volume of composite in the fluid. Furthermore, second-grade fluid type, also known as viscoelastic fluid, is contained in silicone fluids where these fluids show both viscous and elastic properties. Goyal and Bhargava [15] studied the numerical method which is the FEM to solve the viscoelastic solution of flow.

Recently, coating technologies such as wire and fiber coatings are one of the thin-film stretch layer used in industrial applications. The best nature in the process's cycle is controlled by changing the amount of heat in a certain fluid when a sheet is stretched [16]. From the literature review, Carragher and Crane [17], Crane [18] and Sakiadis [19] imposed the theoretical technique to solve the issues of heat transfer in the same geometry. Motivated by the published paper [17-19], the first problem of thin-film flow has been explored by Wang [20]. This author applied the HAM to solve the governing equations analytically. Andersson et. al [21] improved the work of [20] and introduced the dimensionless for temperature.

For the first time, the power-law fluid in a thin film along the stretching sheet was examined in 1996 by Andersson et al. [21]. The heat transfer analysis of Carreau fluid in the same geometry was observed by Tsehla et. al [22], Nagathran et. al. [23] and Khan et. al. [24]. The MHD and thermal radiation effects on thin film in Maxwell nanofluid were observed by Fareesa et. al. [25] and Hameed et. al. [26] joined the MHD and electric field effects on the same fluid. Ali et al [27] investigated the numerical method, fourth-order Runge-Kutta (RK4) of MHD Casson fluid in a liquid film. Rehman et. al. [16] explored the both analytical (HAM) method and numerical method (bvp4c) on the Casson fluid together with slip and suction/injection effects in uniform film thickness over a stretching sheet. Both methods show a good agreement through the calculation of the total average square residual error with the help of MATHEMATICA. The combinations of several effects on MHD Casson film fluid flow have been cross-examined by Vijaya et. al. [28]. Next, the slip parameter on Casson fluid flow affected the film thickness as claimed by Mahmoud and Megahed [29]. Ray et. al [30] added the nanoparticles to the magneto-bioconvection in a similar type of fluid flow. Aziz and Afify [31] investigated the MHD Casson liquid film with the variable of thermal conductivity and thermal radiation. The comparisons of the numerical results with the analytical result from a published paper show an excellent agreement. Megahed [32] investigated the variable of heat flux and viscous dissipation on the fluid flow.

Considering the above broad examinations, it is apparent that the blend of mixture nanoparticles and suction impact in a uniform thin film thickness past an unsteady stretching sheet has not been considered in the writing. This is the fundamental focal point of the current work. The PDEs form of governing equations will be transformed into the simple set of ODEs by employing the similarity transformations from the published paper and then deal with the Keller-box method.

## 2. Method and materials

This study considered an unsteady 2-D incompressible hybrid nanofluid of Casson flow. The film has a uniform thickness $h(t)$ as displays in Figure 1 [20]. The horizontal plane is placed at $x-$axis from the narrow slit. $y-$axis is perpendicular to $x-$axis. The plane is stretching with velocity $U_w(x,t)=bx/(1-\alpha t)$. The constants appearing in the stretched rate are $b>0$ and $\alpha>0$ where $b$ is an initial stretching rate. The fluid is being sucked, $V_w>0$ and injected, $V_w<0$ at the boundary layer with velocity $V_w=(V_w)_0/(1-\alpha t)^{\frac{1}{2}}$ where $(V_w)_0$ is an initial concentration of the reactant. The slit temperature, $T_0$ and reference temperature, $T_{ref}$ at $y=0$ can be defined as $T_w(x,t)=T_0+T_{ref}\left[\dfrac{bx^2}{2\nu}\right](1-\alpha t)^{-1.5}$.

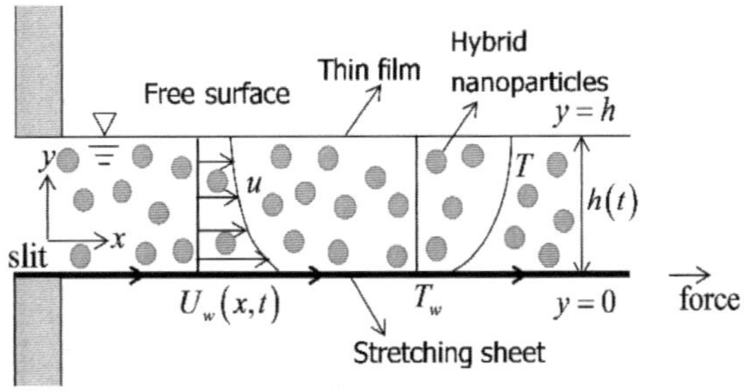


Figure 1: Physical diagram

The rheological logical equation of state for an isotropic and incompressible flow of a Casson fluid can be defined as [31].

$$\tau_{ij}=\begin{cases}2\left(\mu_B+\dfrac{p_y}{\sqrt{2\pi}}\right)e_{ij},\pi>\pi_e\\ 2\left(\mu_B+\dfrac{p_y}{\sqrt{2\pi_e}}\right)e_{ij},\pi<\pi_e\end{cases}$$

Where $\mu_B$ is the plastic dynamic viscosity on the Non-newtonian, $P_y$ is the yield stress of fluid, $\pi$ is the product of the component of deformation rate with itself, namely, $\pi=e_{ij}e_{ij},e_{ij}$ is the $(i,j)$ component of the deformation rate and $\pi_e$ is the critical values of $\pi$ based on the non-Newtonian model. The governing equations can be written as [33]

$$u_x+u_y=0 \tag{1}$$

$$\rho_{hnf}\left(u_t+uu_x+vu_y\right)=\mu_{hnf}\left[1+\frac{1}{\beta}\right]u_{yy} \tag{2}$$

$$\left(\rho c_p\right)_{hnf}\left(T_t+uT_x+vT_y\right)=k_{hnf}T_{yy} \tag{3}$$

Where velocity segments along $x$ and $y$ direction is $(u,v)$. $k_{hnf}$ represents the conduction by the heat of nanodispersion, the density for the hybrid nanofluids is $\rho_{hnf}$, $\left(c_p\right)_{hnf}$ indicates the energy capacity of nanodispersion and $\mu_{hnf}$ demonstrates an effective dynamic viscosity and $\beta$ is Casson fluid parameter. This is subjected to

$$y=0;u=U,v=V_w,T=T_s \tag{4}$$

$$y=h;u_y=T_y=0,v=\frac{dh}{dt} \tag{5}$$

where $U$ and $T_s$ denote the velocity and temperature. Introducing the following transformations vector $f(\eta)$ and $\theta(\eta)$ [20],

$$\psi = \left[ vb(1-\alpha t)^{-1} \right]^{\frac{1}{2}} x\delta f(\eta)$$
$$T = T_0 - T_{ref}\left(\frac{bx^2}{2v}\right)(1-\alpha t)^{-\frac{3}{2}}\theta(\eta) \quad (6)$$
$$\eta = \left(\frac{b}{v}\right)^{\frac{1}{2}}(1-\alpha t)^{-\frac{1}{2}}\delta^{-1} y$$

Where $\eta$ is the similarity variable, $\delta$ is an undefined constant which denotes the film's dimensionless thickness and stream function, $\psi(x,y,t)$. The first Equation (6) automatically satisfies the mass conservation equation (1). At the free surface, $\eta = 1$ and from the last equation (6), when $y = h(t)$ we have:

$$\delta = \left(\frac{b}{v}\right)^{\frac{1}{2}}(1-\alpha t)^{-\frac{1}{2}} h(t) \quad (7)$$

Which gives

$$\frac{dh}{dt} = -\frac{1}{2}\alpha\delta\left(\frac{v}{b}\right)^{\frac{1}{2}}(1-\alpha t)^{-\frac{1}{2}} \quad (8)$$

The equations (1), (2), and (3) with the constraints (4) and (5) can be written in the new variables $f(\eta)$, $\theta(\eta)$ and $\eta$ as a series of ODEs as follows:

$$\left(\frac{A_1}{A_2}\right)\left(1+\frac{1}{\beta}\right)(f'''(\eta)) + \gamma\left[(f(\eta)f''(\eta)) - (f'(\eta))^2 - S\left(f'(\eta) + \frac{1}{2}\eta f''(\eta)\right)\right] = 0 \quad (9)$$
$$\left(\frac{A_3}{A_4}\right)\theta''(\eta) + \Pr\gamma\left[-S\left(\frac{3}{2}\theta(\eta) + \frac{1}{2}\eta\theta'(\eta)\right) - 2(f'(\eta)\theta(\eta)) + (f(\eta)\theta'(\eta))\right] = 0 \quad (10)$$
$$f'(0) = 1, f(0) = w, f(1) = 0.5S, f''(1) = 0 \quad (11)$$
$$\theta'(1) = 0, \theta(0) = 1 \quad (12)$$

Where $A(i = 1,2,3,4)$ is the hybrid nanofluid constants, $\gamma = \delta^2$ is thin film thickness, unsteadiness of the dimensionless measure is $S = \frac{\alpha}{b}$, Prandtl number, $\Pr = \frac{(\mu c_p)_{bf}}{k_{bf}}$ and suction/injection parameter is $w = -\frac{(V_w)_0}{\beta\sqrt{vb}}$. The hybrid nanofluid constants are demonstrated in [7] and the characteristics of the hybrid nanofluids, copper and alumina as well as blood as base fluids are described in [34] and [7].

## 3. Numerical procedure

The boundary value problem in the form of ODEs, Eqs (9) and (10) with related boundary conditions, Eqs (11) and (12) are solved with the help of the numerical method via the Keller-box method in MATLAB software. The book of Cebeci and Bradshaw [35] explained the details of this method. The Keller box method that implements the finite difference method (FDM) is extremely powerful to get the estimated results to an arrangement of a nonlinear differential system like other FDM. Among various other mathematical strategies, the FDM is more versatile for the clarification that underlying approximations control the convergence rate. The four fundamental advances are included to get the mathematical arrangements through the Keller-box technique and are:

a) Reduce the governing equations (Eqs 9 and 10) into a first-order equations' system;
b) The first-order system is written as a finite difference method by applying central differences;
c) Linearize the subsequent equations by using the Newton technique and utilized on the coefficient of a matrix of the finite difference equations;
d) The obtained linear system is understood utilizing the block elimination method by employing a block-tridiagonal factorization scheme.

The solutions can gain when imposing an excellent preliminary guess with a uniform step size $\Delta h = 0.001$. Then, the Prandtl number for blood, $\Pr = 30$ [36] has been fixed throughout the computational process with convergence criterion $10^{-5}$ is used to obtain the solutions.

## 4. Results and discussion

The behavior of the hybrid Casson nanofluid flow and heat transfer in a thin-film is obtained by separating into two cases; nanofluid of Cu/blood and hybrid nanofluid of Cu-$Al_2O_3$/blood through a porous medium when the sheet is stretching. The numerical solutions are obtained for various physical parameters values to explain in-depth the pattern of the physical problems. Validation of the numerical code is observed by comparing the numerical results of the present study with Wang [20], which solved the problem using HAM. Table 1 provides a comparison of the numerical findings with previous publications [20] in the absence of nanoparticles volume fraction $\phi_1 = \phi_2 = 0$, Casson parameter $\beta \to \infty$, suction parameter $w = 0$, and unsteadiness parameter $S = 0.8$ and $S = 1.2$ for different values of Prandtl number. A good agreement was found in a comparison with them despite different methods.

**Table 1.** Comparison of $\theta(\eta)$ when $S = 0.8$ and $S = 1.2$ for different Pr.

| Pr | $\theta(\eta)$ | | | |
|---|---|---|---|---|
| | $S = 0.8$ | | $S = 1.2$ | |
| | Wang [20] | Present study | Wang [20] | Present study |
| 0.01 | -0.090474 | -0.090478 | -0.037734 | -0.037734 |
| 0.1 | -0.756162 | -0.756163 | -0.343931 | -0.343931 |
| 1 | -3.595970 | -3.595574 | -1.999590 | -1.999429 |

Figure 2 until Figure 5 exhibit the velocity profile on the investigated parameters listed as unsteadiness parameter of $S$, nanoparticles volume fraction parameter of $\phi_2$, Casson parameter of $\beta$ and suction parameter of $w$. Figure 2 unveils the variations of the $S$ towards $f'(\eta)$. The velocity of fluid for Cu/blood is higher than Cu-$Al_2O_3$/blood with increasing the values $S$ from 1.0 to 2.5. It causes a decrement of the thickness momentum boundary layer of the fluid flow in a thin film for nanofluid and hybrid nanofluid. The diminishing of fluid speed because of the transverse force makes a drag force when the presence of unsteadiness alongside the stretching sheet. The influence of $\phi_2$ against velocity profile is demonstrated in Figure 3. The increase of the copper concentration ($\phi_2 = 0.01$ to $\phi_2 = 0.04$) leads to a decline in the velocity profile. However, the velocity profile with the presence of hybrid nanofluid is more advanced than regular nanofluid. The momentum boundary layer is more approached to $\eta = 0$. Furthermore, the momentum boundary layer for nanofluid is thinner compared to hybrid nanofluid when the fluid behaves like Newtonian as increases $\beta$ as noticed in Figure 4. Usually, the raising of $\beta$ would boost the volume of plastic dynamic viscosity which made the resistance in the fluid motion. Therefore, the lessening of the velocity profile due to the slow movement of the fluid. Figure 5 and indicates the same result as in Figure 4 for various $w$. The intensity of suction increases as increases the value of $w$ at the surface of a permeable stretching sheet. Enhancement of the $w$ was slow down the molecules in the fluid regime and improve the slow movement of the fluid on the stretching sheet. Therefore, it reduces the momentum boundary layer.

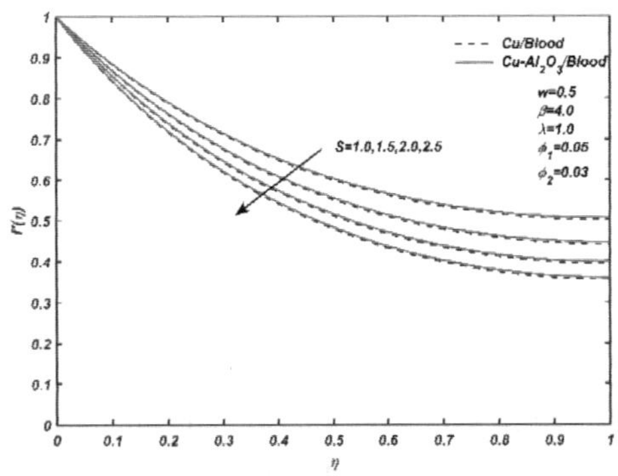


**Figure 2.** Variation of $S$ against $f'(\eta)$

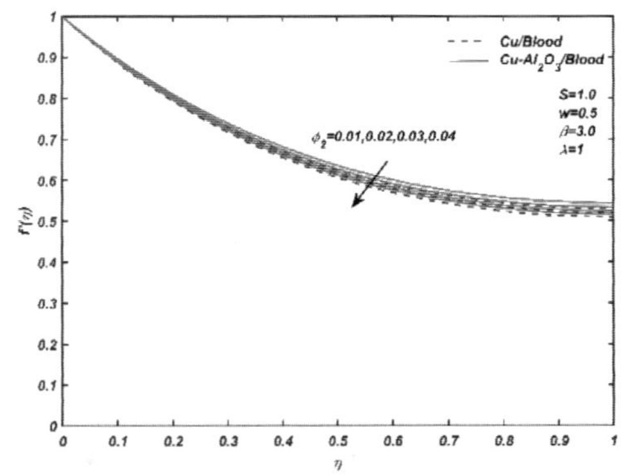


**Figure 3.** Variation of $\phi_2$ against $f'(\eta)$

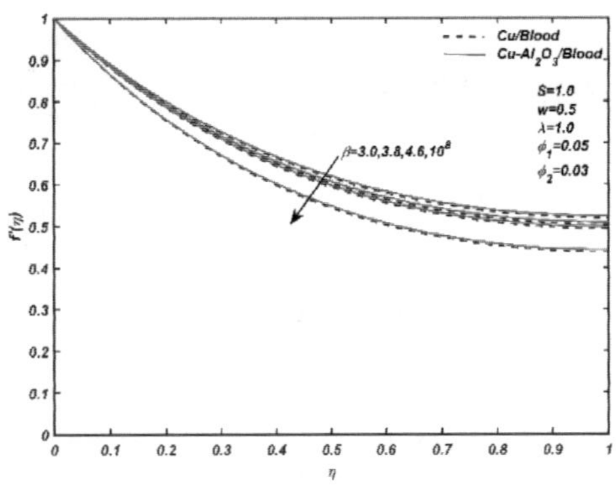


**Figure 4.** Variation of $\beta$ against $f'(\eta)$

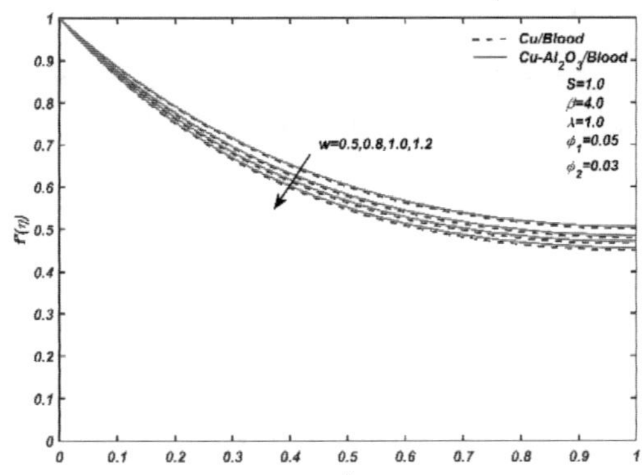


**Figure 5.** Variation of $w$ against $f'(\eta)$

Next, Figure 6 to Figure 9 express the variation of the governing parameters with respect to $\theta(\eta)$ when the sheet is stretched. The Figs illustrate that the temperature profile for hybrid nanofluid is higher when contrasted with regular nanofluid. The temperature profile in Figure 6 declines with decreases in the value of the $S$ of the fluid and it produces a thinner thermal boundary layer. The effect of $\phi_2$ towards temperature profile is viewed in Figure 7. Physically, the nanoparticles scatter energy as warmth. In this way, more energy will be utilized when the expansion of nanoparticles in the fluid and simultaneously causes an increment in temperature and thickness of the thermal boundary layer. The more noteworthy the estimation of a non-Newtonian fluid which is the Casson fluid parameter, $\beta$ the higher the temperature and the thicker the thermal boundary layer as shown in Figure 8. Figure 9 demonstrates the opposite pattern for temperature profile as increasing the intensity of suction, $w$. The upsurge in the suction attracts the amount of fluids particles into the wall then reduces the thermal boundary layer of the fluid in the thin film.

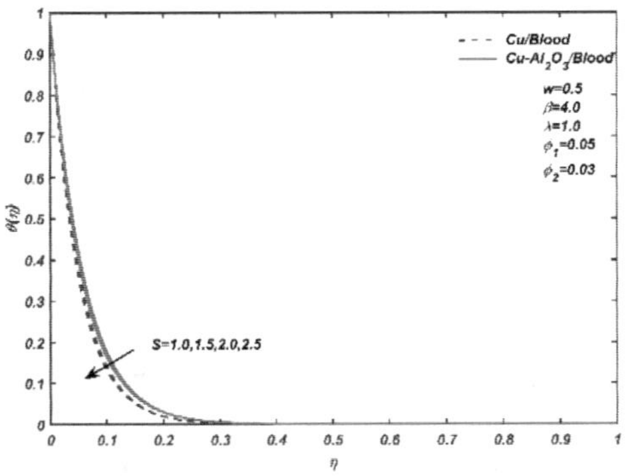

**Figure 6.** Variation of $S$ against $\theta(\eta)$

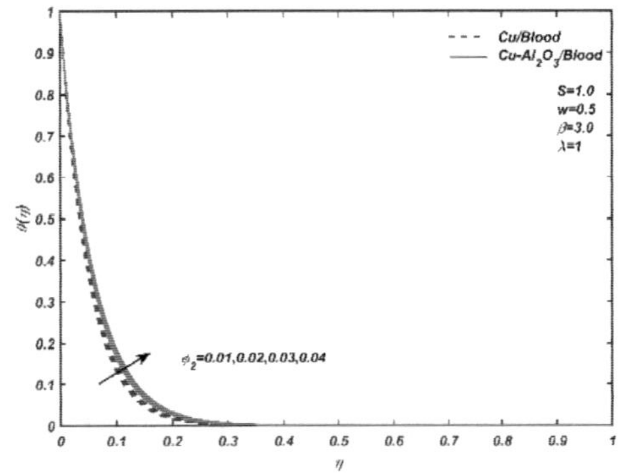

**Figure 7.** Variation of $\phi_2$ against $\theta(\eta)$

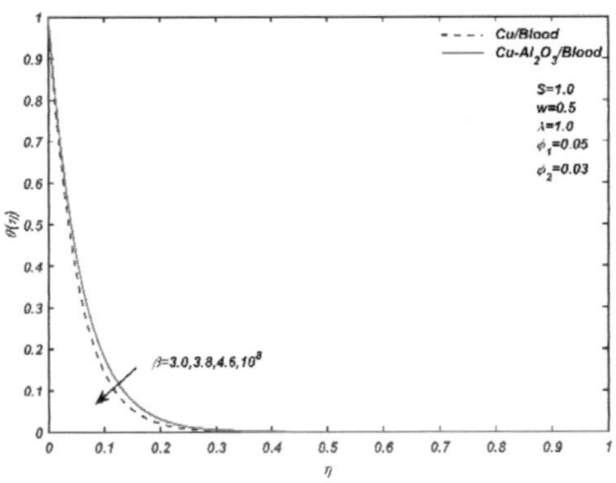

**Figure 8.** Variation of $\beta$ against $\theta(\eta)$

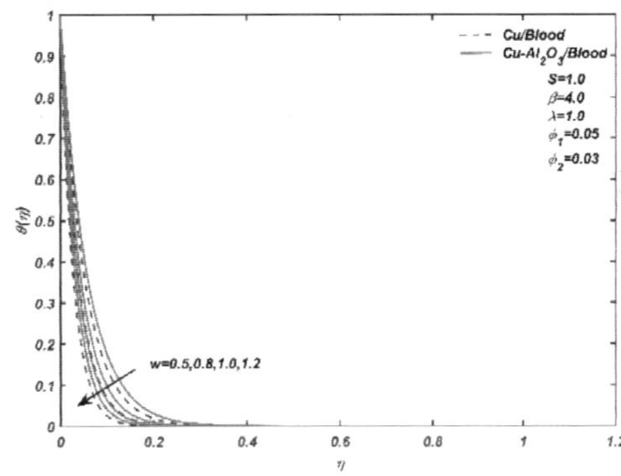

**Figure 9.** Variation of $w$ against $\theta(\eta)$

## 5. Conclusion

The problem of Casson hybrid nanofluid in a thin film over an unsteady stretching sheet has been investigated numerically by using a collocation technique. The model is used together with the suction effect and hybrid nanoparticles. The velocity profile diminishes as raises the investigated parameters. Besides, it found that the increment of the unsteadiness parameter and intensity of suction leads to a decline in the temperature field. The opposite behavior is displayed for nanoparticle volume fraction and Casson parameter.

**Acknowledgments**

This work was supported by research grants vote 5F251 from the Ministry of Higher Education (MOHE) and vote 17J52 from the Research Management Center-UTM, Universiti Teknologi Malaysia (UTM).

**References**

[1] Anitha S, Thomas T, Parthiban V and Pichumani M 2019 What dominates heat transfer performance of hybrid nanofluid in single pass shell and tube heat exchanger? *Advanced Powder Technology* **30** 3107-17

[2] Bai M-j, Liu J-l, He J, Li Z-y, Wei J-j, Chen L-x, Miao J-y and Li C-m 2020 High efficiency heat transfer and antifriction characteristics of SMWCNTs nanofluids *Diamond and Related Materials* **105** 107792

[3] Hashimoto S, Kurazono K and Yamauchi T 2020 Anomalous enhancement of convective heat transfer with dispersed SiO2 particles in ethylene glycol/water nanofluid *International Journal of Heat and Mass Transfer* **150** 119302
[4] Hussien A, Md Yusop N, Abdullah M Z, Al-Nimr M and Janvekar A A 2017 *Heat Transfer Enhancement using Hybrid Nanofluids*
[5] Izadi M, Mohebbi R, Karimi D and Sheremet M A 2018 Numerical simulation of natural convection heat transfer inside a ⊥ shaped cavity filled by a MWCNT-Fe3O4/water hybrid nanofluids using LBM *Chemical Engineering and Processing - Process Intensification* **125** 56-66
[6] Waini I, Ishak A and Pop I 2019 Unsteady flow and heat transfer past a stretching/shrinking sheet in a hybrid nanofluid *International Journal of Heat and Mass Transfer* **136** 288-97
[7] Waini I, Ishak A and Pop I 2020 MHD flow and heat transfer of a hybrid nanofluid past a permeable stretching/shrinking wedge *Applied Mathematics and Mechanics* **41** 507-20
[8] Ali H 2020 *Hybrid Nanofluids for Convection Heat Transfer*
[9] Yazid M N A W M, Sidik N A C and Yahya W J 2017 Heat and mass transfer characteristics of carbon nanotube nanofluids: A review *Renewable and Sustainable Energy Reviews* **80** 914-41
[10] Abbas Z, Sheikh M and Motsa S S 2016 Numerical solution of binary chemical reaction on stagnation point flow of Casson fluid over a stretching/shrinking sheet with thermal radiation *Energy* **95** 12-20
[11] Raju C S K and Sandeep N 2017 Unsteady Casson nanofluid flow over a rotating cone in a rotating frame filled with ferrous nanoparticles: A numerical study *Journal of Magnetism and Magnetic Materials* **421** 216-24
[12] Rawi N A, Ilias M R, Lim Y J, Isa Z M and Shafie S 2017 Unsteady mixed convection flow of Casson fluid past an inclined stretching sheet in the presence of nanoparticles. In: *Journal of Physics: Conference Series,*
[13] Samrat S P, Sulochana C and Ashwinkumar G P 2019 Impact of Thermal Radiation on an Unsteady Casson Nanofluid Flow Over a Stretching Surface *International Journal of Applied and Computational Mathematics* **5**
[14] Zuhra S, Khan N S and Islam S 2018 Magnetohydrodynamic second-grade nanofluid flow containing nanoparticles and gyrotactic microorganisms *Computational and Applied Mathematics* **37** 6332-58
[15] Goyal M and Bhargava R 2013 Numerical Solution of MHD Viscoelastic Nanofluid Flow over a Stretching Sheet with Partial Slip and Heat Source/Sink *ISRN Nanotechnology* **2013** 931021
[16] Rehman S, Shah R A, Khan A and Khan Z 2020 Free surface film flow of an unsteady second grade fluid over a stretching sheet with surface tension linearly varies with temperature *Physica A: Statistical Mechanics and its Applications* 123956
[17] Carragher P and Crane L 1982 Heat transfer on a continuous stretching sheet *ZAMM- Journal of Applied Mathematics and Mechanics/Zeitschrift für Angewandte Mathematik und Mechanik* **62** 564-5
[18] Crane L J 1970 Flow past a stretching plate *Zeitschrift für angewandte Mathematik und Physik ZAMP* **21** 645-7
[19] Sakiadis B C 1961 Boundary-layer behavior on continuous solid surfaces: II. The boundary layer on a continuous flat surface *AIChE Journal* **7** 221-5
[20] Wang C 2006 Analytic solutions for a liquid film on an unsteady stretching surface *Heat and Mass Transfer* **42** 759-66
[21] Andersson H, Aarseth J, Braud N and Dandapat B 1996 Flow of a power-law fluid film on an unsteady stretching surface *Journal of Non-Newtonian Fluid Mechanics* **62** 1-8
[22] Tshehla M 2011 The flow of a Carreau fluid down an incline with a free surface *International Journal of Physical Sciences* **6** 3896-910

[23] Naganthran K, Hashim I and Nazar R 2020 Non-uniqueness solutions for the thin Carreau film flow and heat transfer over an unsteady stretching sheet *International Communications in Heat and Mass Transfer* **117** 104776
[24] Khan N S, Gul T, Kumam P, Shah Z, Islam S, Khan W, Zuhra S and Sohail A 2019 Influence of inclined magnetic field on Carreau nanoliquid thin film flow and heat transfer with graphene nanoparticles *Energies* **12** 1459
[25] Tahir F, Gul T, Islam S, Shah Z, Khan A, Khan W and Ali L 2017 Flow of a nano-liquid film of Maxwell fluid with thermal radiation and magneto hydrodynamic properties on an unstable stretching sheet *Journal of Nanofluids* **6** 1021-30
[26] Khan H, Haneef M, Shah Z, Islam S, Khan W and Muhammad S 2018 The combined magneto hydrodynamic and electric field effect on an unsteady Maxwell nanofluid flow over a stretching surface under the influence of variable heat and thermal radiation *Applied Sciences* **8** 160
[27] Khan N S, Islam S, Gul T, Khan I, Khan W and Ali L 2018 Thin film flow of a second grade fluid in a porous medium past a stretching sheet with heat transfer *Alexandria Engineering Journal* **57** 1019-31
[28] Vijaya N, Sreelakshmi K and Sarojamma G 2016 Effect of magnetic field on the flow and heat transfer in a Casson thin film on an unsteady stretching surface in the presence of viscous and internal heating *Open Journal of Fluid Dynamics* **6** 303-20
[29] Mahmoud M A A and Megahed A M 2017 MHD Flow and Heat Transfer Characteristics in a Casson Liquid Film Towards an Unsteady Stretching Sheet with Temperature-Dependent Thermal Conductivity *Brazilian Journal of Physics* **47** 512-23
[30] Ray Atul K, Vasu B, Anwar Beg O, Gorla R S R and Murthy P V S N 2019 Magneto-bioconvection flow of a casson thin film with nanoparticles over an unsteady stretching sheet: HAM and GDQ computation *International Journal of Numerical Methods for Heat & Fluid Flow* **29** 4277-309
[31] El-Aziz M A and Afify A A 2016 Effects of Variable Thermal Conductivity with Thermal Radiation on MHD Flow and Heat Transfer of Casson Liquid Film Over an Unsteady Stretching Surface *Brazilian Journal of Physics* **46** 516-25
[32] Megahed A 2015 Effect of slip velocity on Casson thin film flow and heat transfer due to unsteady stretching sheet in presence of variable heat flux and viscous dissipation *Applied Mathematics and Mechanics* **36** 1273-84
[33] Sulochana C, Samrat S P and Sandeep N 2018 Magnetohydrodynamic radiative liquid thin film flow of kerosene based nanofluid with the aligned magnetic field *Alexandria Engineering Journal* **57** 3009-17
[34] Kamyar A, Saidur R and Hasanuzzaman M 2012 Application of Computational Fluid Dynamics (CFD) for nanofluids *International Journal of Heat and Mass Transfer* **55** 4104-15
[35] Cebeci T and Bradshaw P 1988 *Physical and computational aspects of convective heat transfer* (New York: Springer)
[36] Zakharov M and Sadovsky M 2013 The role of blood circulatory system in thermal regulation of animals explained by entropy production analysis

# Optimization of laser melting parameter to enhanced surface properties in hot press forming die

**B Norhafzan[1,2], S N Aqida[2,3,*] and C M Khairil[1]**

[1]Politeknik Muadzam Shah, Lebuhraya Tun Razak, 26700 Muadzam Shah, Pahang, Malaysia
[2]College of Engineering, University Malaysia Pahang,26600 Pekan, Pahang, Malaysia
[3]Automotive Engineering Centre, Universiti Malaysia Pahang, 26600 Pekan, Pahang, Malaysia

*Corresponding author's email: norhafzan23@gmail.com

**Abstract.** This paper presents laser melting process with various parameter to enhanced surface properties of AISI H13 tool steel. The design of experiment (DOE) optimisation was conducted to obtain significant model using regression analysis. DOE were analysed using response surface method (RSM) with Box-Behnken design approach. Design Expert 7 software was used to design parameter of laser melting process. The laser melting processes was conducted using Nd:YAG laser system with pulsed mode at a constant average power of 100 W, overlapping rate of 30 to 50%, peak power of 1700 to 2500 W and pulse repetition frequency (PRF) of 50 to 70 Hz respectively. The responses were characterised for sub-surface hardness, melt depth and surface roughness. The results show that optimum parameter of 2500 W peak power, 60 Hz PRF and 30% overlapping rate produced highest surface hardness of 793.7 $HV_{0.1}$ with 0.21 mm melted depth and 3.89 μm surface roughness from experimental data. This finding was important to enhance properties of AISI H13 tool steel for hot press forming die using laser process.

**Keywords:** Laser Melting; DOE optimization; Processing Parameter.

## 1. Introduction

Laser process is preferable due to its high precision, localised surface melting and solidification. Laser modified surface have been shown to have enhanced hardness up to 4 times of the substrate due to grain refinement, secondary carbides and hard non-equilibrium microstructure formation [1, 2]. Control of laser parameter setting yields different effects on the modified surface properties [3]. Optimization of laser parameter is a very important factor to enhanced surface properties in laser surface modification due to cracking tendency at laser modified layer during laser surface melting process [4]. Control of overlapping laser spots yields different effects on the modified surface properties due to additional heating of melted layer. The laser power and scan speed or material-laser interaction time have previously been shown by the applicant to have a strong influence on the resultant temperature profile, modified microstructure, chemical composition, and modified layer depth [5]. Varied processing parameter such as scanning speed and laser power in similar based material produce different microstructure and microhardness [6].

Selection of proper processing parameter is necessary to prevent the high residual stress concentration [7]. Laser parameter play a crucial role in determining the microstructures and surface topography of alloyed zone [8]. To decrease the production time, high processing speeds should be aimed. Furthermore, the surface quality of generated parts has to be improve [9]. Combination of interaction time and power density essential to improve surface properties. Surface properties improved due to short interaction time with high power density caused grain structure refined [10]. Laser surface processing provides a unique tool for high precision surface modification using the heat source to enhance surface integrity of used die.

Major causes of die failures during operations are high thermal shock, mechanical strain, cyclic loading and corrosion which resulting in heat checking, wear, plastic deformation and fatigue [11-13]. Besides, during hot working processes the prolonged contact period between die surfaces and the working material leads to plastic deformation of the die along with the high pressure and temperature [14]. The high temperatures and cyclic loading caused the HPF dies surface to lose strength, hardness and lead to failure and reduce quality product produced from HPF process [11]. To obtained optimum production and reduce production cost, the properties on die surface such as hardness and thermal conductivity must be improved [15]. Die material require for higher hardness caused the typical hardness for dies today is in the range of 44 to 50 HRC. Increase the surface hardness of AISI H13 die is significant to endure wear at elevated temperature during applications [16]. In this study, surface of AISI H13 die insert were modified using laser melting process to improve die properties in hot press forming environment.

## 2. Experimental

A preliminary study was conducted to determine the lower and upper range of peak power, spot overlap and PRF prior to DOE development. Laser surface melting process was designed using Box-Behnken design with three factors. The Box-Behnken design used in this study to select the combination of factor level that produced optimal response with minimal numbers of experiment. Laser process was conducted at a constant average power of 100 W, spot overlap of 30 to 50%, peak power of 1700 to 2500 W and PRF of 50 to 70 Hz respectively. The responses were sub-surface hardness, melt depth and surface roughness which were analysed using response surface methodology (RSM). The RSM analysis was to select optimised parameters for laser melting of ASSAB 8407 (AISI H13) steel surface. Laser processing setup is shown by a schematic diagram in Figure 1. Hardness properties and melted depth of the sample were characterized using Wilson Hardness Tester. The surface roughness was measured using marSurfPS1 roughness tester.

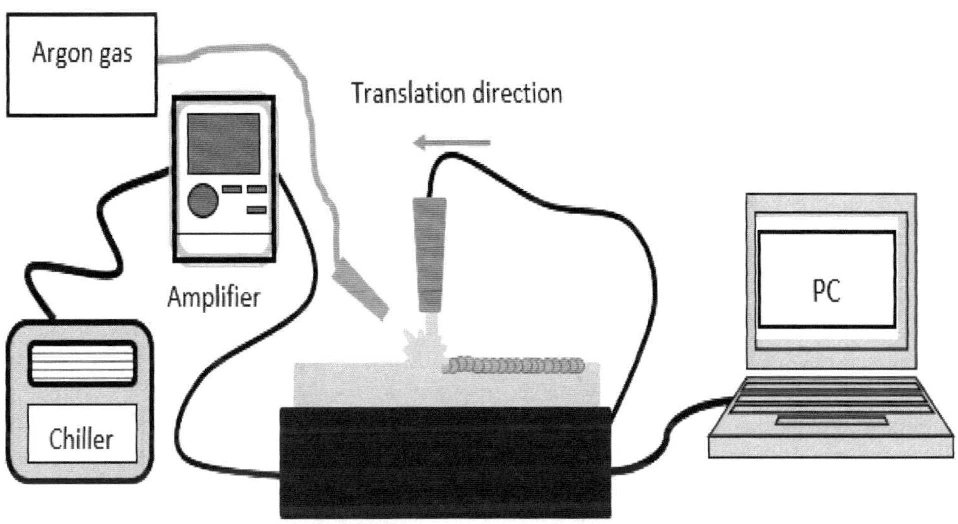


**Figure 1.** Schematic diagram of laser surface melting process setup.

IOP Conf. Series: Materials Science and Engineering 1078 (2021) 012020 doi:10.1088/1757-899X/1078/1/012020

Design of experiment (DOE) optimisation was conducted to obtain significant model using regression analysis. The model reduction using stepwise regression performed to get fit model with eliminates automatically insignificant term. The analysis variance (ANOVA) and optimisation were performed for hardness, roughness and melted depth response. The optimisation processes start with set desired goal of model factor and response. The goal for response were set to maximum for hardness, in range for roughness and minimize for melted depth. The lower and upper limit of factor and response was set and result for optimisation solution and desirability graph were displayed.

## 3. Result and discussion

Set of laser processing parameter were generated using Design Expert Software. The experimental data were analysed using response surface method (RSM) with Box-Behnken technique to analysed surface response on the melted surface. DOE analysis shows the effect of peak power, pulse repetition frequency (PRF) and overlapping percentage on hardness properties, surface roughness and melted depth of the samples. The five centre point was meant for obtaining a precise model and investigate random error of the model. The goal, limit and importance level for setting and outcome parameter from DOE are shown in Table 1. The goals of peak power, PRF, and overlapping rates were set within the appropriate range of is in range. The hardness properties goals were set to maximize, while melted depth was set to minimize and surface roughness set is in range. The importance level for factor parameters were set at Level 3 for all parameters. The importance levels of hardness were set at Level 5, while the roughness was set at Level 3 and melted depth was set at Level 2. The changes on the importance level would affect the desirable solution.

**Table 1.** Constrain selected in design optimization of laser melted processes.

| Name | Goal | Lower Limit | Upper Limit | Lower Weight | Upper Weight | Importance |
|---|---|---|---|---|---|---|
| Peak Power | is in range | 1700 | 2500 | 1 | 1 | 3 |
| PRF | is in range | 50 | 70 | 1 | 1 | 3 |
| Overlapping rate | is in range | 30 | 70 | 1 | 1 | 3 |
| Hardness | maximize | 684.4 | 793.7 | 1 | 1 | 5 |
| Melted depth | minimize | 0.10 | 0.33 | 1 | 1 | 2 |
| Roughness | Is in range | 1.89 | 5.20 | 1 | 1 | 3 |

Design solution in Table 2 shows 23 solution of parameter setting. The highest desirability factor was 0.847 and the lowest solution was 0.416. As for optimisation analysis, the surface hardness was set to maximum limit, while the surface roughness was set to in range limit, and melted depth set to minimize to ensure the highest hardness with better surface roughness after laser melting process. The prediction of setting parameters included peak power of 2500 W, PRF of 58.27 Hz, overlapping rate of 30% produce maximum surface hardness of 793.7 $HV_{0.1}$, roughness of 4.08 μm, and melted depth of 0.2 mm at the highest desirability factor of 0.847. Table 2 shown solution number 1 and 9 with peak power of 2500 W, PRF of 58.27 and 60.1 Hz with overlapping rate of 30% similar with parameter used during experimental process with (peak power of 2500, PRF of 60 Hz and overlap rate of 30%).

The generated optimal condition criteria in the desirability contour plot are shown in Figure 2. The desirability function served to optimise the experimental conditions for maximised surface hardness. The best combination of factors included peak power of 2500 W, PRF of 58.27 Hz, and overlapping rate of 30%. The plotted desirability contour graph of peak power and PRF at the overlapping rate of 30% revealed that desirability factor increased with the increase of peak power. The predicted desirability in the contour plot after model reduction recorded the value of 0.847. The contour plot of optimised surface hardness, melted depth and roughness of melted surface.

**Table 2.** Design solution for maximised surface hardness and melted depth.

| Number | Peak power | PRF | Overlap rate | Hardness | Melted depth | Roughness | Desirability |
|---|---|---|---|---|---|---|---|
| 1 | 2500.00 | 58.27 | 30.00 | 793.7 | 0.20 | 4.08 | 0.847 |
| 2 | 2496.63 | 58.18 | 30.00 | 793.7 | 0.20 | 4.08 | 0.846 |
| 3 | 2499.90 | 58.02 | 30.32 | 793.7 | 0.20 | 4.12 | 0.845 |
| 4 | 2500.00 | 57.80 | 30.61 | 793.7 | 0.20 | 4.16 | 0.844 |
| 5 | 2398.23 | 54.60 | 30.00 | 793.7 | 0.22 | 4.12 | 0.815 |
| 6 | 1864.27 | 70.00 | 30.00 | 773.2 | 0.14 | 2.10 | 0.812 |
| 7 | 1861.26 | 70.00 | 30.00 | 773.1 | 0.14 | 2.10 | 0.812 |
| 8 | 1850.11 | 70.00 | 30.00 | 772.7 | 0.14 | 2.08 | 0.812 |
| 9 | 2500.00 | 60.10 | 30.00 | 785.2 | 0.20 | 3.91 | 0.810 |
| 10 | 2496.75 | 53.34 | 30.00 | 814.2 | 0.23 | 4.48 | 0.795 |
| 11 | 2500.00 | 60.79 | 30.00 | 781.9 | 0.19 | 3.83 | 0.794 |
| 12 | 2184.03 | 68.95 | 30.00 | 765.2 | 0.18 | 2.63 | 0.715 |
| 13 | 1803.07 | 50.00 | 30.25 | 760.9 | 0.19 | 3.69 | 0.670 |
| 14 | 2076.59 | 50.00 | 30.00 | 770.4 | 0.23 | 3.70 | 0.666 |
| 15 | 1846.89 | 70.00 | 63.74 | 761.1 | 0.21 | 1.89 | 0.644 |
| 16 | 1886.24 | 70.00 | 65.18 | 761.5 | 0.21 | 1.89 | 0.643 |
| 17 | 1941.68 | 70.00 | 66.89 | 760.7 | 0.21 | 1.89 | 0.636 |
| 18 | 1750.79 | 70.00 | 53.49 | 752.0 | 0.18 | 2.17 | 0.623 |
| 19 | 2005.62 | 70.00 | 68.45 | 757.9 | 0.21 | 1.89 | 0.622 |
| 20 | 2009.76 | 70.00 | 68.53 | 757.7 | 0.21 | 1.89 | 0.620 |
| 21 | 2046.41 | 69.99 | 69.23 | 755.0 | 0.21 | 1.89 | 0.607 |
| 22 | 2500.00 | 51.74 | 61.33 | 762.5 | 0.25 | 4.53 | 0.581 |
| 23 | 2251.86 | 61.97 | 70.00 | 726.4 | 0.21 | 2.80 | 0.416 |

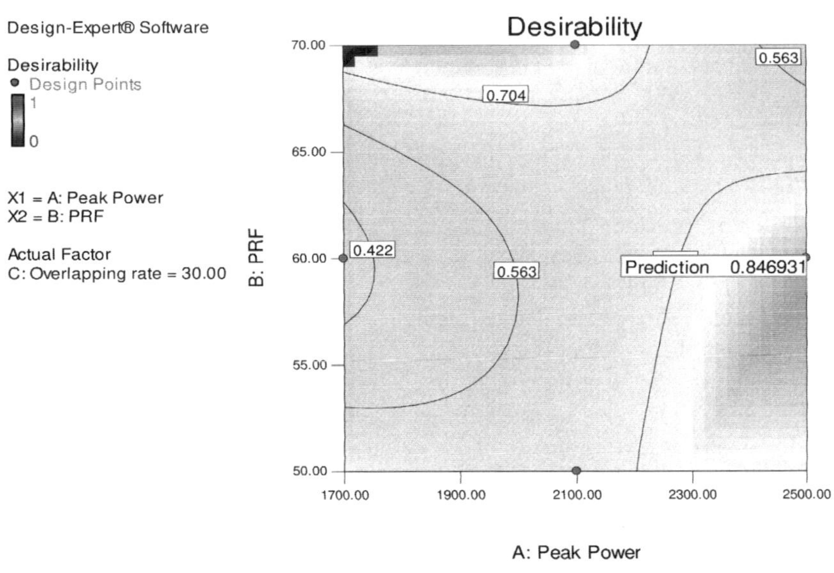


**Figure 2.** Desirability Contour plot of melted surface with highest desirability factor of 0.847.

Based on the result from DOE optimization of laser melted die insert, the optimum parameter of 2500 W peak power, 58.27 PRF and 30% overlapping rate with maximum surface hardness of 793.7 $HV_{0.1}$, melted depth of 0.2 mm and 4.08 μm surface roughness with 0.847 desirability factor. While, the experimental data shown highest surface hardness were 793.7 $HV_{0.1}$ with 0.21 mm melted depth and 3.89 μm surface roughness using laser processing parameter of 2500 W peak power, 60 Hz PRF and 30% overlapping rate. The validation processes done to assess amount of the 95% confidence interval between the predicated value from DOE optimisation and experimental result. The 95% confidence interval for surface hardness and roughness was calculated by using the formula as shown;

$$95\%\ confidence\ interval = \frac{Predicted\ value - Actual\ value}{Predicted\ value} \times 100 \quad (1)$$

The amount of the 95% confidence interval between the predicated value from DOE optimisation and experimental at 2500 W peak power, 60 Hz PRF and 30% overlapping shown hardness of predicted value was 785.2 $HV_{0.1}$ while the experiment value was 793.7 $HV_{0.1}$ with 95% confidence interval of 1.1%. While the surface roughness value from experiment yield 3.89 mm and the predicted value was 3.91 with 0.5% of 95% confidence interval. It can be validated, that the laser parameter used can be produce the expected properties within the investigated range.

## 4. Conclusion

The response surface analysis was performed on the experimental data to examine the hardness, melted depth and roughness properties using Box-Behnken design which involved the quadratic model with model reduction (to improve the model). The box-Behnken analysis was used to identify the independent variables or factors that affect the process and product, and also to study the effects of dependent variables or responses. The optimisation analysis shows optimum laser parameter to produce suitable die insert for HPF process is at 2500 W peak power, 58.27 Hz PRF and 30% overlapping rate with higher surface hardness of 793.7 $HV_{0.1}$. These findings were important to produce optimum parameters for enhanced surface properties in laser melting process.

## Acknowledgements

The authors would like to acknowledge the financial support from University Malaysia Pahang under RDU182203 UMP Flagship research grant, University Tun Hussein Onn Malaysia for providing the laser facility and Politeknik Sultan Mizan Zainal Abidin for providing advanced machining facility.

## References

[1] Aqida S, Brabazon D and Naher S 2011 Designing pulse laser surface modification of H13 steel using response surface method

[2] Fauzun F, Aqida S N and bin Wahab M 2013 Laser Surface Modification of AISI 1025 Low Carbon Steel Using Pulsed Nd: YAG Laser for Enhance Surface Properties. In: *Key Engineering Materials*: Trans Tech Publ) pp 596-602

[3] Norhafzan B, Aqida S, Fazliana F, Reza M, Ismail I and Khairil C 2018 Laser melting of groove defect repair on high thermal conductivity steel (HTCS-150) *Applied Physics A* **124** 169

[4] Petrovič D S and Šturm R 2014 Fine-structured Morphology of Silicon Steel Sheet after Laser Surface Alloying of Sb Powder *Strojniški vestnik-Journal of Mechanical Engineering* **60** 5-11

[5] Aqida S, Brabazon D and Naher S 2013 An investigation of phase transformation and crystallinity in laser surface modified H13 steel *Applied Physics A* **110** 673-8

[6] Elhamali S M, Etmimi K and Usha A 2013 The Effect of Laser Surface Melting on the Microstructure and Mechanical Properties of Low Carbon Steel. In: *Proceedings of World Academy of Science, Engineering and Technology*: World Academy of Science, Engineering and Technology (WASET)) p 447

[7] Farahmand P and Kovacevic R 2014 An experimental–numerical investigation of heat distribution and stress field in single-and multi-track laser cladding by a high-power direct diode laser *Optics & Laser Technology* **63** 154-68
[8] Nath S, Pityana S and Majumdar J D 2012 Laser surface alloying of aluminium with WC+ Co+ NiCr for improved wear resistance *Surface and Coatings Technology* **206** 3333-41
[9] Meier H and Haberland C 2008 Experimental studies on selective laser melting of metallic parts *Materialwissenschaft und Werkstofftechnik* **39** 665-70
[10] Bariman N, Aqida S N and Fauzun F 2017 Laser Melting of High Thermal Conductivity Steel (HTCS) Surface. In: *Materials Science Forum,*
[11] Jhavar S, Paul C and Jain N 2013 Causes of failure and repairing options for dies and molds: A review *Engineering Failure Analysis* **34** 519-35
[12] Altan T and Deshpande M 2011 Selection of die materials and surface treatments for increasing die life in hot and warm forging *ERC for Net Shape Forming, Paper* 1-32
[13] Jia Z-x, Liu Y-w, Li J-q, Liu L-J and Li H-l 2015 Crack growth behavior at thermal fatigue of H13 tool steel processed by laser surface melting *International Journal of Fatigue* **78** 61-71
[14] Papageorgiou D, Medrea C and Kyriakou N 2013 Failure analysis of H13 working die used in plastic injection moulding *Engineering Failure Analysis* **35** 355-9
[15] Valls I, Casas B, Rodríguez N and Paar U 2010 Benefits from using high thermal conductivity tool steels in the hot forming of steels *La Metallurgia Italiana*
[16] Aqida S, Maurel M, Brabazon D, Naher S and Rosso M 2009 Thermal stability of laser treated die material for semi-solid metal forming *International Journal of Material Forming* **2** 761

# Preparation of metal-organic frameworks (MOFs) based Chromium 2,6-Naphtalenedicarboxylic Acid (MIL-101 NDC) for $CO_2$ adsorption application

**Fayza Yulia[1], V J Utami[1], Raden Nanda[1], Nasruddin[1,*], Muhammad Arif Budiyanto[1] and A Zulys[2]**

[1] Department of Mechanical Engineering, Faculty of Engineering, Universitas Indonesia, Depok, 16424, Indonesia
[2]Department of Chemistry, Faculty of Mathematics and Natural Sciences, Universitas Indonesia, Depok, 16424, Indonesia

*Corresponding author's e-mail: nasruddin@eng.ui.ac.id

**Abstract.** Metal-organic frameworks are believed to be one of the solutions in reducing $CO_2$ emissions. Ligand modification, which is one of the methods in optimizing the MOF structure, provides a good opportunity for the development of functional materials with better properties. Ligand functionality in gas separation is important because it provides an insight of the influence of the material topology. MIL-101 (Cr), which is a type of MOF, is composed of metal chromium and terephthalic acid as the linker. This MOF has been extensively researched for $CO_2$ capture applications. In this work, the functionalization of MIL-101 with 1,4-Benzenedicarboxylic acid (BDC) as the ligand was carried out by changing into 2,6-Napthalenedicarboxylic acid (NDC) ligand. The design and architecture of NDC, which has a double hydrocarbon ring, is expected to increase the pore and surface area of the MOF so that it can increase gas adsorption. The synthesis of the modified MOF was performed using solvothermal method and tested for the porous properties, morphology, structure, chemical functionalities and thermal stability by $N_2$ adsorption / desorption data, scanning electron microscopy (SEM), X-ray diffraction (XRD), Fourier transform infrared spectroscopy (FTIR), and thermogravimetric analysis (TGA). The results show that MIL-101 NDC has better thermal stability, as it can maintain its structure up to the temperature of 350°C. The pore diameter measurement results also show a good diameter up to 4nm. The discovery structure of MIL-101 NDC which has excellent stability provides a good prospective adsorbent in $CO_2$ capture applications.

**Keywords:** 2,6-Napthalenedicarboxylic acid (NDC); MIL-101; $CO_2$ Capture; Adsorption; Adsorbent Characterization.

## 1. Introduction
Carbon dioxide capture has been an important research topic regarding its significant amount as greenhouse gas, causing global warming. There are three existing method to reduce carbon dioxide emission through carbon capture: pre-combustion capture, oxy-fuel combustion, and post-combustion capture [1-3]. The industrial and personal use of combustion engine might not be fully replaced anytime soon. Thus, post-combustion capture of carbon dioxide is being considered as the viable solution since this method easily used in retrofit power plant. Many porous materials are used to

capture gasses. The process includes either chemical absorption or physical adsorption [4-6]. Since the implementation of absorption using amine solvents consumes a lot of energy up to 70% by weight, it has an impact of energy consumption escalation and can reduce the efficiency of the power plant [7]. Thus, the adsorption method is the viable solution to overcome this problem.

Most of the research conducted in adsorption is investigating the efficient adsorbents to capture $CO_2$ molecules [8-10]. Some materials which have already been researched are zeolite, metal-organic framework (MOF), and activated carbon [11]. Large surface area, high pore volume, and low density are some of the parameters to be achieved in adsorbents study. Metal-organic framework recently showed up with advantageous characteristic such as high porosity, high selectivity, thermal stability, and moisture resistance which made it suitable for $CO_2$ capture [12]. The ease of modification through pre and post synthetic by functionalizing ligand and the incorporation of open metal sites to increase the properties make MOFs a highly suitable adsorbent for gas storage [13].

MIL-101(Cr) is one of the potential MOF for $CO_2$ adsorption, considering that not only it has large surface area but also it has the combination of high thermal and chemical stability [14]. MIL-101(Cr) is composed by 1,4-benzenedycarboxylic acid (BDC) as its linker and chromium (III) nitrate as its metal. In addition to the density and the large surface area of this MOF, thermal stability is also a parameter which has made this adsorbent becoming the most researched. It is reported that MIL-101 has good thermal stability which is testified by heating it with steam containing nitrogen to a temperature of 325 °C [15]. Another stability experiment is heating it up to100 °C for 7 days on closed vessel and this MOFs can bear up in that condition [16]. This outstanding characteristic makes this MOFs being a subject to various investigations.

In this work, we have modified MIL-101 by ligand functionalization. The adsorbent is modified by replacing the linker from 1,4-benzenedycarboxylic acid (BDC) to 2,6-naphtalenedicarboxylic acid (NDC). The NDC ligand has two hydrocarbon ring which support the hypothetic that this larger structure will make the future MOF be better in surface area and porosity, so it can capture more $CO_2$ molecules [1, 17]. However, the modification in the linker affected some reaction condition such as the temperature reaction, molar ratio, solvent, and time reaction [18]. Therefore, the investigation of MOF MIL-101 NDC was carried out and studied for a variety of reaction condition. Then, the adsorbent which have been synthesized would be characterized by $N_2$ adsorption/desorption data, X-ray diffraction (XRD), Fourier transform infrared spectroscopy (FTIR), scanning electron microscopy (SEM), and thermogravimetric analysis (TGA) to know the porous properties, crystallinity, chemical functionalities, morphology, and thermal stability. This result is expected to become further research that will be carried out as a potential $CO_2$ capture application.

## 2. Methods and materials

### *2.1 Material synthesis*

MIL-101 NDC was synthesized using previously reported procedure with some modification, which scaled up to 50 times [7]. 4 grams of $Cr_2(NO_3)_3.9H_2O$ was dissolved in 10 ml of demineralized water. Next, the mixture was mixed with 1,14 ml $CH_3COOH$, 2.16 grams of NDC, and 47.7 ml of demineralized water. The final mixture was stirred using magnetic stirrer for 30 minutes, put in a Teflon container, and then placed in an autoclave. The sealed autoclave then heated from room temperature to 200°C in 3 hours, then hold for 4 hours in that temperature, and then cooled down to room temperature in 3 hours. The synthesized product was later separated from the liquid using centrifugal spinner. The reaction product was activated three times at 80°C for 5 hours with EtOH. Finally, The synthesized MOF was dried overnight at a temperature of 80°C.

### *2.2 Adsorbent characterization*

The experimental method is carried out by referring to various literatures [9, 19-21]. The presence of functional group was tested by Spectrophotometer FTIR (Fourier Transform Infra-Red) method using Thermo Scientific™ Nicolet™ iS™5. To determine the structure and crystallinity, XRD (X-Ray

Diffraction) was performed using PANanalytical XPERT PRO with step size of 0.002 step/second using a monochromatic CuKα radiation (wavelength ¼ 0.154 nm) operated at 40 kV and 30 mA. The surface area and pore volume were estimated by using $N_2$ adsorption/desorption data of 77.4 K. The adsorbent was regenerated at 160 °C for 6 h before experimental uptakes. SEM (Scanning Electron Microscopy) were carried out using FEI Inspect F50 to estimate its morphology. TGA (Thermo Gravimetric Analysis) was carried out by heating powder sample up to 550°C under oxygen flow for understanding its thermal stability.

## 3. Result and discussion

*3.1 Synthesis*

The material MIL-101 NDC was obtained by hydrothermal synthesis method. The synthesis resulted in greenish powder material as illustrated in
Figure **1**. The pore volume and surface area of the material highly depended on the purification process since there are plenty amount of unreacted linker left either in the outer or inner pores. Besides molarity ratio, temperature, time, and solvents used in the synthesis process, purification also plays an important role in affecting the surface area of MOF, as few unreacted linkers remain on the surface of MOF [22]. Therefore, in this experiment, we purified the direct dissolution method with organic solvents (DMF) and ethanol with second step purification. In the final purification, MOF was washed three times using EtOH to gain highly porous MIL-101 NDC.

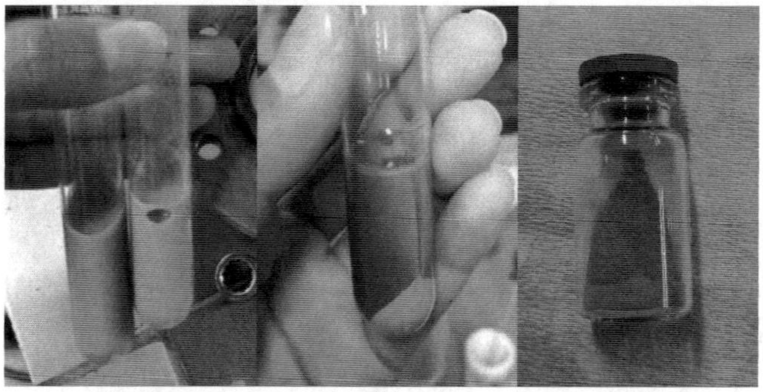

(a) (b) (c)

**Figure 1**. The Synthesized MIL-101 NDC just after reaction, (b) after separation using centrifugal spinner, and (c) after purification.

*3.2 FTIR analysis*

FTIR analysis was performed to identify the functional groups contained in MIL-101 NDC. The measurement of MIL-101 pattern was conducted before isothermal adsorption process. Figure 2 illustrated IR spectroscopy of MIL-101 NDC. Apex existed at wave number 3369.19 $cm^{-1}$ indicated the presence of vibration due to (O-H) molecule representing the existence of $H_2O$ molecule at MIL-101 NDC [23]. Band observed at 1614,51 $cm^{-1}$ indicated the vibration of (C=O) molecule [24]. The forming of band at 1411.93 $cm^{-1}$ means 2,6-Naphthalenedicarboxylic acid was found in MIL-101 NDC [25] . Lastly, vibration of (C-H) molecule observed at spectrum 1142,95, 922,15, dan 782,83 $cm^{-1}$.

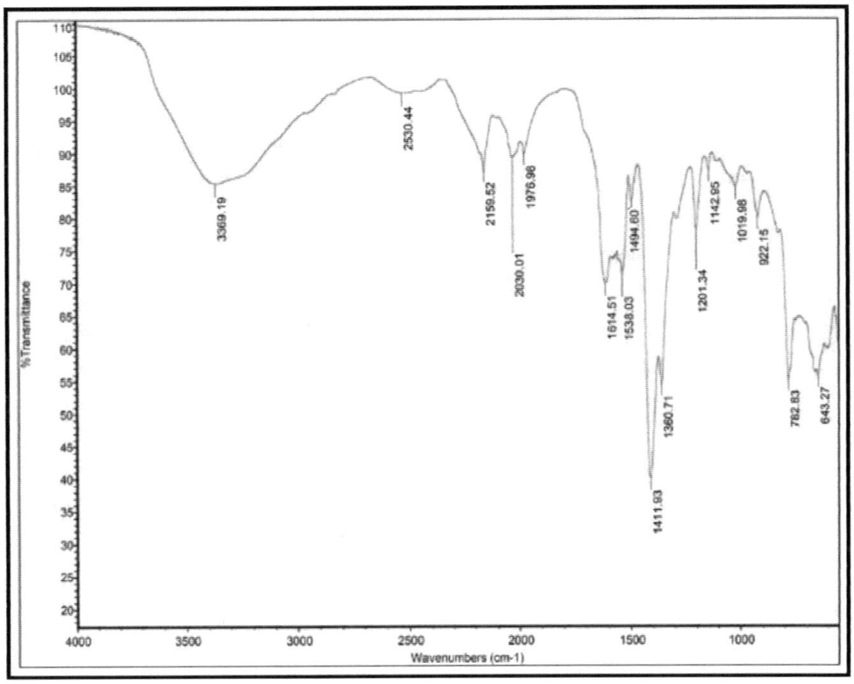


**Figure 2.** FTIR pattern of MIL-101 NDC.

*3.3 Phase identification*

The material MIL-101 NDC was analyzed by X-Ray Diffraction (XRD) to identify its structure and crystallinity. The XRD pattern of synthesized MIL-101 NDC is presented in Figure 3. The appearance of intensive peaks at small angle ($2\theta$) indicated that the material has great amounts of pores in its structure. The executed test using Match! resulted in Tsumgallite as the candidate phase with highest figure of merit. There are sharp peaks at $2\theta = 18^0$, $25^0$, and $28^0$ that identify MIL-101 NDC crystallinity has been quite good.

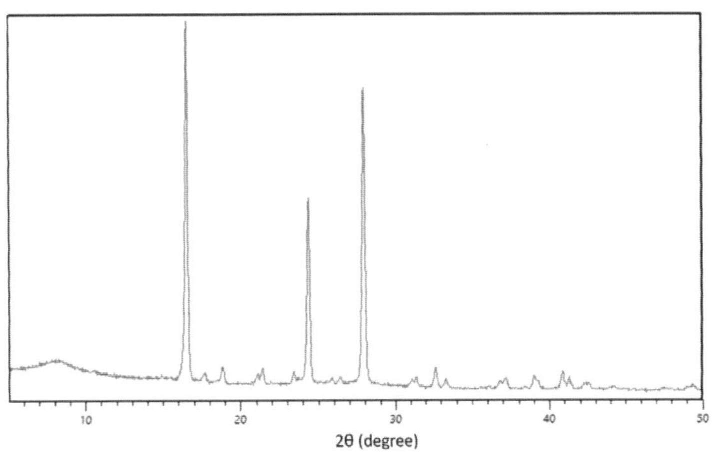


**Figure 3.** XRD pattern of synthesized MIL-101NDC.

*3.4 Surface area and porosity*

The surface area and porosity of materials was investigated using nitrogen sorption experiment. The experiment was conducted at 77 K with pore-filling steps at relative pressure (p/p0) 0.00 to 0.98 in order to avoid chemical adsorption. The materials were degassed at 160°C for 6 hours and the amount of dehydrated materials is 0.22 grams. Nitrogen sorption method was used to evaluate the surface area. The BET and Langmuir surface area result was 298.38 $m^2/g$ and 465.31 $m^2/g$, respectively and the pore volume was 0.169 $cm^3/g$. The corresponding isotherm, which is presented in Figure 4, has the same trends with the parent MIL-101 previous research [26-28]. The pore size distribution was also evaluated using BJH method with the majority of 4 nm diameter variation as shown in Figure 5.

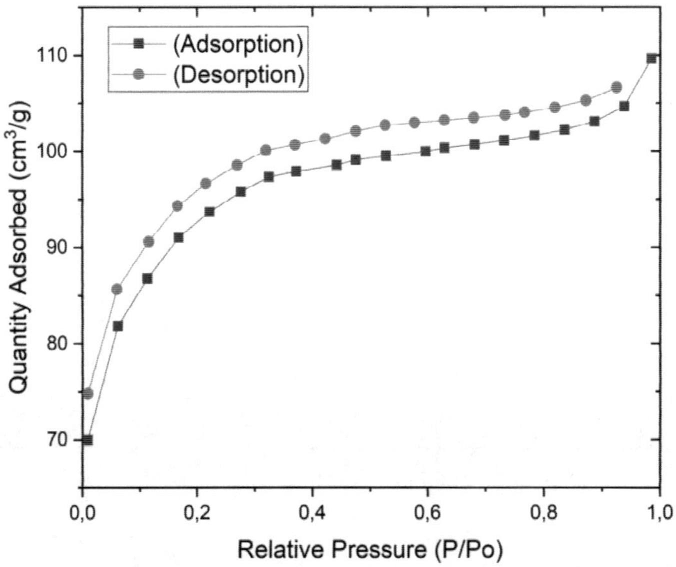


**Figure 4.** Isotherm linear plot for Nitrogen corption.

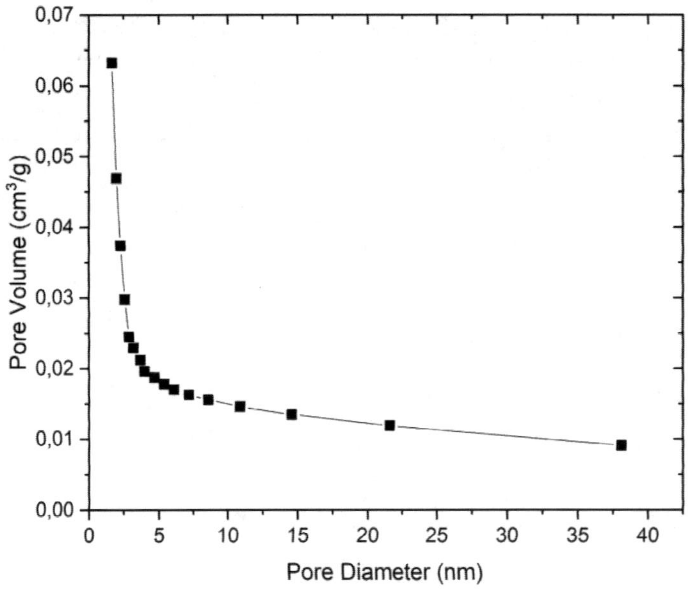


**Figure 5.** Pore size distribution MIL-101 NDC.

*3.5 Morphology*
SEM micrograph which are presented in Figure 6 showed the morphology of synthesized particle MIL-101 NDC for magnificiation of 1,000x and 10,000x. The morphology of these particles shows

the porous and dense surface of MIL-101 NDC. At the magnification of 10,000, the morphology of MIL-101 NDC looks like a block and chunk with little agglomeration, though the uniformity has not been achieved. This MOF microstructure is similar to several MOFs that have been reported [29-31]. From the results, it can be seen clearly that there is a brighter side than the other as illustrated in Figure 6a. It indicates that there are organic ligand residuals that do not react in the formation of MOF. In addition, this bright side also indicates the presence of an organic solvent used in MOF purification step, namely DMF. Thus, to get better properties, the optimization of the MOF structure needs to be improved.

**Figure 6**. SEM (Scanning Electro Micrograph) Test Result at The Magnification of (a) 1000x (b) 10,000x.

*3.6 Thermal stability*

Thermal stability was conducted in order to understand the thermal stability of MIL-101 NDC. The test was carried out using Thermogravimetric Analysis (TGA) in temperature up to 550 °C under oxygen flow. Characteristics of MIL-101 NDC which has been synthesized shows hydrophobic properties and the weight loss significantly decreased at temperatures ranging from 300°C as presented in Figure 7. Furthermore, the greatest weight reduction occured in the temperature of 400-450°C where there is degradation up to 85% from the linker of 2,6-Naphthalenedicarboxylic acid which produces $Cr_2O_3$ residue. The characteristic TGA curve of MIL-101 NDC which shows its hydrophobic properties is similar to MOF Zr- [32]. From this point, it can be seen that the functionalization of the ligand can affect the properties of MOF. Changing the BDC ligand to NDC makes MOF MIL-101 NDC has better thermal stability. This analysis shows the thermal stability of MOF MIL-101 NDC, as this sample should be heated to a temperature of more than 250°C during the regeneration process [33].

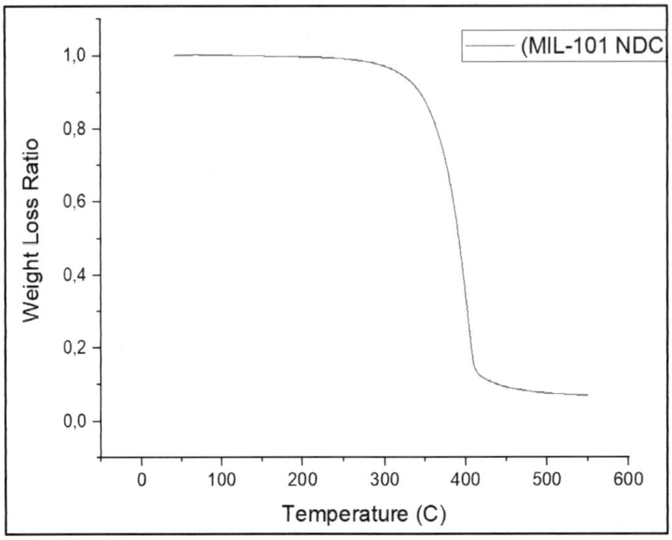


**Figure 7**. Thermogravimetric Analysis Result of MIL-101 NDC

## 4. Conclusion

In this study, the investigation of the ligand functionalization of MIL-101 with BDC into NDC was carried out. To increase the porosity of MOF, the double ring hydrocarbon effect is expected to capture more $CO_2$ molecules. MOF was synthesized by solvothermal reaction by two-step purifcation. The adsorbent characteristics evaluated by various techniques represent an initial successful step in the synthesis of MIL-101 NDC. The results of this study indicate that ligand changes have an impact on increasing the thermal stability of the MOF. Although MIL-101 NDC has shown good results from the FTIR, TGA and crystallinity results which is evaluated by XRD, it is still necessary to increase the porosity of the MOF. Therefore, the optimization of MIL-101 NDC structure should to be continued. The results obtained in this work provide a useful first step for the fabrication of MOF MIL-101 NDC for $CO_2$ capture applications in the future.

## Acknowledgements

This study is funded by PIT 9 research program (NKB-0069/UN2.R3.1/HKP.05.00/2019) held by Universitas Indonesia, Osaka Gas Foundation, and PMDSU Grant (number NKB-445/UN2.RST/HKP.05.00/2020) from Ministry of Research, Technology, and Higher Education, Republic of Indonesia

## References

[1] Mukhtar A, Ullah S, Inayat A, Saqib S, Mellon N B, Assiri M A, Al-Sehemi A G, Niazi M B K, Jahan Z and Bustam M A 2020 Synthesis-structure-property relationship of nitrogen-doped porous covalent triazine frameworks for pre-combustion CO2 capture *Energy* 119230

[2] Sohaib Q, Muhammad A, Younas M and Rezakazemi M 2020 Modeling pre-combustion CO2 capture with tubular membrane contactor using ionic liquids at elevated temperatures *Separation and Purification Technology* **241** 116677

[3] Bairamijamal F 2020 Process for enrichment of the co2 concentration in the co2-streams from the post-combustion and pre-combustion stationary sources of co2 emission upstream of further processing. Google Patents)

[4] Chen Y, Xiang Z, Wang D, Kang J and Qi H 2020 Effective photocatalytic degradation and physical adsorption of methylene blue using cellulose/GO/TiO 2 hydrogels *RSC Advances* **10** 23936-43

[5] Abd A A, Naji S Z, Hashim A S and Othman M R 2020 Carbon dioxide removal through Physical Adsorption using Carbonaceous and non-Carbonaceous Adsorbents: A review *Journal of Environmental Chemical Engineering* 104142
[6] Vega F, Baena-Moreno F, Fernández L M G, Portillo E, Navarrete B and Zhang Z 2020 Current status of CO2 chemical absorption research applied to CCS: Towards full deployment at industrial scale *Applied Energy* **260** 114313
[7] Andreas Sonnauer F H, Michael Froba, Lorenz Kienle, Viola Duppel, Matthias Thommes, Christian Serre, Gerard Ferey, Norbert Stock 2009 Giant Pores in a Chromium 2,6-Naphtalenedicarboxylate Open-Framework Structure with MIL-101 Topology *Angwandte Chemie* 3849-52
[8] Gautam S and Cole D 2020 CO2 Adsorption in Metal-Organic Framework Mg-MOF-74: Effects of Inter-Crystalline Space *Nanomaterials* **10** 2274
[9] Ullah S, Bustam M A, Assiri M A, Al-Sehemi A G, Gonfa G, Mukhtar A, Kareem F A A, Ayoub M, Saqib S and Mellon N B 2020 Synthesis and characterization of mesoporous MOF UMCM-1 for CO2/CH4 adsorption; an experimental, isotherm modeling and thermodynamic study *Microporous and Mesoporous Materials* **294** 109844
[10] Ullah S, Bustam M A, Al-Sehemi A G, Assiri M A, Kareem F A A, Mukhtar A, Ayoub M and Gonfa G 2020 Influence of post-synthetic graphene oxide (GO) functionalization on the selective CO2/CH4 adsorption behavior of MOF-200 at different temperatures; an experimental and adsorption isotherms study *Microporous and Mesoporous Materials* **296** 110002
[11] Jian-Rong Li Y M, M. Colin McCarthy, Julian Sculley, Jiamei Yu, Hae-Kwon Jeong, Perla B. Balbuena, Hong-Cai Zhou 2011 Carbon dioxide capture-related gas adsorption and separation in metal-organic frameworks *Coordination Chemistry Review* 1791-823
[12] Farha O K, Eryazici I, Jeong N C, Hauser B G, Wilmer C E, Sarjeant A A, Snurr R Q, Nguyen S T, Yazaydın A O z r and Hupp J T 2012 Metal–organic framework materials with ultrahigh surface areas: is the sky the limit? *Journal of the American Chemical Society* **134** 15016-21
[13] Anderson S L and Stylianou K C 2017 Biologically derived metal organic frameworks *Coordination Chemistry Reviews* **349** 102-28
[14] How Wei Benjamin Teo A C, Sibnath Kayal 2017 Evaluation of CH4 and CO2 adsorption on HKUST-1 and MIL-101(Cr) MOFs employing Monte Carlo simulation and comparison with experimental data *Applied Thermal Engineering* 891-900
[15] Gao L, Li C-Y V, Yung H and Chan K-Y 2013 A functionalized MIL-101 (Cr) metal–organic framework for enhanced hydrogen release from ammonia borane at low temperature *Chemical Communications* **49** 10629-31
[16] Huang C-Y, Song M, Gu Z-Y, Wang H-F and Yan X-P 2011 Probing the adsorption characteristic of metal–organic framework MIL-101 for volatile organic compounds by quartz crystal microbalance *Environmental science & technology* **45** 4490-6
[17] Sher F, Iqbal S Z, Albazzaz S, Ali U, Mortari D A and Rashid T 2020 Development of biomass derived highly porous fast adsorbents for post-combustion CO2 capture *Fuel* **282** 118506
[18] Bonneau M, Lavenn C, Ginet P, Otake K-i and Kitagawa S 2020 Upscale synthesis of a binary pillared layered MOF for hydrocarbon gas storage and separation *Green Chemistry* **22** 718-24
[19] Sibnath Kayal B S, Anutosh Chakraborty 2015 Study of metal-organic framework MIL-101(Cr) for natural gas (methane) storage and compare with other MOFs (metal-organic framework) *Energy* 772-81
[20] Zulys A, Yulia F, Buhori A, Muhadzib N, Ghiyats M and Saha B B 2020 Synthesis and characterization of a novel microporous lanthanide based metal-organic framework (MOF) using napthalenedicarboxylate acid *Journal of Materials Research and Technology* **9** 7409-17

[21] Yulia F, Utami V J and Nasruddin A Z 2019 Synthesis, Characterizations, and Adsorption Isotherms of CO2 on Chromium Terephthalate (MIL-101) Metal-organic Frameworks (Mofs) *Synthesis* **10**

[22] Hwang Y K, Hong D Y, Chang J S, Jhung S H, Seo Y K, Kim J, Vimont A, Daturi M, Serre C and Férey G 2008 Amine grafting on coordinatively unsaturated metal centers of MOFs: consequences for catalysis and metal encapsulation *Angewandte Chemie International Edition* **47** 4144-8

[23] Kayal S and Chakraborty A 2018 Activated carbon (type Maxsorb-III) and MIL-101 (Cr) metal organic framework based composite adsorbent for higher CH4 storage and CO2 capture *Chemical Engineering Journal* **334** 780-8

[24] Kayal S, Sun B and Chakraborty A 2015 Study of metal-organic framework MIL-101 (Cr) for natural gas (methane) storage and compare with other MOFs (metal-organic frameworks) *Energy* **91** 772-81

[25] Sonnauer A, Hoffmann F, Fröba M, Kienle L, Duppel V, Thommes M, Serre C, Férey G and Stock N 2009 Giant Pores in a Chromium 2, 6- Naphthalenedicarboxylate Open- Framework Structure with MIL- 101 Topology *Angewandte Chemie International Edition* **48** 3791-4

[26] Anılır G, Sert E, Yılmaz E and Atalay F S 2020 Preparation and performance of functionalized metal organic framework, MIL-101, for Knoevenagel reaction *Journal of Solid State Chemistry* **283** 121138

[27] Alamgholiloo H, Rostamnia S, Hassankhani A, Liu X, Eftekhari A, Hasanzadeh A, Zhang K, Karimi-Maleh H, Khaksar S and Varma R S 2020 Formation and stabilization of colloidal ultra-small palladium nanoparticles on diamine-modified Cr-MIL-101: Synergic boost to hydrogen production from formic acid *Journal of Colloid and Interface Science* **567** 126-35

[28] Wu W, Yao T, Xiang Y, Zou H and Zhou Y 2020 Efficient removal of methyl orange by a flower-like TiO 2/MIL-101 (Cr) composite nanomaterial *Dalton Transactions* **49** 5722-9

[29] Ko M, Aykanat A, Smith M K and Mirica K A 2017 Drawing sensors with ball-milled blends of metal-organic frameworks and graphite *Sensors* **17** 2192

[30] Samuel M S, Subramaniyan V, Bhattacharya J, Parthiban C, Chand S and Singh N D P 2018 A GO-CS@MOF [Zn(BDC)(DMF)] material for the adsorption of chromium(VI) ions from aqueous solution *Composites Part B: Engineering* **152** 116-25

[31] Ejeromedoghene O, Oderinde O, Kang M, Agbedor S, Faruwa A R, Olukowi O M, Fu G and Daramola M O 2020 Multifunctional metal-organic frameworks in oil spills and associated organic pollutant remediation *Environmental Science and Pollution Research* **27** 42346-68

[32] Zhang W, Huang H, Liu D, Yang Q, Xiao Y, Ma Q and Zhong C 2013 A new metal–organic framework with high stability based on zirconium for sensing small molecules *Microporous and mesoporous materials* **171** 118-24

[33] Mihad Ibrahim R S, Ghaleb A. Husseini 2018 Synthesis of Metal-organic Framework from Iron Nitrate and 2,6-Naphtalenedicarboxylic Acid and Its Application as Drug Carrier *Journal of Nanoscience and Nanotechnology* 1-8

# Microstructure analysis and mechanical properties of dissimilar AA6061-AA7075 laser brazing with prefixed ER5356 filler

**Haneef M S[1], Aiman M H[1,*], M N M Salleh[1], M M Quazi[1] and M Ishak[1]**

[1]Joining, Welding and Laser Processing (JWL) Lab, Faculty of Mechanical and Automotive Engineering Technology, Universiti Malaysia Pahang (UMP), 26600 Pekan, Pahang, Malaysia.

*Corresponding author's e-mail: haneef.mosabki@gmail.com

**Abstract.** Laser brazing uses a filler metal for joining without melting the base material. This process is a versatile joining technique for a wide range of applications including automotive, aerospace, and medical field because of the ability in joining dissimilar metal and resulted good quality in surface's joint. This study consists of aluminium alloy of AA6061 and AA7075 as a base material with aluminium based ER5356 as a wire filler. Laser brazing was performed using 1.2 kW of laser power, while wire filler was let to be prefixed without using the wire feeder because of using the existing laser welding machine. Microstructure of the joints were studied using 3D measuring laser microscope OLS5000. Besides that, mechanical properties of the joints were evaluated by performing tensile test and hardness test. Microstructures of the brazed joints show the differences in the grain structure followed by the difference's hardness value on each region. The brazed joint shows the average ultimate tensile strength reached 154.71 MPa which was 50% of joint efficiency. However, there is porosities at the fracture surface of the joint.

**Keywords:** Laser Brazing; Aluminium Alloy; Dissimilar Metal; Joining Process; Automotive.

## 1. Introduction

In the last few years, significant development has been made in the automotive sector, particularly in the manufacturing of dissimilar thin sheet metals, especially when joining the body part of a car [1]. The application of joining dissimilar thin sheet metals is typically found as an alternative way to weight reduction in automotive, aircraft and electronic devices [2]. There are many types of joining method in manufacturing processes such as solder, adhesive, bolt, welding and brazing technique [3]. In all types of joining, brazing gives a better joining appearance, especially in certain parts of the body car assembles. It is one of the techniques that have been applied crucially towards metal in manufacturing industries for joint goods. As a promising method of joining thin sheets of various materials, laser brazing has been explored.

Nowadays, luxurious brand in automotive industries has started to use this latest technology to produce a smooth surface joint and therefore is often used in the outer car skin areas when the joints must be invisible to the eye. Brazing is a method of joining metals in which base metals are joined using a braze filler metal without exceeding the base metal's melting point [4]. Therefore, minimal thermal distortion and high precision heat input can be achieved through this process [5]. Brazing is commonly used for joining dissimilar metal including different aluminium alloy. Laser brazing has several advantages compared to conventional welding and laser welding which is minimal heat affected zone, higher degree of mechanical stability and the high aesthetic quality of the joint.

For laser brazing process, only filler metal will be melted and almost not affected the base materials [6]. This process aims to minimize from mixing of the two materials. Meanwhile for laser welding, it is a mechanism that induces material coalescence with the heat obtained from the intense coherent light beam that affects the surfaces to be joined [7]. In certain part of assembling a car body, laser brazing will be more suitable to be applied compared to laser welding as the joining need to be smoother and better appearance such as trunk lid, roof panel and side panel. Technically, both processes use the same power source which is fiber laser as a heat source. In a fiber laser, the laser light is produced in an active fiber and directed through a flexible delivery fiber that acts as a "light guide" to the work piece [8]. Therefore, with some improvement and modification, laser brazing could be run from the existing conventional laser welding machine. Generally, for laser brazing machine, wire feeder was attached to the laser head and the laser beam will focus on filler to melt. Apart from that, scanning speed will be proportional to the wire feed rate. However, for this recent study, conventional laser welding setup will be used as this method can reduce cost and easy to setup. Therefore, prefixed wire filler will be placed directly towards the sheet metal that need to be joined.

During the laser brazing process, there are a few input parameters that determine the brazing quality, which was defined in terms of properties such as mechanical properties and filler spreadability [9]. One of the parameters that need to be highlighted is the wire filler position during the process. With the proper controlled for those parameters, microstructure, and mechanical properties of the joint could be improved [10]. Generally, the microstructure will be related to the reaction of filler metal towards base metal. This is because the growth of the reaction layer will affect the tensile strength and hardness of the joint.

Commonly, joint strength and hardness of the joint can be determined and will be analysed through a tensile test and Vickers hardness test. Meanwhile, microstructure and fracture surface will be analysed through 3D laser measuring and the optical microscope, respectively. The reason for the difference in microstructure might be due to the different experimental set up used. The experimental set up of laser brazing process towards existence laser welding machine is mainly studied. Thus, the case study for this research is to analyse the joint that will be brazed with the prefixed wire filler set up.

## 2. Materials and methods

### *2.1. Materials used*

For this study, aluminium alloy, AA6061-T6 and AA7075-T6 with thickness of 2 mm will be used as a base material. Meanwhile, ER5356 was selected as a wire filler with 2 mm diameter. The filler wire ER5356 is rich in Mg element, which has good thermal cracking resistance and high strength [11]. table 1 shows the physical and mechanical properties for both base materials.

**Table 1.** Physical and mechanical properties of base materials.

| Alloy | Melting point (°C) | Vickers hardness (HV) | Tensile strength (MPa) |
|---|---|---|---|
| AA6061-T6 | 652 | 107 | 310 |
| AA7075-T6 | 635 | 175 | 572 |

**Table 2**. Chemical composition for alloys used.

| Material | Chemical composition (wt.%) | | | | | | | |
|---|---|---|---|---|---|---|---|---|
| AA6061-T6 | Al | Mg | Si | Cr | Mn | Ti | Cu | Zn |
| | 95.8 - 98.6 | 0.8 - 1.2 | 0.4 - 0.8 | 0.04 - 0.35 | Max 0.15 | Max 0.15 | 0.15 - 0.4 | Max 0.25 |
| AA7075-T6 | Al | Mg | Si | Cr | Mn | Ti | Cu | Zn |
| | 87.1 - 91.4 | 2.1 - 2.9 | Max 0.40 | 0.18 - 0.28 | Max 0.30 | Max 0.20 | 1.20 – 2.0 | 5.1 - 6.1 |
| ER5356 | Si | Fe | Co | Mg | Zi | Ti | Cr | Al |
| | 0.25 | 0.40 | 0.10 | 4.60 | 0.10 | 0.15 | 0.15 | Bal |

*2.2. Experimental setup*

The schematic diagram for laser brazing experimental set up was shown in figure 1. This process was performed using Ytterbium Fiber laser (IPG YLS 2000) with continue wave mode operation and capable of delivering maximum 2kW output power. The laser beam is perpendicular 90° to the joint and will move horizontally.

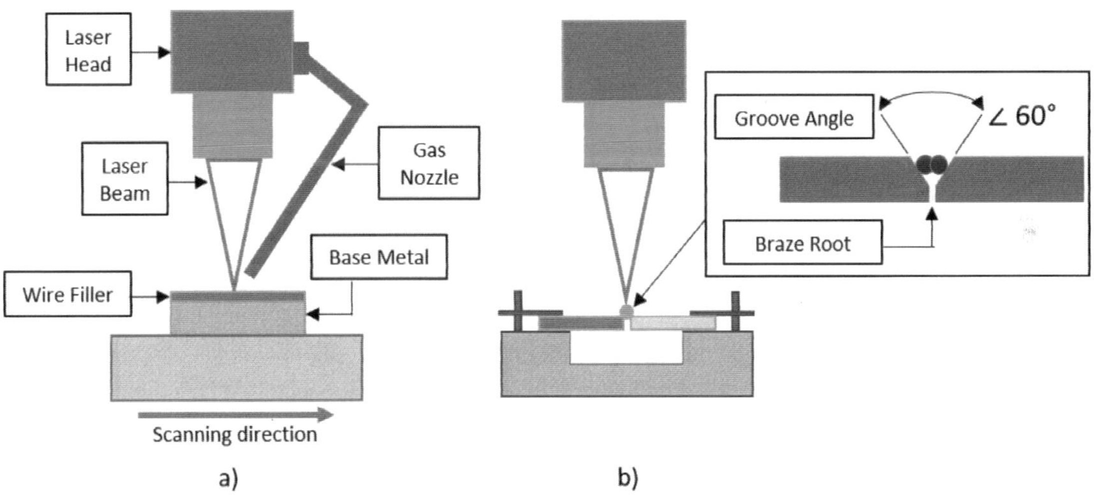


**Figure 1.** Schematic diagram of laser brazing setup (a) front view (b) side view.

*2.3. Specimen preparation and experimental procedure*

In this present study, base materials were mechanically cut into small size which is 150 x 100 mm. The plate will be grinded with 320 grit sandpaper and cleaned with acetone to remove any burr and residual oil during cutting section. After that, the edge of the plate that will be joint need to be shape a groove to let the filler melting through the joint more efficient. Then, two pieces of 1 mm diameter wire filler has been combined and will be placed manually at the centre of the groove before the experiment start. The plate will be brazed in butt joint configuration. Argon gas was the type of shielding gas used in this experiment in order to prevent oxidation towards the welded aluminium alloy [12]. Argon gas provides an inert atmosphere in which welded metals will not oxidise. Laser brazing parameters applied in this experiment are listed in table 3.

**Table 3.** Laser brazing process parameter.

| Process parameter | Laser brazing |
|---|---|
| Scanning speed (mm/min) | 1200 |
| Argon Flow rate(L/min) | 10 |
| Laser Power (kW) | 1.3 |
| Defocused length (mm) | +5 |

After the process was done, a successful joint specimen will be trimmed into ASTM-E8 standard by EDM wire cut machine. Then, the sample will proceed to tensile test and fracture analysis. In addition, cross section was selected at the centre of the joint region will apply on microstructure analysis and micro-hardness test as shown in figure 2.

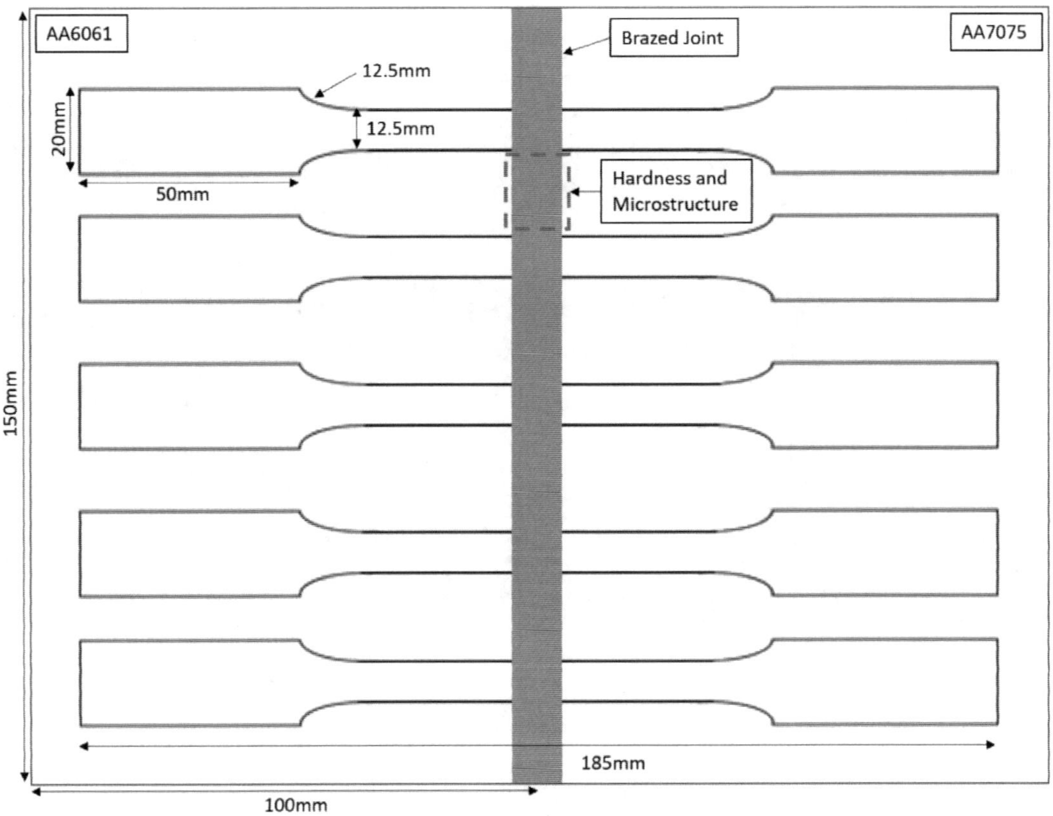


**Figure 2.** Sample for analysis.

The cross-section of the joints was mounted and chemically etched with Keller's Reagent (2.5 ml HNO3 + 1.5 ml HCl + 1 ml HF + 95 ml water) for about 15 seconds. The etched specimens were then examined for their grain size of microstructural structure using 3D Measuring Laser Microscope OLS5000. Hardness measurements were carried at the cross-section of laser weld-brazed joints using a Vickers hardness machine (MMT X7) with a load of 500 gf and dwell time of 15 s through ASTM E-384 standard. Besides that, tensile test and fracture analysis were performed by using INSTRON Universal Testing Machine and MEIJI TECHNO EM2-13TR optical microscope, respectively.

## 3. Results and discussion

*3.1 Microstructure of the cross-section joint area*

The microstructure from the cross-section brazed joint was examined under the 3D measuring laser microscope with the scale of 300μm while the macrostructure was observed by using optical microscope at applied power laser of 1300 W and scan speed of 1200 mm/min. Typically, the brazed joint could be divided into five phases region: Base Metal (BM) AA6061, Interface (IF) AA6061, Filler Zone (FZ) ER5356, IF AA7075 and BM AA7075 as can be shown in figure 3.

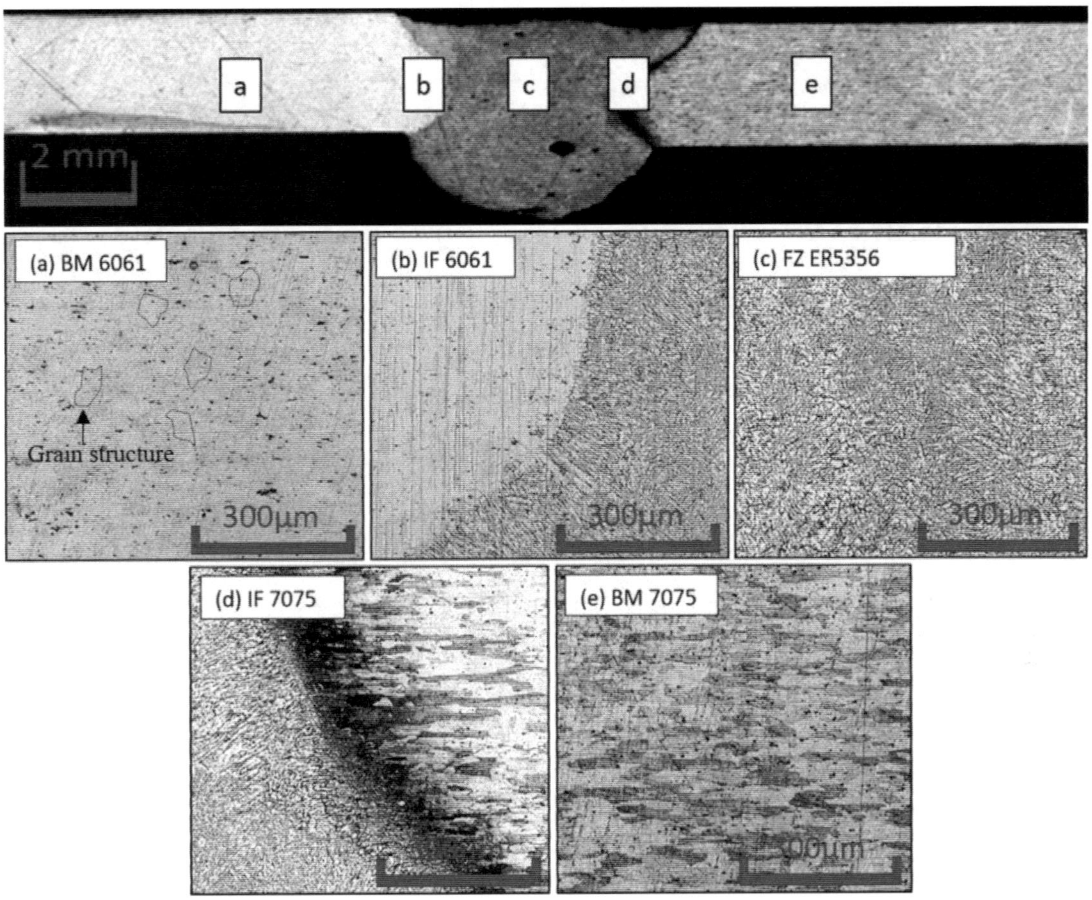


**Figure 3.** Brazed joint and microstructure of the joint: (a) BM AA6061, (b) IF AA6061, (c) FZ ER5356, (d) IF AA7075, (e) BM AA7075.

Based on the figure 3, grain structure for every region are different because of its own chemical composition for each material. Besides that, figure 3(b), shows clearer observation especially on the reaction layer between the brazed alloy than the reaction layer in figure 3(d). The results shown that the filler wire (ER5356) is more compatible towards AA7075 than AA6061 based on their similarity in chemical composition. Even though the reactions showed up are good but the reaction itself can also bring defect toward the material such as increasing their porosity and obviously will reduce the tensile strength of the materials. For the grain size of the base material, there are two differences which are the size between the base material far from the joint (unaffected) region and base material near the joint (interface) region since this region was affected by heat from laser beam during the laser brazing process [13]. For region AA7075, the average region for interface and unaffected base material were 90.60 μm and 148.27 μm, respectively. Meanwhile, the average grain for region

AA6061 interface and AA6061 BM were resulted 81.25 µm and 101.38 µm. Last but foremost, a very fine grain structure was observed in the filler region as shown in figure 3(c). The average grain size in this region ranged between 10 to 20 µm. Subsequently, the changes in grain size will affect the mechanical properties of the joint [14]. The grain sizes were measured by its diameter and the average was calculated and tabulated in table 3.

**Table 4.** Average grain size for the brazed region.

| Brazed region | Average Grain Size (µm) | Std. Deviation |
|---|---|---|
| BM AA6061 | 101.38 | 4.88 |
| IF AA6061 | 81.25 | 9.08 |
| FZ ER5356 | 15.37 | 2.70 |
| IF AA7075 | 90.60 | 7.84 |
| BM AA7075 | 148.27 | 7.38 |

### *3.2. Micro-hardness of the cross-section brazed joint*

The analysis will be conducted by using Vickers hardness test in order to differentiate the laser-brazed region and the unaffected base material for justifying the hardness distribution along the cross-section area. The same parameters were used in all the regions, which is using load of 500 gf and 15 seconds of dwell time. Indentations were to be carried out in every region for identify the average microhardness value. The distance between the indentation was varied fixed at 0.5mm in each region.

From the figure 4, it shows that the filler region (ER5356) has the lowest value of Vickers hardness compare to another region which is calculated to be 69.00 HV. This is due to the direct heat into that region and chemical composition that also can give affect towards the microhardness of the material [15]. In addition, the results show a huge difference with the hardness values between an unaffected base metal and the interface of base metal. Therefore, this is due to the affection of heat from the brazed joint area. From the result, the hypothesis that supporting the evidence which said that the finer grain size of microstructure will increasing the hardness value of the brazed joint is not valid for this study.

### *3.3. Tensile strength*

Table 5 shows the tensile strength result for the brazed joint specimen. The highest tensile strength of the brazed sample was 178.56 MPa while the lowest strength was resulted at 132.52 MPa. All the samples show in different strength even though there were all in same specimen since this method uses a prefixed filler wire setup. Therefore, the inconsistency of heating the wire filler occurs during this process.

$$\text{Welding efficiency (\%)} = \frac{UTS\ of\ welded\ sample}{UTS\ of\ base\ metal} \quad (1)$$

The average UTS of the brazed sample achieved 154.71 MPa that is equivalent to 50% of the joint efficiency. The UTS of AA6061 was 310 MPa as mentioned above. Moreover, standard deviation for this tensile result was 19.38MPa, approximately 12.52%, indicating that the tensile values along the brazed line are consistent. Figure 5 shows the fracture surface of the brazed joint. All the tensile samples having a fracture at the brazed area, which show that the brazed joint was not strong enough.

**Table 5.** Tensile strength result for the laser-brazed joint.

| Samples | Ultimate Tensile Strength (MPa) |
|---|---|
| 1 | 144.09 |
| 2 | 132.52 |
| 3 | 147.16 |
| 4 | 178.56 |
| 5 | 171.20 |
| Average, UTS | 154.71 |
| Std. Deviation | 19.38(12.52%) |
| Joint efficiency (%) | 50% |

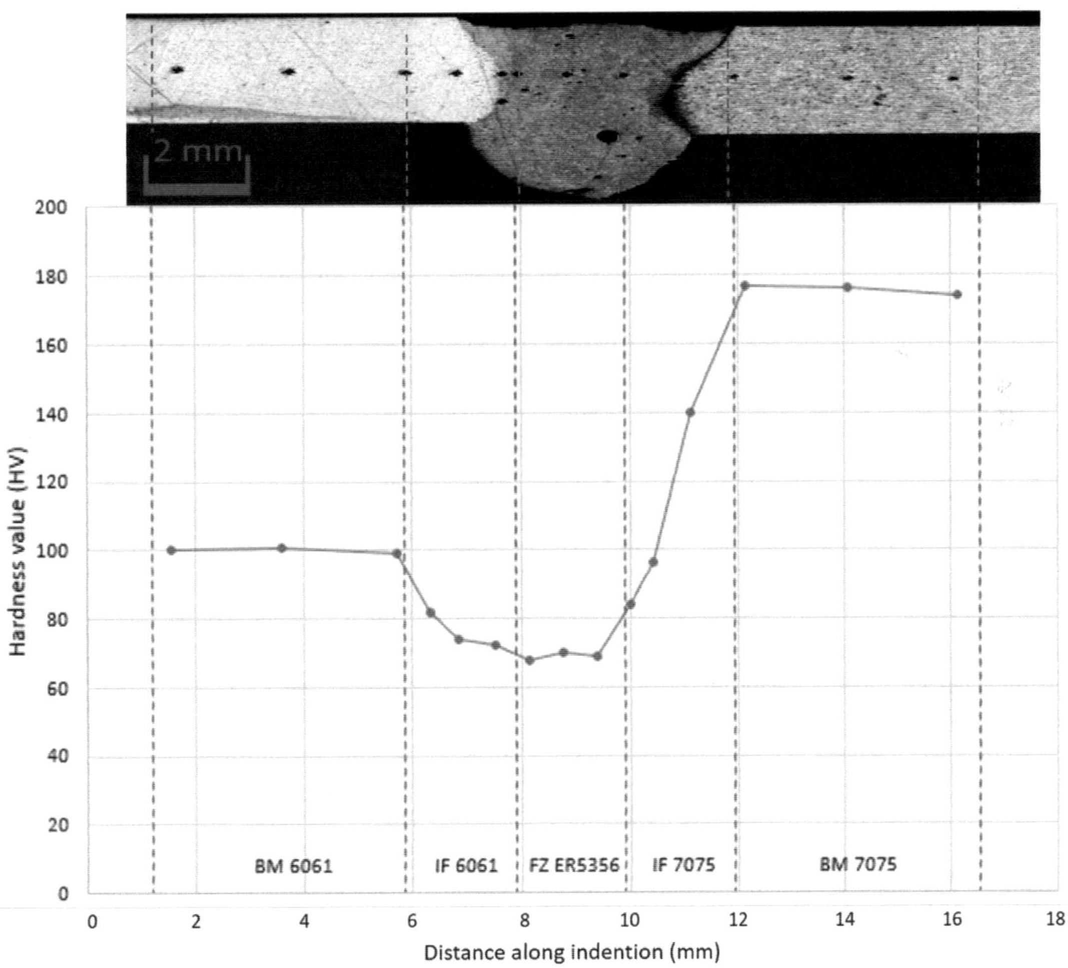


**Figure 4.** Graph of microhardness for all brazed region.

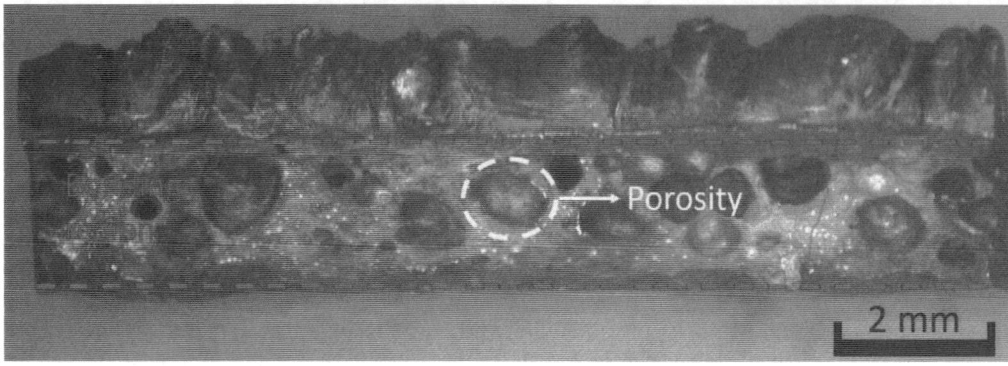


**Figure 5**. Fracture surface of the brazed joint.

Besides, there were several large porosities along the fracture surface which will affect the mechanical properties of the joint. This defect shows that the laser beam was not uniformly heating during the process. Apart from that, this makes the wire filler not consistently melt and affected the solidification process.

## 4. Conclusion

As the experimental work has been done successfully, which is laser brazing process for dissimilar metal butt joint with prefixed wire filler mechanism, some conclusion has been stated:

a) Laser beam scanning changed the microstructure at the filler and interface region. Besides, there is a difference in size grain at the interface region compare to the base metal region since the filler reaction occurs in the area.
b) The Vickers hardness value (HV) at the filler region resulted in the lowest hardness value, which is 69.00 HV. This is due to the direct heat into that region and chemical composition also can give an affect towards the microhardness of the material.
c) All the tensile samples were fractured at the centre of the joint area. The average UTS of the brazed joint was 154.71 MPa which was 50% of the joint efficiency.

Overall, the result was quite good since this experiment was still at the beginning of the research, and the wire filler was set up in prefixed condition as mentioned early. This process has been successfully done by conducting it on the existence of laser welding machine. By further modification and improvement on the experimental setup, the better result will be gained in the future. It can be concluded that the position and feed rate of the wire filler plays a significant role in laser brazing process.

## Acknowledgements

The authors would like to thank the Ministry of Education Malaysia and University Malaysia Pahang for the financial support given through research grant RDU192608 (RACER/1/2019/TK03/UMP//3) and RDU1903118.

## References

[1] Lu Y, Mayton E, Song H, Kimchi M and Zhang W 2019 Dissimilar metal joining of aluminum to steel by ultrasonic plus resistance spot welding-microstructure and mechanical properties *Materials & Design* **165** 107585.
[2] Krishnaja D, Cheepu M and Venkateswarlu D 2018 A review of research progress on dissimilar laser weld-brazing of automotive applications. In: *IOP Conference Series: Materials Science Engineering.*
[3] Rzasinski R and Kochanski L 2018 Joining methods in car body construction. In: *IOP Conference series/materials Science and Engineering,* pp 1-8.

[4] Gatzen M, Radel T, Thomy C and Vollertsen F 2016 Wetting and solidification characteristics of aluminium on zinc coated steel in laser welding and brazing *Journal of Materials Processing Technology* **238** 352-60.
[5] Zaifuddin A, Aiman M, Quazi M, Ishak M and Ariga T 2020 Effect of Laser Surface Modification (LSM) on laser energy absorption for laser brazing. In: *IOP Conference Series: Materials Science and Engineering*: IOP Publishing) p 012013.
[6] Kim Y, Park K and Kwak S 2016 A review of arc brazing process and its application in automotive *International Journal of Mechanical Engineering and Robotics Research* **5** 246.
[7] Sun J, Feng K, Zhang K, Guo B, Jiang E, Nie P, Huang J and Li Z 2017 Fiber laser welding of thick AISI 304 plate in a horizontal (2G) butt joint configuration *Materials & Design* **118** 53-65.
[8] Salleh M, Ishak M, Romlay F and Aiman M 2016 A study on bead-on-plate welding of AA7075 using low power fiber laser *Journal of Mechanical Engineering and Sciences* **10** 2065-75.
[9] Hissyam W, Aiman M, Ishak M and Ariga T 2019 Effect of copper based filler composition on the strength of brazed joint *Journal of Mechanical Engineering and Sciences* **13** 5090-103.
[10] Ye Z, Huang J, Gao W, Zhang Y, Cheng Z, Chen S and Yang J 2017 Microstructure and mechanical properties of 5052 aluminum alloy/mild steel butt joint achieved by MIG-TIG double-sided arc welding-brazing *Materials & Design* **123** 69-79.
[11] Xu F, Chen L, Lu W and He E 2017 Effect of different filler wires on weld formation for fiber laser welding 6A02 Aluminum alloy. In: *IOP Conference Series: Materials Science and Engineering*: IOP Publishing) p 012041.
[12] Shiganov I N, Misyurov A I, Trushnikov A N, Kholopov A A and Blinkov V V 2017 Hybrid laser-arc welding of aluminium alloys *Welding International* **31** 67-70.
[13] Salleh M, Ishak M, Aiman M, Quazi M and Hanafi A 2019 Weld Geometry Investigation on Dissimilar Boron Steel Laser Welded for TWB Application *International Journal of Automotive and Mechanical Engineering* **16** 7364-74.
[14] Salleh M, Ishak M, Aiman M, Zaifuddin Q and Quazi M 2020 The effect of laser surface hardening on the surface hardness of mild steel. In: *IOP Conference Series: Materials Science and Engineering*: IOP Publishing) p 012014.
[15] Sadeghian M, Shamanian M and Shafyei A 2014 Effect of heat input on microstructure and mechanical properties of dissimilar joints between super duplex stainless steel and high strength low alloy steel *Materials & Design* **60** 678-84.

# Leak diagnosis of pipeline based on empirical mode decomposition and support vector machine

**A Faysal[1], M S N A Adhreena[2], E Vorathin[2], Z M Hafizi[2] and W K Ngui[1,*]**

[1]College of Engineering, Universiti Malaysia Pahang (UMP), 26300 Gambang, Pahang, Malaysia.
[2]Advanced Structural Integrity and Vibration Research (ASIVR), Faculty of Mechanical & Automotive Engineering Technology, Universiti Malaysia Pahang (UMP), 26600 Pekan, Pahang, Malaysia.

*Corresponding author's e-mail: wkngui@ump.edu.my

**Abstract.** The pipeline is used as a medium of transportation in global gas and oil industries, providing the most efficient, convenient and transportation method for natural gas and oil from downstream to upstream production of the economical mode of the power station, refineries, and domestic needs. However, the pipeline leakages become a major concern as their failure may contribute to operational and economic loss as well as environmental pollution. This paper proposed a system to detect pipe fault at different locations. Empirical Mode Decomposition (EMD) was applied for feature extraction using energy and kurtosis. The one-against-one (OAO) and one-against-all (OAA) multiclass SVM with radial basis function (RBF), polynomial and sigmoid kernel functions were implemented in order to classify the multiple fault locations from the extracted features. RBF kernel function recorded the highest classification accuracy for both OAO and OAA approaches with 97.77% and 96.29%, respectively, followed by slightly reduced accuracy for sigmoid whereas significantly low accuracy for the polynomial kernel. The outputs were further analysed to justify the performance of the classifiers. From all the cases, it was observed that OAO-SVM with RBF kernel performed the best for pipe fault diagnosis.

**Keywords:** Fault Diagnosis; Condition Monitoring; Support Vector Machine.

## 1. Introduction

The use of pipelines for fluid transportation has been widely utilised across the world. However, over time, the pipelines are susceptible to leakage due to several factors such as corrosion, degradation of the material, poor quality of fittings and many more [1]. Leakage in pipelines will not only consequence in the operational loss but also to environmental pollution. Therefore, effective leak detection in the pipeline is essential. Over the years, leak detection using vibration analyses are the most widely reported.

Karkulali *et al.* [2] reported on the use of a piezo thin-film sensor to capture the leak signals and Fast Fourier Transform (FFT) signal processing was used to analyse the leak. As a result, the FFT spectrum capable of distinguishing between no leak and leak conditions. However, the authors did not further report on the classification of the leak in terms of size and distance. Wijayanto *et al.* [3] utilised the same signal processing method to determine the size of the leak in a pipeline. However, the results showed that the FFT spectrum could not differentiate between no leak and small leak conditions. The frequency-domain was only significant when the leak was massive. According to Bentoumi *et al.* [4], the used of FFT signal processing was also associated with drawbacks in terms of noise, fault geometry and fluid nature. However, for various limitation of frequency domain analysis for vibration signals, time-frequency adaptive signal processing can effectively mitigate the discrepancies.

An unconventional, relatively new and a robust time-frequency analysis, EMD was first presented in 1998. Since then, EMD has been applied to stationery and nonstationary signals in numerous fault diagnosis research [5, 6]. The principle of EMD is based on the signal's local characteristics in the time scale. The time scale is divided into a set of orthogonal components known as intrinsic mode function (IMF). IMFs are determined by the signal, not some kernel. Therefore, EMD is self-adaptive and can be applied efficiently to any signal. Since EMD is so well fit for any signal, it has been applied affluently in condition monitoring and fault diagnosis research. Jiang *et al.* [7, 8] used an improved EMD for rolling element bearing fault diagnosis where the results showed that the proposed method can not only diagnose known faults but also monitor unknown faults with strong, robust performance. Lei *et al.* [7, 8] used locomotive rolling element bearings to demonstrate the performance of EMD. In this paper, EMD is applied to the pipe fault signal for effective fault feature extraction.

An effective fault diagnosis method would have the adaptability to accommodate to signal variations and different system. Machine learning algorithms meet these criteria and that makes them a helpful tool in fault diagnosis research. SVM is a powerful supervised learning algorithm which analyses data for regression analysis or non-probabilistic classification. SVM uses hyperplane to maximise the distance between the two classes. Gao *et al.* [9] applied wavelet packet transform (WPT) to extract features and used SVM for classification of the samples in reciprocating pumps fault diagnosis. Yuan and Chu [10] successfully applied SVM for fault diagnosis of the turbo-pump rotor. However, SVM has many kernel function and approach for multiclass classification where the hyperparameter tuning dominates the classification accuracy. So, it is important to investigate which multiclass SVM approach with the right kernel function performs the best with the feature extraction algorithm. In this paper, EMD was used for feature extraction, while SVM was used for classification.

## 2. Theoretical background

### *2.1. EMD*

EMD is a new time-frequency adaptive method which uses the shifting process to decompose the signal into IMFs. The shifting process is conducted by connecting the local maxima and minima to form the upper and lower envelopes. The sifting process is repeated until the following two conditions satisfy:

(a) The number of extrema in the whole time series must differ with the zero-crossing value by at most one.

(b) The mean value of the upper and lower envelopes at any given location is zero.

An IMF, $c_1$, is produced when the above-mentioned conditions are satisfied. The residue is obtained by subtracting the $IMF_1$ from the main signal which is denoted by $r_1$ and is used to obtain the next IMFs. The decomposition is performed n times until the residue gets monotonic. By adding the IMFs and the remaining residue, the original signal can be obtained as follows:

$$X(t) = \sum_{i=1}^{n} c_i + r_n \tag{1}$$

where X(t) is the original signal, $c_i$ is the $i$th IMF, and $r_n$ is the $n$th residue.

### 2.2. SVM

SVM uses hyperplane to maximise the distance between the two classes. An optimal hyperplane is obtained that represents the margin of the vectors known as support vectors. This type of SVM uses a linear decision boundary and is called linear SVM and expressed as follows:

$$w.x + b = 0 \tag{2}$$

which indicates

$$y_i(w.x + b = 0) \geq 1, i = 1, \cdots, N \tag{3}$$

The goal of SVM is to obtain the highest separating hyperplane by minimizing $||w||$ using the conditions. Here $||w||$ the Euclidean norm of w, and $2/||w||$ denotes the distance between the hyperplane and the nearest data points of each class. By introducing Lagrange multipliers $\alpha_i$, the SVM is trained to solve a convex quadratic problem (QP). A unique globally optimized result is the solution, which contains the following properties:

$$w = \sum_{i}^{N} \alpha_i y_i x_i \tag{4}$$

Only if corresponding $\alpha_i > 0$, these $x_i$ are called support vectors.

The decision function of the trained SVM can be expressed as:

$$f(x) = sign\left(\sum_{i=1}^{N} \alpha_i y_i (x \cdot x_i) + b\right) \tag{5}$$

When the observations are non-linearly separable, SVM performs a nonlinear mapping of the input vector x from the input space $R^N$ into a higher dimensional Hilbert space, where the mapping is determined by the kernel function. Different types of kernels are suitable for different classification problems to reach the optimal classification error.

## 3. Experimental procedures and results

The test rig setup for the collection of the vibration signal from the pipe is shown in figure 1. An air storage tank with the length and diameter of 120 cm and 11.43 cm was used to replicate the leaks in a pipeline. A total of 5 plug fittings with a hole diameter of 1 mm were welded along the length of the pipe. The distance between each plug was fixed at 20 cm. A ball valve was fitted to the plug to control the leak. The storage tank was pressurised by using centralised compressed air system. The pressure inside the tank was monitored by using a pressure gauge. An accelerometer with the sensitivity of 100 mV/g was used to capture the leak signals. The accelerometer was attached at the distance of 15 cm away from leak 1. The leak signals were acquired by using National Instruments data acquisition model NI 9234. The signals were captured at a sampling rate of 25.6 kHz. Initially, the storage tank was pressurised with a pressure of 0.2 MPa and the signal was recorded for 1 minute which is labelled as no leak condition. Next, the ball valve at 15 cm was opened to induce the leak and the signals were recorded for 1 minute. The same procedures were repeated for leaks at 35 cm, 55 cm, 75 cm and 95 cm away from the accelerometer. The total leak type considered in this study is illustrated in table 1.

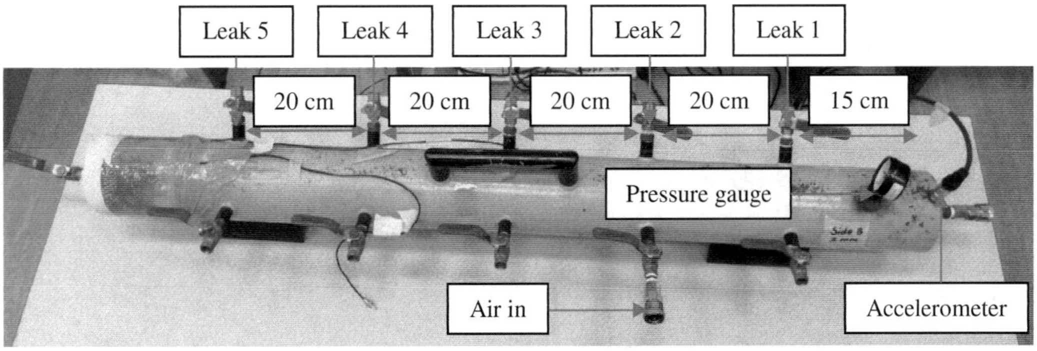


**Figure 1.** Experimental setup of pipe fault test rig.

**Table 1.** Pipe leak at different locations from the sensor.

| Leak type | No leak | Leak 1 | Leak 2 | Leak 3 | Leak 4 | Leak 5 |
|---|---|---|---|---|---|---|
| Fault class | C1 | C2 | C3 | C4 | C5 | C6 |

### *3.1. Classification modelling*
For feature extraction, 6250 data points were taken per sample. For each condition, 75 samples were obtained and for all six different conditions, the total number of samples is 450. EMD was applied to decompose the vibration signals into several IMFs. It is to be noted that the lower order IMFs carry the most information and for higher order, the information decreases. So, taking all the IMFs into consideration would increase the computational burden and decrease the accuracy. In this paper, only the first five IMFs are considered [11].

In this paper, two statistical parameters, energy and kurtosis are used to obtain the feature vector. The summation of the energy can obtain the total energy of a signal in each sub-band and energy feature in each band. Kurtosis is a dimensionless statistical measure that characterises the flatness of a signal's probability density function defined as the fourth-order moment of signal data and kurtosis of the sub-band coefficient [10.1007/s00500-013-1055-1]. From each condition, out of five IMFs, ten features were extracted which consist of energy and kurtosis for each IMF. Table 2 represents one feature vector row from each fault class chosen randomly for visualization of the features. The final size of the feature vector is 450x10 for the entire feature set. This feature vector was normalised between 0 to 1 and fed into the SVM for classification. OAO SVM and OAA SVM were implemented to classify the faults at different locations. In SVM, kernel function is crucial to get desired classification accuracy. In this study, the radial basis function (RBF), polynomial and sigmoid kernel functions are used. To evaluate the performance of the classifier, 70% of the data was used to train the classifier and the rest 30% to test the accuracy. So, the number of samples obtained for training and testing were 315 and 135, respectively. The data was partitioned randomly which results in uneven distribution of the number of samples in different fault class.

## 4. Results and discussion
All three combinations of the kernel were applied to OAO and OAA approach of SVM. From table 3, it is observed that the RBF kernel obtained the highest classification accuracy for both approaches, whereas the polynomial kernel obtained the lowest. The testing accuracy with RBF in OAO was slightly higher than OAA. In both cases of sigmoid, the testing accuracy was slightly lower than RBF, which is 94.81% and 94.07% for OAO and OAA approach, respectively. Since the accuracy of the polynomial kernel in both OAO and OAA was significantly low, no further analysis was conducted with it. The confusion matrix with precision, recall and F1 score for RBF and Sigmoid from both OAO and OAA approach are provided in table 4 and table 5.

**Table 2.** Representation of the feature vector for different class.

| Fault | IMF1 | | IMF2 | | IMF3 | | IMF4 | | IMF5 | |
|---|---|---|---|---|---|---|---|---|---|---|
| Class | Energy | Kurtosis | Energy | Kurtosis | Energy | Kurtosis | Energy | Kurtosis | Energy | Kurtosis |
| No leak | 0.001228 | 0.000434 | 0.000214 | 0.000176 | 0.000148 | 2.3732 | 3.0511 | 3.2202 | 2.8091 | 4.0188 |
| Leak 1 | 2.9025 | 0.48943 | 0.10441 | 0.030102 | 0.009964 | 2.3998 | 3.1116 | 3.8018 | 3.2697 | 2.7446 |
| Leak 2 | 4.7834 | 0.66713 | 0.14955 | 0.039948 | 0.019658 | 2.3546 | 3.1181 | 3.7326 | 3.4483 | 2.381 |
| Leak 3 | 2.1343 | 0.88299 | 0.069758 | 0.019215 | 0.008126 | 2.2583 | 3.0966 | 3.5044 | 3.7247 | 3.9072 |
| Leak 4 | 2.3039 | 0.42037 | 0.050061 | 0.012056 | 0.006248 | 2.3876 | 3.0483 | 3.7251 | 2.9731 | 4.3754 |
| Leak 5 | 2.6432 | 1.7895 | 0.11299 | 0.033748 | 0.010545 | 2.9254 | 3.033 | 5.2849 | 3.9484 | 4.6836 |

In all results of table 4 and 5, although the number of samples varies for individual class because of the random partition of feature set, the total observation is 135. Among all four cases, since OAO-SVM with RBF kernel obtained the highest accuracy, its precision, recall and F1 score were also the highest with 0.98, 0.97 and 0.98. respectively. In case of OAO-SVM, the no fault condition could obtain the perfect accuracy for both RBF and sigmoid kernel which is 1 for precision, recall and F-1 score. Only class C2 and C6 in OAO-SVM with RBF kernel could obtain the perfect accuracy too. As expected, the precision, recall and F1 score for the OAA-SVM with RBF kernel was the second highest which were 0.96 for all of them. The performance of sigmoid was lower for both SVM approach where OAO-SVM with sigmoid kernel was slightly more accurate than OAA-SVM with Sigmoid kernel. From table 5(a), it is observed that the first three classes obtained the perfect accuracy whereas in table 5(b) C3 and C6 could obtain it. These differences show why the classifier with RBF kernel was better. In all cases, OAO-SVM performed better than OAA-SVM for the same kernel whereas OAO-SVM with RBF kernel was found to be the best performing classifier in this study.

**Table 3.** SVM accuracy using different kernel functions.

| Approach | Kernel | Test accuracy (%) |
|---|---|---|
| OAO | RBF | 97.77 |
| | Polynomial | 74.81 |
| | Sigmoid | 94.81 |
| OAA | RBF | 96.29 |
| | Polynomial | 75.55 |
| | Sigmoid | 94.07 |

**Table 4(a).** Confusion matrix with precision, recall and F1 score for OAO-SVM with RBF kernel.

| Class | C1 | C2 | C3 | C4 | C5 | C6 | Total | Precision | Recall | F1 |
|---|---|---|---|---|---|---|---|---|---|---|
| C1 | 27 | 0 | 0 | 0 | 0 | 0 | 27 | 1 | 1 | 1 |
| C2 | 0 | 20 | 0 | 0 | 0 | 0 | 20 | 1 | 1 | 1 |
| C3 | 0 | 0 | 21 | 0 | 1 | 0 | 22 | 1 | 0.95 | 0.98 |
| C4 | 0 | 0 | 0 | 17 | 1 | 0 | 18 | 0.94 | 0.94 | 0.94 |
| C5 | 0 | 0 | 0 | 1 | 20 | 0 | 21 | 0.91 | 0.95 | 0.93 |
| C6 | 0 | 0 | 0 | 0 | 0 | 27 | 27 | 1 | 1 | 1 |
| Total | | | | | | | 135 | 0.98 | 0.97 | 0.98 |

**Table 4(b).** Confusion matrix with precision, recall and F1 score for OAO-SVM with Sigmoid kernel.

| Class | C1 | C2 | C3 | C4 | C5 | C6 | Total | Precision | Recall | F1 |
|---|---|---|---|---|---|---|---|---|---|---|
| C1 | 27 | 0 | 0 | 0 | 0 | 0 | 27 | 1 | 1 | 1 |
| C2 | 0 | 20 | 0 | 0 | 0 | 0 | 20 | 0.91 | 1 | 0.95 |
| C3 | 0 | 1 | 21 | 0 | 0 | 0 | 22 | 0.91 | 0.95 | 0.93 |
| C4 | 0 | 0 | 0 | 16 | 2 | 0 | 18 | 0.94 | 0.89 | 0.91 |
| C5 | 0 | 1 | 0 | 1 | 19 | 0 | 21 | 0.90 | 0.90 | 0.90 |
| C6 | 0 | 0 | 2 | 0 | 0 | 25 | 27 | 1 | 0.93 | 0.96 |
| Total | | | | | | | 135 | 0.94 | 0.95 | 0.94 |

**Table 5(a).** Confusion matrix with precision, recall and F1 score for OAA-SVM with RBF kernel.

| Class | C1 | C2 | C3 | C4 | C5 | C6 | Total | Precision | Recall | F1 |
|---|---|---|---|---|---|---|---|---|---|---|
| C1 | 24 | 0 | 0 | 0 | 0 | 0 | 24 | 1 | 1 | 1 |
| C2 | 0 | 21 | 0 | 0 | 0 | 0 | 21 | 1 | 1 | 1 |
| C3 | 0 | 0 | 19 | 0 | 0 | 0 | 19 | 1 | 1 | 1 |
| C4 | 0 | 0 | 0 | 20 | 2 | 0 | 22 | 0.91 | 0.91 | 0.91 |
| C5 | 0 | 0 | 0 | 2 | 17 | 1 | 20 | 0.89 | 0.85 | 0.87 |
| C6 | 0 | 0 | 0 | 0 | 0 | 29 | 29 | 0.97 | 1 | 0.98 |
| Total | | | | | | | 135 | 0.96 | 0.96 | 0.96 |

**Table 5(b).** Confusion matrix with precision, recall and F1 score for OAA-SVM with Sigmoid kernel.

| Class | C1 | C2 | C3 | C4 | C5 | C6 | Total | Precision | Recall | F1 |
|---|---|---|---|---|---|---|---|---|---|---|
| C1 | 18 | 0 | 0 | 0 | 0 | 1 | 19 | 1 | 0.95 | 0.97 |
| C2 | 0 | 24 | 0 | 1 | 0 | 0 | 25 | 1 | 0.96 | 0.98 |
| C3 | 0 | 0 | 29 | 0 | 0 | 1 | 29 | 1 | 1 | 1 |
| C4 | 0 | 0 | 0 | 14 | 4 | 0 | 18 | 0.82 | 0.78 | 0.80 |
| C5 | 0 | 0 | 0 | 2 | 17 | 0 | 19 | 0.77 | 0.89 | 0.83 |
| C6 | 0 | 0 | 0 | 0 | 0 | 25 | 25 | 1 | 1 | 1 |
| Total | | | | | | | 135 | 0.93 | 0.93 | 0.93 |

## 5. Conclusion

Throughout this study, EMD and SVM were successfully developed for early detection and localisation of small leaks occurred along the pipeline. The OAO and OAA SVM approaches were implemented to classify the multiple fault location with applying the kernel function of RBF, polynomial and sigmoid functions. Compared to others, RBF kernel function recorded the highest classification accuracy in both OAO and OAA approaches with a percentage of 97.77% and 96.29% respectively followed by a slightly lower performance of sigmoid, 94.81% and 94.07%, respectively. On the other hand, the performance of polynomial kernel was significantly poor in both SVM approach. The classification results with RBF and sigmoid were further analysed using confusion matrix, precision, recall and F1 score. It was observed that OAO-SVM with RBF kernel is the best combination since it can obtain the highest accuracy with the best precision, recall and F1 score.

**Acknowledgments**

The authors would like to thank the Faculty of Mechanical & Automotive Engineering Technology, Universiti Malaysia Pahang (http://www.ump.edu.my/) for providing the laboratory facilities. Finally, special thanks to the UMP Research and Innovation Department for providing the internal research grant under grant no. RDU1903120.

**References**

[1] Ranginkaman M H, Haghighi A and Vali Samani H M 2016 Inverse frequency response analysis for pipelines leak detection using the particle swarm optimisation *Iran University of Science & Technology* **6** 1-12.

[2] Karkulali P, Mishra H, Ukil A and Dauwels J 2016 Leak detection in gas distribution pipelines using acoustic impact monitoring *IECON 2016-42nd Annual Conference of the IEEE Industrial Electronics Society* 412-6.

[3] Wijayanto A P and Miasa I M 2019 Effects of variations in pipe leakage diameter and pipe leakage distance to the FFT diagrams *Journal of Physics: Conference Series* **1351** 012055.

[4] Bentoumi M, Chikouche D, Mezache A and Bakhti H 2016 Wavelet dt method for water leak-detection using a vibration sensor: an experimental analysis *IET Signal Processing* **11** 396-405.

[5] Lei Y, Lin J, He Z and Zuo M J 2013 A review on empirical mode decomposition in fault diagnosis of rotating machinery *Mechanical systems and signal processing* **35** 108-26.

[6] He Q, Li P and Kong F 2012 Rolling bearing localised defect evaluation by multiscale signature via empirical mode decomposition *Journal of vibration and acoustics* **134.**

[7] Jiang F, Zhu Z, Li W, Chen G and Zhou G 2013 Robust condition monitoring and fault diagnosis of rolling element bearings using improved EEMD and statistical features *Measurement Science and Technology* **25** 025003.

[8] Lei Y, Liu Z, Ouazri J and Lin J 2017 A fault diagnosis method of rolling element bearings based on CEEMDAN *Proceedings of the Institution of Mechanical Engineers, Part C: Journal of Mechanical Engineering Science* **231** 1804-15.

[9] Gao J, Shi W, Tan J and Zhong F 2002 Support vector machines based approach for fault diagnosis of valves in reciprocating pumps *IEEE CCECE2002. Canadian Conference on Electrical and Computer Engineering. Conference Proceedings (Cat. No. 02CH37373)* **3** 1622-7.

[10] Pan S J and Yang Q 2009 A survey on transfer learning *IEEE Transactions on knowledge and data engineering* **22** 1345-59.

[11] Cheng Y, Wang Z, Chen B, Zhang W and Huang G 2019 An improved complementary ensemble empirical mode decomposition with adaptive noise and its application to rolling element bearing fault diagnosis *ISA transactions* **91** 218-34.

# Investigation of design space in manufacturing meta-biomaterials by additive manufacturing

**Siti Rohaida Mohamed[1,*] and Saiful Anwar Che Ghani[1]**

[1]Faculty of Mechanical and Automotive Engineering Technology, Universiti Malaysia Pahang, 26600, Pekan, Pahang, Malaysia.

*Corresponding author's e-mail: sitirohaida0603@gmail.com

**Abstract.** Topology features such as interconnectivity, pore shape and size, porosity, struts thickness, and used materials play the key roles for mechanical and biological properties of meta-biomaterials structures. However, the influences of morphological geometries on the mechanical and biological properties are not certainly intuitive. This paper develops parametric model that use to visualize the morphological geometries of unit cell of meta-biomaterials on design space that governing the manufacturing limitation, mechanical and biological requirements. The selected samples within design spaces tested to determine manufacturing accuracy and effective elastic modulus by finite element analysis. The geometries discrepancies between designed models and manufactured samples obtained percentage of average errors of 13% for diamonds structures and 21% for square structures. The proposed technique yielded average error reduced to 74.4% for diamond structures and 44.4% for square structures of effective elastic modulus from theoretical calculation. The approach and the implications of the results discussed in the context of mechanical and biological criteria with highlight of advantages and limitations of meta-biomaterials manufactured by additive manufacturing for orthopaedic implants.

**Keywords:** Meta-biomaterials; Additive Manufacturing; Finite Element Analysis; CoCrMo.

## 1. Introduction

Meta-biomaterials known as a unique class of materials that exhibited unprecedented combination of mechanical, mass transport and biological properties made from biomaterials for biomedical applications [1, 2]. Metamaterials are design with a repeating unit cell and used for other applications including optical [3, 4] and acoustic [5, 6] given to micro-architecture level. Bone-mimicking meta-biomaterials are progressively being studied which are particularly relevant for bone substitution since the implants offer a high surface area to promote bone in-growth, facilitate transportation of nutrient components, expedite tissue regeneration as well as improve implant longevity [7]. In addition, the mechanical properties of meta-biomaterials ideally tailored to have comparable elastic modulus with human bone as to reduce stress-shielding phenomenon that influence bone resorptions and revision surgeries.

The reliability to manufacture meta-biomaterials with highly controlled micro geometry architectures is vital to the great achievement of orthopaedic implants. Conventional techniques for example, space holder [8, 9] and direct metal foaming [10, 11] provide restricted freedom in fabricating complex structures with desired porosity and homogenous pore distribution and interconnectivity that can fulfill specific requirements of mechanical properties and biological response [12-14]. Recently, additive manufacturing (AM) technology such as electron beam melting (EBM) and selective laser melting (SLM), bring versatility in three-dimensional (3D) layer-by-layer processes of patient-specific implant devices [15-17]. Additionally, gradient of pore size and porosity are introduced into the micro-architecture to improve mechanical behaviour and biological response with the respect of anatomical location of local bone.

The micro-architectures of the unit cell of meta-biomaterials customized with rationally controlled pore shape, pore size, and porosity to provide sufficient mechanical stability and optimum structural suitability for bone-implant interactions [18, 19]. From previous studies, topological design of unit cell meta-biomaterials, relative density and used materials influenced the compressive and fatigue properties of the structures [20-22]. For meta-biomaterials, there are presently no quantitative standards of specifying pore parameters and porosity volume constraints for cells in-growth and mechanical stability of the implants. Furthermore, there is lack of study that discussed the effects of unit cell size, porosity percentage, and strut thickness on the mechanical and biological properties of meta-biomaterials particularly for orthopaedic implant applications.

Many recent studies are focusing on selection of unit cell topologies of porous biomaterials with no systematic approach to obtain designed structures that are manufacturable for implants [23, 24]. However, this method does not provide a full perspective of the precise viable design space to define each unit cell topological parameters. Additionally, this technique frequently leads to design and produce porous implants with geometrical parameters higher than the ideal range for biological properties requirements. Any insight into how the topologies design, biological and manufacturing requirements influenced the acceptable parameter design range of a unit cell of meta-biomaterials. In essence, an extensive characterization of the developed design space for meta-biomaterials is crucial to comprehend the relationship of mechanical properties and biological requirements with the geometrical design boundaries.

This paper presents systematic procedure for understanding the interaction relationship between geometrical parameters with the mechanical and biological properties. The chart model allows graphically examining the acceptable porosity percentage, unit cell size, strut thickness and pore size that are possible for fabricating process. This work also goals to investigate the purpose of design space in manufacturing meta-biomaterials by SLM, as well as to comprehend the relationship of with geometrical constraints inflicted by biological necessities, mechanical requirements and additive manufacturing drawback.

## 2. Methodology

### *2.1. Development of cell topologies design domains*

The parametric geometrical of diamond and square unit cell types were applied to develop the design domains. Both unit cells have modulus matrix with cubic symmetry. The designed unit cells exhibited almost isotropic mechanical properties. From the given geometric topology, computer-aided design (CAD) models generated via SolidWorks software. The overall unit cell type controlled by two parameters, which were strut thickness (*t*) and unit cell size (*l*). These parameters combination regulated each unit cell to obtain resultant of desired porosity and pore size. In this study, the pore size defined by the largest circle correspond that can permit through adjacent cells in the samples. The porosity derived as in equation 1.

$$Porosity\ (\%) = \left(1 - \frac{Vp}{Vs}\right) \times 100 \quad (1)$$

where $v_p$ refers to void volume of unit cell and $v_s$ refers to the volume of solid unit cell. The pore sizes and porosities plotted in contour design spaces where on the *x*-axis represented pore size and the *y*-axis represented strut thickness. Meanwhile, isometric lines illustrated the value of cell size and porosity of the structures. From contour area, the accompanying biomechanical requirements and fabrication limits overlaid to feature the allowable design maps as in figure 1 and figure 2. The according vital criteria for generating design space as follows:

- Biological requirement: Pore size range of 50 µm and 800 µm with porosity more than 50% are favourable for optimum biological response. The red lines showed the boundaries of these values in the design space maps.
- Manufacturing constraint: Additive manufacturing (AM) technologies restricted to manufacture a minimal strut thickness of 300 µm of meta-biomaterials depending to selected process parameters. The constraint illustrated in the *x-axis* design space maps as in red line.

The design domain maps develops with borders as in the blue region represents the geometric parameters with respect to appropriate for bone in-growth and manufacture capability of AM technology. All the designs of meta-biomaterials falling within the domains area were satisfactory solutions that emerge the bone in-growth requirements and AM limitations.

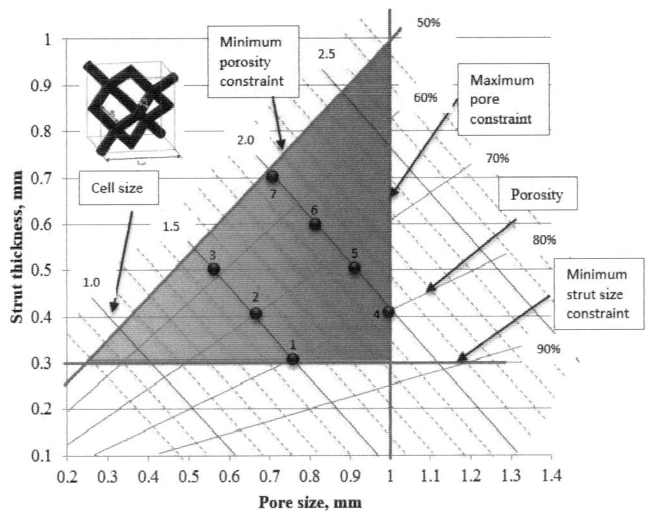


**Figure 1.** Admissible design space for diamond type.

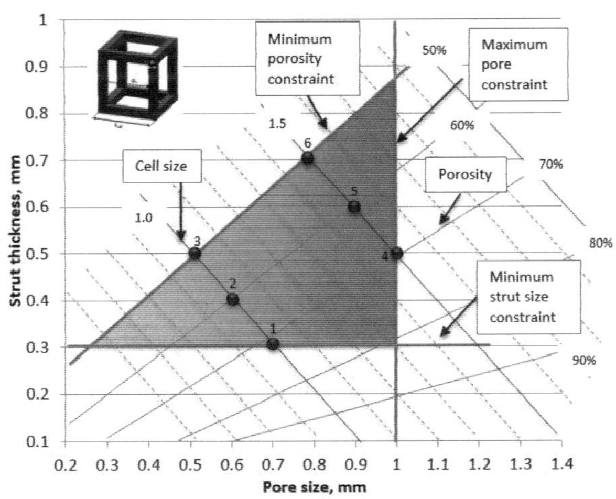


**Figure 2.** Admissible design space for square type.

*2.2. Design parameters of selected samples*

The representative samples (in black) at the boundary of the design spaces as in figure 1 and figure 2 selected for further investigation. Seven points of diamond type designated with the porosity in the range of 50% to 80% (Diamond #1-7). Meanwhile, for square unit cell, six points were selected in the range of porosity of 50% to 70% (Square #1-6). These points considered direct correlation of the consequence of geometrical parameters with morphology corresponding to ideal requirement of orthopaedic implants. Table 1 summarises the morphological details for the selected points.

**Table 1.** Geometrical details of the representative samples.

| Unit cell type | # | Unit cell size (mm) | Strut thickness (mm) | Pore size (mm) | Porosity (%) |
|---|---|---|---|---|---|
| Diamond | 1 | 1.5 | 0.3 | 0.76 | 82 |
| | 2 | 1.5 | 0.4 | 0.66 | 70 |
| | 3 | 1.5 | 0.5 | 0.56 | 58 |
| | 4 | 2.0 | 0.4 | 1.1 | 80 |
| | 5 | 2.0 | 0.5 | 0.91 | 74 |
| | 6 | 2.0 | 0.6 | 0.81 | 66 |
| | 7 | 2.0 | 0.7 | 0.71 | 54 |
| Square | 1 | 1.0 | 0.3 | 0.70 | 70 |
| | 2 | 1.0 | 0.4 | 0.60 | 60 |
| | 3 | 1.0 | 0.5 | 0.50 | 50 |
| | 4 | 1.5 | 0.5 | 1.0 | 70 |
| | 5 | 1.5 | 0.6 | 0.9 | 60 |
| | 6 | 1.5 | 0.7 | 0.80 | 50 |

*2.3. Material and fabrication process*

The selected geometries fabricated by SLM process (SLM® 125HL) out of cobalt chrome molybdenum (CoCrMo) alloys with powder particles had average size of approximately 22 μm. The manufacturing process used 300 W laser and laser spot diameter of 80 μm with energy density 119 J/mm$^3$. The samples manufactured at intervals of 30 μm layer thicknesses and hatch spacing of 120 μm with layer-by-layer exposure on a stainless steel base plate. The build chamber was flushed with argon gas to prevent contamination and oxidation. The samples experienced thermal stress-relieved at temperature of 1050 °C for two (2) hours in an argon atmosphere. Electrical discharge machining (EDM) used to detach the build samples from the base plate.

*2.4. Morphology evaluations*

Each design points were randomly selected and morphological evaluated in order to measure as manufactured strut thickness and pore size. The strut size of produced samples investigated and analyzed in addressing the manufacturability and accuracy of SLM performance of the produced meta-biomaterials using optical microscope (Dino-lite Digital Microscope). These average values measured as ten dimensional values at random points for every measurement. The results were compared to the as CAD designed strut thickness and pore size.

*2.5. Prediction of effective elastic modulus*

A model proposed by Gibson and Ashby [25] was employed in order to predict the effective elastic modulus of designed samples as in equation 2. Elastic modulus of solid Co-Cr-Mo was consumed as 220 GPa [26].

$$E^*/E_s = C(\rho^*/\rho_s)^2 \tag{2}$$

where $E^*$ = Elastic modulus of meta-biomaterials;
$E_s$ = Elastic modulus of solid;
$C$ = Constant;
$\rho^*$ = Density of meta-biomaterials; and
$\rho_s$ =Density of solid

From above equation, elastic modulus demonstrated that was dependent on the relevant density. Equation 3 used to determine the relevant density that associated to porosity volume [25].

$$(\rho^*/\rho_s)=1-\varphi \tag{3}$$

where $\varphi$ was the porosity. Then, equation 3 rewritten as equation 4 in order to determine the effective elastic modulus [25].

$$E_{eff}= E_s\left(1-\varphi)^2\right) \tag{4}$$

where $E_{eff}$ = effective elastic modulus

Gibson and Ashby model was employed in this study for prediction effective elastic modulus ($E_{eff}$) of three-dimensional porous structure when the structure loaded vertically upon unit cell geometry. ANSYS 18.0 also used to predict the effective elastic modulus of the samples. In order to explore what impact this has on the effective stiffness, the study considered the effect of variable within unit cell parameters. Finite element model was designed with a structure that having six layers with one unit cell in each layer. SOLID 185 used to mesh each model by four node tetrahedral elements with size of element of 0.1 mm. Under the materials elastic limit, a uniform 0.1% compressive strain applied to the top surface of the model, while fixed at the bottom. No boundary constraints added to the sides of the model. Due to its impact on simulation accuracy and computational time, the influence of mesh configuration was a vital consideration in structural analysis [27, 28]. Results of effective elastic modulus obtained from mesh sensitivity compared with theoretical CoCrMo materials properties as 200 GPa. The mesh element size from 0.05 mm to 0.5 mm and quadrilateral element used to study the mesh independent.

## 3. Results and discussion

### *3.1. Physical manufacturability*

Table 2 summaries the measurement and averages of errors of struts thickness and pore size of produced meta-biomaterials. From the table, the averages strut thickness error for diamonds structures is 13% and 21% for square structures. Meanwhile, average error of the pore sizes for diamond structures is 1.8% and 1.4% for square structures. The struts thickness and pore sizes measured for the diamond structures are in good agreement with the as-designed strut sizes for sample D#3, D#5 - D#7. Interestingly, the good manufacturing agreement with the self-supporting feature demonstrated in diamond structures due to angle of inclination between the two neighbouring strut layers. The inclination angle of strut has the potential to support the output of next layer after the previous scanned layer during the manufacturing activity [29, 30]. However, smaller strut sizes resulted in inaccuracy and shrinkage owed to loss connectivity among neighboring layers of unit cell. Hence, the strut size of diamond types with 0.3 and 0.4 mm are too thin for fabrication process. Furthermore, the strut angle for diamond is lower than 45° from the horizontal plane or overhanging angle. Thus, deformation occurred during manufacturing process where the struts build on loose powder led to the defect on products [31, 32].

Meanwhile, the strut thickness of square type was higher than designed strut related to the initial measurement for the models. It caused the pore sizes results smaller than compared to as designed models for square structures. The increased of the strut thickness was the outcome of the partially melted powder particles added the strut core and melt pool size on the strut boundary. The average actual strut sizes were higher and pore sizes were smaller than CAD models. The overhang phenomenon occurred in the unsupported region on square unit cells as stresses led to dross formation and the partially melted powder particles added the average values of product's strut sizes. The short interaction of powder bed and heat source triggered by the scanning speed of laser beam led to rapid heating during the manufacturing process [34, 35]. Then, the melting phase followed significantly. The shrinkage in the components triggered this phenomenon. The results indicate that manufacturing errors were more susceptible to the strut thickness close to the manufacturing limits at high porosity volume.

**Table 2.** Manufacturability of meta-biomaterials.

| Sample | Strut thickness (mm) | As-manufactured strut thickness (mm) | Error (%) | Pore size (mm) | As-manufactured pore size (mm) | Error (%) |
|---|---|---|---|---|---|---|
| D#1 | 0.3 | 0.29 ± 0.006 | 3.3 | 0.76 | 0.75 ± 0.006 | 1.32 |
| D#2 | 0.4 | 0.39 ± 0.003 | 2.5 | 0.66 | 0.65 ± 0.003 | 1.52 |
| D#3 | 0.5 | 0.50 ± 0 | 0 | 0.56 | 0.56 ± 0 | 0 |
| D#4 | 0.4 | 0.29 ± 0.006 | 3.3 | 1.1 | 0.99 ± 0.005 | 10 |
| D#5 | 0.5 | 0.50 ± 0 | 0 | 0.91 | 0.91 ± 0 | 0 |
| D#6 | 0.6 | 0.60 ± 0 | 0 | 0.81 | 0.81 ± 0 | 0 |
| D#7 | 0.7 | 0.70 ± 0 | 0 | 0.71 | 0.71 ± 0 | 0 |
| S#1 | 0.3 | 0.31 ± 0.006 | 3.3 | 0.70 | 0.69 ± 0.006 | 1.42 |
| S#2 | 0.4 | 0.41 ± 0.003 | 2.5 | 0.60 | 0.59 ± 0.006 | 1.67 |
| S#3 | 0.5 | 0.51 ± 0.006 | 2 | 0.50 | 0.49 ± 0.003 | 2 |
| S#4 | 0.5 | 0.51 ± 0.006 | 2 | 1.0 | 0.99 ± 0.006 | 1 |
| S#5 | 0.6 | 0.61 ± 0.001 | 1.6 | 0.9 | 0.89 ± 0.006 | 1.11 |
| S#6 | 0.7 | 0.71 ± 0.006 | 1.4 | 0.80 | 0.79 ± 0.01 | 1.25 |

### *3.2. FEA of effective elastic modulus*

Originally, the developed Gibson and Ashby model used to predict the rigidity of the three-dimensional (3D) open cell structure when applied a load to unit cell structures that related approach used in this study [25]. Previous studies demonstrating the used of the Gibson and Ashby models were in harmony with the values obtained from mechanical testing [35, 36]. The effective elastic modulus derived from the mathematical calculation for each designed meta-biomaterials.

As the results, the effective elastic modulus of diamond type was ranging from 6.5 GPa to 42 GPa, and range of 18 GPa to 50 GPa for square type. Interestingly, the elastic modulus for higher volume porosity were met a good agreement as the values closer to the human bone properties (1-30 GPa). While further study mechanical properties through experimental was required, meta-biomaterials possessed load-bearing potential that allowed the structures to withstand severe cyclic loading condition. In addition, geometrical design of meta-biomaterials inside the design domain had capacity for tailored porosity to imitate the mechanical properties of the host bone properties.

Mesh element size from 0.1 mm to 0.5 mm resulted effective elastic modulus of 169.23 GPa with accuracy 15.39% error. Mesh independent treatment was introduced to eliminate the geometric discretization error in complex morphologies of structures [37]. From finite element analysis, the effective modulus of diamond type ranging from 3.01 GPa to 4.79 GPa. Meanwhile, the effective modulus of square type ranging from 6.34 GPa to 28.75 GPa. Both diamond and square samples in the range of cancellous bone properties (1-15 GPa). Samples of square type of S#3, S#5 and S#6 exhibited elastic modulus similar to cortical bone properties (15-30 GPa) except for sample S#1 and S#4 which were in the range of cancellous bone. The results showed the pore structure could tailor the elastic modulus from the porosity value and different pore shape to produce accurate prediction of mechanical compressive behavior. Figure 3 illustrates the comparison of effective elastic modulus from Gibson and Ashby model and finite element analysis. The proposed technique yields accurate predictions for effective elastic modulus with decreasing of average error to 74.4% for diamond structures and 44.4% for square structures from theoretical calculation of Gibson and Ashby equation.

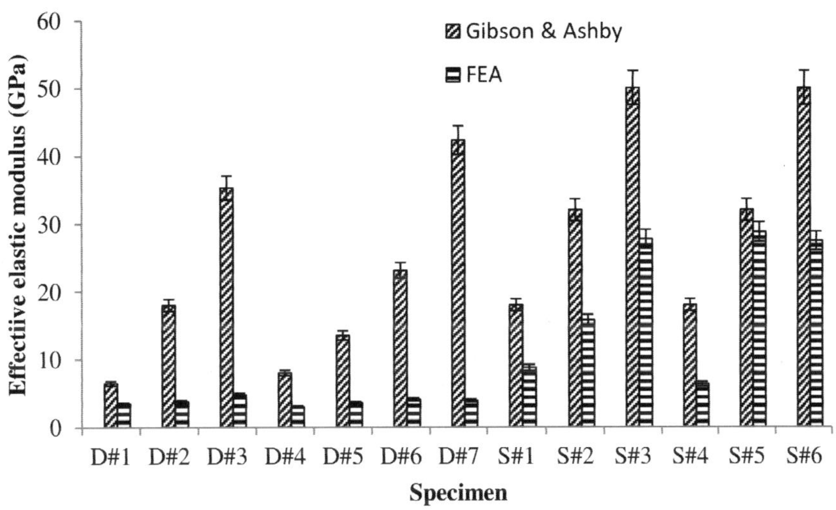


**Figure 3.** Effective elastic modulus of meta-biomaterials.

The FEA model showed substantially homogeneous deformation, beginning with buckling for both types and followed by complete compaction due to interaction between adjacent struts core. Figure 4 shows the deformation behaviours of the selected design structures predicted by FEA. It worth to note that FEA model was built for a fundamental analysis of deformation in porous structures. Particularly, this analysis emphasized the default options available for structural module in ANSYS software. In addition, the designed model based on controlled topologies where the strut surfaces can be irregular. This initiated flaws in manufacturing the structures as stress concentrators or added material to the struts by disputing compressive load. In addition, the mechanical properties of the meta-biomaterials might influenced by porosity and surface roughness, but those factors were not taken into account in the finite element analysis.

Both type of structures with different volume fractions deformed with same failure mechanism. Deformation mechanism represented plasticity for square and diamond type structures with 10%. From observation, the deformation commenced from the top and then, the structure collapsed by each layer of the cell continuously. The homogeneity of failure mechanism indicated the potential of high impact resistance from the struts of the samples [38]. Buckling micro-struts predicted to result in high-energy releasing at respectively failure stage since the struts able to endure axial distortion due to their equivalent direction of load [39].

Failure of diamond type structures followed by shearing of the microscale struts led the structure to collapse as bending dominated deformation. Crushing struts to the center of plates contributed to a continuing shear fracture in 45°. Thus, in contrast with the buckling deformation of square structures, less energy released during collapse of diamond structure owing to its strut inclination angle. The findings indicated that orientation of strut play a key role in the mechanism of deformation and stress-strain comportments of the meta-biomaterials [39].

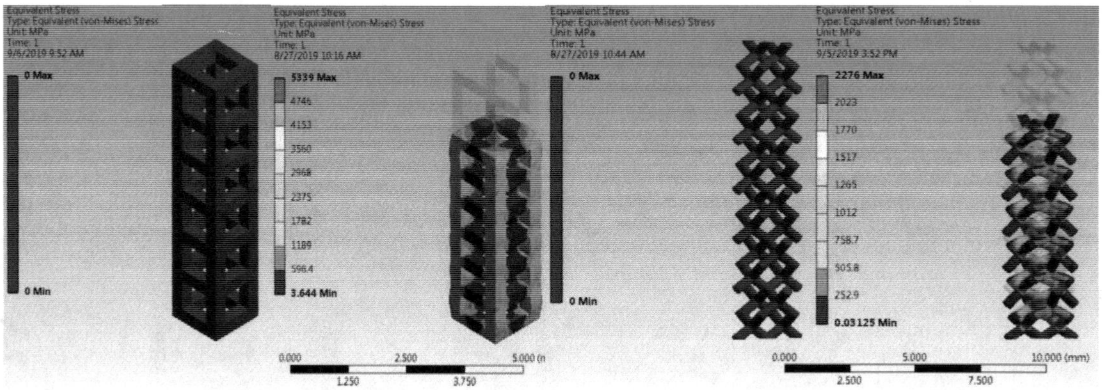


(a) (b)

**Figure 4.** Deformation mechanism of (a) square and (b) diamond structures.

## 4. Conclusion

This paper presents two distinct meta-biomaterials topologies that are dependable for orthopaedic implants. This paper also successfully presented a technique for visually understanding the function of porosity and pore size as design variables of meta-biomaterials. In addition, the design space model showed how biological and mechanical requirements and manufacturing limitation can clearly incorporated in design topologies constraints that allow a holistic knowledge of the relationship of unit cell geometries on mechanical and biological requirements and manufacturing drawback. This method can help to clarify the factors that can influence the effectiveness of meta-biomaterials as orthopedic implants.

Morphological investigation had shown geometries deviations between CAD models and manufactured meta-biomaterials. The averages error of the strut thickness is 13% for diamonds structures and 21% for square structures. Meanwhile, average error of the pore sizes is 1.8% for diamond structures and 1.4% for square structures. The discrepancies predominantly attributed to the overhang phenomenon and partially melted powder particles. Manufacturing intolerance leads to reduction of pore size and shrinkage of the components. The prediction of elastic modulus confirmed that samples at higher porosity exhibit stiffness comparable to human bone with mean error decreased to 74.4% for diamond structures and 44.4% for square structures from theoretical calculation.

**Acknowledgments**
The authors gratefully acknowledge the financial support supported by the Malaysian Ministry of Higher Education and Universiti Malaysia Pahang (www.ump.edu.my) for Postgraduate Research Grants Scheme (PGRS1903164) and RDU200301.

## References

[1] Kolken H M, Janbaz S, Leeflang S M, Lietaert K, Weinans H H, Zadpoor A A 2018 *Materials Horizons* **5** 28-35.
[2] Kolken H M A, Lietaert K, van der Sloten T, Pouran B, Meynen A, Van Loock G, et al. 2020 *Journal of the Mechanical Behavior of Biomedical Materials* **104** 103658.
[3] Mahmud S, Islam S S, Mat K, Chowdhury M E H, Rmili H, Islam M T 2020 *Results in Physics* **18** 103259.
[4] Forouzeshfard M R, Ghafari S, Vafapour Z 2021 *Journal of Luminescence* **230** 117734.
[5] Ma K, Tan T, Yan Z, Liu F, Liao W-H, Zhang W 2020 *Nano Energy* 105693.
[6] Ji G, Fang Y, Zhou J 2020 *Extreme Mechanics Letters* **36** 100648.
[7] Ahmadi S M, Hedayati R, Li Y, Lietaert K, Tümer N, Fatemi A, et al. 2018 *Acta Biomaterialia* **65** 292-304.
[8] Rodriguez-Contreras A, Punset M, Calero J A, Gil F J, Ruperez E, Manero J M 2021 *Journal of Materials Science & Technology* **76** 129-49.
[9] Lu X Z, Lai C P, Chan L C 2020 *Materials & Design* **188** 108474.
[10] Kim J, Hong D, Badshah M A, Lu X, Kim Y K, Kim S-m 2018 *Micromachines* **9** 376.
[11] Ahmed N 2019 *Journal of Manufacturing Processes* **42** 167-91.
[12] Pereira R F, Bártolo P J. 10.10 - Recent Advances in Additive Biomanufacturing. In: Yilbas S H F B J V T, editor. Comprehensive Materials Processing. Oxford: Elsevier; 2014. p. 265-84.
[13] Javaid M, Haleem A 2018 *Journal of Clinical Orthopaedics and Trauma* **9** 202-6.
[14] Kim T, See C W, Li X, Zhu D 2020 *Engineered Regeneration* **1** 6-18.
[15] Li N, Huang S, Zhang G, Qin R, Liu W, Xiong H, et al. 2018 *Journal of Materials Science & Technology*.
[16] Mitchell A, Lafont U, Hołyńska M, Semprimoschnig C 2018 *Additive Manufacturing* **24** 606-26.
[17] Bartolomeu F, Costa M M, Alves N, Miranda G, Silva F S 2020 *Optics and Lasers in Engineering* **134** 106208.
[18] Sobral J M, Caridade S G, Sousa R A, Mano J F, Reis R L 2011 *Acta Biomaterialia* **7** 1009-18.
[19] Rahimizadeh A, Nourmohammadi Z, Arabnejad S, Tanzer M, Pasini D 2018 *Journal of the Mechanical Behavior of Biomedical Materials* **78** 465-79.
[20] Che Ghani S A, Mohamed S R, Sha'ban M, Wan Harun W S, Md Noar N A Z 2020 *Procedia CIRP* **89** 79-91.
[21] Xue R, Cui X, Zhang P, Liu K, Li Y, Wu W, et al. 2020 *Extreme Mechanics Letters* **40** 100918.
[22] Mohamed S, Ghani S 2018 *Proceedings of Asia International Conference on Tribology* 2018: Malaysian Tribology Society.
[23] Moussa A, Rahman S, Xu M, Tanzer M, Pasini D 2020 *Journal of the Mechanical Behavior of Biomedical Materials* **105** 103705.
[24] Liu Y J, Ren D C, Li S J, Wang H, Zhang L C, Sercombe T B 2020 *Additive Manufacturing* **32** 101060.
[25] Gibson L J, Ashby M F 1999 *Cellular solids: structure and properties* Cambridge university press).
[26] Nakano T. 3 - Mechanical properties of metallic biomaterials. In: Niinomi M, editor. Metals for Biomedical Devices: Woodhead Publishing; 2010. p. 71-98.
[27] Lim H, Battaile C C, Bishop J E, Foulk J W 2019 *International Journal of Plasticity* **121** 101-15
[28] Huang X, Xie Y M 2007 *Finite Elements in Analysis and Design* **43** 1039-49.
[29] Yan C, Hao L, Hussein A, Young P, Huang J, Zhu W 2015 *Materials Science and Engineering: A* **628** 238-46.
[30] Pérez-Sánchez A, Yánez A, Cuadrado A, Martel O, Nuño N 2018 *Materials & Design* **155** 106-15.
[31] Hardes C, Pöhl F, Röttger A, Thiele M, Theisen W, Esen C 2019 *Additive Manufacturing* **29** 100786.

[32] Zhuang J, Gu D, Xi L, Lin K, Fang Y, Wang R 2020 *Powder Technology* **368** 59-69.
[33] Wang P, Lei H, Zhu X, Chen H, Fang D 2019 *Journal of Alloys and Compounds* **789** 852-9.
[34] Panda B K, Sahoo S 2019 *Results in Physics* **12** 1372-81.
[35] Martin A A, Calta N P, Khairallah S A, Wang J, Depond P J, Fong A Y, et al. 2019 *Nature communications* **10** 1-10.
[36] Jagodzinski D J, Miksch M, Aumann Q, Müller G 2020 *Mechanics Based Design of Structures and Machines* 1-15.
[37] Zhu H, Lin Z, Peng J, Zhang H-Q, Zhu J, Zhang J 2020 *Heat Transfer Summer Conference and the ASME 2020 18th International Conference on Nanochannels, Microchannels, and Minichannels*: American Society of Mechanical Engineers Digital Collection.
[38] Rashed M, Ashraf M, Mines R, Hazell P J 2016 *Materials & Design* **95** 518-33.
[39] Kadkhodapour J, Montazerian H, Darabi A C, Anaraki A P, Ahmadi S M, Zadpoor A A, et al. 2015 *Journal of the Mechanical Behavior of Biomedical Materials* **50** 180-91.

# Effect of thread profile variation on pullout and bending strength of a pedicle screw

**Rosdi Daud[1,2,*], W Y Kae[2], H Mas Ayu[3] and A Shah[4]**

[1]Automotive Engineering Centre, Universiti Malaysia Pahang, 26600 Pekan, Pahang, Malaysia.
[2]Faculty of Mechanical & Automotive Engineering Technology, Universiti Malaysia Pahang, 26600 Pekan, Pahang, Malaysia.
[3]Faculty of Manufacturing & Mechatronic Engineering Technology, Universiti Malaysia Pahang, 26600 Pekan, Pahang, Malaysia.
[4]Faculty of Technical and Vocational,Universiti Pendidikan Sultan Idris, 35900 Kuala Kubu Bharu, Perak, Malaysia.

*Corresponding author's e-mail: rosdidaud@ump.edu.my

**Abstract.** Pedicle screw is an important instrument in treatment of spinal degeneration or disease. However, the pedicle screw failure still occurs due to the screw loosening, fracture and pullout. There are few factors that affected the pedicle screw performance as reported by previous research but still lacking study related to the pedicle screw thread profile. Thus, the aim of this research is to investigate the effect of variation of thread pitch on the pullout and bending strength of pedicle screw. The research is carried out by constructing 3D pedicle screw models (model 1-6), importing the models into ANSYS, meshing and post processing analysis. The equivalent or Von-Mises stress used to compare the bending and pullout performance of the pedicle screws. Based on the obtained finite element analysis result, the single thread pedicle screw (model 3) has the optimum performance in bending while the model 4 is the optimum in pullout performance. While, for the dual threaded pedicle screw, the model 6 which has coarse thread pitch at screw tip is better than model 5 (fine thread). Thus, it can be conclude that both the single and dual threaded pedicle screw with coarse thread pitch has lower maximum equivalent stress than fine thread pitch, which is means it has better bending and pullout performance.

**Keywords:** Pedicle Screw; Finite Element Analysis; Pedicle Screw Thread Profile; Loosening.

## 1. Introduction

Pedicle screw is a surgical implant that used to stabilize the spinal segments in the spinal fusion surgery. A pedicle screw has played an important role in the treatment of spinal degeneration to limit the movement of the spine parts to assist the affiliation between broken bones by providing support and strength to the bony structures since the 1960s [1-4]. However, the failure of pedicle screw due to the fracture, loosening and pullout of screw still occurs although the pedicle screw contributes in long-term human lumbar spine segments stability in more than 90% cases, stated by W Qi *et al.* [5]. Many studies did on different pedicle screw designs to prevent screw loosening. Various designs such as screw with different outer diameter or length of the body of screw, thread profiles, cylindrical or conical core, expanding screws and cannulated screws are studied constantly [6-10]. The improvement of this medical device is getting more important in the medical industry.

The mechanical performance of the pedicle screw can evaluated by Finite Element Analysis (FEA) software as practised by other researchers [1-10]. The real condition and performance of pedicle screw during implantation can simulated by the FEA software. It allows the designers to predict and optimize the physical properties and implantation of the pedicle screw before applied on the patients in the surgery especially for the elderly patients. Nowadays the number of the population of elderly is rising gradually in the whole world. Biswas *et al.* (2019) stated that the spinal degeneration caused by the age becomes a serious problem for the older generation and brings intense pain to them. Fortunately, the spinal degeneration problem can reduced via the spinal surgery with the help of the pedicle screw [2]. However, screw loosening and pullout may happen due to insufficient interfacial strength between the surfaces of bone and screw.

Therefore, there are many studies done on the different pedicle screw designs to improve the structure of screw and to reduce the probability of screw loosening [1-10]. The designs on the outer diameter, length, thread profiles, core of the screw are studied. Although lot of studies related to the pedicle screw are done but the nightmare of pedicle screw (loosening) still happened till this date which shown that there is still no gold standard to overcome this main problem of pedicle screw performance. Thus the aim of this study are to investigate the effect of thread pitch variation on the pullout strength of single threaded pedicle screw using FEA and to investigate the difference in the bending strength of solid pedicle screws designed with different thread pitch by FEA.

## 2. Methodology

This project began with constructed two 3D models of the pedicle screw which it dimensions based on the published journal using SolidWorks software [6]. Next, the models were simulated using ANSYS software with the boundary conditions and load referred from the published journals. They then validated by comparing the results in the published journal to check whether the model can be used as the references for other modified models. After that, the proposed design of the pedicle screw (focused on thread profile) were constructed using SolidWorks software too. The models then imported into the ANSYS software to carry out simulation with the same boundary conditions and load. The table 1 below is the summarised dimensions of the constructed pedicle screws.

**Table 1.** Structures of pedicle screws.

| Parameters | Pedicle Screw from Previous Research by Chao *et al.* [6] | | Proposed Design of Pedicle Screw | | | |
|---|---|---|---|---|---|---|
| | Model 1 | Model 2 | Model 3 | Model 4 | Model 5 | Model 6 |
| Length | 45 | | | | | |
| Outer Diameter (mm) | 6.5 | | | | | |
| Inner Diameter at Screw Tip (mm) | 3.9 | 4.9 | 3.9 | | | |
| Thread Pitch (mm) | 2.8 | | 1.8 | 3.8 | -Screw tip to half-length of screw body: 1.8<br>-Rest of screw body length: 3.8 | -Screw tip to half-length of screw body: 3.8<br>-Rest of screw body length: 1.8 |
| Thread Width (mm) | 0.2 | | | | | |
| Proximal Half Angle (degree) | 14 | | | | | |
| Distal Half Angle (degree) | 25 | | | | | |
| Conical Angle from the Screw Tip (degree) | 1.655 | 1.018 | 1.655 | | | |

Figure 1 shows the important structures/parameters of a pedicle screw.

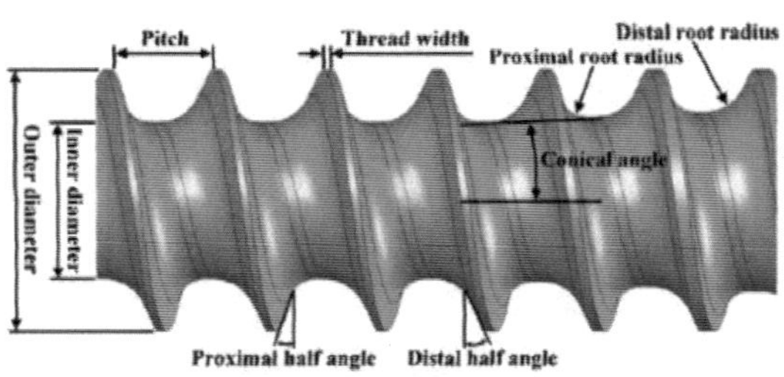


**Figure 1.** Structures of the pedicle screws (thread part).
Source: Chao *et al.* [6].

While figure 2 till figure 5 shows the pedicle screws from the previous published research (model 1 and 2) and the proposed design of pedicle screw (model 3-5) [6].

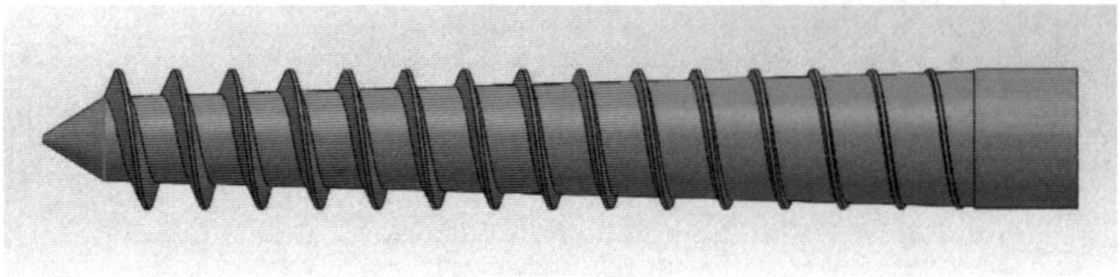

**Figure 2.** Model 1.

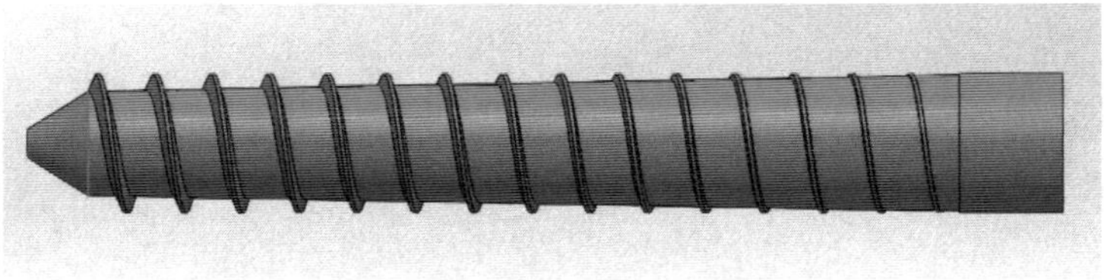

**Figure 3.** Model 2.

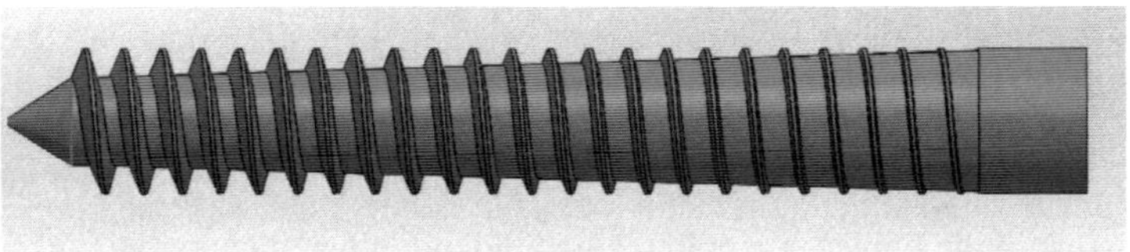

**Figure 4.** Model 3.

**Figure 5.** Model 4.

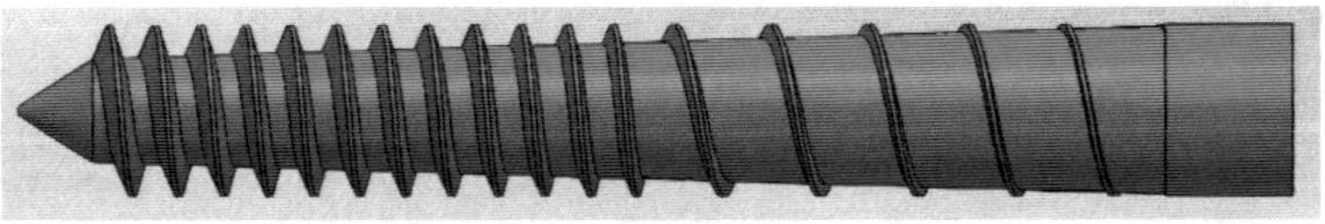

**Figure 6.** Model 5.

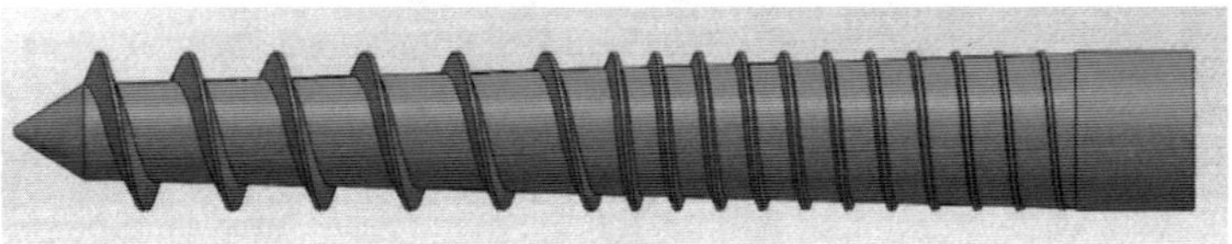

**Figure 7.** Model 6.

The models of the pedicle screw then saved as IGES file for imported to ANSYS software for finite element analysis. The material that used in finite element analysis for the pedicle screws is titanium alloy (Ti-6Al-4V) according to the specification of American Standard of Tested Materials (ASTM) F136-96. Chao *et al.* stated that two important parameters to describe the mechanical properties of the titanium alloy are elastic modulus (E) and Poisson's ratio (v) [6]. The titanium alloy chosen because of its superior biocompatibility and mechanical strength if it is compared to stainless steel based on the statement of Shea *et al.* [12]. According to the Chao *et al.*, the titanium alloy has the Young's modulus of 114 GPa and the Poisson ratio of 0.3

[6]. Therefore, in ANSYS, the Ti-6Al-4V in the additive manufacturing materials added into the contents of engineering data. The material then added to the body of the pedicle screws.

### *2.1 Boundary Condition*

There are two situations of screw for the simulation: 1) the screw is bent; 2) the screw was pull by given load. Therefore, the boundary conditions of the pedicle screw in both situations are different. For the situation when the screw is bent, the screw head fixed in all direction while a force of 330N is applied to the body of the screw in negative y-direction as shown in figure 8. Next, for the situation when the screw is pull, the screw head also fixed in all direction but the displacement of 0.01mm applied to the screw tip in positive z-direction as shown in figure 9.

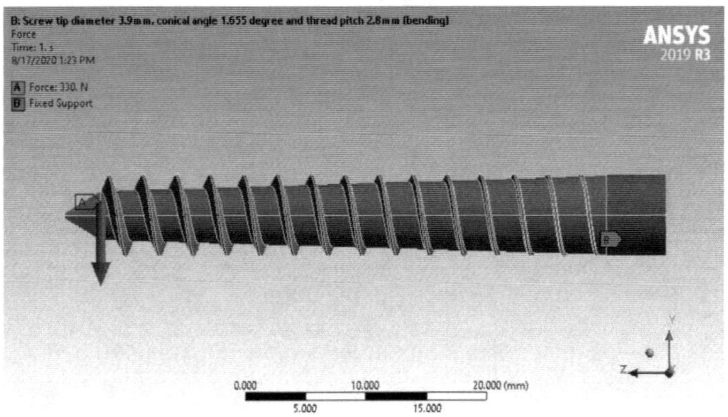


**Figure 8.** Boundary conditions of pedicle screw (bending).

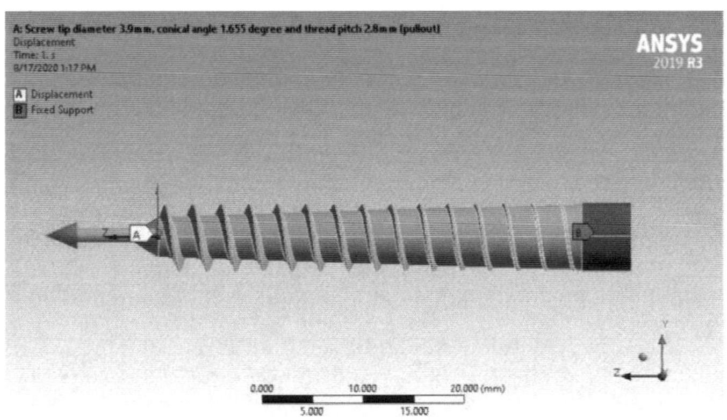


**Figure 9.** Boundary conditions of pedicle screw (pullout).

### *2.2 Meshing*

The titanium alloy has been add as the material of the pedicle screws and the 3D meshes were then generated at the body of the pedicle screw. The size of mesh on the body of the pedicle screw is controlled using edge sizing method as shown in figure 10. All of the edge sizing are using the type of number of divisions. For the first three-edge sizing, the number of divisions is set at 30. For edge sizing 4, the number of divisions is set at 10 while the number of divisions for edge sizing 5 is set at 5. The details of number of nodes and elements of each models can be refer to the table 2. One of assumption made in our Finite Element Analysis (FEA) was that the mesh used was not too course or too fine but it is still converged. A mesh with more nodes would have improved the accuracy of the computational results because it would account for more points on the sample. This was not done however, because more elements means more calculations for the program to do and there is limited time in lab. Some other assumptions that did in our FEA such as materials were consider as linear and forces were apply slowly and did not change direction in time.

**Figure 10.** Meshing on the body of pedicle screw.

**Table 2.** Number of nodes and elements for each model.

| Model No. | 3D Models of Pedicle Screw<br>Conditions | Nodes | Elements |
|---|---|---|---|
| 1 | Screw tip diameter 3.9mm, conical angle 1.655 degree and thread pitch 2.8mm | 19618 | 10570 |
| 2 | Screw tip diameter 4.9mm, conical angle 1.018 degree and thread pitch 2.8mm | 20740 | 11272 |
| 3 | Screw tip diameter 3.9mm, conical angle 1.655 degree and thread pitch 1.8mm | 19421 | 10414 |
| 4 | Screw tip diameter 3.9mm, conical angle 1.655 degree and thread pitch 3.8mm | 17165 | 9395 |
| 5 | Screw tip diameter 3.9mm, conical angle 1.655 degree and thread pitches 1.8mm (half-length of screw body starting from screw tip) and 3.8mm (rest of length) | 20514 | 11287 |
| 6 | Screw tip diameter 3.9mm, conical angle 1.655 degree and thread pitches 3.8mm (half-length of screw body starting from screw tip) and 1.8mm (rest of length) | 20086 | 10796 |

IOP Conf. Series: Materials Science and Engineering 1078 (2021) 012025 doi:10.1088/1757-899X/1078/1/012025

## 3. Results and discussion

Finite element analysis result (maximum equivalent stress) of pedicle screw model 1 shown in figure 11 and 12 for both situation bending and pullout. The maximum equivalent stress of other models are as shown in table 3 and 4.

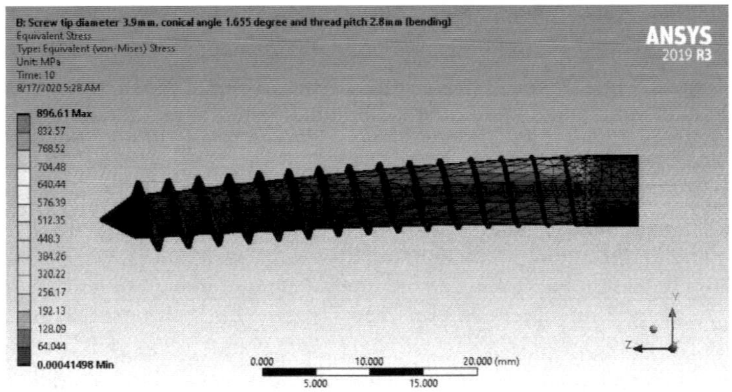


**Figure 11.** Equivalent (von-Mises) stress (bending) for model 1.

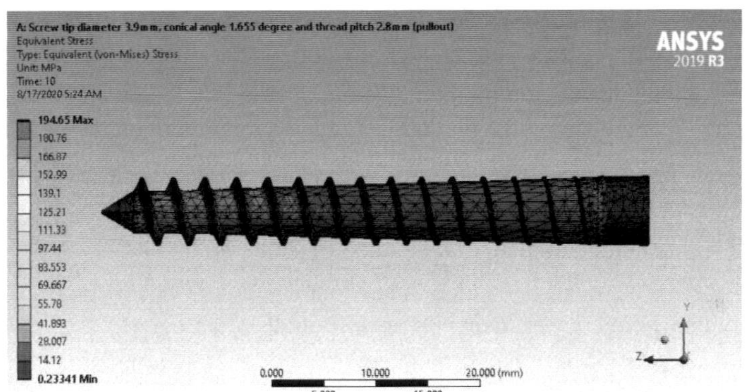


**Figure 12.** Equivalent (von-Mises) stress (pullout) for model 1.

**Table 3.** Maximum equivalent stress (bending).

| Model | Maximum Equivalent Stress (MPa) |
|---|---|
| Reference model 1 [6] | 896.61 |
| Reference model 2 [6] | 892.65 |
| Proposed model 3 | 1097.4 |
| Proposed model 4 | 709.05 |
| Proposed model 5 | 1067.9 |
| Proposed model 6 | 732.91 |

**Table 4.** Maximum equivalent stress (pullout).

| Model | Maximum Equivalent Stress (MPa) |
|---|---|
| Reference model 1 [6] | 194.65 |
| Reference model 2 [6] | 150.62 |
| Proposed model 3 | 172.47 |
| Proposed model 4 | 186.05 |
| Proposed model 5 | 198.65 |
| Proposed model 6 | 176.32 |

When the force applied on the body of pedicle screw models to let it bend, the model 4 has the lowest maximum equivalent stress, which is 709.05 MPa while model 3 has the highest maximum equivalent stress, 1097.4 MPa (table 3-4). Based on the theoretical, the higher maximum equivalent stress the easier to fail. In the bending test, model 3 fail first while model 4 (with the highest value of thread pitch) has the best bending performance compared to other pedicle screw models. Thus, it is proof that the thread pitch affects the performance of the pedicle screw in bending and has effect on pullout performance as proof by Gausepohl *et al.* [13]. The higher value of thread pitch of pedicle screw has better performance in bending. However further investigation especially by experiment should be done to verify this finding as did by Lee *et al.* [14].

When the load displacement is applied on the screw tip of the pedicle screw models, the model 5 has the highest maximum equivalent stress which is 198.65 MPa but model 2 (reference model) has the lowest equivalent stress, 150.62 MPa as shown in table 4. Thus, model 5 show the worst pullout performance. Based on the stress value in table 4, we found that the model 3 (a single threaded pedicle screw) has the optimum pullout performance among the proposed design of the pedicle screw. It is also better than the model 1(reference model). While, for the dual threaded pedicle screw, the model 6 is better than model 5 in term of pullout performance.

Model 1, 3, 4, 5 and 6 has the same inner diameter but various thread pitch. Based on table 3 and 4, when comparing the bending and pullout performance of the models which have same inner diameter but different thread pitch, model 4 and model 6 have lower maximum equivalent stress than model 1 which is a reference model in both bending and pullout tests. The model 1 fail faster than model 4 and 6 when bending and pullout are applied. Therefore, it can be conclude that the model 4 and 6 is the optimum design of pedicle screw thread since it is better than the model 1(reference model that currently available in the market).

## 4. Conclusion

As a conclusion, the pedicle screw with lower maximum equivalent stress has better performance in bending and pullout. Comparing the model 1, 3, 4, 5 and 6, model 4 is the best in bending performance since it has the lowest maximum equivalent stress. For pullout, the model 3 has the optimum performance even though it has the worst bending performance. To compare all proposed pedicle screw models with the reference model (model 1) in both bending and pullout aspects, model 4 and model 6 are better than model 1. Therefore, it can be conclude that the pedicle screw with coarse thread pitch has better pullout and bending performance. For the dual threaded pedicle screw, the pedicle screw with coarse thread pitch at the screw tip is better than the pedicle screw with fine thread pitch at the screw tip. This findings have potential in assisting the optimum design of pedicle screw in the future although still need to do further investigation to support this finding.

## Acknowledgments

We would like to thank Universiti Malaysia Pahang through research grant RDU190327 and RDU180395 for fully support the facilities and resources for this study. The authors also would like to acknowledge research grant FRGS/1/2018/TK03/UMP/02/5 which provided by Ministry of Higher Education Malaysia for support the resources for this study.

## References

[1] Bianco, R. J., Arnoux, P. J., Wagnac, E., Mac-Thiong, J. M., & Aubin, C. É. 2017 *In Clinical Spine Surgery* **30** 3.
[2] Biswas, J. K., Sahu, T. P., Rana, M., Roy, S., Karmakar, S. K., Majumder, S., & Roychowdhury, A. 2019 *Biocybernetics and Biomedical Engineering* **39** 1.
[3] Biswas, J., Karmakar, S., Majumder, S., Chowdhury, A. 2012 *Indian Journal of Biomechanics* **3** 1-2.
[4] Boos, N., & Webb, J. K. 1997 *European Spine Journal* **6** 1.

[5] Qi, W., Yan, Y. bo, Zhang, Y., Lei, W., Wang, P. jie, & Hou, J. 2011 *Orthopaedic Surgery*, **3** 1.
[6] Chao, C. K., Hsu, C. C., Wang, J. L., & Lin, J. 2008 *Journal of Spinal Disorders and Techniques*, **21** 2.
[7] Karakaşli, A., Acar, N., Özcanhan, M. H., & Ertem, F. 2016 *Eklem Hastaliklari ve Cerrahisi* **27** 3.
[8] R Daud, A Asyraf, HM Ayu, A Shah, SH Tomadi 2020 *TEST Engineering & Management* **83** 1097-1103.
[9] Giacaglia, G. E. O., & Lamas, W. de Q. 2015 Pedicle screw rupture: A case study. Case Studies.
[10] Hsieh, Y. Y., Chen, C. H., Tsuang, F. Y., Wu, L. C., Lin, S. C., & Chiang, C. J. 2017 *Clinical Biomechanics* **43** 115–120.
[11] Kiyak, G., Balikci, T., Heydar, A. M., & Bezer, M. 2018 *Asian Spine Journal*, **12** 1.
[12] Shea, T. M., Laun, J., Gonzalez-Blohm, S. A., Doulgeris, J. J., Lee, W. E., Aghayev, K., & Vrionis, F. D. 2014 *BioMed Research International.*
[13] Gausepohl, T., Möhring, R., Pennig, D., & Koebke, J. 2001 *Injury* **32** SUPPL. 4.
[14] Lee, E. S., Goh, T. S., Heo, J. Y., Kim, Y. J., Lee, S. E., Kim, Y. H., & Lee, C. S. 2019 *Applied Sciences* **9** 3.

IOP Conf. Series: Materials Science and Engineering 1078 (2021) 012026 doi:10.1088/1757-899X/1078/1/012026

# Effect of thread profile on bending strength of pedicle screw

**Rosdi Daud[1,*], S J Kennedy[1], H Mas Ayu[2] and A Shah[3]**

[1]Faculty of Mechanical & Automotive Engineering Technology, Universiti Malaysia Pahang, 26600 Pekan, Pahang, Malaysia.
[2]Faculty of Manufacturing & Mechatronic Engineering Technology, Universiti Malaysia Pahang, 26600 Pekan, Pahang, Malaysia.
[3]Faculty of Technical and Vocational,Universiti Pendidikan Sultan Idris, 35900 Kuala Kubu Bharu, Perak, Malaysia.

*Corresponding author's e-mail: rosdidaud@ump.edu.my

**Abstract.** Current pedicle screw designs have not achieved strong bending strength due to varying bone density among patients. Osteoporosis further presents complications in screw fixation and instrumentation failure due patients having less dense bones. We conducted this study to investigate the effect of thread profile on the bending strength of pedicle screw. The study also focus on the bending performance comparison between single threaded and dual threaded pedicle screw via finite element method. The finite element analysis (FEA) was apply to investigate the effect of thread design on the bending strength of pedicle screw (dual threaded, double dual threaded and the dual lead dual threaded screw designs) under normal and osteoporotic bone conditions. The FEA results obtained show that the thread profile does influence the bending strength of a pedicle screw in normal and osteoporotic bones. The dual lead dual threaded pedicle screw showed an improvement of 1.88% in its deflection due to bending. While for the osteoporotic condition, the dual lead dual threaded pedicle screw showed an improvement of 32.3% in term of deflection. Thus, this finding have potential in assisting the optimum pedicle screw design in future although further investigation needed to support this finding.

**Keywords:** Bending Strength; Pedicle Screw; Optimization; Osteoporosis.

## 1. Introduction

As our society becomes more advanced, humans are beginning to have longer lifespans. However, with this comes issues pertaining to caring for their health, as the human body ages our mental and physical abilities start to deteriorate and as many developed countries recently figuring out a lot of energy and care is required to take care of the elderly. One of the common issues faced by elderly people is osteoporosis in which is a bone disease that occurs when the bone density decreases. As a result, the bone strength decreases and may break from falling or, in more profound cases, from sneezing or minor bumps. It is indeed is a grim reality and many of the elderly do indeed have to go through procedures to ensure that they have a decent quality of life. Spinal fusion is one of the most common procedures used by the elderly. This is a procedure for combining two or more vertebrae into a single structure [1-4].

It works by preventing movement between the joined vertebrae, which reduces or prevents back pain. After the bones have fused, they do not move any more as they used to. This prevents the patient from stretching ligaments and nerves of surrounding muscles, which may have caused pain. Pedicle screws, which are often use in the spinal fusion process to provide additional support and strength to the fusion when healing. Pedicle screws are insert into the vertebrae, and a rod is use to connect the fused screws together. There have been issues on pedicle screw usage in surgery, which has become a standard for spinal surgery. Since its first usage in a surgery in the 1990s. Common problems that are face by this adjacent segment disease and instrumentation failure, which might be cause by biomechanical problems [5-7]. As mentioned earlier the decreasing bone density in the elderly has resulted in poorer performance of pedicle screws in spinal surgery. Current studies are gearing towards finding a more optimum screw design, which will have better bending strength [8-14]. One of the proposed designs is a dual threaded screw.

Pedicle screws commonly used to treat a variety of spinal conditions, such as fracture, tumour, infection, or spine degeneration. As society continues to enjoy longer life spans, surgeons commonly face complicating cases in treating spinal problems such as where patients are not only suffering osteoporosis which results weak bones but also require pedicle screw fixation for treatment [1-14]. Current issues faced by the industry are the current design of pedicle screws that still have not accomplished a strong bending strength due to the fact, that the density of the bone varies from person to person and the bone is significantly less dense in patients suffering osteoporosis. Thus, the purpose of this research is to investigate the bending strength of single and dual threaded pedicle screw, the effect of thread profile design on the bending strength of a pedicle screw either in normal bones or in osteoporotic bones.

## 2. Methodology

This project began with constructed 3D model of pedicle screw which focusing its thread profile in detail. The thread construction based on the parameters listed in table 1 and referred from report by Kao *et al.* [13].

**Table 1**. Thread design parameter.

| Parameter | Value |
|---|---|
| Proximal Root Radius | 0.4mm |
| Distal Root Radius | 1.2mm |
| Thread Width | 0.1mm |
| Proximal Half Angle | 5° |
| Distal Half Angle | 25° |

To create the thread profile, we select the thread profile folder and an existing default thread profile is then edit to the specifications from table 1. Figure 1 shows the dimension of the thread profile while figure 2 shows the thread profile.

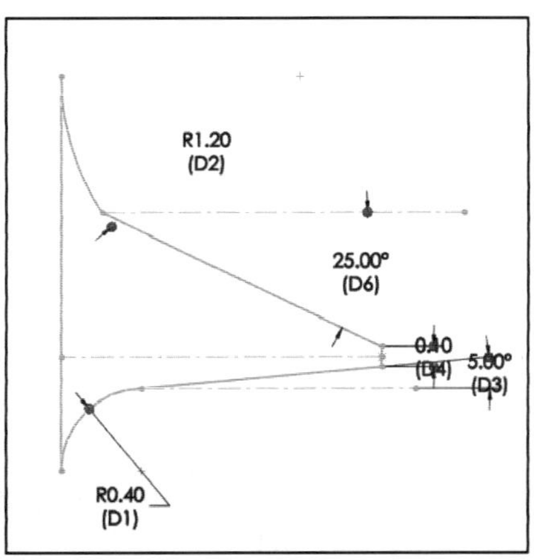


**Figure 1**. Thread profile dimensions for single threaded screw.

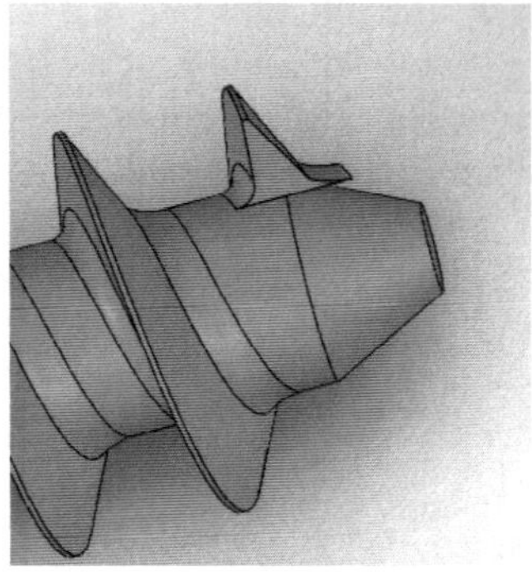

**Figure 2.** Thread profile of pedicle screw.

After the module has been selected the data for the material that represents the bone and screw must be inputted. The mechanical properties listed in the table 2 where the shear modulus and bulk modulus were derive by the software from the elastic modulus.

**Table 2**. Mechanical properties of the model.

| Type | Screw (Titanium alloy) | Bone |
|---|---|---|
| Elastic Modulus | 110 GPa | 2.6 GPa |
| Poisson Ratio | 0.3 | 0.3 |
| Shear Modulus | 4.2308 x 10^10 Pa | 1 x 10^9 Pa |
| Bulk Modulus | 9.16667 x 10^10 Pa | 2.1667 x 10^9 Pa |

*2.1 Finite element model (FEA) validation*
In order to validate the FEA model, the maximum deflection and the tensile stress of the single threaded screw compared to the FEA result obtained by Kao *et al.* [13]. The detail of the comparisons are as shown in table 3.

**Table 3**. FE model validation.

| Type | Reference [12] | Our Model | Percentage Error (%) |
|---|---|---|---|
| Maximum Deflection (mm) | 2.13 | 2.11 | 0.93 |
| Maximum Tensile Stress (MPa) | 2726 | 3028.7 | 11.10 |

Based on the low percentage error (0.93 & 11.10%), our FE model can be considered valid and acceptable. Once the model was validate, the finite element analysis was run under normal and osteoporotic bone conditions for the proposed models (dual threaded, double dual threaded and the dual lead dual threaded screw designs). This FEA done to investigate the effect of thread design on the bending strength. For the bending test, the head of the screw was fully constrained and the contact between the screw and bone was set as frictionless. A force of 220N was set at distance of 40mm from the head of the screw. The finite element analysis results such as the maximum principal stress and deformation obtained was use to represent the bending strength performance of the proposed pedicle screw. Proposed pedicle screw are consisted of three models which is the dual threaded, double dual threaded and the dual lead dual threaded screw designs. While for the reference model is single threaded screw (currently available in the market).

*2.2 Dual threaded screw design*
The parameters of the screw were maintained as the same as the single threaded screw the only difference being that the shaft screw was divided into two portions of 19mm each not including the tip of the screw. The pitch was 3mm and 1.95mm portions as shows in in figure 3.

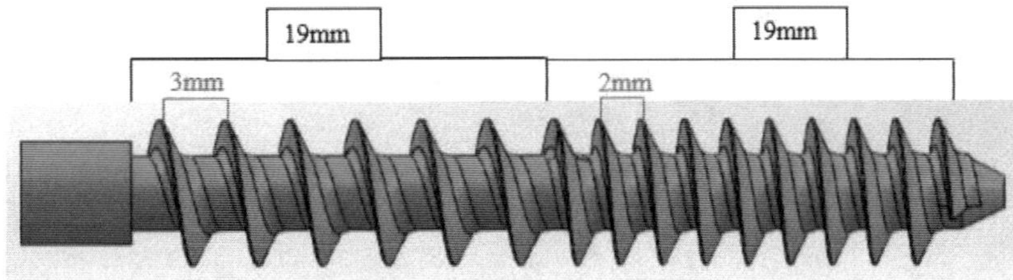


**Figure 3.** Dual threaded pedicle screw design.

*2.3 Double dual threaded screw design*

In a double dual threaded screw, there are two different thread profiles (buttress thread and V thread profile). Figure 4 shows the double dual threaded pedicle screw. The upper portions use the same thread profile as used in the single threaded screw while the lower portion uses the thread seen in figure 4. The thread was construct using the holes wizard feature in SOLIDWORKS and aligned with each other. The bone model was construct the same way as the dual threaded pedicle screw.

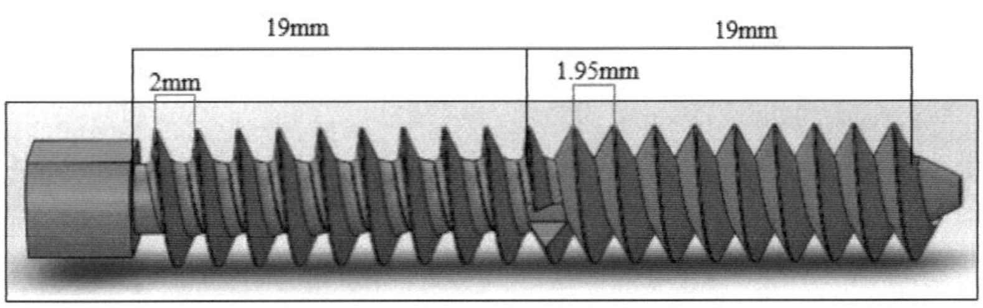


**Figure 4.** Dual threaded pedicle screw design.

*2.4 Dual lead dual threaded screw design*

The third proposed screw design is dual lead dual threaded; the screw uses the buttress thread as shown in figure 5, the upper portion is a dual lead with a pitch of 3mm while the lower portion is 1.7mm. The thread was create using the holes wizard feature and made to align each other using Solidworks software. This proposed design is like a combination between the proposed design 1 (dual threaded pedicle screw) and proposed design 2 (dual threaded pedicle screw).

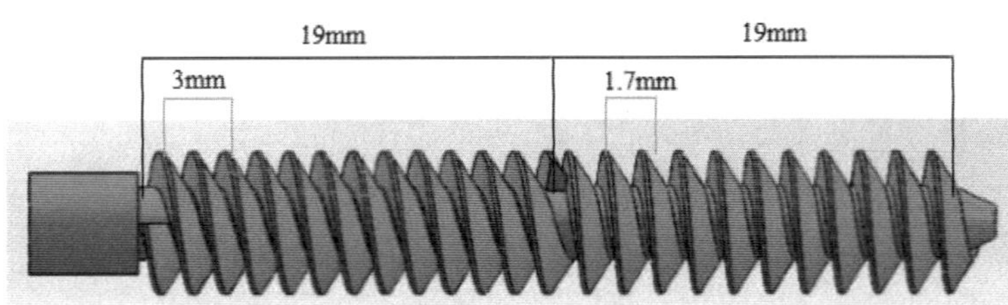


**Figure 5.** Dual lead dual threaded pedicle screw design.

*2.5 Osteoporotic bone condition*

To simulate the conditions of an osteoporotic bone the value for the elastic modulus is changed to mimic the loss of bone density from aging and a reduction from mechanical performance. The value of the young's modulus is choose to be half of the amount of the normal healthy bone base on the published journal by Jimenez *et al.* [5]. Therefore, since the elastic modulus for the normal healthy bone is assumed to be 2.6 GPa, the osteoporotic condition is set at 1.3 GPa to represent the decline in mechanical performance, the Poisson ratio is maintained constant at 0.3.

*2.6 Meshing*

The finite element mesh used in the present study consisted of solid elements (hexahedral and tetrahedral) with reduced integration. The models (three proposed design of pedicle screw models) are comprised of 9000 to 11000 elements and 17000 to 21000 nodes. One of assumption made in our Finite Element Analysis (FEA) was that the mesh used was not too course or too fine but it is still converged. Other assumptions that did in our FEA are such as materials were consider as linear and forces are being applied slowly and did not change direction in time.

## 3. Results and discussion

*3.1 Comparison of maximum deflection of pedicle screw in normal bone*

Based on the table 4, the proposed design 3 (dual lead dual threaded design) shows the best improvement since its maximum deflection, 1.94mm compared to the single threaded which is 2.11mm. This indicates that the proposed design 3 can withstand a higher force before starting to undergo deformation. Figure 6 shows the deflection of the dual lead dual threaded screw.

**Table 4.** Maximum deflection in normal bone.

| Type | Single Threaded | Dual Threaded | Double Dual Threaded | Dual Lead Dual Threaded |
|---|---|---|---|---|
| Maximum Deflection (mm) | 2.11 | 2.13 | 2.02 | 1.94 |

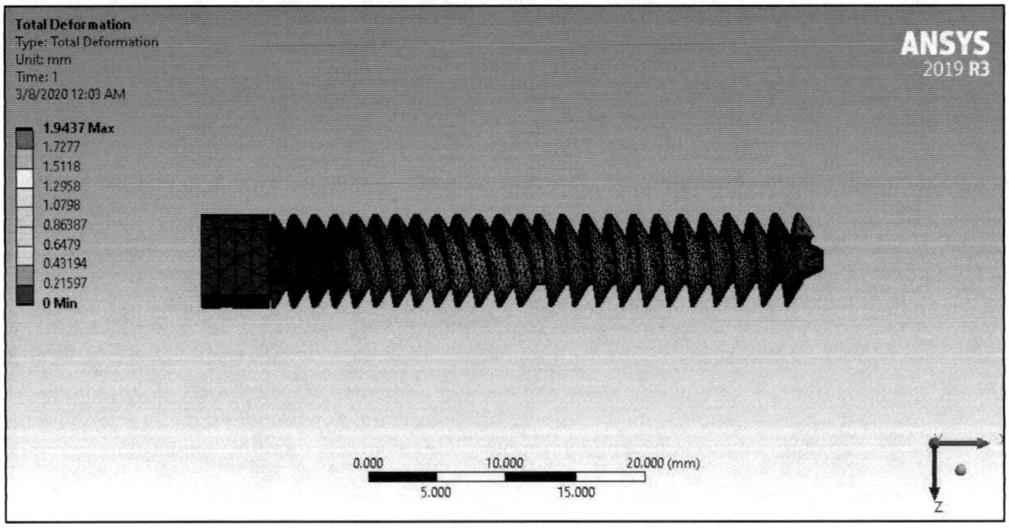


**Figure 6.** Deflection of dual lead dual threaded pedicle screw in normal bone.

*3.2 Comparison of maximum deflection of pedicle screw in osteoporotic bone*

For the osteoporotic bone condition in which the mechanical properties of the bone were greatly the test were conduct in the exact same condition. Figure 7 shows the deflection of the single threaded pedicle screw in osteoporotic bone. The maximum deflections of the dual threaded, double dual threaded and dual lead dual threaded pedicle screw are listing in table 5.

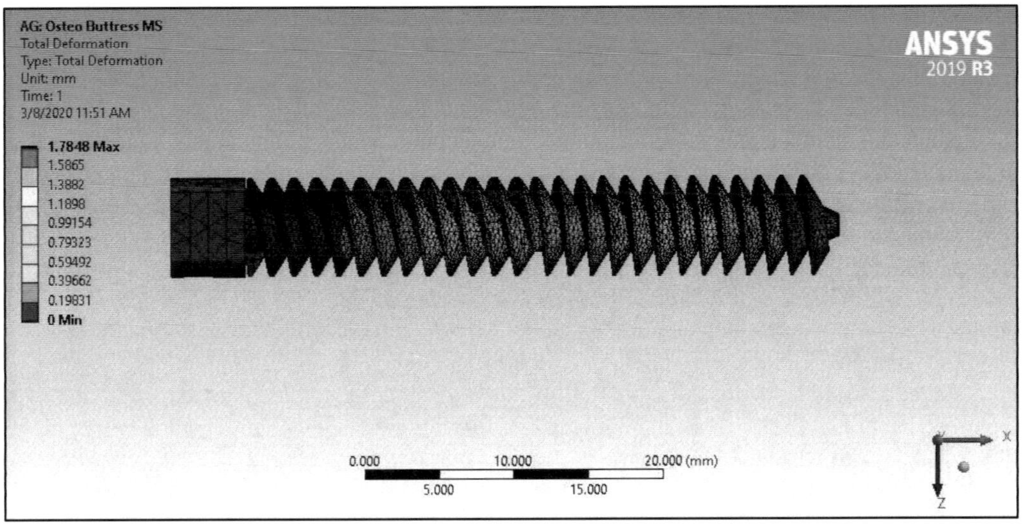


**Figure 7.** Deflection of dual lead dual threaded pedicle screw in osteoporotic bone.

**Table 5.** Maximum deflection in osteoporotic bone.

| Type | Single Threaded | Dual Threaded | Double Dual Threaded | Dual Lead Dual Threaded |
|---|---|---|---|---|
| Maximum Deflection (mm) | 2.63 | 2.65 | 2.52 | 1.78 |

The dual lead dual threaded design shows the lowest maximum deflection in the osteoporotic condition and 32.31% improvement when compared to the single threaded design (reference thread profile of pedicle screw that currently available in market). While the dual threaded did not show any significant improvements over the single threaded in terms of bending performance. While the double dual threaded performed slightly better than the single threaded design with an improvement of 4.18%. After comparing the performance of the three proposed screw design to the single threaded screws performance, dual lead dual threaded shows the most improvement over the single threaded pedicle screw performance. In the normal bone condition, it showed an improvement of 8.4% in terms of maximum deflection and improvement of 32.31% in osteoporotic bones.

## 4. Conclusion

Based on the finite element analysis result, the dual threaded pedicle screw did provide significant improvements over the single threaded pedicle screw in terms of bending performance in both normal and osteoporotic bone. The thread design seem to have an effect on the bending strength of the pedicle screw in which the dual lead dual threaded screw showed an improvement of 8.4% in the maximum deflection. In the osteoporotic bone, the dual lead dual threaded screw also showed an improvement 32.31% improvement. Thus, this finding have potential in assisting the optimum pedicle screw design in future although further investigation needed to support this finding.

## Acknowledgements

We would like to thank Universiti Malaysia Pahang through research grant RDU180395 and RDU190327 for fully support the facilities and resources for this study. The authors also would like to acknowledge research grant FRGS/1/2018/TK03/UMP/02/5 which provided by Ministry of Higher Education Malaysia for support the resources for this study.

## References

[1] Deyo, R. A., Nachemson, A., & Mirza, S. K. 2004 *The Spine Journal*, **4** 5.
[2] Yu, C., Ou, Y., Xie, C. 2020 *J Orthop Surg Res* **15** 1.
[3] Webb, J. K., & Boos, N. 1997 *European Spine Journal* **6** 2-18.
[4] Ian A. Harris Adrian Traeger Ralph Stanford Christopher G. Maher Rachelle Buchbinder. 2018 *Internal Medicine Journal* **48** 1430-1434.
[5] Jimenez-Palomar, I., Shipov, A., Shahar , R., & Barber Asa, H. 2015 *Frontiers in Materials* **9**.
[6] Mummaneni, P. V., Haddock, S. M., Liebschner, M. A., Keaveny, T. M., & Rosenberg, W. S. 2002 *Journal of Spinal Disorders & Techniques* 64-68.
[7] Salama, A.A., Amin, M.A., Soliman, A.Y. 2019 *Egypt J Radiol Nucl Med* **50** 57.
[8] Liu MY, Tsai TT, Lai PL, Hsieh MK, Chen LH, Tai CL. 2020 PLoS One 15 2.
[9] Brasiliense, L. B., Lazaro, B. C., Phillip, R. M., Newcomb, A. G., Turner, J. L., Crandall, D. G., & Crawford, N. R. 2013 *The Spine Journal*, **13** 8.
[10] Chatzistergos, P. E., Magnissalis, E. A., & Kourkoulis, S. K. 2010 *Medical Engineering & Physics* **32** 2.
[11] Claeson, A., Gandhi, A., & Mehbod, A. A. 2019 *The Spine Journal* **19** 9.
[12] Jazini, E., Petraglia, C., Moldavsky, M., Tannous, O., Weir, T., Saifi, C.Ludwig, S. C. 2017 *The Spine Journal* **17** 574-578.
[13] Kao, S.-S., Hsu, C., Hou, S.-M., Yu, S.-C., & Liaw, C.-K. 2015 *Medical Engineering and Physics* 879-884.
[14] Shen, F., Kim, H. J., Kang, K. T., & Yeom, J. S. 2019 *Applied Sciences (Switzerland).*
[15] Louis, R. 1986 *Clinical Orthopedics and Related Research* **203** 18-33.

# The effect of quick response manufacturing in factory layout: Assessment using Witness software

**M Y M Yusoof[1,*], N M Z N Mohamed[1], Nelfiyanti[1,2] and M Z I A Hadi[1]**

[1]College of Engineering, Universiti Malaysia Pahang, Lebuhraya Tun Razak, 26300 Kuantan, Pahang, Malaysia
[2]Department of Industrial Engineering, Faculty of Engineering, Universitas Muhammadiyah Jakarta

*Corresponding author's email: yusrizal.yusoof@daimler.com

**Abstract.** Nowadays, in a new era of industrial revolution 4.0, there are still many manufacturing companies that tend to neglect the knowledge of optimizing the factory layout. By doing so, the productivity can be increase and lessen the burden of the operator while making sure the quality and also the number of products are also at optimum. The objective of this study is to conduct an analysis on the actual arrangement assembly line using witness software based on QRM in V model assembly and also to propose a new arrangement of the assembly line using Witness software. The data has been collected and recorded through some observation during the visit to X Company. The current layout of the V model assembly line is simulated in the Witness software to get its actual process, percentage of idle, and also the setup time. The software is also used for the new improvement of the assembly line layout to get the comparison for the current layout and after the improvement. Based on the current result, there is some station that has a high percentage of blockage occur in the layout and after the improvement, the percentage of blockage has been reduced which help in increase the production of the model per day. The data that has been recorded from the current is compared with the data after the improvement has been made to the assembly line.

**Keywords:** Quick Response Manufacturing (QRM); Automotive Industry; Line Balancing.

## 1. Introduction

As for the automotive industry, it is the industry that covers a wide range of companies involved in the design, development, manufacturing, marketing, and selling of motor vehicles towed vehicles, motorcycles, and also its parts. The automotive industry is one of the most demanding market sectors, in which flexibility and response capability constitute the basis for success [1]. The automotive industry plays a significant role in transforming Malaysia into a more industrial country, which helps to increase economic activities, improved the standard of living of the people as well as higher-paying jobs. As the sector continues to advance, more high-value jobs will be available and these include industrial engineers, technicians as well as product, process, and tool designers. From this statement, the concept of Quick Response Manufacturing (QRM) was introduced. It was first introduced by the late Rajan Suri who was a professor at the University of Wisconsin-Madison at that time in the 1980s. It is quite similar to Lean and also the Theory of Constraints (TOC). The difference between QRM than the other is it emphasizes time or specifically reducing the lead time in a manufacturing process. This strategy helps the company that supplies high variation, less number, and custom-engineered products or parts for customers. For example, Engineer-to-Order (ETO) manufacturers tend to follow the exact time taken for the delivery according to their customers after the order [2]. Time and responsiveness have become strategic factors for the competitiveness of businesses that face the challenge of competing in the dynamic and unpredictable markets of globalized trade [3]. Thus, flexibility and adaptability are vital factors for organizations to respond quickly enough to the demands of twenty-first-century customers, which are primarily characterized by small volume and a high variety of customized products [4, 5].

The importance of reducing lead time was initially recognized in the Toyota Production System (TPS), coined Lean Manufacturing in the occident. For the TPS, non-value added time is one of the seven types of wastes to be removed from production processes in order to maximize customer value and improve efficiency. However, most Lean Manufacturing implementations do not primarily focus directly on lead time reduction but instead on the well-known seven wastes (overproduction, unnecessary transportation, inventory, motion, defects, overprocessing and waiting) which is different from QRM where all of the improvement efforts are directed towards lead time reduction [6]. The POLCA (Paired-cell Overlapping Loops of Cards with Authorization) system implementation [7-9]. In a recent review of the literature on lead time reduction within the context of Time-Based Competition and QRM,[10] noted that currently available literature on practical studies of lead time reduction, particularly QRM, is quite limited.

## 2. Background

Manufacturing in the automotive industry Increase in societal demand for sustainability has resulted in attention to sustainable manufacturing [11]. Produce vehicles is a big business, but the automotive industry is also a demanding industry for the customer. The industry is made up of companies and workers who manufacture and deliver cars, trucks, and other vehicles to companies that sell them. Companies in the automotive industry fall into one of two primary segments which are car manufacturers and car parts manufacturers. Today's vehicles are more complex and involve many more parts and electronics compare to past years. These changes lead to an increase in the number of parts and the number of manufacturers.The automotive industry can also improve the economy of the country as there are an increase in jobs and position for people in the industries. At a manufacturing factory, the position that available for them are line workers or operators, supervisors, quality control workers or inspectors, managers, engineers, designers, safety engineers, and executives. Besides, some people provide the routine business operations any organization needs, such as accounting, marketing and advertising, and human resources. On the supplier side, the same types of workers are required. Sales professionals also play an important role in these companies.

### *2.1 Assembly line*

While many people think that Henry Ford invented the automotive assembly line, it was invented by Ransom Eli Olds. Olds had worked on cars for most of his life, including steam-powered cars in the 1880s and 1890s. His assembly line allowed him to be the first mass-producer of cars in the United States, and he dominated the American automotive industry from 1901 to 1904. Assembly line or also known as a progressive line is a manufacturing process where the parts are assembled part by part and from a station to another station sequentially until it becomes a finished product. The line is usually equipped with a conveyor or similar handling system that helps in moving the semi-finished part to another station. Every line has its own cycle time at each station. They need to finish the process according to or below the time given to fulfill the customer's satisfaction. The product can be done faster with less labor needed by using this process because less number of workers carry the parts. Usually, the assembly line consists of workers, tools, and machines that need to be arranged so that it is easy for the workers to do the operation and at the same time shorten the time taken for the products to be made. There some models of assembly line such as single model assembly line (SMAL), batch model assembly line (BMAL), and mixed-model assembly line (MMAL) which differ according to its product. For automotive industries that produce numbers of identical products, they tend to used SMAL because it is more efficient compare to others.

*2.2 Line balancing*

The concept of line balancing according to Parvez, et. al is to "design a smooth production flow by allotting processes to workers to allow each worker to complete the allotted workload within a given time [12]. Assembly line balancing is a family of combinatorial optimization problems that have been widely studied in the literature due to its simplicity and industrial applicability [13]. Line balancing is a method where the workload, machines, and also labor are evenly arranged according to the type of work that needs to be done in each work station to ensure that every station can complete the task in the designated time.

It is important to reduce the processing time or buffer, remove any bottlenecks that can occur in the assembly line, and also reduce any worker to overwork. Since the installation of the assembly line is a long-term decision and highly cost-intensive, there is a proper need of designing the assembly line and balancing the workload at the workstations [14]. To calculate for line balancing, firstly we need to have some data for the number of operators, operation, performance, number of products per hour, time to achieve the target, capacity, and also number of the product ordered by the customer.

*2.3 Bottleneck process*

A bottleneck is a point of congestion in a production system such as an assembly line that occurs when workloads arrive too quickly for the production process to handle. This problem often creates delays in the line and increases the production cost. The manufacturing industry tends to face this problem especially if they produce new products or using a new production line because there may be some flaws from it. The bottleneck may also happen when the demand from the customer is high unexpectedly exceeds the production from the manufacturer. The processor constraints of the bottleneck in the line are defined by the total cycle time in the line. The line capacity is decided by this bottleneck cycle time. Line Capacity is Source of Bottleneck Cycle Time (C / t) and Total Available Time, if Bottleneck C / T < Current Time then demand from the customer has been fulfilled. With the past projected production delivery or from the expected future demand, the takt time is identified for the manufacturing system. With the established takt time, the bottleneck process is defined from the Value Stream Mapping (VSM) and the difference between capacity and demand is determined [15]. For Company X, a bottleneck always happens when there is some model that does not have enough parts such as a seat or its cockpit. If that problem happened, the model is put under Work-in-Project (WIP) as they need to wait for the supplier to send the missing part. The model will still need to be assembled with the available part and go through all the work station before the model is in the store at the end of an assembly line.

*2.4 Lead time, cycle time, and takt time*

Practitioners and academics are aware of the bad effect of lead times on production and distributions system [16]. Lead time is the time taken between the ordering of products by customers and receiving them back by the manufacturer. Lead-time varies for every company and depends on the products that have been ordered, the number of workers and also the work station and even the location of suppliers for the parts. Lead- time should be considered a variable to be controlled by management rather than a constant. Lead time is important for the satisfaction of the customer as customer want the product as quickly as possible. The shorter the lead time, the more satisfaction the customer gets. The company needs to have accurate forecasting, planning, and schedule on how to reduce it.

Cycle time has various definitions in the literature [17]. The definition of cycle time is the time taken from the beginning of the process until it is finished. Cycle time includes process time for a unit in each work station and also the delay time or buffer which is the waiting time for a unit to move to the next work station. The formula of cycle time is:

$$\text{cycle time} = \frac{\text{Time taken for a unit to produce}}{\text{Number of units required}} \quad (1)$$

Tact time from the German word' taktzeit' means' measure," loop' or' pulse.' This refers to the 1930's march, pacing, and speed control. The Axis Powers collaborated with Germany and Japan. Instead, Japan uses this term to unite during World War 2. This refers to the maximum time required to produce a product to meet the demand of customers. It is important to take time to ensure that all business potential is prepared and used and still meets the overall demand of the customer. This means that the right product is delivered to the consumers in the right amount at the right time. Tact time is a commonly used method within the lean production lines to ensure the most productive flow through each workstation. Take-up time refers to the amount of time needed to produce a product from raw materials to finished products to satisfy customer demand. Tact time will greatly help manufacturing companies achieve higher efficiency. This tool helps in the process to eliminate as much waste as possible.

$$Takt\ time = \frac{Available\ production\ time\ per\ day}{Customer\ demand\ per\ day} \quad (2)$$

*2.5 Quick response manufacturing (QRM)*

Quick response manufacturing (QRM) is one of the lead time reduction methods that suitable for a company that produces in a low volume but varies in types. QRM is a management concept that focuses on time as the key factor in competitive manufacturing, especially in manufacturing environments for customer-oriented engineering and ordering [18]. A literature review identifies that QRM is based on four concepts which are the power of time, enterprise-wide application, and organization structure and system dynamics.

First, the power of time. Traditionally, the focus of the manufacturing company is on scale and cost management strategies resulting in a high degree of labor specialization and departmental hierarchy. QRM reveals why this conventional system adversely affects lead times and generates a lot of hidden costs. Next, enterprise-wide application. QRM extends time-based management concepts to all aspects of the company, including office activities, inventory scheduling, quality control, supply chain, and new product launch.

For organization structure, QRM provides cells in high mix, low volume, customized environments with principles and tools. They are designed around the Focused Target Market Segment, which is a market segment where shorter lead times provide maximum benefits to the company. They apply to the workplace and office. Lastly, system dynamics. A thorough understanding of system dynamics specifically tailored to high mix conditions complements the cellular structure of QRM. The application of the principles of common process dynamics leads to better capacity planning and optimized batch sizes for short lead times.

It is the difference with lead time as lead time is suitable for a company that produces a higher volume of production but less varies in types or models. While in Make-to-Order environments lean manufacturing can work well, it does only when the demand is very flat and constant. Lean uses a 'pull' System that does not stagnate but produces an additional stock if Lean cannot produce an optimal flow. Lean also uses this system. QRM is different because the monitoring of order flows is done by POLCA (Paired-cell overlapping cards with permission). POLCA is a powerful and simple flow controller that guarantees the speed of orders without any human intervention within the business.

*2.6 Witness simulation software*

WITNESS Simulation software is a software developed by the Lanner Group Ltd. WITNESS has a simple and interactive interface that enables the users to easily navigate through the simulation. Thompson et al. in their paper states that Witness simulation software is a discrete, systematic occurrence and continuous simulation of processes. This is designed to allow a person who is familiar with the process under study to debug rapidly, incrementally, and accurately. Include complex models and check them. It has a grid interface that displays the toolbar, window modeling, and window structure support. Models are built by using the elements in the model element window tabs.

## 3. Methodology

This research was conducted in the automotive manufacturing environment located in Malaysia. The methodology of the study is based according to Figure 1. It was started by doing a literature review on QRM, then continue with a tour and internship at the automotive assembly line to learn and learn about the work processes performed by employees while working. This assembly line consists of the Trim Assembly, Mechanical Assembly, and Final of Line Assembly Line. Next, identify the problem during the arrangement of the layout After that, the data from the factory on the assembly line document was collected and the study proceeded to study the selected tools that had already been applied at the automotive manufacturing company X.

The Assembly line situation was analyzed. This research was focused on the factory's low-output assembly line in the V model assembly line number 1. This assembly line has 20 different process stations for each station such as wiring, engraving, and assembly line body preparation station. The overview of methods used to complete the study is generally reviewed in this section. The data was collected here for the simulation. The development of the new model or concept is continued with the Witness simulation program, following the current update to the data. The existing data are analyzed and improved in this section. The setup time is studied and the setup time is improved and improvements are made to the appropriate situation to improve the process.

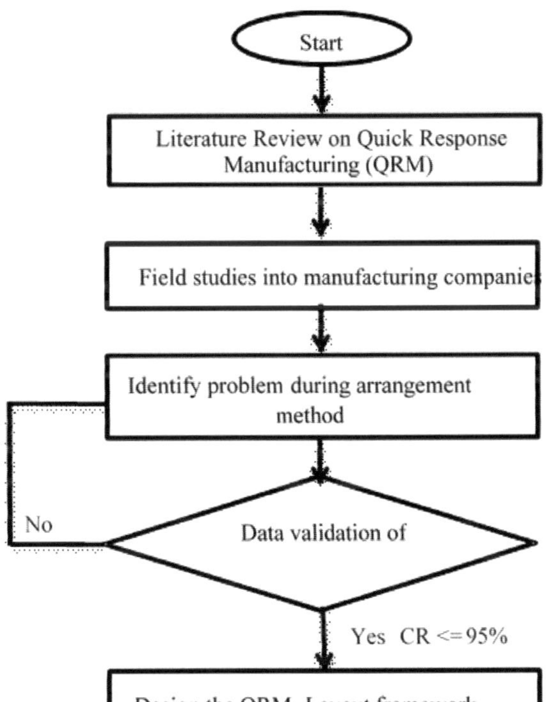


**Figure 1.** Flow diagram of research methodology.

Data validation is made to get a quality result from the outcome of the simulation. If there is an addition to information or data that is needed for this project, a discussion will be made with the supervisor before going to the company for data collection. The result from both before and after the simulation is recorded and put in the report. The report must be organized according to the requirement to help in understanding it better. The report will be validated by the supervisor before submitting it.

*3.1 Data analysis*

This research was conducted on company X and focused on just one assembly line that is the V model assembly line. This assembly line consists of 20 workstations and operator numbers with different processes for each station as shown in figure 2. Each station has its number of the operator according to the difficulties of the process. The process that required less work will have less number of operators compared to the heavier work process to reduce any unbalance of manpower in the assembly line. Every assembly line has one person in charge for every three-station which is called multi while the person in charge for the whole line is called PENGHULU. Balancing the number of operators at each station is a must as balancing the line helps in reducing the lead time. It will also remove any buffer in the assembly line. The percentage of blocked parts will also be reduced.

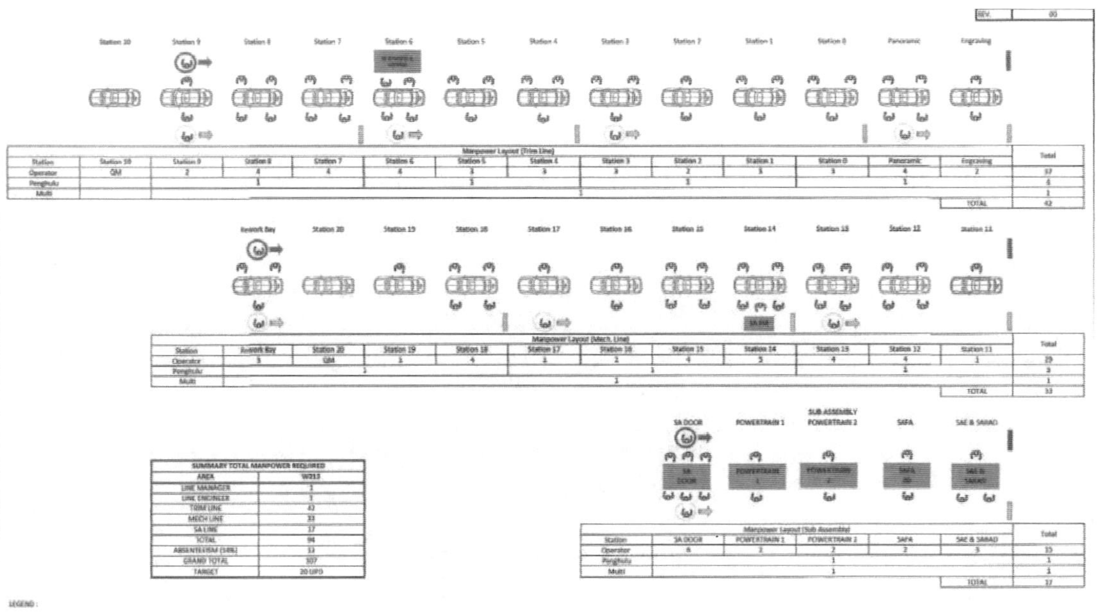

**Figure 2.** Manpower mapping layout for V Model assembly line.

Each station also had its cycle and tact time. The tact time that the X Company had stated needs to be followed by the operator to get the target output for the day. If there is any station that exceeds the tact time, it will create a buffer as the assembly line is in a single line model. Thus, it will reduce the number of cars produced on that day and increase the lead time of the V model.

From the table 1 shows that a few stations exceed the takt time. Thus, the lead time is also affected. The station that has the problem is the station that we need to tackle to balance it and make sure it is under the takt time. The takt time is had been made by the company to achieve its daily target output. From the current layout, the number of operations or output is 24 units per day. All of the data has been taken during the visit to the assembly line with the help of staff in that company. The collected data will then be used in the simulation to get the initial or current view of the assembly line.

From the data, the takt time as the cycle time for each station in the Witness software. It will be representative of the current layout cycle time. After the analysis, the research begins the improvement of the layout based on the analysis of the current layout.

IOP Conf. Series: Materials Science and Engineering 1078 (2021) 012027 doi:10.1088/1757-899X/1078/1/012027

**Table 1.** Time study for V model assembly line.

| Station | Operator A | Operator B | Operator C | Operator D | Highest time | Manpower | Takt time/unit |
|---|---|---|---|---|---|---|---|
| Door Dissemble/ Engraving | 10.00 | 10.00 | - | - | 10.00 | 2 | 20 |
| Panoramic/ Roof | 15.00 | 15.00 | 15.00 | 15.00 | 15.00 | 4 | 20 |
| Station 0 | 17.00 | 19.00 | - | - | 19.00 | 2 | 20 |
| Station 1 | 17.00 | 16.00 | 17.50 | - | 17.50 | 3 | 20 |
| Station 2 | 15.00 | 15.00 | - | - | 15.00 | 2 | 20 |
| Station 3 | 23.00 | 22.00 | 20.00 | 22.00 | 23.00 | 4 | 20 |
| Station 4 | 21.00 | 19.00 | 15.00 | 17.00 | 21.00 | 4 | 20 |
| Station 5 | 13.00 | 14.00 | 13.00 | - | 14.00 | 3 | 20 |
| Station 6 | 17.00 | 19.00 | 18.00 | - | 19.00 | 3 | 20 |
| Station 7 | 15.00 | 21.00 | 20.00 | 20.00 | 21.00 | 4 | 20 |
| Station 8 | 20.00 | 18.00 | 16.00 | - | 20.00 | 3 | 20 |
| Station 9 | 20.00 | 18.00 | - | - | 20.00 | 2 | 20 |
| Station 11 (MER) | 4.60 | 4.6 | - | - | 4.6 | 2 | 20 |
| Station 12 (MER) | 20.37 | 20.37 | 20.37 | 20.37 | 20.37 | 4 | 20 |
| Station 13 (MER) | 20.91 | 20.91 | 20.91 | - | 20.91 | 3 | 20 |
| Station 14 | 26.20 | 27.50 | 25.60 | 25.60 | 27.50 | 4 | 20 |
| Station 15 | 24.50 | 23.70 | 22.10 | 22.10 | 24.50 | 4 | 20 |
| Station 16 | 20.00 | 19.50 | - | - | 20.00 | 2 | 20 |
| Station 17 | 22.00 | - | - | - | 22.00 | 1 | 20 |
| Station 18 | 20.00 | 22.10 | 22.10 | - | 22.10 | 3 | 20 |
| Station 19 | 24.30 | - | - | - | 24.30 | 1 | 20 |

## 4. Results and discussion

Production productivity is likely to be largely affected by the production line process. Consequently, the assembly line must not be wasteful of action for each process and operation, since the production output is carried out every minute. The current configuration of the V model line is examined to obtain more data on the process setup and to obtain essential data for increasing process and flow, such as cycle time.

IOP Conf. Series: Materials Science and Engineering 1078 (2021) 012027 doi:10.1088/1757-899X/1078/1/012027

The exchanges that take place between the mechanic line and the trim line play a major role in the productivity output that has been wasted as much time there. X company has 3 lines each with different models of car. In this study, we only focused on line 1 or V model assembly line. The improvement is needed to be done to this line to increase the productivity of X company. The arrangement of the layout is important in the improvement.

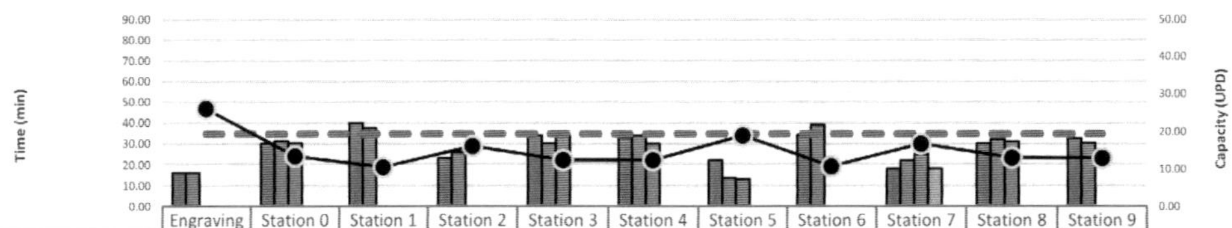


| | Engraving | Station 0 | Station 1 | Station 2 | Station 3 | Station 4 | Station 5 | Station 6 | Station 7 | Station 8 | Station 9 |
|---|---|---|---|---|---|---|---|---|---|---|---|
| Operator A | 16.00 | 30.10 | 40.00 | 23.00 | 34.00 | 33.00 | 22.10 | 34.20 | 18.00 | 30.10 | 32.50 |
| Operator B | 16.00 | 31.20 | 37.50 | 26.00 | 30.10 | 34.00 | 13.40 | 39.10 | 22.00 | 32.20 | 30.30 |
| Operator C | 0.00 | 30.30 | 0.00 | 0.00 | 34.00 | 30.00 | 13.00 | 0.00 | 25.00 | 31.10 | 0.00 |
| Operator D | 0.00 | 0.00 | 0.00 | 0.00 | 0.00 | 0.00 | 0.00 | 0.00 | 18.00 | | 0.00 |
| Operator E | 0.00 | 0.00 | 0.00 | 0.00 | 0.00 | 0.00 | 0.00 | 0.00 | 0.00 | 0.00 | 0.00 |
| Takt time/unit (12upd) | 34.71 | 34.71 | 34.71 | 34.71 | 34.71 | 34.71 | 34.71 | 34.71 | 34.71 | 34.71 | 34.71 |
| Capacity (unit/day) | 26.03 | 13.35 | 10.41 | 16.02 | 12.25 | 12.25 | 18.85 | 10.65 | 16.66 | 12.93 | 12.82 |

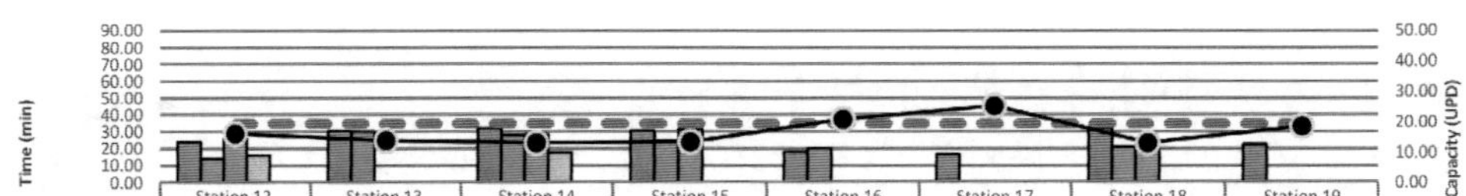


| | Station 12 | Station 13 | Station 14 | Station 15 | Station 16 | Station 17 | Station 18 | Station 19 |
|---|---|---|---|---|---|---|---|---|
| Operator A | 24.00 | 30.40 | 32.00 | 30.50 | 18.00 | 16.53 | 32.00 | 22.53 |
| Operator B | 14.00 | 30.20 | 28.00 | 24.50 | 20.00 | 0.00 | 21.00 | 0.00 |
| Operator C | 26.00 | 0.00 | 28.32 | 31.40 | 0.00 | 0.00 | 20.00 | 0.00 |
| Operator D | 16.00 | 0.00 | 17.62 | 0.00 | 0.00 | 0.00 | 0.00 | 0.00 |
| Operator E | 0.00 | 0.00 | 0.00 | 0.00 | 0.00 | 0.00 | 0.00 | 0.00 |
| Takt time/unit (12upd) | 34.71 | 34.71 | 34.71 | 34.71 | 34.71 | 34.71 | 34.71 | 34.71 |
| Capacity (unit/day) | 16.02 | 13.70 | 13.02 | 13.26 | 20.83 | 25.20 | 13.02 | 18.49 |

**Figure 3.** Cycle time V Model before improvement.

From Figure 3 shows that some station has exceeded its takt time which will affect the production output at the end. Each station has its required task and the task of assembling part is depending on the machine at the station and the number of operators allocated for every station. This data was taken when the manpower well training at every station. Figure 4 shows that the current layout using Witness Software. 20 stations arranged in a line and there was a different number of labor's in each station according to the task, process, and machine use. Each operator is responsible for finishing the required task during the process. Each operator must complete the process at all stations under takt time which is 20 minutes. It is different between the trim line and the mechanical line. Trim needed less time as it was only involved in wiring and small part while a mechanical line is where all the important or big components are assembled. The company was set all stations to be the same as some stations required conveyor and some stations are not. So, to avoid any restrictions or further blockage occur, they decide to sync all stations with maximum tact time which is 20 minutes. From the simulation, the situation of the assembly includes the percentage blocked, the percentage of busy and idle, and also the number of operations or output.

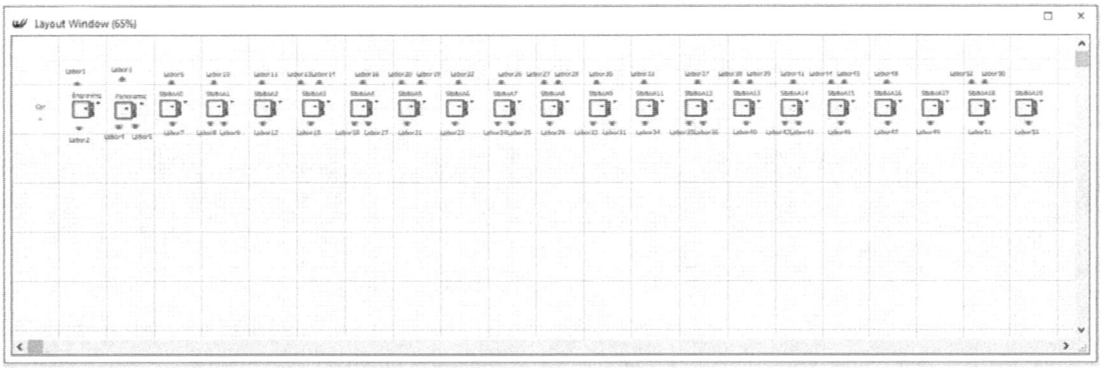

**Figure 4.** Layout before improvement and cycle time.

**Table 2.** Machine statistic for V Model.

| Name | % Idle | % Busy | % Filling | % Emptying | % Blocked | % Cycle Wait Labor | % Setup | % Setup Wait Labor | % Broken Down | % Repair Wait Labor | No. Of Operations |
|---|---|---|---|---|---|---|---|---|---|---|---|
| STN ENGR | 0 | 100 | 0 | 0 | 0 | 0 | 0 | 0 | 0 | 0 | 24 |
| STN00 | 0 | 100 | 0 | 0 | 0 | 0 | 0 | 0 | 0 | 0 | 24 |
| STN ROOF | 0 | 100 | 0 | 0 | 0 | 0 | 0 | 0 | 0 | 0 | 24 |
| STN01 | 0 | 100 | 0 | 0 | 0 | 0 | 0 | 0 | 0 | 0 | 24 |
| STN02 | 0 | 100 | 0 | 0 | 0 | 0 | 0 | 0 | 0 | 0 | 24 |
| STN03 | 0 | 100 | 0 | 0 | 0 | 0 | 0 | 0 | 0 | 0 | 24 |
| STN04 | 0 | 100 | 0 | 0 | 0 | 0 | 0 | 0 | 0 | 0 | 24 |
| STN05 | 0 | 100 | 0 | 0 | 0 | 0 | 0 | 0 | 0 | 0 | 24 |
| STN06 | 0 | 100 | 0 | 0 | 0 | 0 | 0 | 0 | 0 | 0 | 24 |
| STN07 | 0 | 100 | 0 | 0 | 0 | 0 | 0 | 0 | 0 | 0 | 24 |
| STN08 | 0 | 100 | 0 | 0 | 0 | 0 | 0 | 0 | 0 | 0 | 24 |
| STN09 | 0 | 100 | 0 | 0 | 0 | 0 | 0 | 0 | 0 | 0 | 24 |
| STN10 | 0 | 100 | 0 | 0 | 0 | 0 | 0 | 0 | 0 | 0 | 24 |
| STN11 | 0 | 100 | 0 | 0 | 0 | 0 | 0 | 0 | 0 | 0 | 24 |
| STN12 | 0 | 100 | 0 | 0 | 0 | 0 | 0 | 0 | 0 | 0 | 24 |
| STN13 | 0 | 100 | 0 | 0 | 0 | 0 | 0 | 0 | 0 | 0 | 24 |
| STN14 | 0 | 100 | 0 | 0 | 0 | 0 | 0 | 0 | 0 | 0 | 24 |
| STN15 | 0 | 100 | 0 | 0 | 0 | 0 | 0 | 0 | 0 | 0 | 24 |
| STN16 | 0 | 100 | 0 | 0 | 0 | 0 | 0 | 0 | 0 | 0 | 24 |
| STN17 | 0 | 100 | 0 | 0 | 0 | 0 | 0 | 0 | 0 | 0 | 24 |
| STN18 | 0 | 100 | 0 | 0 | 0 | 0 | 0 | 0 | 0 | 0 | 24 |
| STN19 | 11.18 | 75 | 0 | 0 | 0 | 0 | 0 | 0 | 0 | 0 | 24 |

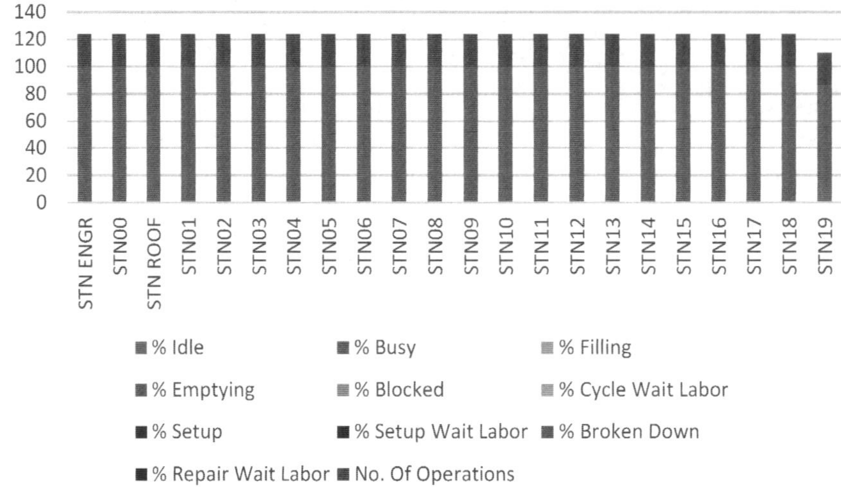


**Figure 5.** Graph percentage of machine for the current layout.

From Table 2 and Figure 5, it shows that the percentage of busy and idle of the manpower and also the machine in the current layout at the V model assembly line. In this current layout, each operator is responsible for their task. The operator does not do the work for another station. Station 19 has 75% busy while the other has 100%. Station 19 is also the only station that recorded the percentage of idle which is 25% while the others are null.

### *4.1 Improvement with added buffer area*

To reduce the time of the operation can be made. Reviewed and then make a changed or alteration in the mechanism by which the system has been reordered, as in the new interface was implemented. With each station, the process changed and there was a workload imbalance. Therefore, added the buffer area or storage for any parts that finished before the next station took it to continue the assembly process. From the current layout, the percentage idle at station 19 while the others have 0% idle. Moreover, to overcome it by adding or placing a buffer area or storage area between station 9 and station 11 which this station is between the trim line and mechanical line. The number of cars that can be put in the area is 1. By having this area, the percentage of busy is reduced while the percentage of idle is increased for all station.

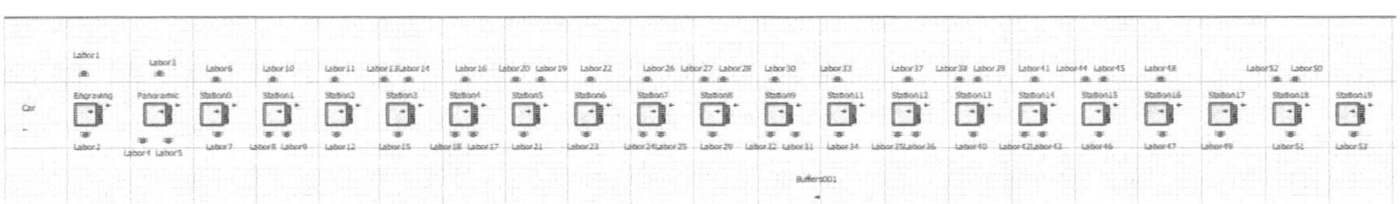

**Figure 6.** Proposed layout after adding buffer area.

| Name | Total In | Total Out | Now In | Max | Min | Avg Size | Avg Time | Avg Delay C | Avg Delay Ti | Min Time | Max Time |
|---|---|---|---|---|---|---|---|---|---|---|---|
| Buffers001 | 23 | 23 | 0 | 1 | 0 | 0.05 | 1.00 | 0.00 | 0.00 | 1.00 | 1.00 |

**Figure 7.** Buffer statistics.

The results, shows that the buffer area will have the same amount of parts for in and out which is 23, and help in reducing the lead time rather than have to wait for the station to finish before transfer it to the next station.

**Table 3.** Machine statistic after added 1 buffer area.

| *Name* | *% Idle* | *% Busy* | *% Filling* | *% Emptying* | *% Blocked* | *% Cycle Wait Labor* | *% Setup* | *% Setup Wait Labor* | *% Broken Down* | *% Repair Wait Labor* | *No. Of Operations* |
|---|---|---|---|---|---|---|---|---|---|---|---|
| *STN ENGR* | *1.67* | *98.33* | *0* | *0* | *0* | *0* | *0* | *0* | *0* | *0* | *23* |
| *STN00* | *1.67* | *98.33* | *0* | *0* | *0* | *0* | *0* | *0* | *0* | *0* | *23* |
| *STN ROOF* | *1.67* | *98.33* | *0* | *0* | *0* | *0* | *0* | *0* | *0* | *0* | *23* |
| *STN01* | *1.46* | *98.54* | *0* | *0* | *0* | *0* | *0* | *0* | *0* | *0* | *23* |
| *STN02* | *1.67* | *98.33* | *0* | *0* | *0* | *0* | *0* | *0* | *0* | *0* | *23* |
| *STN03* | *1.67* | *98.33* | *0* | *0* | *0* | *0* | *0* | *0* | *0* | *0* | *23* |
| *STN04* | *1.46* | *98.54* | *0* | *0* | *0* | *0* | *0* | *0* | *0* | *0* | *23* |
| *STN05* | *1.67* | *98.33* | *0* | *0* | *0* | *0* | *0* | *0* | *0* | *0* | *23* |
| *STN06* | *1.67* | *98.33* | *0* | *0* | *0* | *0* | *0* | *0* | *0* | *0* | *23* |
| *STN07* | *1.46* | *98.54* | *0* | *0* | *0* | *0* | *0* | *0* | *0* | *0* | *23* |
| *STN08* | *1.67* | *98.33* | *0* | *0* | *0* | *0* | *0* | *0* | *0* | *0* | *23* |
| *STN09* | *1.67* | *98.33* | *0* | *0* | *0* | *0* | *0* | *0* | *0* | *0* | *23* |
| *STN11* | *1.46* | *98.54* | *0* | *0* | *0* | *0* | *0* | *0* | *0* | *0* | *23* |
| *STN12* | *1.67* | *98.33* | *0* | *0* | *0* | *0* | *0* | *0* | *0* | *0* | *23* |
| *STN13* | *1.67* | *98.33* | *0* | *0* | *0* | *0* | *0* | *0* | *0* | *0* | *23* |
| *STN14* | *1.46* | *98.54* | *0* | *0* | *0* | *0* | *0* | *0* | *0* | *0* | *23* |
| *STN15* | *1.67* | *98.33* | *0* | *0* | *0* | *0* | *0* | *0* | *0* | *0* | *23* |
| *STN16* | *1.67* | *98.33* | *0* | *0* | *0* | *0* | *0* | *0* | *0* | *0* | *23* |
| *STN17* | *1.46* | *98.54* | *0* | *0* | *0* | *0* | *0* | *0* | *0* | *0* | *23* |
| *STN18* | *1.67* | *98.33* | *0* | *0* | *0* | *0* | *0* | *0* | *0* | *0* | *23* |
| *STN19* | *1.67* | *98.33* | *0* | *0* | *0* | *0* | *0* | *0* | *0* | *0* | *23* |

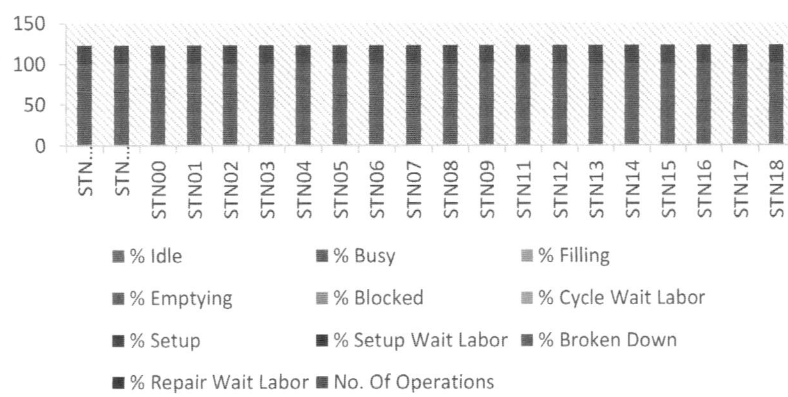


**Figure 8.** Graph percentage of the machine after adding one buffer.

The simulation to install more number of a buffer area in to improve that the result can help in reducing the percentage of idle and busy while increasing in production. Unfortunately, the result is the same as the previous one. From the simulation, the percentage of idle and busy for each station is the same for install one buffer area or two buffer areas as shown in Table 4 and Figure 9.

**Table 4.** Machine statistic after added 2 buffer area.

| *Name* | *% Idle* | *% Busy* | *% Filling* | *% Emptying* | *% Blocked* | *% Cycle Wait Labor* | *% Setup* | *% Setup Wait Labor* | *% Broken Down* | *% Repair Wait Labor* | *No. Of Operations* |
|---|---|---|---|---|---|---|---|---|---|---|---|
| *STN ENGR* | *1.67* | *98.33* | *0* | *0* | *0* | *0* | *0* | *0* | *0* | *0* | *23* |
| *STN00* | *1.67* | *98.33* | *0* | *0* | *0* | *0* | *0* | *0* | *0* | *0* | *23* |
| *STN ROOF* | *1.67* | *98.33* | *0* | *0* | *0* | *0* | *0* | *0* | *0* | *0* | *23* |
| *STN01* | *1.46* | *98.54* | *0* | *0* | *0* | *0* | *0* | *0* | *0* | *0* | *23* |
| *STN02* | *1.67* | *98.33* | *0* | *0* | *0* | *0* | *0* | *0* | *0* | *0* | *23* |
| *STN03* | *1.67* | *98.33* | *0* | *0* | *0* | *0* | *0* | *0* | *0* | *0* | *23* |
| *STN04* | *1.46* | *98.54* | *0* | *0* | *0* | *0* | *0* | *0* | *0* | *0* | *23* |
| *STN05* | *1.67* | *98.33* | *0* | *0* | *0* | *0* | *0* | *0* | *0* | *0* | *23* |
| *STN06* | *1.67* | *98.33* | *0* | *0* | *0* | *0* | *0* | *0* | *0* | *0* | *23* |
| *STN07* | *1.46* | *98.54* | *0* | *0* | *0* | *0* | *0* | *0* | *0* | *0* | *23* |
| *STN08* | *1.67* | *98.33* | *0* | *0* | *0* | *0* | *0* | *0* | *0* | *0* | *23* |
| *STN09* | *1.67* | *98.33* | *0* | *0* | *0* | *0* | *0* | *0* | *0* | *0* | *23* |
| *STN11* | *1.46* | *98.54* | *0* | *0* | *0* | *0* | *0* | *0* | *0* | *0* | *23* |
| *STN12* | *1.67* | *98.33* | *0* | *0* | *0* | *0* | *0* | *0* | *0* | *0* | *23* |
| *STN13* | *1.67* | *98.33* | *0* | *0* | *0* | *0* | *0* | *0* | *0* | *0* | *23* |
| *STN14* | *1.46* | *98.54* | *0* | *0* | *0* | *0* | *0* | *0* | *0* | *0* | *23* |
| *STN15* | *1.67* | *98.33* | *0* | *0* | *0* | *0* | *0* | *0* | *0* | *0* | *23* |
| *STN16* | *1.67* | *98.33* | *0* | *0* | *0* | *0* | *0* | *0* | *0* | *0* | *23* |
| *STN17* | *1.46* | *98.54* | *0* | *0* | *0* | *0* | *0* | *0* | *0* | *0* | *23* |
| *STN18* | *1.67* | *98.33* | *0* | *0* | *0* | *0* | *0* | *0* | *0* | *0* | *23* |
| *STN19* | *1.67* | *98.33* | *0* | *0* | *0* | *0* | *0* | *0* | *0* | *0* | *23* |

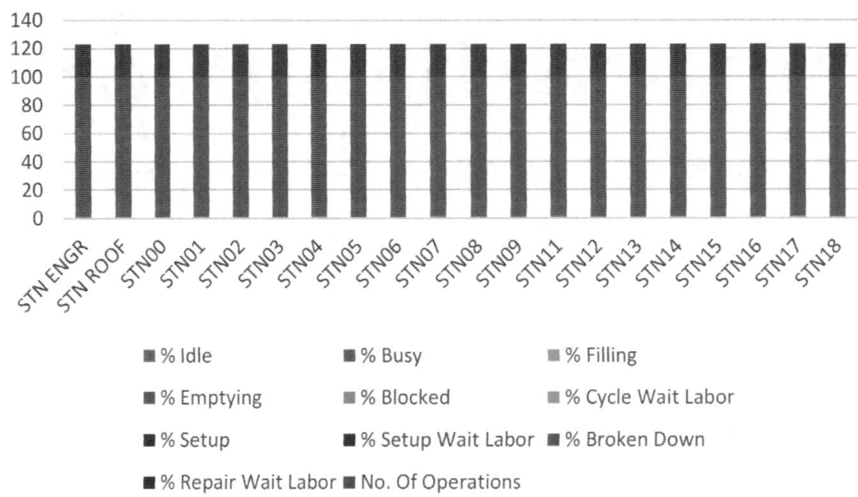


**Figure 9.** Graph percentage of the machine after adding two buffer.

*4.2 Combining station*

Moreover, there have proposed another method which is by combining some stations that is suitable according to its process. By combining these stations, this research reduces some buffer time or reducing any time waste as it will avoid any blockage from the previous stations. From this research to combine any station starting from station 11 until station 19 or known as the mechanical line. It is due to its process which involves assembling the components of the car such as the engine, cockpit, and also exhausting. All of this part is big and has its machine that helps in assembling it. So, this research suggests that to combine stations from the trim line which are station door dissemble and engraving with station panoramic or roof. For the door dissemble and engraving station, the labor or worker task is to disassemble the door before the car is a move to the next station to ease the operation. They also need to engrave the chassis number for the car in this station. As for the panoramic or roof station, the operator needs to disassemble the sunroof of the car. The other station that we combine are station 4 and station 5 as the scope of work is quite similar. Both stations involve processes such as assembling the cockpit or dashboard and also installing all the cables or wires around the interior of the car. Two more stations that were combined are station 7 and station 8. The process that involves in that station is assembling the rear window and also installing the rear-view mirror. All the stations as involves in assembling small parts and also do not depend on the process from previous stations. For the stations in the mechanical line, it is quite impossible to combine as there are many big components and huge assembly machines required to use to continue the process.

**Figure 10.** Pictures of station 4, 5, 7 and 8.

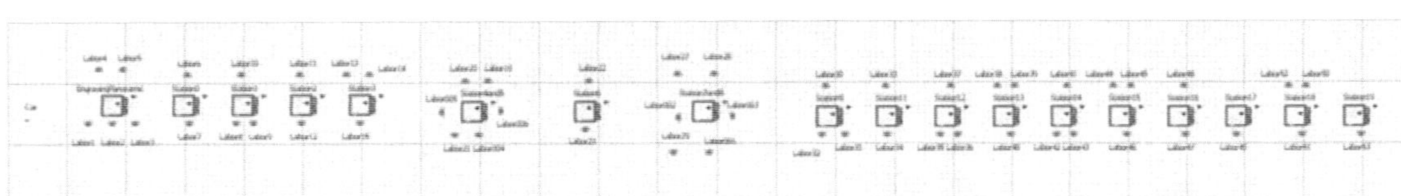

**Figure 11.** Proposed layout after combine station.

IOP Conf. Series: Materials Science and Engineering 1078 (2021) 012027 doi:10.1088/1757-899X/1078/1/012027

**Table 5.** Machine statistic after combine station.

| Name | % Idle | % Busy | % Filling | % Emptying | % Blocked | % Cycle Wait Labor | % Setup | % Setup Wait Labor | % Broken Down | % Repair Wait Labor | No. Of Operations |
|---|---|---|---|---|---|---|---|---|---|---|---|
| *STN00* | *0* | *100* | *0* | *0* | *0* | *0* | *0* | *0* | *0* | *0* | *24* |
| *STN01* | *0* | *100* | *0* | *0* | *0* | *0* | *0* | *0* | *0* | *0* | *24* |
| *STN02* | *0* | *100* | *0* | *0* | *0* | *0* | *0* | *0* | *0* | *0* | *24* |
| *STN03* | *0* | *100* | *0* | *0* | *0* | *0* | *0* | *0* | *0* | *0* | *24* |
| *STN04 & 05* | *0* | *100* | *0* | *0* | *0* | *0* | *0* | *0* | *0* | *0* | *24* |
| *STN06* | *0* | *100* | *0* | *0* | *0* | *0* | *0* | *0* | *0* | *0* | *24* |
| *STN07 & 08* | *0* | *100* | *0* | *0* | *0* | *0* | *0* | *0* | *0* | *0* | *24* |
| *STN09* | *0* | *100* | *0* | *0* | *0* | *0* | *0* | *0* | *0* | *0* | *24* |
| *STN11* | *0* | *100* | *0* | *0* | *0* | *0* | *0* | *0* | *0* | *0* | *24* |
| *STN12* | *0* | *100* | *0* | *0* | *0* | *0* | *0* | *0* | *0* | *0* | *24* |
| *STN13* | *0* | *100* | *0* | *0* | *0* | *0* | *0* | *0* | *0* | *0* | *24* |
| *STN14* | *0* | *100* | *0* | *0* | *0* | *0* | *0* | *0* | *0* | *0* | *24* |
| *STN15* | *0* | *100* | *0* | *0* | *0* | *0* | *0* | *0* | *0* | *0* | *23* |
| *STN ENGRV & ROOF* | *0* | *100* | *0* | *0* | *0* | *0* | *0* | *0* | *0* | *0* | *24* |
| *STN16* | *0* | *100* | *0* | *0* | *0* | *0* | *0* | *0* | *0* | *0* | *24* |
| *STN117* | *0* | *100* | *0* | *0* | *0* | *0* | *0* | *0* | *0* | *0* | *24* |
| *STN18* | *0* | *100* | *0* | *0* | *0* | *0* | *0* | *0* | *0* | *0* | *24* |
| *STN19* | *0* | *100* | *0* | *0* | *0* | *0* | *0* | *0* | *0* | *0* | *24* |
| *STN18* | *0* | *100* | *0* | *0* | *0* | *0* | *0* | *0* | *0* | *0* | *24* |
| *STN19* | *0* | *100* | *0* | *0* | *0* | *0* | *0* | *0* | *0* | *0* | *24* |

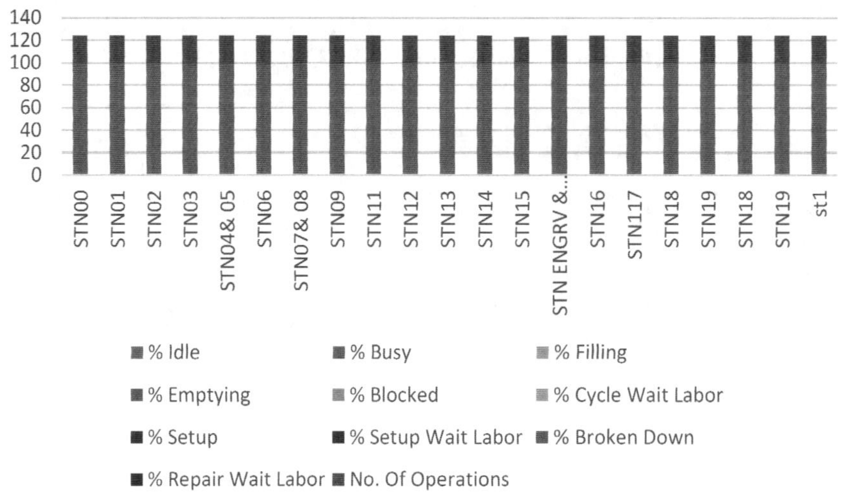


**Figure 12.** Graph percentage of the machine after combine station.

*4.3 Discussion*

From the research, as to conclude that the best improvement that can be made is by combining the station compare to adding a buffer area between the trim and mechanical line. The results show and are compared with each other. Combining the station in the trim line is better.

**Table 6.** Comparison between current, adding buffer and combine layout.

| Name | % Idle | % Busy | No. Of Operations |
|---|---|---|---|
| Current | 1.190476 | 98.8095 | 24 |
| Buffer | 1.61 | 98.39 | 23 |
| Combine | 0 | 100 | 24 |

From the table, adding a buffer area between station 9 and station 11 will not improve its output or percentage of busy. The current output is 24 while after improvement is 23. The percentage of busy is also reduced from 98.8095% to 98.39%. As for the percentage of idle, the current layout will give 1.190476% while adding the buffer area will increase the percentage of idle to 1.61%.

This buffer area is only some suggestion that can be made but cannot be done directly as the factory does not have enough space to provide it. This research needs to consult with the responsible person in the company to execute the process. The current layout of the factory is quite close to another assembly line so it will be difficult to have some space for buffer or storage area between stations.

By combining the station, the production output is constant with 24 units per day but the percentage idle will be eliminated from 1.190476%. The percentage of busy will also increase to 100% of 98.8095%. After combining the selected station, the time for one unit to be assembled is faster compare to the current one. This method will balance the time for all stations so there will be no stations that will have any lead time.

For the current layout,

$$21\ stations\ (20minutes) = 420\frac{minutes}{unit} \tag{3}$$

After combining the station,

$$18\ stations\ (20minutes) = 360\frac{minutes}{unit} \tag{4}$$

According to the QRM method, it is important to reduce the lead time as it will affect the productivity of the company. QRM is a suitable method for this company as this company also supplies high variation and fewer numbers of products for customers. For example, this company has 3 classes of the car with 3 different assembly line but all of the lines only produced below 100 units per day.

The best solution for improvement by combining the stations. Even though the output is still the same and cannot be increasing as the percentage of busy has reached 100%, it still can eliminate the percentage of idle that affected the lead time. After combining the stations, the lead time can be reduced and the time taken for one unit to be producing will be shortened.

## 5. Conclusion

The objective of this research is to analyses the current layout and also propose a new arrangement of the assembly line by using the Witness software. At first, the simulation of the current layout of the V model assembly line has been set up using Witness software. According to the result that we have from the simulation, station 19 is the most critical area in the line. It is because station 19 does not achieve 100% of busy while another station achieves it. The percentage of idle from station 19 is 25%. We need to reduce the percentage of idle as it will increase the time for production. This extra time is called lead time and we need to reduce it.

This problem brought to the next objective which is to propose a new arrangement of the assembly line. From the current layout, the proposed two new possible arrangement or improvement that can be made by using the Witness software. Firstly, by adding a buffer area or storage between station 9 and station 11 as it is the intersection of the trim and mechanical line but the result does not reach the research expectation. Thus, the output is decreased to 23 units per day from 24 units per day. The next arrangement by combining the stations in the trim line. The results achieved the target by eliminating the percentage of idle in station 19 and make the percentage of busy 100%. Even though the output is still the same, the reduction of the time for one unit to be produced from this improvement. This method was selected as the result is better to compare to the previous one.

**Acknowledgements**
The authors would like to be obliged to the Ministry of Higher Education, Malaysia and Universiti Malaysia Pahang for providing financial assistance under project no. FRGS/1/2019/TK03/UMP/02/20.

**References**

[1] Rosa C, Silva F, Ferreira L P and Campilho R 2017 SMED methodology: The reduction of setup times for Steel Wire-Rope assembly lines in the automotive industry *Procedia Manufacturing* **13** 1034-42
[2] Siong B C, Eng C K and Teknikal U 2018 Implementing Quick Response Manufacturing to Improve Delivery Performance in an ETO Company *International Journal of Engineering & Technology* **7** 38-46
[3] STALK G 1989 TIME-THE NEXT SOURCE OF COMPETITIVE ADVANTAGE (REPRINTED FROM HARVARD BUSINESS REVIEW JULY AUGUST, 1988) *Quality Progress* **22** 61-8
[4] Suri R 2010 *It's about time: the competitive advantage of quick response manufacturing*: CRC Press)
[5] Jitpaiboon T, Gu Q and Truong D 2016 Evolution of competitive priorities towards performance improvement: a meta-analysis *International Journal of Production Research* **54** 7400-20
[6] Godinho Filho M, Marchesini A G, Riezebos J, Vandaele N and Ganga G M D 2017 The application of Quick Response Manufacturing practices in Brazil, Europe, and the USA: An exploratory study *International Journal of Production Economics* **193** 437-48
[7] Chinet F S and Godinho Filho M 2014 POLCA System: literature review, classification, and analysis *Gestão & Produção* **21** 532-42
[8] Germs R and Riezebos J 2010 Workload balancing capability of pull systems in MTO production *International Journal of Production Research* **48** 2345-60
[9] Vandaele N, Van Nieuwenhuyse I, Claerhout D and Cremmery R 2008 Load-based POLCA: an integrated material control system for multiproduct, multimachine job shops *Manufacturing & Service Operations Management* **10** 181-97
[10] Godinho Filho M and Saes E V 2013 From time-based competition (TBC) to quick response manufacturing (QRM): the evolution of research aimed at lead time reduction *The International Journal of Advanced Manufacturing Technology* **64** 1177-91
[11] Stoycheva S, Marchese D, Paul C, Padoan S, Juhmani A-s and Linkov I 2018 Multi-criteria decision analysis framework for sustainable manufacturing in automotive industry *Journal of Cleaner Production* **187** 257-72

[12] Parvez M, Amin F and Akter F 2017 Line Balancing Techniques To Improve Productivity Using Work Sharing Method *IOSR Journal of Research & Method in Education (IOSRJRME)* **7** 07-14
[13] Álvarez-Miranda E and Pereira J 2019 On the complexity of assembly line balancing problems *Computers & Operations Research* **108** 182-6
[14] Nilakantan J M, Ponnambalam S and Nielsen P 2017 *Handbook of Neural Computation*: Elsevier) pp 239-67
[15] Rahani A and Al-Ashraf M 2012 Production flow analysis through value stream mapping: a lean manufacturing process case study *Procedia Engineering* **41** 1727-34
[16] Godinho Filho M, Marchesini A G, Riezebos J, Vandaele N and Ganga G M D 2017 The extent of knowledge of Quick Response Manufacturing principles: an exploratory transnational study *International Journal of Production Research* **55** 4891-911
[17] Ohno T 1988 *Toyota production system: beyond large-scale production*: crc Press)
[18] Yasir A and Mohamed N 2018 Assembly line efficiency improvement by using WITNESS simulation software. In: *IOP Conference Series: Materials Science and Engineering*: IOP Publishing) p 012004

# Various risks and safety analysis to reduce fire in oil refinery plant

**Md Zahangir Alam[1], Md Nazmus Sakib[2,3,*], Minarul Islam[4] and Mohammad Jalal Uddin[3]**

[1]Department of Mechanical Engineering, Faculty of Mechanical Engineering, Khulna University of Engineering & Technology, Khulna, 9203, Bangladesh.
[2]Faculty of Manufacturing and Mechatronic Engineering Technology, Universiti Malaysia Pahang (UMP), 26600, Pekan, Pahang, Malaysia.
[3]Department of EEE, Faculty of Science & Engineering, International Islamic University Chittagong (IIUC), Chittagong-4318, Bangladesh.
[4]Faculty of Electrical and Electronic Engineering Technology, Universiti Malaysia Pahang (UMP), 26600, Pekan, Pahang, Malaysia

*Corresponding author's email: sakibme@gmail.com

**Abstract.** It is quite hazardous to produce the final product of oil or gas in a petrochemical refinery plant due to its flammable or combustible and explosive materials. Small mistakes can cause massive damage to life, property, pollution, injury, ecosystem, and business by fire. The entire system is challenging to manage. Therefore, fire risk assessment and forecasting are necessary to overcome personal, environmental, and refinery plants' hazard situations. There have been four main threats in any refinery facility: electrical, mechanical, civil, and chemical issues, maximum cases its result burning. This research aimed to study and assess fire risk in the refinery plant by using a multi-stage early warning system and reducing the fire. The fire hazard safety layer technique would be used for our petrochemical process. Some equipment is set in place to forecast the danger and execution before and after it happens. Geographic information systems (GIS), remote sensing (RS) are some techniques for fire incidents tracking. Flame detectors, heat detectors, and gas detectors are used to maintain good contact in the entire risk analyzer portion. Various techniques and monitoring have been proposed to operate the plant efficiently and safely, like controlling, predicting, and pre-warning strategic planning.

**Keywords:** Refinery Plant Risk; Fire Reduction; Personal Safety; Plant Safety.

## 1. Introduction

An oil refining plant is a factory whereby various functions extract various oil types from petroleum products[1]. It's not a simple feature of the refining system. The explanation for the selection of extremely flammable and explosive materials is very complicated. The danger is the most critical issue for an oil refining facility. There are many threats that not only directly but also indirectly occur in oil refining plants. Also, the plant's waste leakage is flammable, explosive, and very dangerous [2]. Any of the waste goes to the river and the atmosphere. It pollutes the river's habitat and water [3]. For all animals living in the atmosphere and rivers, it is hazardous. These are the indirect risks involved. Electrical, electronic, civil, and chemical threats are directly involved [1,4]. There will be firing for these direct threats, and workers may be hurt. The entire plant can even be lost. Some years ago, in China, the petrochemical firms had fire & explosion incidents [5]. The overall upward trend and situation remained severe, such as the Jilin petrochemical company fire & explosion accident at the benzene factory [5–7]. The heavy casualties and loss of land and a wide area of water contamination in the Songhua River have caused adverse effects in the country and the international community. Petro China Lanzhou Petrochemical Company No.316 Tank Farm Fire and Explosive Accident on January 7th, 2010 [8]. That's why the serial explosion of about eleven tanks one after another. Dalian Dagushan Pipeline Explosion and Fire in Petro-China occurred on July 16th, 2010 [9]. And that is why the Farm No.103 tank fire caused several oil spills and the creation of nearly 60000 square meters of floor flowing fire field, leading to a wide area of water contamination near the Dalian Sea area. Therefore, it is essential to prevent these accidents and protect all things affected by those accidents above or like them, suggesting that some direct and indirect technologies need to be set up. There are various forms of immediate risk use of technology: Monitoring system, voltage control valve, walkie-talkie instead of cell phone, to resolve the electrical risk, the RS system is used. The lubricating fluid measurement scale for regularly checking the amount of lubricating fluid to determine mechanical risk, monitoring the machinery after a period of time and replacing, if necessary post-welding heat treatment, non-destructive heat treatment (UV test, X-ray test, UST test, magnetic test, flange joint, and gasket), heat detector, flame detector, gas detector, etc. [10–12]. It is essential to develop plant-based model to resolve the civil danger. On the actual design and take enough protection, monitoring at a regular time, pre-commissioning test, joint flame test, and GIS. The components of crude oil such as hydrocarbon (CH), sulfur-di-oxide (SO2), sulfur (S) are needed to maintain an acceptable percentage to avoid chemical danger [13]. The waste is re-used as bituminous in liquid form for indirect risk, and gaseous waste is used as a heating source that pollutes the river and the atmosphere.

## 2. Risk of oil refinery plant

The danger is the likelihood of damage or harm caused by any threat. Adverse effects on humans, plants, or costs are also reported. A qualitative risk evaluation typically includes a matrix of the probability versus the risk acquisition implications from which a risk management decision is made [14]. In oil refinery plants, there are primarily four forms of danger that are followed. Those such as.

### *2.1. Electrical risk*

Most of the machines and equipment in oil refining plants are electro-mechanical and are powered by electrical power. There is, therefore, a greater risk of facing electrical problems. Such as short circuits, networks, high voltage. Shooting is the most likely cause of these issues [15].

### *2.2. Mechanical risk*

An oil refining facility builds many weld joints about 50,000 or higher. So, there is more likely to leak inside or outside the joint for looseness or over the joint's tightness. Increases in high temperature and pressure can result in an explosion. These are referred to as mechanical threats [16].

### *2.3. Civil risk*

The plant is built predominantly by civil design. The soil of the Earth consumes the entire plant load. The plant will break if the soil is not perfect for making the plant because more vibration comes when the plant runs. Any plant is influenced by the environment surrounding it. The bulk of the plant is located next to the river [17]. And in that location, it can face some dangerous storm and Could break the plant. If a plant fails, some problems must be faced, such as if staff may injure or die, production may hamper, and any system may damage, leakage may occur while shooting.

### *2.4. Chemical risk*

Many flammable and active substances are made up of crude and refinery oil, both organic and inorganic. It often produces vapor from liquid organic oil [18]. If the temperature increases, fire is likely.

IOP Conf. Series: Materials Science and Engineering 1078 (2021) 012028 doi:10.1088/1757-899X/1078/1/012028

Sulfur and sulfur dioxide, both flammable and very active compounds, are gasoline's main components. As it is so active, when it goes through it interacts with the pipe and is responsible for eroding the pipe.

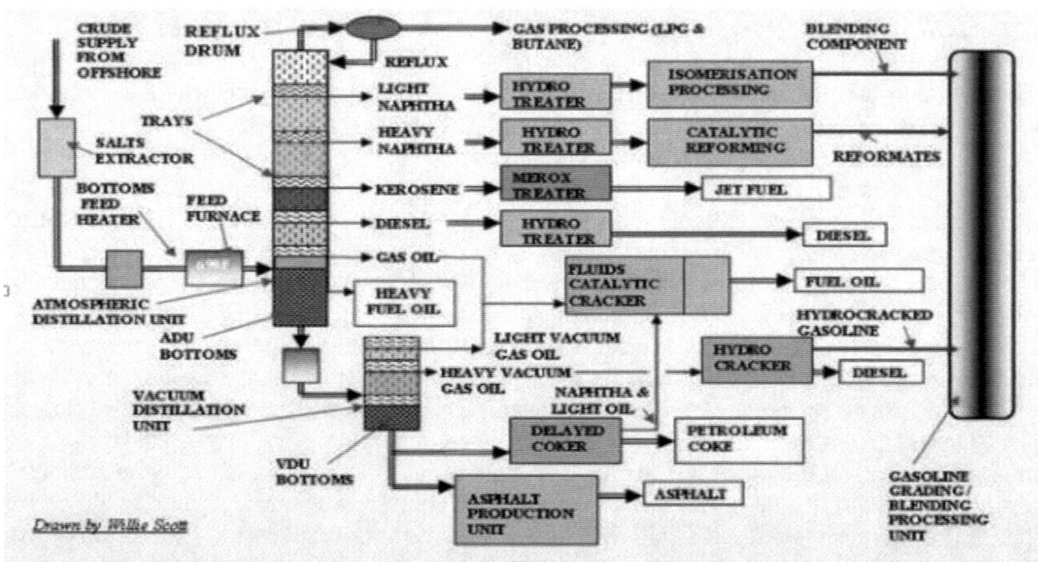


**Figure 1.** Schematic diagram of oil refinery plant [19].

## 3. Safety assessment of refinery plant

A person's safety can be described as the depreciation of human-hazard contact and is primarily concerned with the prevention of physical harm (injury) to a person." [20]. Plant safety means 'secure operation of all facilities, safe working people's activities, safe storage of raw materials and goods.' More generally, safety refers to rules and efforts to keep people unharmed, safeguard properties, provide uninterrupted output, and protect the environment in an oil refining plant; there are primarily two types of safety.

### *3.1. Personnel Safety*

Personal Protective Equipment (PPE) is intended to protect personnel from serious accidents or illnesses in the workplace from interaction with chemical, physical, electrical, mechanical, radiological, or other workplace hazards [21]. By providing a shield against workplace hazards, the Personal Protective Equipment (PPE) program intends to protect workers from the risk of injury. Personal protective equipment is not a replacement for sound engineering or administrative controls, or acceptable work practices. Still, it can be used to ensure employees' safety and health in accordance with these controls. Personal protective equipment will be provided, used, and maintained when it is decided that its usage is required and that the risk of a workplace injury or disease will be minimized [22]. Eye, ears, head, foot, and hand safety are discussed in this curriculum. For respiratory and hearing safety, different systems exist.

### *3.2. Plant Safety*

In an oil refining facility, four types of protection systems are primarily maintained.

Electrical Safety: First the supply of electricity is necessary for the operation of any electrical machine. If any issue arises, the production of the plant will initially stop completely. Therefore, adequate electrical protection needs to be provided, such as;

- Electrical supply machinery control.

- Using the Voltage Control Valve.
- Using walkies-talkies for contact instead of mobile phones and so on.

*3.3. Mechanical Safety:* Most of the machinery in every plant is electro-mechanical, and most of the operating procedures are mechanically handled. In mechanical systems, it is essential to provide protection.

- Using the measuring scale for lubricating fluid and tracking
- The sum of lubricating fluid over a normal period.
- Monitoring for a period and machinery
- Replace it if necessary,
- Ultraviolet test, Xray test, and UST (ultrasonic test: post welding and heat treatment test
- Use the flange joint and gasket to indicate and transfer signal into the control house using the flame detector while the flame is produced anywhere in the factory.

*3.4. Civil safety:* Construct the plant-based on the actual design and take enough factor safety. There is also a need to monitor the plant regularly, and a pre-commissioning test and a joint flame test are needed. Painting: Near coastal and salty regions, the maximum oil plants are located. There could even be a cement factory next to it. The structure of the plant can be affected by different dust and salt forms. That's why any part of the factory, mostly located in the atmosphere, is painted over.
Wall: If it affects the oil, oil is stored in the tank and can stream over the region and pollute the water. To protect the atmosphere, a wall is constructed around the tank with the same tank volume, and the plant needs to be washed.

*3.5. Chemical safety*: At acceptable limits, the excess sulfur, excess hydrocarbon, and excess sulfur dioxide are retained. Internal chemical safety: You can extract sulfur by using a catalyst. The removal of hydrogen sulfide is also required. Humans can die if the amount of H2S exceeds 50 ppm.

## 4. Result and Discussion

*4.1. Data collection to risk measurement*
A total of 23 samples of data were collected from a reputated Oil Refinery plant of Bangladesh. The data collection was done with two months of observation (July-August, -2019). All data were collected with a written consent form from the industry general manager. From 23 data samples for mechanical lubrication system failure, mechanical device damage and a mechanical short circuit were 162, 272, and 408y. Correspondingly, electrical, civil, and chemical risk are shown in detail in Table 1.

*4.2. Data collection on safety measurement*
On the other hand, safety measurement data were collected from the same refinery plant, which is shown in detail in Table 2. From Figure 2. (a) it is concluded that, for risk of analyzed data with excess sulfur 13%, lubrication 7%, the system fails 12%, mechanical 18%, device damage 10%, short circuit 4%, high voltage 5% and network 22% respectively. Networking is the topmost risk factor, whereas the short circuit shows the lowest. However, for safety the Figure 3. (b) shows a helmet for head protection 5%, mask for nose protection 13%, hand gloves 4%, safety shoe 5%, detector 24%, foaming 26%, respectively. Table 3 shows the risk, effect, and solution of the oil refinery plant accordingly.

**Table 1.** Data collection on risk from an oil refinery plant.

| Risk Identification | Mechanical | | Electrical | | | Civil | Chemical | | |
|---|---|---|---|---|---|---|---|---|---|
| | Lubrication System fail | Mechanical Device damage | Short circuit | High voltage | Network | Constructional | Excess sulfur | Excess $So_2$ | Excess CH |
| S1 | 10 | 10 | 30 | 10 | 10 | 5 | 10 | 10 | 10 |
| S2 | 10 | 10 | 20 | 10 | 5 | 5 | 20 | 10 | 10 |
| S3 | 5 | 5 | 10 | 5 | 5 | 5 | 40 | 10 | 10 |
| S4 | 5 | 10 | 10 | 10 | 5 | 5 | 25 | 10 | 20 |
| S5 | 5 | 15 | 20 | 15 | 5 | 5 | 20 | 5 | 10 |
| S6 | 10 | 15 | 15 | 10 | 5 | 5 | 20 | 10 | 10 |
| S7 | 5 | 10 | 15 | 10 | 5 | 5 | 10 | 20 | 20 |
| S8 | 5 | 10 | 20 | 12 | 3 | 10 | 25 | 5 | 10 |
| S9 | 10 | 10 | 20 | 15 | 5 | 5 | 15 | 10 | 10 |
| S10 | 12 | 20 | 20 | 15 | 2 | 6 | 15 | 3 | 8 |
| S11 | 5 | 10 | 15 | 10 | 3 | 5 | 20 | 18 | 14 |
| S12 | 5 | 8 | 20 | 10 | 2 | 7 | 25 | 8 | 15 |
| S13 | 5 | 10 | 25 | 10 | 5 | 3 | 20 | 7 | 15 |
| S14 | 10 | 15 | 15 | 10 | 2 | 4 | 22 | 10 | 12 |
| S15 | 5 | 10 | 15 | 10 | 3 | 5 | 25 | 12 | 15 |
| S16 | 8 | 12 | 15 | 12 | 5 | 5 | 17 | 12 | 14 |
| S17 | 10 | 12 | 20 | 15 | 3 | 4 | 20 | 6 | 10 |
| S18 | 15 | 20 | 15 | 10 | 5 | 5 | 15 | 5 | 10 |
| S19 | 5 | 10 | 15 | 10 | 2 | 3 | 30 | 10 | 15 |
| S20 | 3 | 10 | 12 | 7 | 3 | 5 | 30 | 15 | 15 |
| S21 | 5 | 10 | 30 | 15 | 5 | 5 | 15 | 5 | 10 |
| S22 | 4 | 10 | 16 | 5 | 2 | 5 | 30 | 10 | 18 |
| S23 | 5 | 10 | 15 | 5 | 5 | 5 | 30 | 10 | 15 |
| Total | 162 | 272 | 408 | 241 | 95 | 117 | 499 | 221 | 296 |

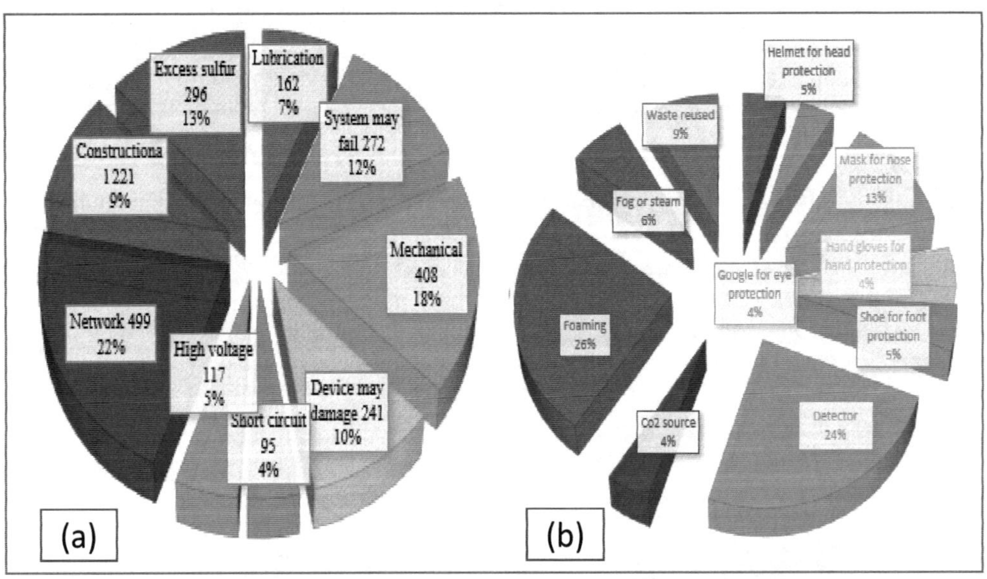


**Figure 2.** (a) Risk and (b) Safety analysis in the pie chart by percentage.

**Table 2.** Data collection on safety measure from an oil refinery plant.

| Safety provides | Personal | | | | | Environ mental | Plant | | | |
|---|---|---|---|---|---|---|---|---|---|---|
| | Hel met | Goog le | Mask | Hand gloves | Sho e | Waste reused | Dete ctor | $Co_2$ source | Foa ming | Fog or steam |
| S1 | 5 | 5 | 10 | 5 | 5 | 9 | 20 | 6 | 25 | 10 |
| S2 | 4 | 2 | 12 | 5 | 5 | 10 | 22 | 5 | 25 | 10 |
| S3 | 5 | 3 | 10 | 5 | 7 | 8 | 25 | 5 | 20 | 12 |
| S4 | 7 | 3 | 15 | 8 | 5 | 8 | 20 | 3 | 25 | 6 |
| S5 | 6 | 4 | 15 | 5 | 6 | 8 | 25 | 5 | 22 | 4 |
| S6 | 5 | 5 | 12 | 5 | 5 | 10 | 22 | 3 | 25 | 8 |
| S7 | 5 | 6 | 10 | 8 | 5 | 4 | 25 | 5 | 28 | 4 |
| S8 | 5 | 5 | 10 | 5 | 5 | 10 | 20 | 5 | 30 | 5 |
| S9 | 4 | 4 | 12 | 4 | 4 | 7 | 27 | 4 | 30 | 4 |
| S10 | 5 | 3 | 15 | 3 | 5 | 6 | 30 | 5 | 24 | 4 |
| S11 | 8 | 3 | 10 | 4 | 5 | 11 | 22 | 4 | 27 | 6 |
| S12 | 5 | 5 | 8 | 3 | 6 | 12 | 25 | 5 | 25 | 6 |
| S13 | 4 | 6 | 12 | 5 | 4 | 10 | 25 | 4 | 27 | 3 |
| S14 | 6 | 3 | 15 | 3 | 5 | 6 | 24 | 3 | 30 | 5 |
| S15 | 5 | 5 | 10 | 5 | 5 | 10 | 22 | 5 | 25 | 8 |
| S16 | 5 | 4 | 12 | 4 | 5 | 11 | 25 | 5 | 25 | 4 |
| S17 | 4 | 3 | 15 | 3 | 5 | 7 | 28 | 4 | 24 | 7 |
| S18 | 5 | 4 | 12 | 5 | 5 | 7 | 30 | 3 | 25 | 4 |
| S19 | 3 | 4 | 16 | 3 | 5 | 10 | 25 | 3 | 25 | 6 |
| S20 | 3 | 3 | 15 | 3 | 3 | 5 | 30 | 3 | 30 | 5 |
| S21 | 5 | 3 | 12 | 3 | 5 | 10 | 22 | 4 | 28 | 8 |
| S22 | 5 | 4 | 18 | 2 | 5 | 11 | 24 | 2 | 24 | 5 |
| S23 | 2 | 2 | 20 | 4 | 3 | 8 | 25 | 4 | 22 | 10 |
| Total | 111 | 89 | 296 | 100 | 113 | 198 | 563 | 95 | 591 | 144 |

### *4.3. Discussion*

According to the data collected from the oil refinery plant, some solutions were listed based on the risk in Table 3. Most of the risk in the refinery plant causes firing and huge damage to life and wealth. Proper maintenance and guideline reduce the risk and environmental hazard and increase plant production and efficiency. Providing adequate safety equipment and training is a must for every person related to the refinery plant. Every counter or section should have a checking system for the safety uniform and have a safe assembly point. While the risk alarm will ring, everybody should assemble a safe place and flow in the direction learn by training and not panic.

### *4.4. Fire hazard prevention and control model*

In the case of firing or prevent firing in the plant, the authors proposed a hazard prevention and control model showen if figure 3. The proposed model will be effective and efficient if the plant authority flows accordingly. Saloua et.all [23] used a hybrid technique (HAZID) for identifying the risk scenarios that can result explosion or fire accidents, theirs causes, location and protection measures. They also used PHAST software to simulate and predict the fire.

IOP Conf. Series: Materials Science and Engineering 1078 (2021) 012028 doi:10.1088/1757-899X/1078/1/012028

**Table 3**. Risk, effect, and solution of oil refinery plant.

| Risk | | Effect | Solution |
|---|---|---|---|
| Electrical | Short Circuit<br>High Voltage<br>Network | Firing | Monitoring<br>Used Voltage Regulating Valve<br>Used okitoki instead of cell phone |
| Mechanical | Lubrication systems fail | | Used lubricating fluid measuring scale and monitoring the level of lubricating fluid in a periodic time. |
| | Mechanical Device fail<br>Temperature Valve<br>Pressure Valve<br>Compressor<br>Pump<br>Engine Function<br>Welding | Temperature rise and firing | Monitoring replace period of machinery.<br>Non-destructive heat treatment.<br>Post welding heat treatment,<br>X-Ray test.<br>Magnetic test<br>UST test<br>Flange Joint & gasket.<br>UV Test. |
| Civil | Construction Fail | Personnel injure or die<br>Device may damage<br>Production may hamper<br>Firing may occur for leakage | Construction based on actual enough safety.<br>Monitoring regular<br>Pre-commissioning flame joint test. |
| Chemical | Excess Sulphur<br>Excess $SO_2$<br>Excess CH | Corrosion in pipe<br>Firing | Reduce it to a reasonable level. |

Security management is an interconnected effort that requires leadership for effective implementation and practice at all levels. Employee interest in formulating the organization's safety policy and its role in enforcing these policies in the workplace is of great importance. Participative management is the modern-day philosophy of handling a business concern. A successful safety committee will improve the organization's general safety standards, thereby avoiding injuries. Ultimately, this establishes a greater understanding between employees and managers, less downtime in output, less maintenance costs, consumer confidence, etc. These directly promote the company's competitiveness and profit. More benefit increases the staff's compensation package and welfare activities. A dynamic activity is the development of a thriving safety culture. But the organization must develop a positive safety culture to achieve the best employee results. The safety culture is a delicate entity-it takes time to develop, but if not properly cared for, it can easily be destroyed. There should be a safety & firefighting department, specifically in refinery organizations, to protect.

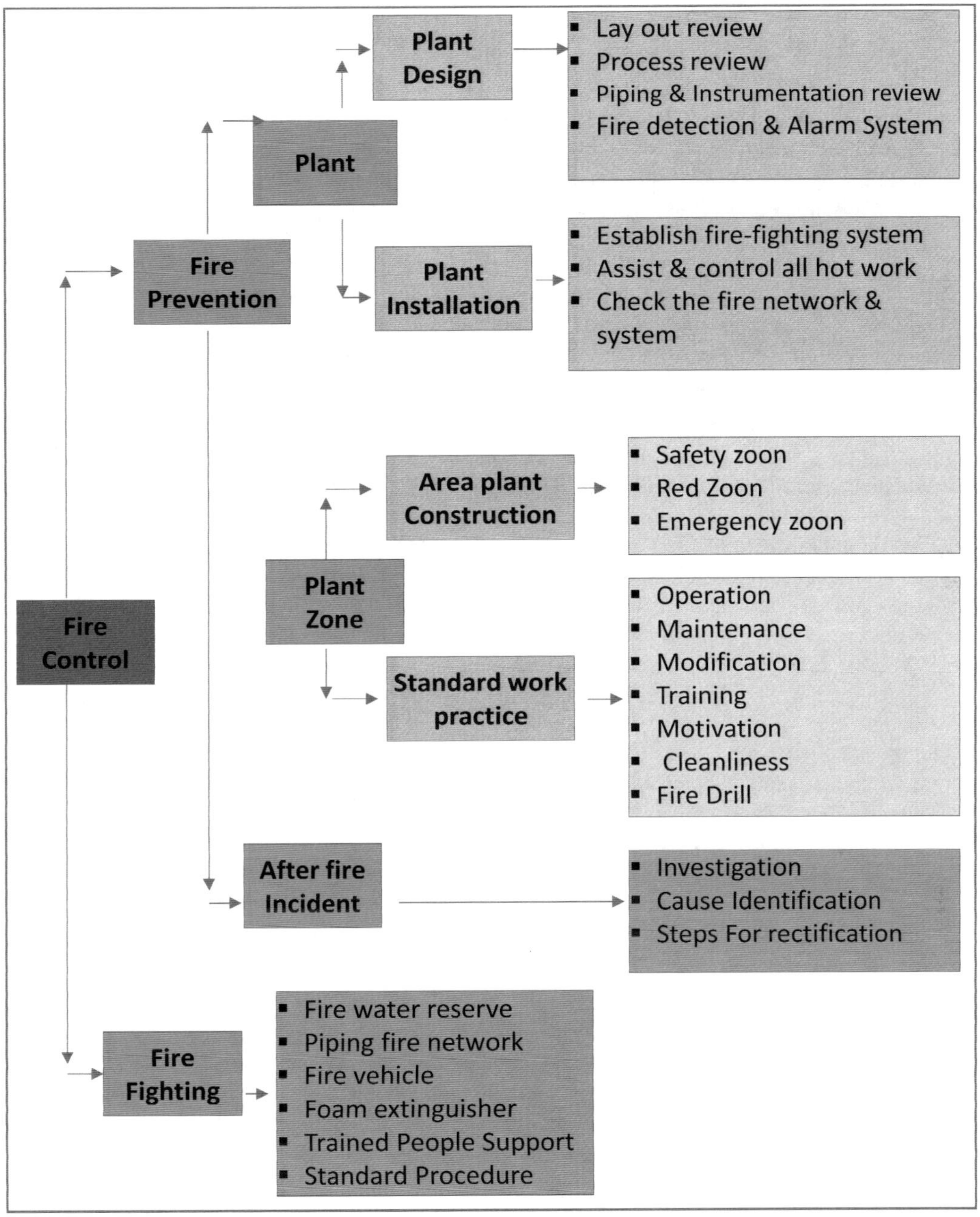


**Figure 3.** Proposed fire hazard prevention and control model.

## 5. Conclusion

In both the domestic and industrial areas, fire is a prevalent occurrence. Due to this case, massive losses are suffered by the industries every year. The size-large and operational-critical were rendered by the present design definition of the industries. If not timely and adequately attended, a minor fire incident will end up in the catastrophic form. No unnecessary occurrence of fire, intent, and effort should be made, and if it occurs, timely, and adequately attended. To ensure the facility's proper protection, each industry should have safety regulations, fire control, emergency preparedness & evacuation plan.

In all facilities/companies, the following emergency protocols should be adopted.

- Establish a fire alarm device, sirens, whistles, etc.
- Present written emergency plans in popular positions and provide all staff with copies.
- Maintain direct access to buildings and locations.
- Warn security staff in case of an alarm to open gates/doors.
- Install simple signs in prominent positions, showing fire access route locations, escape routes and extinguisher positions, other combat facilities, etc.

**Acknowledgments**
Authors are greatfull to Petromax Oil Refinery Limited, Mongla Bagerhat, Bangladesh for their cordial support by providing the facility of data collection and investigation.

**References**

[1] Saloua B, Mounira R and Salah M M 2019 Fire and Explosion Risks in Petrochemical Plant: Assessment, Modeling and Consequences Analysis *J. Fail. Anal. Prev.* **19** 903–16
[2] Moreno-Sader K, Alarcón-Suesca C and González-Delgado A D 2020 Application of environmental and hazard assessment methodologies towards the sustainable production of crude palm oil in North-Colombia *Sustain. Chem. Pharm.* **15** 100221
[3] Katansky A A, Minchenok E E, Zhuravleva N G, Aleksandrova Y Y and Trotsenko A A 2020 Industrial and ecological safety analysis for modeling oil and petroleum products spills in the Arctic waters of the Kola Bay *IOP Conference Series: Materials Science and Engineering* vol 862 (Institute of Physics Publishing) p 062007
[4] Mohammadfam I and Zarei E 2015 Safety risk modeling and major accidents analysis of hydrogen and natural gas releases: A comprehensive risk analysis framework *Int. J. Hydrogen Energy* **40** 13653–63
[5] Cai B, Liu Y, Liu Z, Chang Y, Jiang L, Cai B, Liu Y, Liu Z, Chang Y and Jiang L 2020 Bayesian Network-Based Risk Analysis Methodology: A Case of Atmospheric and Vacuum Distillation Unit *Bayesian Networks for Reliability Engineering* (Springer Singapore) pp 189–216
[6] Luo T, Wu C and Duan L 2018 Fishbone diagram and risk matrix analysis method and its application in safety assessment of natural gas spherical tank *J. Clean. Prod.* **174** 296–304
[7] Zio E and Aven T 2013 Industrial disasters: Extreme events, extremely rare. Some reflections on the treatment of uncertainties in the assessment of the associated risks *Process Saf. Environ. Prot.* **91** 31–45
[8] Liu X Y and Wang X Z 2013 Fire risk forecast and early warning technology for large oil and gas storage & transport tank areas *Advanced Materials Research* vol 726–731 (Trans Tech Publications Ltd) pp 4654–9
[9] Liu X 2012 The petrochemical park fire safety planning study based on fire risk analysis *Advanced Materials Research* vol 518–523 (Trans Tech Publications Ltd) pp 1045–51
[10] Sadeghi-Yarandi M, Karimi A, Ahmadi V, Sajedian A A, Soltanzadeh A and Golbabaei F 2020 Cancer and non-cancer health risk assessment of occupational exposure to 1,3-butadiene in a petrochemical plant in Iran. *Toxicol. Ind. Health* 748233720962238
[11] Raposo M A M, Gomes C J P and Nunes L J R 2020 Selective Shrub Management to Preserve Mediterranean Forests and Reduce the Risk of Fire: The Case of Mainland Portugal *Fire* **3** 65
[12] Onyeabor S and Zalewski P 2020 Best practices in gas flaring and emission reduction - Importance of safety equipment *Society of Petroleum Engineers - SPE Nigeria Annual International Conference and Exhibition 2020, NAIC 2020* (Society of Petroleum Engineers)
[13] Hosseini N, Givehchi S and Maknoon R 2020 Cost-based fire risk assessment in natural gas industry by means of fuzzy FTA and ETA *J. Loss Prev. Process Ind.* **63** 104025
[14] Ibrahim H A and Rao P J 2019 *FIRE RISK ANALYSIS IN FLNG PROCESSING FACILITY USING BAYESIAN NETWORK* vol 14

[15] Lim J W, Baalisampang T, Garaniya V, Abbassi R, Khan F and Ji J 2019 Numerical analysis of performances of passive fire protections in processing facilities *J. Loss Prev. Process Ind.* **62** 103970
[16] Anon Natech Risk Assessment and Management: Reducing the Risk of Natural-Hazard ... - Elisabeth Krausmann, Ana Maria Cruz, Ernesto Salzano - Google Books
[17] Brunoro Ahumada C, Quddus N and Mannan M S 2018 A method for facility layout optimisation including stochastic risk assessment *Process Saf. Environ. Prot.* **117** 616–28
[18] Ouffroukh L A, Chaib R, Ion V and Khochmane L 2018 Analysis of risk and the strengthening of the safety technical barriers: Application of Skikda (Algeria) oil refining complex *World J. Eng.* **15** 99–109
[19] Anon Crude Oil: Refining Process Of Crude Oil
[20] Priatno D H and Sunaryo 2020 Study of Pioneer Sea Transportation's Safety with Formal Safety Assessment Method *IOP Conference Series: Earth and Environmental Science* vol 557 (IOP Publishing Ltd) p 012016
[21] Chen C, Xu T, Cao S and Wang X 2018 Study on risk assessment of trans-regional transmission lines tripping disaster induced by wildfire *IOP Conference Series: Materials Science and Engineering* vol 452 (Institute of Physics Publishing) p 032121
[22] Lakhe P, Prehn E M, Habib T, Lutkenhaus J L, Radovic M, Mannan M S and Green M J 2019 Process Safety Analysis for Ti 3 C 2 T x MXene Synthesis and Processing *Ind. Eng. Chem. Res.* **58** 1570–9
[23] Saloua B, Mounira R and Salah M M 2019 Fire and Explosion Risks in Petrochemical Plant: Assessment, Modeling and Consequences Analysis *J. Fail. Anal. Prev.* **19** 903–16

# A review on 3D printed polymer-based composite for thermal applications

**N A Halim[1,*], J Mogan[1], L Sandanamsamy[1], W S W Harun[2], K Kadirgama[3], D Ramasamy[2] and F Tarlochan[4]**

[1]Institute of Postgraduate Studies,Universiti Malaysia Pahang, 26300 Gambang, Kuantan Pahang, Malaysia.
[2]Department of Mechanical Engineering,College of Engineering, Universiti Malaysia Pahang, 26300 Gambang, Kuantan Pahang, Malaysia.
[3]Faculty of Mechanical and Automotive Engineering Technology, Universiti Malaysia Pahang, 26600 Pekan, Pahang, Malaysia.
[4]Mechanical and Industrial Engineering, Qatar University, Qatar.

*Corresponding author's e-mail: norakmalhalim1925@gmail.com

**Abstract.** Recently, 3D printing techniques for the polymer-based part has become popular in the industry and academic area. Fused deposition modelling (FDM) is one of the most popular methods among 3D printing techniques because of its low cost, ability to fabricate objects with complex structures and geometries, specific functions, ease in processing, safety, reliability, and availability of various thermoplastic materials. This technique's application is widespread on the electronic application, medical application, and rapid tooling applications. This paper aims to review fused deposition modelling (FDM) 3D printed polymer-based composite techniques for thermal management applications. This paper will provide an overview of fused deposition modelling (FDM) techniques in 3D printing, conducting polymer base and conducting nanofiller additive material used, and current trend research in this area.

**Keywords**: Fused Deposition Modelling; Thermal Management; Conducting Polymer; Conducting Nanofiller Additives.

## 1. Introduction

Higher-dimensional accuracy and shorter time to construct prototypes have become the reason to use 3D printed techniques in modern industries. There are several types of 3D printed techniques that have been commonly used in modern industry such as stereolithography (SLA), selective laser sintering (SLS), fused deposition modelling (FDM), laminated object manufacturing (LOM), and direct metal laser sintering (DMLS). The most commonly used 3D printed technique in the industry is fused deposition modelling because of its ability to construct a complex structure in a shorter time and shorter cost. It has been widely used in applications such as medical [1, 2], electrical and electronics [3-7], rapid tooling [8], automotive sector [9], toys [10] and aircraft industries. Traditional FDM systems only used to fabricate parts with thermoplastics. Current FDM systems can produce parts in a wide range of engineering and industrial thermoplastics such as acrylonitrile butadiene styrene (ABS), polylactic acid (PLA), etc.

### *1.1. FDM basic principles*

The FDM system's basic principle is by depositing extruded materials layer by layer through a nozzle of feedstock filament offers the great potential of filament extrusion for a wide range of other materials such as carbon, metals and ceramics as shown in figure 1. FDM system comprises filament winder, drive wheels, filament feedstock, extrusion heads, nozzle, and building platform to depositing layer by layer extruded materials of 3D printing. The 3D printing filament feedstock can be either a single polymer, polymer blends, or composite materials, combining the base polymer and filler materials depending on the preferred application. Metals, ceramics, and carbon are an example of filler materials. The new composites materials from a combination of base polymers and filler materials in feedstock filament can be developed and used in the FDM process with suitable size and properties. A. Dorigato *et al.* [11] has made up filament for FDM feedstock from combinations of acrylonitrile-butadiene-styrene (ABS), and multi-walled carbon nanotubes (MWCNT). Liu W. *et al.* [12] has made a hybrid composite by combining two filler materials, silicon carbide (SiC) / C / PLA as feedstock filament of FDM and produced 3D printed polymer composite parts. FDM technique has been used in improving mechanical properties [13-20], improving thermal properties [11, 20-23], improving electrical conductivity [11], improve process ability [21, 24-26], and fabricate complex parts [27-29].

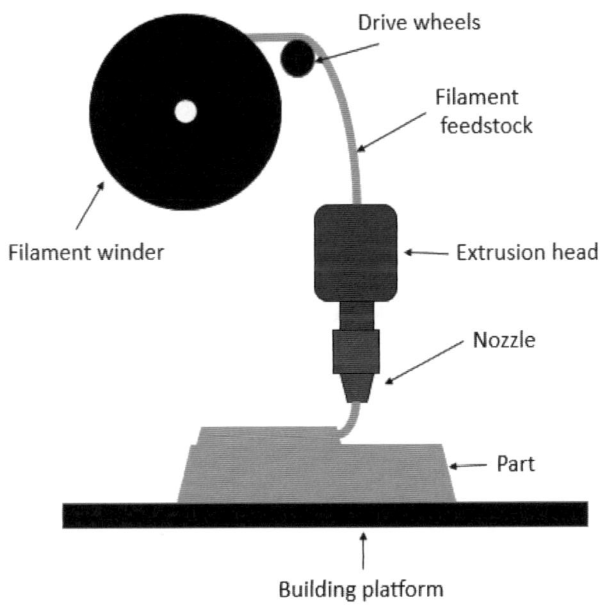


**Figure 1.** Schematic diagram of FDM primary mechanism.

### *1.2. Overview of 3D printed polymer composites*

The classification of 3D printed polymer composite can be divided into a single polymer, poly-polymer, and composite. Single polymer 3D printed materials are the materials that using single polymer for 3D printed application. In contrast, poly-polymer 3D printed materials are the materials that blend two or more polymers together in their research. In composite classification includes single polymer composite, polymer blends, hybrid filler, and base-filler hybrid. Polymer blends composite can be achieved by blending two or more polymers, together with the addition of filler material. A brief schematic of 3D printed polymer composite classification using the FDM technique for thermal application is illustrated in figure 2. In contrast, figure 3 elaborated on the polymer-based composite research trend in thermal application acquired from 2001 to 2020.

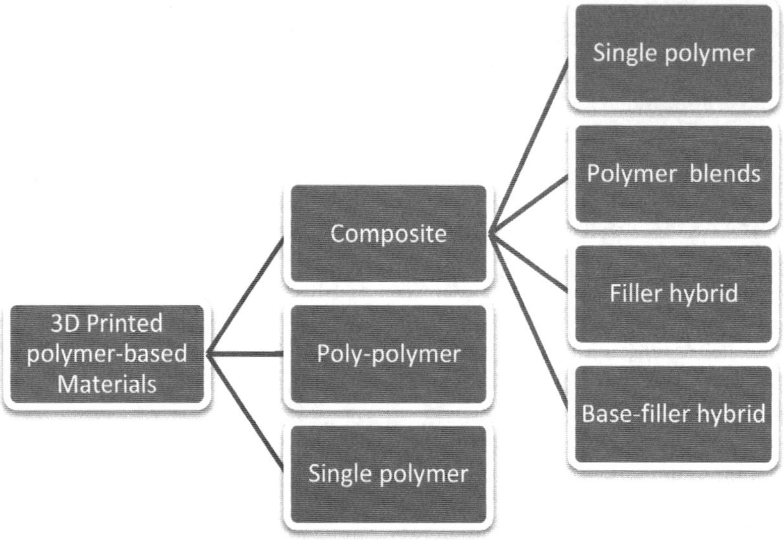


**Figure 2.** Classification of 3D printed polymer-based composite.

The purpose of blends polymer together is improved the processability of the composite. Blanco I. *et al.* [21] have blended two polymers together, polyetherimide and polycarbonate to be used in advanced applications. Single polymer composite uses one type of polymer and one type of filler only while hybrid filler composite is acquired by combining two or more fillers with a polymer. Hybrid composite purposes are to enhance the properties while maintaining the others to the composite materials. To enhance the crystallinity, thermal, electrical, and mechanical properties of composite, Kotsilkova R. *et al.* [20] have made filament feedstock using graphene and carbon nanotubes. A Base-filler hybrid composite is a composite that combining two or more polymers with two or more fillers to become a hybrid composite. Researchers choose composites over single polymers or poly-polymer because composite materials give different characteristics to the materials depending on the polymer and filler used [4, 11, 13, 21, 30].

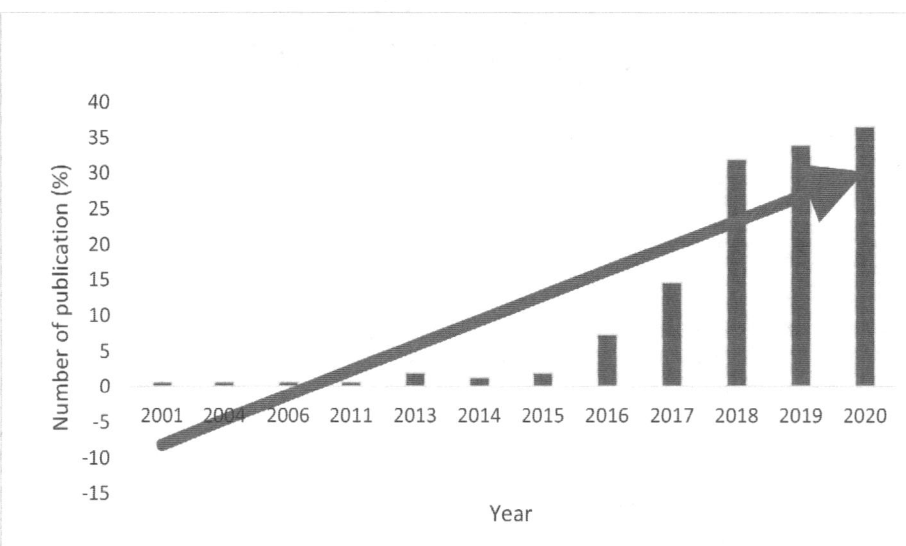


**Figure 3**. Research trend of composite-based polymer in thermal application.

The graph from figure 3 shows the increasing trend of polymer-based composite research for thermal application. The trendline shown is covered from 2001 to 2020, including the data collected from more than 150 research papers, which means that this research topic is an interesting topic to be explored and has a wide range of applications. There has been increasing in the research field of 3D printed polymer composite materials from 2011 until now because of the fast advancement of nanotechnology and microelectronic device that demand new materials for their wide range of application. MXene is an example of new material described in the year 2011 that has become supermaterial with excellent thermal and electrical conductivity, chemical stability, and high mechanical properties which are believed to lead to a drastic increase of polymer composite research in material engineering that can be compared with graphene. Graphene also has excellent properties such as thermal properties and has great potential in energy storage technology and electronic device [31].

In this review, the trend of advances in 3D printed polymer composite for the thermal application was presented. Fused deposition modelling (FDM) of 3D printing techniques has been selected among other methods because of ease of manufacturing, low cost, and fabricating complex shapes. This paper aims to analyze and summarize the processing parameter and response parameter of the FDM technique such as base materials, filler materials, processing temperature, and experimental study. This review outcome outlines the progress in developing materials in 3D printed polymer composite based on recent literature. Current or future trends of this area of study and the limitations of current technologies will also be discussed.

## 2. Thermally conductive nanofiller additive materials

Polymers such as polylactic acid (PLA), ABS, polypropylene (PP), polyethylene (PE), polyetherimide(PE), and others have been widely used in industry for decades because of its flexibility, processability, lightweight, low cost, excellent chemical stability, and excellent mechanical properties. The only technological barrier of these polymers to some applications such as electrical and electronic applications [32], rapid prototyping [33], energy storage [12], and biomaterial, is that polymers have low thermal conductivity. The effort to enhance polymers' thermal conductivity is crucial for these applications that have become an active research topic in the last decades.

### *2.1. Single filler*

Lots of filler materials can be used to enhance thermal conductivity, as shown in table 1, such as metal, ceramic, and carbon filler material. The metal filler used for thermal application such as silver (Ag), copper (Cu), gold, zink (Zn), stainless steel (SS), aluminium (Al), iron (Fe), titanium dioxide, and zirconia but copper has attracted researcher more due to its high thermal conductivity among other metal fillers. Typical high thermal ceramic fillers are aluminium oxide (Al2O3), silicon carbide (SiC), aluminium nitride (AlN), and boron nitride(BN). Ceramic-based fillers are more preferred than metallic-based fillers for high thermal conductivity and electrical insulation properties and thermal stabilities. However, the electrical conductivity of metallic filler can be tailored by surface treatment or oxidation. Ceramic nanofiller such as boron nitride (BN), silicon carbide (SiC), alumina (Al2O3), beryllium oxide (BeO), silica (SiO2), aluminium nitride (AlN), and zinc oxide (ZnO) also can be used to overcome the problem because of high thermal conductivity and high electric insulation properties. Compared to ceramic-based fillers and metallic-based fillers, carbon-based fillers have received more attention due to their high thermal conductivity. Graphene, carbon nanotubes (CNT), carbon black, and carbon fibres are the carbon-based fillers used in thermal application. Even being used at the smallest filler content from 1 wt% to 8 wt%, the addition of CNT still very effective in enhancing the thermal and electrical conductivity compared to neat ABS [11]. However, graphene is regarded as the most promising candidates with high thermal conductivity, high mechanical strength, and chemical stability. As Vijay T. *et al* reported, graphene's addition into ABS enhances the thermal stability with thermal decomposition increased by 4.9% [30].

### 2.2. *Hybrid filler*

The hybrid nanocomposite combines two or more nanofillers that are employed together as reinforcement in polymer composite. The nanofiller combination can be with carbon, ceramic, cellulose, metal, salts, and others. Generally, the purpose of combining two nanofillers is to enhance the properties of the composite such as thermal, mechanical, electrical properties, and morphologies of the composites. One of the researchers combined ceramic and carbon nanofiller [34, 35] in the research as filament feedstock of FDM 3D printing using silicon carbide (SiC) and carbon (C). The other researcher used silicon carbide (SiC) and Aluminum oxide ($Al_2O_3$) [8, 36] as hybrid fillers in their research for rapid tooling application. To enhance thermal stability, electrical conductivity, and crystallinity of nanocomposite, graphene nanoplatelets (GNP) and carbon nanotube (CNT) have been combined as hybrid nanofiller for the feedstock of FDM 3D printing [20]. Another effort for thermal enhancement using ABS as polymer and silicon carbide (SiC) and zink (Zn) as hybrid nanofiller [37] for novel 3D printing feedstock.

## 3. The conducting base polymers

Thermal conductivity is an essential key to the world of research because it characterizes the materials' ability to conduct heat flow when operating with high-temperature procedures. Thermoplastics' composite thermal conductivity is crucially determined by polymer crystallinity that varied from 0.14 W/mK for amorphous polymers such as polystyrene (PS) to 0.53 W/mK for high-density polyethylene (HDPE) which is known as a highly crystalline polymer. Semi-crystalline polymers thermal conductivity, on the other hand, is reported to increase with crystallinity.

### 3.1. *Single polymers*

Polymers can be categorized in many ways such as according to the origin of materials; natural or synthetic, based on their thermal response properties; thermosets or thermoplastics, and polymer chain structure; linear, branched, cross-linked, and network. For thermal application composite materials, polymers are categorized based on their thermal response properties: thermosets and thermoplastics. Thermoplastics polymers composite dominated thermosets polymer composite over 93%. Most of the polymer composite experiments for thermal applications using thermoplastics such as ABS, PLA, PEEK, HDPE, PHA, PVC, PC, and TPU compared to thermosets epoxy. This is due to the brittleness and less tough of thermosets materials which made thermoset are not the choice for composite materials. For thermal application, the commonly used polymers are ABS and PLA, as shown in table 2. The polymers' application includes electrical and electronic, medical, rapid prototyping [21], and aerospace application [28].

### 3.2. *Polymer blends*

It is called polymer blends when a system combines or blends two polymers to improve specific properties to the composite while maintaining others. The examples of polymer blends in FDM 3D printing are polyetherimide (PEI)/ polycarbonate (PC), polylactic acid (PLA)/polyhydroxyalkanoate (PHA), polycaprolactone (PCL)/ hydrolyzed collagen (HC), and polyethylene terephthalate (PET)/ polypropylene (PP). Using polymer blends is to improve the processability [21, 38] for the 3D printing filament. Blanco [21] has shown that polymer blends can lower thermal degradation resistance and improve the filament processability by TGA analysis. Other than that, FDM 3D printing was also used to print pots using polymer blends using PCL and HC for biodegradable items for applications in agriculture and plant nurseries where the observed HC thermal stability showed the suitability to be printed with PCL without having thermal degradation [38]. Polymer blends can also be a cost-effective way to reduce plastic usage by recycling plastics such as PET in water bottle bodies and PP in water bottle caps. The recycled plastics are proved as viable printing stock of FDM 3D printing and ease in the process while its tensile strength comparable with HIPS, lower-end commercial filaments [39].

IOP Conf. Series: Materials Science and Engineering 1078 (2021) 012029 doi:10.1088/1757-899X/1078/1/012029

**Table 1**. Fillers, and their thermal conductivity.

| Fillers Category | Fillers | Composition (wt%) | Thermal Conductivity (W/m.K) | Application | Ref. |
|---|---|---|---|---|---|
| Metal | Silver (Ag) | 40 - 55 | 450 | Electrical and electronic application | [32] |
| | Copper | 5- 40 | 398 | Rapid prototyping | [33, 40] |
| | Copper oxide | 8- 14 | - | - | [40] |
| | Zink | 10 - 30 | - | Rapid prototyping | [8, 37] |
| | Stainless steel | 10 - 23 | 247 | - | [18] |
| | Aluminium oxide | 5- 15 | 30 | Rapid prototyping | [8] |
| | Iron | 5 - 50 | - | Rapid prototyping | [33, 36] |
| | Titanium dioxide | 30 | - | Rapid prototyping | [36] |
| | Zircornia | 30 - 40 | - | Biomaterial application | [41] |
| Carbon | Graphene | 0.2- 8 | 4840-5300 | Rapid prototyping, energy storage, electrical and electronic application | [5, 6, 12, 15, 30, 42] |
| | Graphite | 20-50 | 100-400 | | |
| | Carbon nanotube | 1- 15 | 2000-6000 | Electrical and electronic application | [11, 20, 25, 37] |
| | Carbon black | 1.5 - 32.3 | 6-174 | Electrical and electronic application | [43-45] |
| Ceramic | Hydroxyapatite | 4 | - | Medical application | [46] |
| | Barium titanate | 6- 45 | - | Electrical and electronic application | [35] |
| | Boron nitride | 5- 40 | 250-300 | Electrical and electronic application | [34, 47] |
| | Silicon carbide | 5- 40 | 270 | Rapid prototyping, energy storage | [8, 12] |

**Table 2.** Polymer base materials, and their thermal conductivity.

| Base Category | Base Material | Abbrev. | Thermal Conductivity (W/(mK)) | Application | Ref. |
|---|---|---|---|---|---|
| Thermoplastics | Acrylonitrile-butadiene-styrene | ABS | 0.1 | Rapid prototyping | [48-50] |
| | Polylactic acid | PLA | - | Electronic device | [7, 42] |
| | Polypropylene | PP | 0.11-0.17 | Medical device | [46] |
| | Polycarbonate | PC | 0.19-0.21 | Rapid prototyping | [21] |
| | Poly-ether-ether-ketone | PEEK | 0.25 | Aerospace, medical device | [28] |
| | Polyvinylchloride | PVC | 0.13-0.29 | Medical device | [46] |
| | High-density polyethylene | HDPE | 0.33-0.53 | - | |
| Thermoset | Epoxy resin | - | 0.19 | - | |

## 4. Nanofiller in thermal application

### *4.1. Preparation of nanofiller composite*

The important in preparing the nanofiller composite is the materials such as a base polymer, the fillers, the composition of the materials, and the apparatus used. The materials' choosing process can be done by reviewing the other publications in the applied application. One of the preparations of nanofiller composite is a solution blending method. The solution blending method uses a solvent liquid such as acetone [18, 37, 51] or chloroform [52] to melt or blend the base polymer before mixing with the nanofiller.

### *4.2. FDM 3D printing processing parameter and response parameter*

To ensure a good quality finishing product with enhanced thermal or mechanical properties of FDM 3D printing, the processing parameter choice must be suitable for its application for 3D printing machine, as shown in table 3. Table 3 shown the optimum process setting parameter for FDM 3D printing including nozzle diameter, nozzle diameter, bed temperature, nozzle speed, filament diameter, infill pattern, infill density, and layer height for thermal application. The filament diameter used by the researchers is 1.75mm [23], 2.85 mm [53], and 3 mm [7] diameter filament feedstock but commonly used for 3D printing is 1.75 mm. As for layer height of FDM deposition are 0.1 mm [54], 0.2 mm [25], and 0.3 mm [4]of layer thickness. On the other hand, bed temperature varied from 60°C [27], 110°C [43], and 120°C [55] depends on the melting temperature of the materials used. Temperature control is critical to control residual stress as residual stress accumulated during layer by layer build-up leads to warping and delamination during the 3D printing process [54].

**Table 3.** Printing process parameter for FDM 3D printing.

| Parameter | Value | References |
|---|---|---|
| Nozzle Diameter | 0.4mm | [1, 4, 13, 24, 27, 29, 42, 43, 45, 56-60] |
| Nozzle Temperature | 230°C | [15, 32, 45, 49] |
| Bed Temperature | 60°C | [27, 42] |
| Nozzle Speed | 30mm/s | [13, 55] |
| Filament Diameter | 1.75mm | [4, 5, 10, 12, 13, 15, 17, 21, 23-25, 30, 32, 35, 40, 43, 44, 50, 55, 61-67] |
| Infill Pattern | Linear | [4, 13, 15, 24, 27, 37, 49, 54, 57, 62] |
| Infill Density | 100% | |
| Layer Height | 0.2mm | [13, 25, 29, 35, 43, 49, 61, 66] |

The response parameter includes thermal properties, physical properties, mechanical properties, morphological, and characterization studies. FDM 3D printing's thermal properties are thermogravimetric analysis (TGA) [13, 21, 30], differential scanning calorimetry (DSC) [13, 24, 30], thermal conductivity [4, 11, 13], electrical conductivity [43], heat capacity [11], electrical resistivity [11], and thermal diffusivity [65]. Still, thermal conductivity, DSC and TGA are the most commonly testing conducted in literature for FDM 3D printing for thermal application. The ASTM testing standard for thermal conductivity in literature is ASTM D5470 for filament thermal conductivities [12]. Simultaneously, physical properties include density [11, 68] and dimensional accuracy [55, 69]. On the other hands, morphological and characterization in literature studies are scanning electron microscope (SEM) [4, 13, 24], transmission electron microscope (TEM) [19, 62], and computed tomography (CT) scan [24], x-ray diffraction (XRD) [24, 26, 62], x-ray photoelectron spectroscopy (XPS) [70, 71], and Fourier transform infrared spectroscopy (FTIR) [72, 73]. Several mechanical properties studies were conducted from literature, such as tensile [4, 11, 23, 24], flexural [13, 24, 25], impact [19, 54, 74], and hardness [27]. Still, tensile, flexural, and impact tests are commonly conducted in literature for mechanical properties for FDM 3D printing technique for thermal application. The ASTM standard testing method for tensile test is D638 Type V (63.5mm × 9.53mm × 4mm) [12, 18, 24] , for flexural test is ASTM D790 (127mm × 12.7mm × 3.2 mm) [62, 75] , whereas for impact test is ASTM D256 ( 64mm × 12.7mm × 3.2 mm ) [26].

## 5. Limitations of thermal conductivity of FDM 3D printed composites

The problem with the thermal application for a 3D printed composite is low thermal conductivity. Most of the base polymers have a lower thermal conductivity which is not sufficient for the high heat conduction applications such as electronic application. The addition of heat conductive fillers to polymer-based composite enhances the composite materials' thermal conductivity. They typically heat conductive fillers are introduced into polymers to increase the thermal conductivity such as graphene, carbon nanotube, boron nitride, silicon carbide, and metal particles. Sonsalla *et al.* has experimented with 3D filament contributing ABS as polymer base, Zn flakes, SWCNT, MWCNT, and SCMW as the filler materials, for the enhancement of thermal conductivity of the 3D filament as the sufficient thermal conductivity of filler are more than 100 times than polymer base thermal conductivity to achieve the highest thermal conductivity at any level of filler loading [37]. The addition of graphene nanoplatelets to the base polymer shown increment in glass transition temperature with the increasing graphene volume suggests that thermal stability enhanced with graphene [30].

Other than that, lower filler loading also can lead to low thermal conductivity. However, too much filler can also cause poor composite, low mechanical properties, and high cost. The best option is to use nanofillers such as graphene, silicon carbide, boron nitride, and other, potent materials if only used in a small number of materials. Recent studies of FDM 3D printing have shown that thermal properties can be influenced by the addition of CNT nanofiller to the ABS base polymer by increasing 55% of thermal diffusivity and 30% increment in thermal conductivity 6wt% of CNT filler loading [11]. The viscosity of extruding filament for the 3D printing cannot be too high which will agglomerate and clogged the nozzle suggests that 4wt% of graphene is the optimal value as reported by S. Dul *et al.* [15].

The other issue with the 3D printed composites' limitation is the low interaction between the polymer and the filler itself, leading to high thermal interfacial resistance and reducing the polymer composite's thermal conductivity. By improving the interaction of polymer-filler is believed can increase the thermal conductivity of the overall composite. The coupling agent is one of the solutions to improving polymer filler interaction. The commonly used coupling agent is a silane and titanate coupling agent. The titanate coupling agent can improve the polymer-filler interactions by increasing polymer-filler adhesion and increasing thermal conductivity and lowering thermal expansion [37]. Compatibilizing agent, in contrast, functions to stabilize and modifies interfacial between two polymer or polymer blends. The examples of compatibilizing agents in FDM 3D printing are maleic anhydride and styrene ethylene butylene styrene (SEBS). SEBS was proved to improved bonding between phases and led to an increase in glass transition temperature thus increase the performance of the window materials although not significantly increase the tensile strength [39].

Printer orientation of 3D printing is horizontal, vertical, and flat, which is also taking part in thermal conductivity enhancement in a filament. Horizontal printing orientation of 3D printing is believed to give lower thermal conductivity enhancement than vertical printing orientation of filament. It leads to a larger number of layer junctions, more considerable interlayer resistance, and thermal transport along the shortest dimension of filler particle. While vertical orientation reduces thermal interfaces, filler alignment is in the direction of heat flow, and shear forces align fillers in the print directions which leads to enhancement of thermal and electrical polymer composite proved by Sonsalla *et al.* that filament printed by vertical give 70% thermal enhancement than horizontal print orientation [37].

## 6. Prospect and recommendation
Thermal conductivity enhancement for 3D printing of thermal application was investigated in this article. All of the experimental and theoretical approaches have been considered in this paper. Since nanofiller materials can influence the thermal conductivity, future research can be extended by combining two nanofiller with higher thermal conductivity in producing filament 3D printing. Also, combining two different base polymers with higher thermal conductivity can be applied for 3D printing thermal applications for new materials.

## 7. Conclusion
In conclusion, the processing parameter and response parameter of FDM techniques has been summarized. The processing parameter of FDM 3D printing techniques are the materials used and the processing temperature. The most commonly used base material for FDM 3D printing for thermal application is ABS. In contrast, the filler material is graphene, regarded as the most promising candidates with high thermal conductivity, high mechanical strength, and chemical stability. The processing temperature, such as nozzle temperature and bed temperature, also has been reviewed. The nozzle temperature commonly used is 230 °C, bed temperature used is 60 °C, the nozzle speed is 30mm/s, and layer height is 0.2mm. The response parameter includes thermal properties, physical properties, mechanical properties, morphological, and characterization studies. Thermal properties conducted in literature are DSC, TGA, and thermal conductivity while physical properties are density and dimensional accuracy. Tensile, flexural, and impact tests are the mechanical testing conducted in literature. On the other hand, morphological and characterization studies include SEM, TEM, CT scan, XRD, XPS, and FTIR. The limitations such as low thermal conductivity, the material used such as the base materials and filler materials, and experimental parameters such as nozzle temperature, nozzle diameter, nozzle speed, layer thickness, and others have been discussed through this paper and taking consideration for the prospect of FDM 3D printing for thermal application.

**Acknowledgements**
The authors would like to acknowledge Universiti Malaysia Pahang (www.ump.edu.my) with the facilities and resources for this research. The authors would like to acknowledge their support for the Universiti Malaysia Pahang internal grants (RDU1903134, RDU190354) and Fundamental Research Grant Scheme (FRGS) FRGS/1/2019/TK03/UMP/02/15, provided by the Ministry of Higher Education, Malaysia.

**References**
[1] Ceretti E, Ginestra P, Neto P I, Fiorentino A and Da Silva J V L 2017 Multi-layered Scaffolds Production via Fused Deposition Modeling (FDM) Using an Open Source 3D Printer: Process Parameters Optimization for Dimensional Accuracy and Design Reproducibility *Procedia CIRP* **65** 13-8
[2] Rankin T M, Giovinco N A, Cucher D J, Watts G, Hurwitz B and Armstrong D G 2014 Three-dimensional printing surgical instruments: are we there yet? *Journal of Surgical Research* **189** 193-7
[3] Aslanzadeh S, Saghlatoon H, Honari M M, Mirzavand R, Montemagno C and Mousavi P 2018 Investigation on electrical and mechanical properties of 3D printed nylon 6 for RF/microwave electronics applications *Additive Manufacturing* **21** 69-75
[4] Chen G, Chen N and Wang Q 2018 Preparation of poly (vinyl alcohol)/ionic liquid composites with improved processability and electrical conductivity for fused deposition modeling *Materials & Design* **157** 273-83
[5] Zhang D, Chi B, Li B, Gao Z, Du Y, Guo J and Wei J 2016 Fabrication of highly conductive graphene flexible circuits by 3D printing *Synthetic Metals* **217** 79-86
[6] Singh R, Singh H, Farina I, Colangelo F and Fraternali F 2019 On the additive manufacturing of an energy storage device from recycled material *Composites Part B: Engineering* **156** 259-65
[7] Johnson P R, Copeland P M, Ayodele A O, Tarekegn E N, Bromley S J, Harrell W R, Sosolik C E and Marler J P 2020 In-vacuum performance of a 3D-printed ion deflector *Vacuum* **172** 109061
[8] Singh N, Singh R and Ahuja I P S 2018 Recycling of polymer waste with SiC/Al2O3 reinforcement for rapid tooling applications *Materials Today Communications* **15** 124-7
[9] Jaisingh Sheoran A and Kumar H 2019 Fused Deposition modeling process parameters optimization and effect on mechanical properties and part quality: Review and reflection on present research *Materials Today: Proceedings*
[10] León-Cabezas M A, Martínez-García A and Varela-Gandía F J 2017 Innovative functionalized monofilaments for 3D printing using fused deposition modeling for the toy industry *Procedia Manufacturing* **13** 738-45
[11] Dorigato A, Moretti V, Dul S, Unterberger S H and Pegoretti A 2017 Electrically conductive nanocomposites for fused deposition modelling *Synthetic Metals* **226** 7-14
[12] Liu W, Wu N and Pochiraju K 2018 Shape recovery characteristics of SiC/C/PLA composite filaments and 3D printed parts *Composites Part A: Applied Science and Manufacturing* **108** 1-11
[13] Liao G, Li Z, Cheng Y, Xu D, Zhu D, Jiang S, Guo J, Chen X, Xu G and Zhu Y 2018 Properties of oriented carbon fiber/polyamide 12 composite parts fabricated by fused deposition modeling *Materials & Design* **139** 283-92
[14] Guessasma S, Belhabib S and Nouri H 2019 Understanding the microstructural role of bio-sourced 3D printed structures on the tensile performance *Polymer Testing* **77** 105924
[15] Dul S, Fambri L and Pegoretti A 2016 Fused deposition modelling with ABS–graphene nanocomposites *Composites Part A: Applied Science and Manufacturing* **85** 181-91
[16] Gebisa A W and Lemu H G 2019 Influence of 3D Printing FDM Process Parameters on Tensile Property of ULTEM 9085 *Procedia Manufacturing* **30** 331-8
[17] Zhong W, Li F, Zhang Z, Song L and Li Z 2001 Short fiber reinforced composites for fused deposition modeling *Materials Science and Engineering: A* **301** 125-30

[18] Ryder M A, Lados D A, Iannacchione G S and Peterson A M 2018 Fabrication and properties of novel polymer-metal composites using fused deposition modeling *Composites Science and Technology* **158** 43-50
[19] Berretta S, Davies R, Shyng Y T, Wang Y and Ghita O 2017 Fused Deposition Modelling of high temperature polymers: Exploring CNT PEEK composites *Polymer Testing* **63** 251-62
[20] Kotsilkova R, Petrova-Doycheva I, Menseidov D, Ivanov E, Paddubskaya A and Kuzhir P 2019 Exploring thermal annealing and graphene-carbon nanotube additives to enhance crystallinity, thermal, electrical and tensile properties of aged poly(lactic) acid-based filament for 3D printing *Composites Science and Technology* **181** 107712
[21] Blanco I, Cicala G, Ognibene G, Rapisarda M and Recca A 2018 Thermal properties of polyetherimide/polycarbonate blends for advanced applications *Polymer Degradation and Stability* **154** 234-8
[22] Dehdari Ebrahimi N and Ju Y S 2018 Thermal conductivity of sintered copper samples prepared using 3D printing-compatible polymer composite filaments *Additive Manufacturing* **24** 479-85
[23] Malekipour E, Attoye S and El-Mounayri H 2018 Investigation of Layer Based Thermal Behavior in Fused Deposition Modeling Process by Infrared Thermography *Procedia Manufacturing* **26** 1014-22
[24] Rinaldi M, Ghidini T, Cecchini F, Brandao A and Nanni F 2018 Additive layer manufacturing of poly (ether ether ketone) via FDM *Composites Part B: Engineering* **145** 162-72
[25] Yang L, Li S, Zhou X, Liu J, Li Y, Yang M, Yuan Q and Zhang W 2019 Effects of carbon nanotube on the thermal, mechanical, and electrical properties of PLA/CNT printed parts in the FDM process *Synthetic Metals* **253** 122-30
[26] Wang L and Gardner D J 2017 Effect of fused layer modeling (FLM) processing parameters on impact strength of cellular polypropylene *Polymer* **113** 74-80
[27] Sang L, Han S, Peng X, Jian X and Wang J 2019 Development of 3D-printed basalt fiber reinforced thermoplastic honeycombs with enhanced compressive mechanical properties *Composites Part A: Applied Science and Manufacturing* **125** 105518
[28] Stepashkin A A, Chukov D I, Senatov F S, Salimon A I, Korsunsky A M and Kaloshkin S D 2018 3D-printed PEEK-carbon fiber (CF) composites: Structure and thermal properties *Composites Science and Technology* **164** 319-26
[29] Yang C, Tian X, Li D, Cao Y, Zhao F and Shi C 2017 Influence of thermal processing conditions in 3D printing on the crystallinity and mechanical properties of PEEK material *Journal of Materials Processing Technology* **248** 1-7
[30] Tambrallimath V, Keshavamurthy R, D S, Koppad P G and Kumar G S P 2019 Thermal behavior of PC-ABS based graphene filled polymer nanocomposite synthesized by FDM process *Composites Communications* **15** 129-34
[31] Korkmaz S and Kariper İ A 2020 Graphene and graphene oxide based aerogels: Synthesis, characteristics and supercapacitor applications *Journal of Energy Storage* **27** 101038
[32] Lei Z, Chen Z, Zhou Y, Liu Y, Xu J, Wang D, Shen Y, Feng W, Zhang Z and Chen H 2019 Novel electrically conductive composite filaments based on Ag/saturated polyester/polyvinyl butyral for 3D-printing circuits *Composites Science and Technology* **180** 44-50
[33] Nikzad M, Masood S H and Sbarski I 2011 Thermo-mechanical properties of a highly filled polymeric composites for Fused Deposition Modeling *Materials & Design* **32** 3448-56
[34] Liu J, Li W, Guo Y, Zhang H and Zhang Z 2019 Improved thermal conductivity of thermoplastic polyurethane via aligned boron nitride platelets assisted by 3D printing *Composites Part A: Applied Science and Manufacturing* **120** 140-6
[35] Kim H, Johnson J, Chavez L A, Garcia Rosales C A, Tseng T-L B and Lin Y 2018 Enhanced dielectric properties of three phase dielectric MWCNTs/BaTiO3/PVDF nanocomposites for energy storage using fused deposition modeling 3D printing *Ceramics International* **44** 9037-44
[36] Singh R, Kumar R, Mascolo I and Modano M 2018 On the applicability of composite PA6-TiO2 filaments for the rapid prototyping of innovative materials and structures *Composites Part B: Engineering* **143** 132-40

[37] Sonsalla T, Moore A L, Radadia A D and Weiss L 2019 Printer orientation effects and performance of novel 3-D printable acrylonitrile butadiene styrene (ABS) composite filaments for thermal enhancement *Polymer Testing* **80** 106125

[38] Seggiani M, Altieri R, Puccini M, Stefanelli E, Esposito A, Castellani F, Stanzione V and Vitolo S 2018 Polycaprolactone-collagen hydrolysate thermoplastic blends: Processability and biodegradability/compostability *Polymer Degradation and Stability* **150** 13-24

[39] Zander N E, Gillan M, Burckhard Z and Gardea F 2019 Recycled polypropylene blends as novel 3D printing materials *Additive Manufacturing* **25** 122-30

[40] Hamzah K A, Keat Y C, Leng T P, Noor M M, Sazali S A and Yun A Y 2019 Tensile strength and hardness property of ABS filled CuO prepared via 3D printing approach *Materials Today: Proceedings* **16** 1778-81

[41] Abdullah A M, Tuan Rahim T N A, Mohamad D, Akil H M and Rajion Z A 2017 Mechanical and physical properties of highly ZrO2 /β-TCP filled polyamide 12 prepared via fused deposition modelling (FDM) 3D printer for potential craniofacial reconstruction application *Materials Letters* **189** 307-9

[42] Foster C W, Elbardisy H M, Down M P, Keefe E M, Smith G C and Banks C E 2020 Additively manufactured graphitic electrochemical sensing platforms *Chemical Engineering Journal* **381** 122343

[43] Schmitz D P, Ecco L G, Dul S, Pereira E C L, Soares B G, Barra G M O and Pegoretti A 2018 Electromagnetic interference shielding effectiveness of ABS carbon-based composites manufactured via fused deposition modelling *Materials Today Communications* **15** 70-80

[44] Hashemi Sanatgar R, Campagne C and Nierstrasz V 2017 Investigation of the adhesion properties of direct 3D printing of polymers and nanocomposites on textiles: Effect of FDM printing process parameters *Applied Surface Science* **403** 551-63

[45] Kwok S W, Goh K H H, Tan Z D, Tan S T M, Tjiu W W, Soh J Y, Ng Z J G, Chan Y Z, Hui H K and Goh K E J 2017 Electrically conductive filament for 3D-printed circuits and sensors *Applied Materials Today* **9** 167-75

[46] Sharma R, Singh R, Penna R and Fraternali F 2018 Investigations for mechanical properties of Hap, PVC and PP based 3D porous structures obtained through biocompatible FDM filaments *Composites Part B: Engineering* **132** 237-43

[47] Owais M, Zhao J, Imani A, Wang G, Zhang H and Zhang Z 2019 Synergetic effect of hybrid fillers of boron nitride, graphene nanoplatelets, and short carbon fibers for enhanced thermal conductivity and electrical resistivity of epoxy nanocomposites *Composites Part A: Applied Science and Manufacturing* **117** 11-22

[48] Raut S, Jatti V S, Khedkar N K and Singh T P 2014 Investigation of the Effect of Built Orientation on Mechanical Properties and Total Cost of FDM Parts *Procedia Materials Science* **6** 1625-30

[49] Ning F, Cong W, Qiu J, Wei J and Wang S 2015 Additive manufacturing of carbon fiber reinforced thermoplastic composites using fused deposition modeling *Composites Part B: Engineering* **80** 369-78

[50] Yin J, Lu C, Fu J, Huang Y and Zheng Y 2018 Interfacial bonding during multi-material fused deposition modeling (FDM) process due to inter-molecular diffusion *Materials & Design* **150** 104-12

[51] Palmero E M, Casaleiz D, de Vicente J, Hernández-Vicen J, López-Vidal S, Ramiro E and Bollero A 2019 Composites based on metallic particles and tuned filling factor for 3D-printing by Fused Deposition Modeling *Composites Part A: Applied Science and Manufacturing* **124** 105497

[52] Sharma V, Kapoor S, Goyal M and Jindal P 2020 Enhancement of the mechanical properties of graphene based acrylonitrile butadiene styrene (ABS) nanocomposites *Materials Today: Proceedings* **28** 1744-7

[53] Le Duigou A, Castro M, Bevan R and Martin N 2016 3D printing of wood fibre biocomposites: From mechanical to actuation functionality *Materials & Design* **96** 106-14

[54] Arif M F, Kumar S, Varadarajan K M and Cantwell W J 2018 Performance of biocompatible PEEK processed by fused deposition additive manufacturing *Materials & Design* **146** 249-59
[55] Dawoud M, Taha I and Ebeid S J 2016 Mechanical behaviour of ABS: An experimental study using FDM and injection moulding techniques *Journal of Manufacturing Processes* **21** 39-45
[56] Lebedev S M, Gefle O S, Amitov E T, Berchuk D Y and Zhuravlev D V 2017 Poly(lactic acid)-based polymer composites with high electric and thermal conductivity and their characterization *Polymer Testing* **58** 241-8
[57] Xiang D, Zhang X, Li Y, Harkin-Jones E, Zheng Y, Wang L, Zhao C and Wang P 2019 Enhanced performance of 3D printed highly elastic strain sensors of carbon nanotube/thermoplastic polyurethane nanocomposites via non-covalent interactions *Composites Part B: Engineering* **176** 107250
[58] Dias M V, de Medeiros H S, Soares N d F F, Melo N R d, Borges S V, Carneiro J d D S and Pereira J M T d A K 2013 Development of low-density polyethylene films with lemon aroma *LWT - Food Science and Technology* **50** 167-71
[59] Xiao X, Chevali V S, Song P, He D and Wang H 2019 Polylactide/hemp hurd biocomposites as sustainable 3D printing feedstock *Composites Science and Technology* **184** 107887
[60] Geng P, Zhao J, Wu W, Ye W, Wang Y, Wang S and Zhang S 2019 Effects of extrusion speed and printing speed on the 3D printing stability of extruded PEEK filament *Journal of Manufacturing Processes* **37** 266-73
[61] Lin L, Ecke N, Huang M, Pei X-Q and Schlarb A K 2019 Impact of nanosilica on the friction and wear of a PEEK/CF composite coating manufactured by fused deposition modeling (FDM) *Composites Part B: Engineering* **177** 107428
[62] Weng Z, Wang J, Senthil T and Wu L 2016 Mechanical and thermal properties of ABS/montmorillonite nanocomposites for fused deposition modeling 3D printing *Materials & Design* **102** 276-83
[63] Gonzalez Ausejo J, Rydz J, Musioł M, Sikorska W, Sobota M, Włodarczyk J, Adamus G, Janeczek H, Kwiecień I, Hercog A, Johnston B, Khan H R, Kannappan V, Jones K R, Morris M R, Jiang G, Radecka I and Kowalczuk M 2018 A comparative study of three-dimensional printing directions: The degradation and toxicological profile of a PLA/PHA blend *Polymer Degradation and Stability* **152** 191-207
[64] Kuo C-C, Liu L-C, Teng W-F, Chang H-Y, Chien F-M, Liao S-J, Kuo W-F and Chen C-M 2016 Preparation of starch/acrylonitrile-butadiene-styrene copolymers (ABS) biomass alloys and their feasible evaluation for 3D printing applications *Composites Part B: Engineering* **86** 36-9
[65] Jia Y, He H, Geng Y, Huang B and Peng X 2017 High through-plane thermal conductivity of polymer based product with vertical alignment of graphite flakes achieved via 3D printing *Composites Science and Technology* **145** 55-61
[66] Han P, Tofangchi A, Deshpande A, Zhang S and Hsu K 2019 An approach to improve interface healing in FFF-3D printed Ultem 1010 using laser pre-deposition heating *Procedia Manufacturing* **34** 672-7
[67] Shabani F, Kumar L and Esmaeili A 2015 A modelling implementation of climate change on biodegradation of Low-Density Polyethylene (LDPE) by Aspergillus niger in soil *Global Ecology and Conservation* **4** 388-98
[68] Wang Y-n, Weng Y-x and Wang L 2014 Characterization of interfacial compatibility of polylactic acid and bamboo flour (PLA/BF) in biocomposites *Polymer Testing* **36** 119-25
[69] Alafaghani A a and Qattawi A 2018 Investigating the effect of fused deposition modeling processing parameters using Taguchi design of experiment method *Journal of Manufacturing Processes* **36** 164-74
[70] Feng Y, Han G, Wang B, Zhou X, Ma J, Ye Y, Liu C and Xie X 2020 Multiple synergistic effects of graphene-based hybrid and hexagonal born nitride in enhancing thermal conductivity and flame retardancy of epoxy *Chemical Engineering Journal* **379** 122402
[71] Ren L, Zeng X, Zhang X, Sun R, Tian X, Zeng Y, Xu J-B and Wong C-P 2019 Silver nanoparticle-modified alumina microsphere hybrid composites for enhanced energy density

and thermal conductivity *Composites Part A: Applied Science and Manufacturing* **119** 299-309

[72] An D, Cheng S, Xi S, Zhang Z, Duan X, Ren Y, Li J, Sun Z, Liu Y and Wong C-P 2020 Flexible thermal interfacial materials with covalent bond connections for improving high thermal conductivity *Chemical Engineering Journal* **383** 123151

[73] Morsi M A, Rajeh A and Al-Muntaser A A 2019 Reinforcement of the optical, thermal and electrical properties of PEO based on MWCNTs/Au hybrid fillers: Nanodielectric materials for organoelectronic devices *Composites Part B: Engineering* **173** 106957

[74] Luo M, Tian X, Shang J, Zhu W, Li D and Qin Y 2019 Impregnation and interlayer bonding behaviours of 3D-printed continuous carbon-fiber-reinforced poly-ether-ether-ketone composites *Composites Part A: Applied Science and Manufacturing* **121** 130-8

[75] Heidari-Rarani M, Rafiee-Afarani M and Zahedi A M 2019 Mechanical characterization of FDM 3D printing of continuous carbon fiber reinforced PLA composites *Composites Part B: Engineering* **175** 107147

# Investigation of pipe materials and thermal conductivity of soil on the performance of ground heat exchanger operating under Malaysia climate

**A M Aizzuddin[1], T M Yusof[1,*] and W H Azmi[1]**

[1]College of Engineering, Universiti Malaysia Pahang (UMP), Lebuhraya Tun Razak, 26300 Gambang, Kuantan, Pahang, Malaysia

*Corresponding author's e-mail: myusof@ump.edu.my

**Abstract.** In nature, renewable energy is inherently free and must be implemented. The use of air conditioning and refrigerants that affect global warming is a serious issue. In building applications, renewable energy from the geothermal source, namely ground heat exchanger (GHE), has great potential. The main concept of GHE is utilizing the ground as an infinite thermal reservoir for cooling and heating to the fluid medium. In the GHE system, the air is used as a fluid medium of work. Because of the temperature difference between the air and underground temperature, the air cools in summer and gets heated in winter. In this present work, a study has been conducted to investigate the effect of pipe materials and thermal conductivity of soil on the performance of the GHE. The study acknowledges that the pipe materials do not give a significant effect on the performance of the GHE. Therefore, the lower thermal conductivity of pipe materials with low cost can be used in GHE implementation. The study also revealed that the range of thermal conductivity of soil which gives good ground heat exchanger performance is between 1.5 to 5 W/m·K. Besides, the length of the pipe was reduced from 25 to 10 meters.

**Keyword:** Ground Heat Exchanger, Earth to Air Heat Exchanger, GHE Performance.

## 1. Introduction

Geothermal technologies can result in significant energy savings, mainly related to building heating and cooling [1]. In ancient times, the idea of using the earth as a heat sink was well known. In about 3000 B.C., Iranian architects were used passive cooling through wind towers and underground tunnels [2, 3]. In the 20th century, air conditioning is widely used in tropical climates, not only for industrial production, but also for the comfort of humans. It can be efficiently achieved with vapor compression refrigerant machines. However, the accumulation of chlorofluorocarbons (CFCs) in the atmosphere released by the vapor compression refrigerant machines has reduced the ozone layer. Consequently, the world needs for reduction in the use of vapor compression refrigerant machines. The ground heat exchanger is one such suggestion that can support the reduction of the use of vapor compression refrigerant machines as shown in Figure 1. GHE is used for years in developed countries because of their higher energy efficiency than conventional heat and cooling systems. Their heat exchangers have been a common feature of the system. The use of ground-based heat exchangers is known to lead to an efficient, economical and environmentally friendly design [4]. GHE uses soil as a heat sink and air as the medium of heat transfer for the summer cooling space. When the warm air flows through the GHE pipe, heat passes from the air to the earth. As a result, at the exit of the GHE pipe, the air temperature is much lower than the ambient temperature [5]. GHE is a passive climate control technology that uses the underground soil temperature at a depth of around 2.5 to 3 meters in residential and farm buildings [3]. Previous studies have shown that GHE systems have excellent cooling and heating potential [6]. A study conducted by Gao, Li, Xu, Gang and Yan [7] showed that the use of the GHE system is an excellent solution to meet the Zero Energy Buildings (ZEB) target.

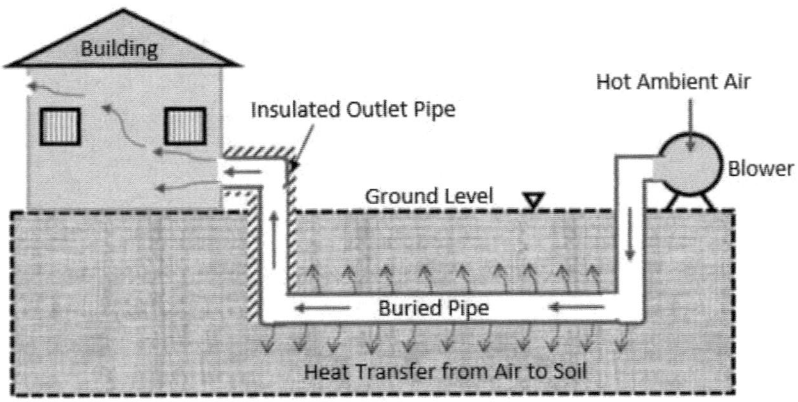


**Figure 1.** GHE system used for summer cooling [8].

Several researchers, such as Sodha, Sharma, Singh, Bansal and Kumar [9], have studied the use of the earth as a heat source and sink. At one of the hospitals in India, they tested a large GHE system intended to provide thermal comfort inside the entire building complex. A parametric model in which different parameters were length of pipe, radius of pipe, air velocity inside the tube and the depth of the buried pipe under the surface of the earth was presented by Mihalakakou, Santamouris, Asimakopoulos and Tselepidaki [10]. The effect of various ground surface boundary conditions on the efficiency of a single and multiple parallel earth to air heat exchanger system was investigated by Mihalakakou, Santamouris, Lewis and Asimakopoulos [11]. The conservation potential of an earth air pipe system coupled with a building without air conditioning was evaluated by Kumar, Ramesh and Kaushik [12]. A thermal model was developed by Ghosal, Tiwari, Das and Pandey [13] to investigate the performance of earth to air heat exchanger (EAHE) which is integrated with green house. The cooling capacity of the earth to air heat exchangers for domestic buildings in a desert climate was studied by Al-Ajmi, Loveday and Hanby [14]. A ground heat exchanger model based on the numerical transient bi-dimensional approach was developed by Badescu [15]. In order to evaluate the effects of the operating parameters for example the length of pipe, pipe radius, depth of ground and air flowrate on the thermal performance and air-pipe cooling capacities, Wu, Wang and Zhu [5] developed a transient and implicit model. The model based on numerical heat transfer and computational fluid dynamics and then implemented it on the CFD (computational fluid dynamics) platform, PHOENICS. A one-dimensional transient analytical model was proposed Cucumo, Cucumo, Montoro and Vulcano [16] to estimate the performance of earth to air heat exchangers, installed at different depths, used for cooling/heating buildings. Finite element numerical model for the simulation of the ground Cui, Yang and Fang [17] have developed a finite element numerical model for the simulation of ground heat exchangers in alternative modes of operation over a short period of time for ground-coupled heat pump applications.

From literature that have been done by the author, it has been found that the range of study for thermal conductivity of soil for the GHE implementation is not comprehensively covered. Since the thermal conductivity of soil and pipe materials have related to each other on the performance of the GHE. Therefore, in the present study, the effect of thermal conductivity of soil and pipe materials on the performance of the GHE will be investigated using simulation analysis. The performance analysis of the

GHE is simulated for cooling during summer only. There are five different types of pipe materials have been used in this study namely PVC, HDPE, stainless steel, aluminium, and copper with different thermal conductivity from low to high thermal conductivity. Meanwhile, the soil thermal conductivity was 0.5, 1.5, 2.5, 5.0 and 10 W/m.K.

## 2. Methodology

Simulation of GHE system is an interesting approach to predict and analyse the performance of the GHE based on available information and input parameters. It requires less time to analyse the performance of the system, repeatable analysis and running at various input settings. In the present work, the simulation uses the Excel software to obtain the temperature of the GHE system. Basically, the performance of GHE strongly depends on the thermal characteristics of the heat exchanger. The effect of pipe materials and soil thermal conductivity on GHE performance is analysed in the simulation. Figure 2 shows the flow chart of the methodology.

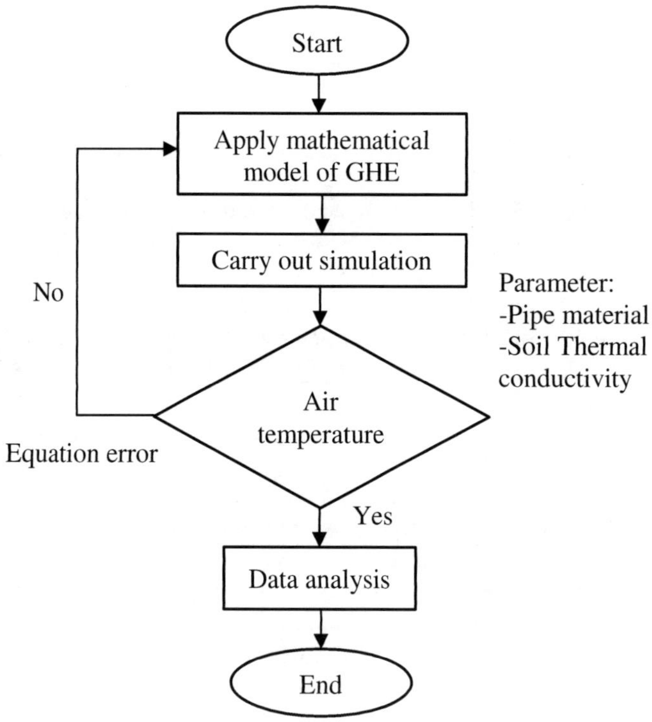


**Figure 2.** Flow chart.

Based on the schematic diagram in Figure 3 [18], the mathematical model of heat transfer from the air inside the GHE pipe to the surrounding ground can be analysed. The figure shows the ambient temperature of air ($T_{air}$) that enters the pipe. At a constant ground temperature ($T_{z,t}$), heat will be transferred to the surrounding ground as the air flows inside the pipe. In this case, several assumptions have been made that i) the heat capacity rate of the ground surrounding is infinity, or treated as a thermal reservoir, ii) the presence of a pipe in the area around the pipe does not affect the temperature profile. Therefore, the temperature in the surface is consistent in y direction, iii) the pipe has a uniform cross sectional area, iv) the air convection flow in the pipe develops hydrodynamic and thermally, and v) thermal properties of the ground around the pipe is homogeneous and its thermal conductivity has a constant value [19, 20].

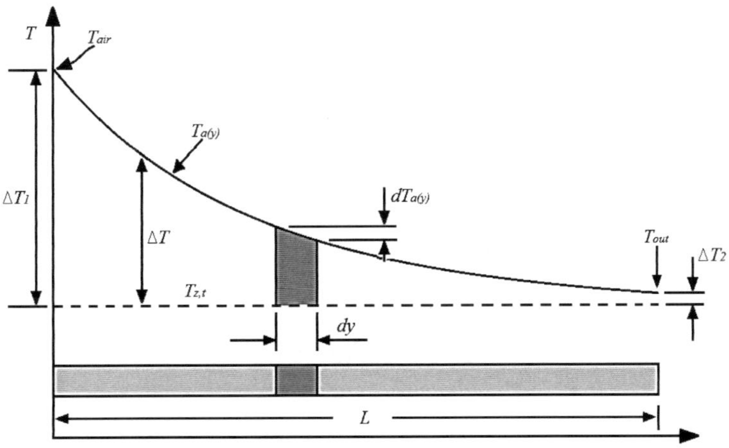


**Figure 3.** Schematic of heat transfer process in GHE pipe.

The hot ambient air at $T_{air}$ enters the pipe, which will decrease at $T_{a(y)}$ with the distance of the pipe from the inlet ($y$) increases. At $T_{out}$, the hot air leaves the pipe which is slightly above ground temperature, $T_{z,t}$. Therefore, the heat transfer rate received by the ground for arbitrary length in the differential section of $dy$ shall be expressed by Equation 1.

$$dQ = U\pi D\left[T_{air(y)} - T_{z,t}\right]dy \tag{1}$$

On the other side of the heat transfer, the heat rate transmitted by the air inside the pipe in the differential section of the $dy$ is shown in Equation 2 [21];

$$dQ = -\dot{m}_{air}c_p\left[dT_{air(y)}\right] \tag{2}$$

The temperature change of the air is a negative quantity, which decrease the temperature as the pipe length increases and a negative sign therefore added to Equation 2, which result in positive heat transfer rate. In theory, the heat transfer within the pipe between the earth and the air is equal to the heat loss as the air flows along the pipe. Therefore, Equation 1 and Equation 2 are then equated to yield Equation 3 as follow.

$$U\pi D\left[T_{air(y)} - T_{z,t}\right]dy = -\dot{m}_{air}c_p\left[dT_{air(y)}\right] \tag{3}$$

Rearranging Equation 3 in order to solve for $T_{a(y)}$ yields;

$$\frac{dT_{air(y)}}{T_{air(y)} - T_{z,t}} = -\frac{U\pi D}{\dot{m}_{air}c_p}dy \tag{4}$$

By integrating both sides of Equation 4, for $T_{air(y)}$ from $T_{air(y)}$,0 to $T_{air(y),L}$ and for $y$ from 0 to $L$. Hence;

$$\int_{T_{air(y)}=T_{air(y),0}}^{T_{air(y),L}} \frac{1}{T_{air(y)} - T_{z,t}}dT_{air(y)} = -\frac{U\pi D}{\dot{m}_{air}c_p}\int_{y=0}^{L} dy$$

$$ln\frac{T_{air(y),L} - T_{z,t}}{T_{air(y),0} - T_{z,t}} = -\frac{U\pi Dy}{\dot{m}_{air}c_p}\Bigg|_{y=0}^{y=L}$$

$$T_{air(y),L} - T_{z,t} = \left(T_{air(y),0} - T_{z,t}\right)\left(e^{-UA_{sur}/\dot{m}_{air}c_p}\right)$$

$$T_{air(y),L} = T_{z,t} + \left(T_{air(y),0} - T_{z,t}\right)\left(e^{-UA_{sur}/\dot{m}_{air}c_p}\right) \tag{5}$$

Therefore, Equation 5 can be written as Equation 6 for $y = L$.

$$T_{air.out} = T_{z,t} + \left(T_{air.in} - T_{z,t}\right)\left(e^{-UA_{sur}/\dot{m}_{air}c_p}\right) \tag{6}$$

where $T_{air}$ out is the outlet air temperature, $T_{z,t}$ is the ground temperature at depth $z$ and time $t$, $T_{air}$ in is the inlet air temperature, $U$ is the overall heat transfer coefficient, $y$ is the length of the pipe, $\dot{m}$ is the mass flowrate and $c_p$ is specific heat of air. By definition;

$$NTU = \frac{UA_{sur}}{\dot{m}_{air}c_p} \tag{7}$$

Therefore Equation 6 can be written as a function of NTU as shown below;

$$T_{air.out} = T_{z,t} + \left(T_{air.in} - T_{z,t}\right)e^{-NTU} \tag{8}$$

Table 1 shows the physical and thermal parameters used in the simulation to investigate the effect of pipe materials and thermal conductivity of soil on the GHE performance. There are five pipe materials with different thermal conductivity which are polyvinyl chloride (PVC), high-density polyethylene (HDPE), stainless steel, aluminium and copper were considered in this simulation. Meanwhile, five different of thermal conductivity of soil that has been used in this simulation that is 0.5, 1.5, 2.5, 5 and 10 W/m·K for each pipe materials. Table 2 tabulates conditions of all the pipe materials and its surrounding for the input of the simulation. Simulation is conducted based on same pipe diameters for all pipe materials with 25 m length. Besides, the velocity of air, flowrate, ground temperature and air inlet temperature also set to be constant in this simulation. The thermal conductivity value of the ground is considered as homogenous [22, 23].

**Table 1.** Physical and thermal parameters used in simulation.

| Materials | Thermal conductivity of pipe, $k_p$ (W/m·K) | Thermal conductivity of soil, $k_s$ (W/m·K) |
|---|---|---|
| PVC | 0.18 | 0.5, 1.5, 2.5, 5.0 and 10 |
| HDPE | 0.47 | 0.5, 1.5, 2.5, 5.0 and 10 |
| Stainless steel | 13 | 0.5, 1.5, 2.5, 5.0 and 10 |
| Pure aluminium | 220 | 0.5, 1.5, 2.5, 5.0 and 10 |
| Pure copper | 386 | 0.5, 1.5, 2.5, 5.0 and 10 |

**Table 2.** Conditions of pipe and its surrounding for the input of the simulation.

| Input name | Value | Units |
|---|---|---|
| Outer diameter of pipe, OD | 100 | mm |
| Inner diameter, ID | 101.8 | mm |
| Pipe length, L | 25 | m/s |
| Velocity of air, V | 10.73 | m/s |
| Flowrate, ṁ | 0.1 | kg/s |
| Thermal diffusivity, α | 0.046 | $m^2$/day |
| Ground temperature, $T_{z,t}$ | 24 | °C |
| Air inlet temperature, $T_{air.in}$ | 35 | °C |

## 3. Results and discussion

Analysis of the output air temperature has been conducted for all the pipe materials and five different soil thermal conductivity. In this analysis, there is two parameters variation which are pipe materials and soil thermal conductivity while, other parameters are fixed. The aim of this analysis is to investigate the effect of pipe materials and soil thermal conductivity on the performance of GHE. The output air temperature has been recorded for all pipe materials with five different soil thermal conductivity. Table 3 shows the output air temperature of the GHE for PVC pipe and five different soil thermal conductivity. The data was recorded from 0 to 25 m length of pipe with an increment of 1 m. Meanwhile, other materials of pipe are HDPE, stainless steel, aluminium, and copper were analysed the same as PVC pipe materials.

Data from the analysis for all the pipe materials have been manipulated in the form of graphs as shown in Figure 4 which represents pipe length against air temperature for five different soil thermal conductivity. From the graph, the same trend has been obtained throughout length against air temperature for all pipe materials. However, their air temperature is different for all pipe materials at the same thermal conductivity of soil. This is due to all pipe materials namely PVC, HDPE, stainless steel, aluminium and copper has different thermal conductivity which are 0.18, 0.47, 13, 220 and 386 W/m.K respectively. Regarding the thermal conductivity of soils, the variation of air temperature decreased along the length of pipe from 0 to 25 m as the thermal conductivity of soil increased for all pipe materials. The thermal conductivity of soil range from 0.5 to 1.5 W/m.K has a significant effect on the air temperature drop along the length of pipe for all pipe materials. In the meantime, the thermal conductivity of soil from range 1.5 to 10 W/m.K has no significant effect on the air temperature drop along the length of pipe for all the pipe materials as shown in Figure 4. Therefore, it can be concluded that the good range of soil thermal conductivity was between 1.5 to 10 W/m.K.

**Table 3.** Simulation data of GHE temperature output for PVC pipe with five different of soil thermal conductivity PVC.

| **Pipe length, L** | **Thermal conductivity of soil, $k_s$** | | | | |
|---|---|---|---|---|---|
| | 0.5 | 1.5 | 2.5 | 5 | 10 |
| | $T_{out}$, °C | $T_{out}$, °C | $T_{out}$, °C | $T_{out}$, °C | $T_{out}$, °C |
| 0 | 35.00 | 35.00 | 35.00 | 35.00 | 35.00 |
| 1 | 32.90 | 32.04 | 31.77 | 31.54 | 31.41 |
| 2 | 31.21 | 29.88 | 29.49 | 29.17 | 29.00 |
| 3 | 29.84 | 28.29 | 27.88 | 27.55 | 27.37 |
| 4 | 28.72 | 27.14 | 26.74 | 26.43 | 26.27 |
| 5 | 27.82 | 26.29 | 25.94 | 25.67 | 25.53 |
| 6 | 27.10 | 25.68 | 25.37 | 25.14 | 25.03 |
| 7 | 26.51 | 25.23 | 24.97 | 24.78 | 24.69 |
| 8 | 26.03 | 24.90 | 24.68 | 24.54 | 24.47 |
| 9 | 25.64 | 24.65 | 24.48 | 24.37 | 24.32 |
| 10 | 25.33 | 24.48 | 24.34 | 24.25 | 24.21 |
| 11 | 25.08 | 24.35 | 24.24 | 24.17 | 24.14 |
| 12 | 24.87 | 24.26 | 24.17 | 24.12 | 24.10 |
| 13 | 24.71 | 24.19 | 24.12 | 24.08 | 24.07 |
| 14 | 24.57 | 24.14 | 24.09 | 24.06 | 24.04 |
| 15 | 24.46 | 24.10 | 24.06 | 24.04 | 24.03 |
| 16 | 24.37 | 24.07 | 24.04 | 24.03 | 24.02 |
| 17 | 24.30 | 24.05 | 24.03 | 24.02 | 24.01 |
| 18 | 24.25 | 24.04 | 24.02 | 24.01 | 24.01 |
| 19 | 24.20 | 24.03 | 24.02 | 24.01 | 24.01 |
| 20 | 24.16 | 24.02 | 24.01 | 24.01 | 24.00 |
| 22 | 24.11 | 24.01 | 24.01 | 24.00 | 24.00 |
| 25 | 24.06 | 24.00 | 24.00 | 24.00 | 24.00 |

Figure 5 shows a graph of the pipe length against air temperature for all pipe materials at constants thermal conductivity which is 10 W/m·K. By referring the Figure 4, HDPE pipe materials give a more significant effect compared with PVC pipe materials while, stainless steel, aluminium and copper pipe materials give a larger effect to the performance of GHE system compared with HDPE and PVC pipe materials. This variation happened due to the different of thermal conductivity of pipe materials. However, stainless steel, aluminium and copper pipe materials give almost the same performance as shown in Figure 5 even the copper pipe materials have the highest thermal conductivity compared with stainless steel and aluminium pipe materials. Therefore, this analysis can be concluded that the pipe materials does not give a significant effect on the performance of GHE system which similar with previous studies [24, 25]. The low-cost pipe materials with low thermal conductivity can be used in GHE system instead of high-cost pipe materials with high thermal conductivity. In addition, the drop of air temperature remains consistent at a range of pipe length between 10 m to 25 m for PVC pipe. Hence, the length of pipe that can be used was 10 m instead of 25 m. It can be reducing the space of pipe installation and at the same the cost of installation also can be reduced.

IOP Conf. Series: Materials Science and Engineering 1078 (2021) 012030 doi:10.1088/1757-899X/1078/1/012030

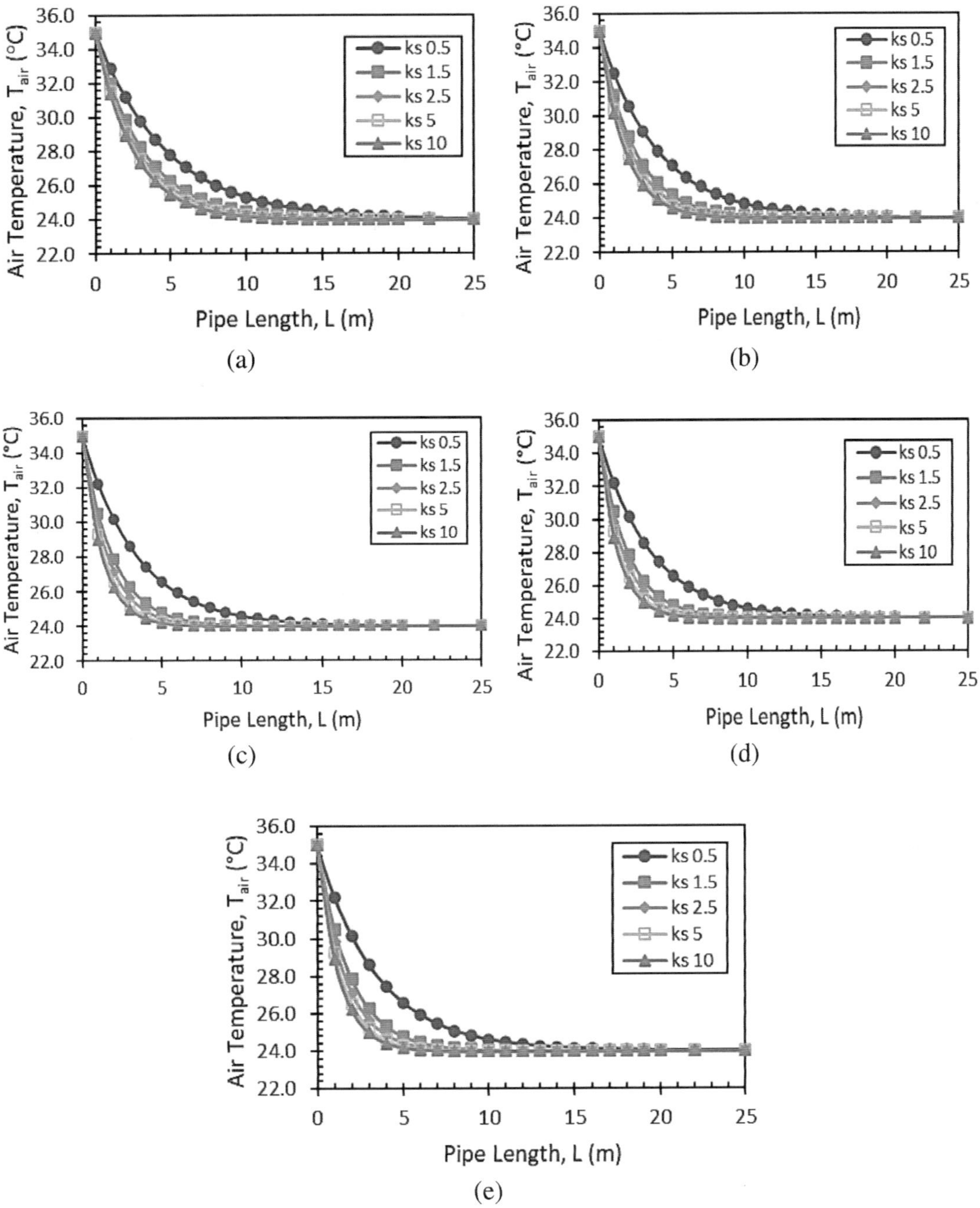


**Figure 4.** Pipe length against air temperature for five different of soil thermal conductivity (a) PVC, (b) HDPE, (c) Stainless steel, (d) Aluminium and (e) Copper.

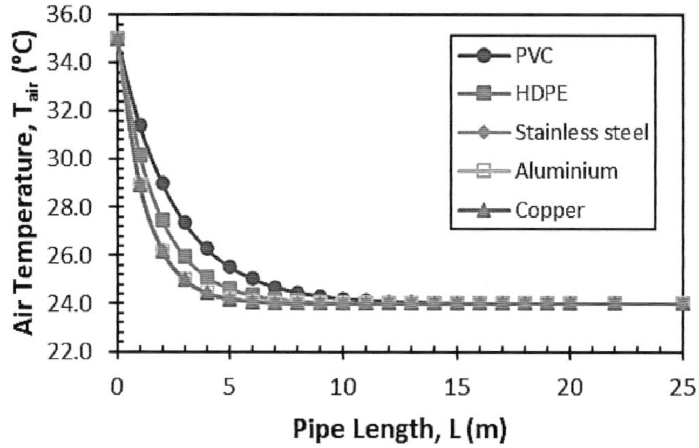


**Figure 5.** Pipe length against air temperature for five different materials of pipe at 10 W/m·K of soil thermal conductivity.

## 4. Conclusion

The influence of pipe materials and thermal conductivity of soil on the performance of the ground heat exchanger (GHE) system has been determined by a simulation study. There are five different types of pipe materials which are PVC, HDPE, stainless steel, aluminium, and copper with different thermal conductivity 0.18, 0.47, 13, 220 and 386 W/m·K respectively. Meanwhile, there are five different soil thermal conductivity that has been used in this study, which is 0.5, 1.5, 2.5, 5 and 10 W/m·K. The soil thermal conductivity from 1.5 to 10 W/m·K is a good range for the performance of GHE system. in the meantime, from this analysis the pipe materials do not give a significant effect on the performance of GHE system even the thermal conductivity of the pipe materials is lower. Therefore, it can be concluded that the low cost of pipe material can be used in GHE system instead of high-cost pipe materials. besides, the optimum length of pipe is about 10 m for PVC pipe materials at range 1.5 to 10 W/m·K of soil thermal conductivity.

## Acknowledgments

The authors thanked the Ministry of Higher Education (MOHE) for financially supported this study through Fundamental Research Grant Scheme (FRGS/1/2018/TK03/UMP/03/4) and Universiti Malaysia Pahang (www.ump.edu.my) for providing facilities to conduct this study under research grant RDU190143.

## References

[1] D'Agostino D and Mazzarella L 2019 What is a Nearly zero energy building? Overview, implementation and comparison of definitions *Journal of Building Engineering* **21** 200-12
[2] Scott N 1965 Analysis and performance of an earth-air heat exchanger
[3] Goswami D and Ileslamlou S 1990 Performance analysis of a closed-loop climate control system using underground air tunnel
[4] Soni S K, Pandey M and Bartaria V N 2015 Ground coupled heat exchangers: A review and applications *Renewable and Sustainable Energy Reviews* **47** 83-92
[5] Wu H, Wang S and Zhu D 2007 Modelling and evaluation of cooling capacity of earth–air–pipe systems *Energy Conversion and Management* **48** 1462-71
[6] Li H, Ni L, Liu G and Yao Y 2019 Performance evaluation of Earth to Air Heat Exchange (EAHE) used for indoor ventilation during winter in severe cold regions *Applied Thermal Engineering* **160** 114111
[7] Gao J, Li A, Xu X, Gang W and Yan T 2018 Ground heat exchangers: Applications, technology integration and potentials for zero energy buildings *Renewable Energy* **128** 337-49

[8] Agrawal K K, Misra R and Agrawal G D 2020 Improving the thermal performance of ground air heat exchanger system using sand-bentonite (in dry and wet condition) as backfilling material *Renewable Energy* **146** 2008-23
[9] Sodha M, Sharma A, Singh S, Bansal N and Kumar A 1985 Evaluation of an earth—air tunnel system for cooling/heating of a hospital complex *Building and Environment* **20** 115-22
[10] Mihalakakou G, Santamouris M, Asimakopoulos D and Tselepidaki I 1995 Parametric prediction of the buried pipes cooling potential for passive cooling applications *Solar Energy* **55** 163-73
[11] Mihalakakou G, Santamouris M, Lewis J and Asimakopoulos D 1997 On the application of the energy balance equation to predict ground temperature profiles *Solar Energy* **60** 181-90
[12] Kumar R, Ramesh S and Kaushik S 2003 Performance evaluation and energy conservation potential of earth–air–tunnel system coupled with non-air-conditioned building *Building and environment* **38** 807-13
[13] Ghosal M, Tiwari G, Das D and Pandey K 2005 Modeling and comparative thermal performance of ground air collector and earth air heat exchanger for heating of greenhouse *Energy and buildings* **37** 613-21
[14] Al-Ajmi F, Loveday D and Hanby V I 2006 The cooling potential of earth–air heat exchangers for domestic buildings in a desert climate *Building and Environment* **41** 235-44
[15] Badescu V 2007 Simple and accurate model for the ground heat exchanger of a passive house *Renewable energy* **32** 845-55
[16] Cucumo M, Cucumo S, Montoro L and Vulcano A 2008 A one-dimensional transient analytical model for earth-to-air heat exchangers, taking into account condensation phenomena and thermal perturbation from the upper free surface as well as around the buried pipes *International Journal of Heat and Mass Transfer* **51** 506-16
[17] Cui P, Yang H and Fang Z 2008 Numerical analysis and experimental validation of heat transfer in ground heat exchangers in alternative operation modes *Energy and Buildings* **40** 1060-6
[18] Derbel H B J and Kanoun O 2010 Investigation of the ground thermal potential in tunisia focused towards heating and cooling applications *Applied thermal engineering* **30** 1091-100
[19] Mongkon S, Thepa S, Namprakai P and Pratinthong N 2014 Cooling performance assessment of horizontal earth tube system and effect on planting in tropical greenhouse *Energy Conversion and Management* **78** 225-36
[20] Benhammou M and Draoui B 2015 Parametric study on thermal performance of earth-to-air heat exchanger used for cooling of buildings *Renewable and Sustainable Energy Reviews* **44** 348-55
[21] Naili N, Hazami M, Kooli S and Farhat A 2015 Energy and exergy analysis of horizontal ground heat exchanger for hot climatic condition of northern Tunisia *Geothermics* **53** 270-80
[22] Cuny M, Lin J, Siroux M, Magnenet V and Fond C 2018 Influence of coating soil types on the energy of earth-air heat exchanger *Energy and Buildings* **158** 1000-12
[23] Shojaee S M N and Malek K 2017 Earth-to-air heat exchangers cooling evaluation for different climates of Iran *Sustainable Energy Technologies and Assessments* **23** 111-20
[24] Bansal V, Misra R, Agrawal G D and Mathur J 2010 Performance analysis of earth–pipe–air heat exchanger for summer cooling *Energy and Buildings* **42** 645-8
[25] Bansal V, Misra R, Agrawal G D and Mathur J 2009 Performance analysis of earth–pipe–air heat exchanger for winter heating *Energy and Buildings* **41** 1151-4

# A review on 3D printing bio-based polymer composite

**L Sandanamsamy[1], J Mogan[1], N A Halim[1], W S W Harun[2,*], K Kadirgama[3] and D Ramasamy[2]**

[1]Institute of Postgraduate Studies, Universiti Malaysia Pahang, 26300 Gambang, Kuantan, Pahang, Malaysia
[2]Department of Mechanical Engineering, College of Engineering, Universiti Malaysia Pahang, 26300 Gambang, Kuantan, Pahang, Malaysia
[3]Faculty of Mechanical and Automotive Engineering Technology, Universiti Malaysia Pahang, 26600 Pekan, Pahang, Malaysia

*Corresponding author's email: sharuzi@ump.edu.my

**Abstract.** Polymers play a vital role in our daily lives. In various fields such as medical, food industry and automotive applications, the use of biopolymers is commonly used. The most widely used polymers and fillers among biopolymers are polylactic acid (PLA) and cellulose, which are biocompatible and biodegradable due to their eco-friendly properties. Extensive usage of cellulose in various forms has been applied in combination to PLA but there is only a few research that has been done by using the 3D printing method. This paper covers the types of biodegradable biopolymer materials, types of coupling agents and plasticizers, mechanical properties and applications. This paper discusses the types of cellulose ranging from micro to nano, including other types and sources of cellulose that have been researched and are compatible with PLA. In order to generate biocompatible polymers with stronger and better mechanical properties, the findings of these experiments are all tied together. These biopolymers are commonly used in the biomedical industry and are expected to improve their benefits in this field.

**Keywords:** Biopolymers; Polylactic Acid (PLA); Cellulose; 3D Printing; Mechanical Properties; Applications.

## 1. Introduction

Additive manufacturing is a technique that applies consecutive content layers to form a 3D model that requires the use of a computer-based 3D model as well as a 3D printer and a process to develop a physical model that is based on the original model concept[1, 2]. In the past years, additive manufacturing (AM) has been frequently utilized as an effective way of prototyping and generating parts with a high level of fracture surface[1]. It offers the facility to create goods at a higher pace and with dimensional precision than most modelling approaches [3]. 3D printing controls the principle of additive manufacturing (AM) which is widely termed as rapid prototyping (RP) and free-form manufacturing that involves a range of methods that uses a layer by layer process to develop parts or items typically of small size, limited quantities and customized design [2]. 3D Printing (3DP) is an extremely versatile fabrication system which can be enforced on various materials, such as ceramics, metals, polymers as well as other materials [1]. FDM is commonly utilized in additive manufacturing processes that produces practical prototypes of many thermoplastics because of its ability to safely build high-quality complex goods. Besides, different additive manufacturing techniques are available, for example, FDM (fused deposition modelling), DMD (direct metal deposition), 3D printing, SLS (selective laser sintering), IJM (inkjet modelling) and SLA (stereo-lithography). Each technique is different in terms of the method and materials that are used for production [3]. While there are many approaches available linked to 3D printing, a technique that is used widely is fused deposition modelling (FDM). This additive manufacturing method allows us to obtain products with better durability, dimensional precision, surface roughness [4]. Materials such as acrylonitrile-butadienestyrene and polylactic acid are frequently used polymer materials for additive processing filament [5].

During the manufacturing phase of a product, the usage of 3D printing has reduced the extra costs encountered. Besides, with comparatively low prices, additive manufacturing can 3D print limited amounts of personalized products [6]. Besides, over the years, the resolution, precision, usability, and repeatability of 3D printed products have increased. The rising number of 3D printers and quick access to applications has resulted in a decrease in the cost of the production process. [7].

3D printing is driving major modernization in a wide variety of fields involving medical equipment, biotechnology and so on [8]. Oprea (2020) suggested that more works need to be done using 3D printing when involving cellulose in order to obtain biomaterials. The FDM process is therefore commonly used in aerospace, automotive manufacturing, bio-medicines, smart houses, stationaries, training aid and artistic gifts [9]. On the other hand, this process has been used for many years to manufacture products for aerospace, medical, mould design and industrial applications from a wide variety of materials, such as plastics, metal powder, ceramics and composites [10]. It is possible to ensure the consistency of FDM 3D printed components that are specially designed by altering printing variables such as the thickness of a layer, orientation of printing and air gap [11].

Graphical description of various biopolymers origin is represented in Figure 1 [12] as polymers play a crucial role in our everyday lives. The assortment of materials are based on the characteristics are defined as the word 'biopolymers'. However, the monomeric units of the biochemical composition and polymers can be classified into two major groups which consist of polymers that are produced biologically and synthetically[12, 13]. Synthetic and natural biopolymers and the subdivision are shown in Table 1 [12]. Therefore, biologically produced polymers are examples of natural polymers derived from animals, microorganisms and plants whereas oils, sugars and amino acids are chemically synthesized polymers that are synthetically processed biopolymers are extracted from biological material. Polyesters, polylactic and polycarbonates are synthetic biodegradable polymers, while polyethylene and polypropylene are some examples of synthetic non-biodegradable [12]. Metabolic processes involving enzyme-catalysis and activated monomers through chain-growth polymerization reactions produce natural bio-degradable polymers[14]. The usage of biopolymers is wide such as they are used in the medical, food industry, which involves additives and packaging materials [12]. This review paper aims to discuss the studies related to bio-based polymer composites which are biodegradable by using 3D printing. Particular focus will be given to Polylactic acid (PLA) reinforced cellulose.

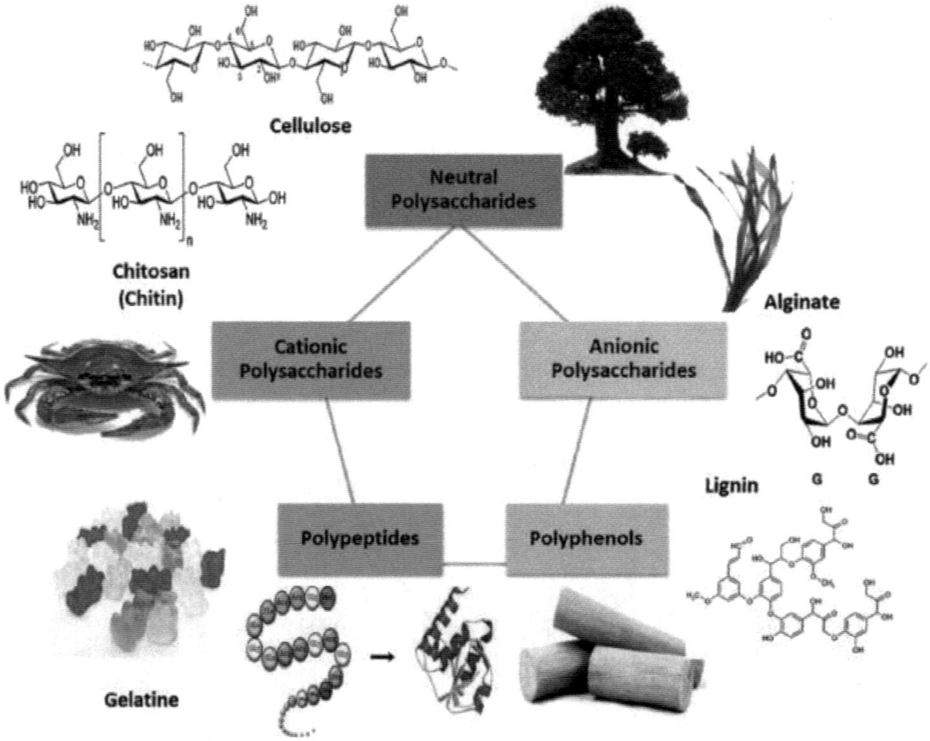


**Figure 1.** Biopolymers from different sources.

**Table 1.** Biopolymer categorization.

| | | Biopolymers | Reference |
|---|---|---|---|
| Natural | Protein | Soy protein, collagen, fibrinogen, silk, gelatin | [13] |
| | Polysaccharide | Cellulose, starch, chitin, Hyaluronic acid | [5,18] |
| Synthetic | Biodegradable | Polylactic acid, Polyglycolide, Polycaprolactane, Polyvinyl alcohol, Polyhydroxybutyrate, Polyglutamic acid | [5,7,18,21] |
| | Non-biodegradable | Polyolefins, Nylon, Polyurethane, Polypropylene, Polyethylene, Polyamides | [7,21] |

## 2. Materials

*2.1. Polylactic Acid*

PLA has made tremendous strides due to its biodegradability within the 3D printing sector. Generally, for example, corn starch is extracted from biomaterials and is biodegradable in compost, PLA is more renewable and presently in use which has been authorized for numerous biomedical applications [15, 16]. Also, there are hardly have any thermal warping or curling problems with 3D printed PLA parts [16]. Since the PLA is a semi-crystalline structure, thus, the mechanical properties and crystalline structure of PLA are significantly affected by thermal background. Therefore, thermal annealing therapy is a widely used method aimed at removing the impact of the thermal background and rebuilding the crystallinity, crystallite size, and mechanical properties of PLA [16].

In addition, thermal annealing is intended to increase the crystallinity of semi-crystalline polymers by adding thermal heat in between the transition temperature of the polymer (Tg) as well as the melting temperature (Tm). Thermal annealing stimulates the mechanism of polymer nucleation, which further enables existing crystallites to expand [17]. Furthermore, plant-derived fibres like, kenaf, hemp, flax and bamboo fibres were utilized to improve PLA. Micropowders resulting from agricultural by-products have also been packed with PLA. The composite modulus was typically increased, although the elongation at break reduced relative to tidy PLA. A good effect of natural fibres and man-made cellulose fibres on the mechanical properties of roller card-oriented PLA-based composites. [18]. In addition, utilizing nanofillers, nanocomposites can be treated very easily, but the composite's biodegradability was impaired [19].

### 2.2. Types of fillers (cellulose)

Natural resource-based green polymers are an environmentally sustainable alternative to traditional materials, and its use will minimize the vast amount of non-biodegradable waste created on a daily basis by the pharmaceutical sector. Thus, among these polymers, owing to its unique properties and biocompatibility, cellulose, the key component found in the structure of plant wall and also developed by some microorganisms, is intensively studied in form of aerogels, hydrogels, fillers or films for biomedical application [20]. With its outstanding qualities such as biodegradability low density, toxicological harmlessness, eco-friendliness and robust thermal, cellulose, bio-based polymer, shows tremendous potential uses in automobile, plastics and packaging sectors. In addition to reducing the overall cost of the material from the pure polymer, these cellulosic fibres may also act as composite material reinforcements. It is commonly used for the preparation of (partially) bio-based and biodegradable composites. For instance, using cellulosic fibers, the mechanical characteristics of plastics like PLA, polycaprolactone and polyamide have been modified. Cellulose has an almost ideal inerratic form and robust mechanical properties as an organic raw resource. It was used in composite materials as a fiber improver, and its scale, morphology, and structure governed the enhancing effect [15].

Based on Figure 2, from the studies conducted, many researchers have used nanocellulose as the fillers to be reinforced in PLA compared to microcellulose and other cellulose. Properties of fiber-reinforced composites are highly affected by reactions between the materials. It's well recognized that the use of nanometer-sized fillers helps one to improve the surface contact of a filler/matrix and change properties at relatively low loading speeds [18]. Therefore, nanocellulose consists of many types of fillers such as nanocrystalline cellulose, cellulose nanofibers (CNF), cellulose nanocrystals (CNC), carbonized cellulose nanofibers (CCNF), man-made cellulose and cellulose nanowhiskers. As strengthening fillers for different natural polymeric matrices, for example, polylactic acid (PLA), polyhydroxyalkanoates (PHAs), poly(vinyl alcohol) (PVA), natural rubber or starch, CNCs, CNFs and BC are often utilized [20]. Majorly from plants or vegetation, cellulose nanofibers (CNFs) are obtained. Due to its exceptional mechanical, physicochemical properties, CNF has been widely used as a filler material for polymer resins. However, when combined with hydrophobic polymers, for example, PLA, the existence of CNF-strong hydrogen bonds typically creates dispersion problems [16].

As for microcellulose, it consists of micro fibrillated cellulose and microcrystalline cellulose. Besides, other cellulose is categorized as natural cellulose, natural fiber, cellulose fiber, man-made cellulose and abaca fibers. A complete "biocomposite "can be produced by using natural fibres only. The 'lightweight capacity' of natural fibres (NF) is one of the key benefits. Venkatarajan (2020) stated that polymer composites strengthened with natural cellulose fibres have recently gained substantial recognition among researchers for their eco-friendliness and strong physical as well as mechanical properties. Due to their desirable advantages, such as renewability, availability, low cost, low density and adequate mechanical properties, natural cellulose fibres are known as a possible alternative to synthetic fibres in polymer matrix composites [21]. Only natural fibres of technological consistency ensure adequate reproducibility of mechanical characteristics [19]. In addition to plant fibres, PLA may be reinforced by man-made fibres dependent on recycled raw materials, for example, viscose, rayon (Cordenka) or Lyocell [22].

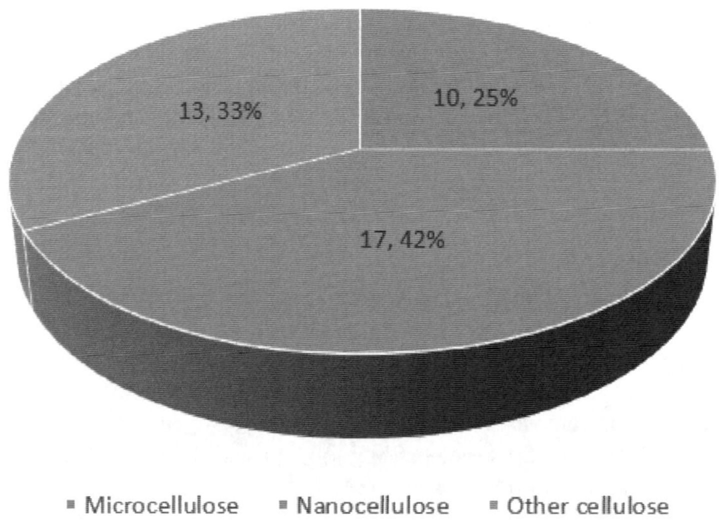


**Figure 2.** Types of fillers that are suitable for cellulose.

*2.3. Types of coupling agent & plasticizer*

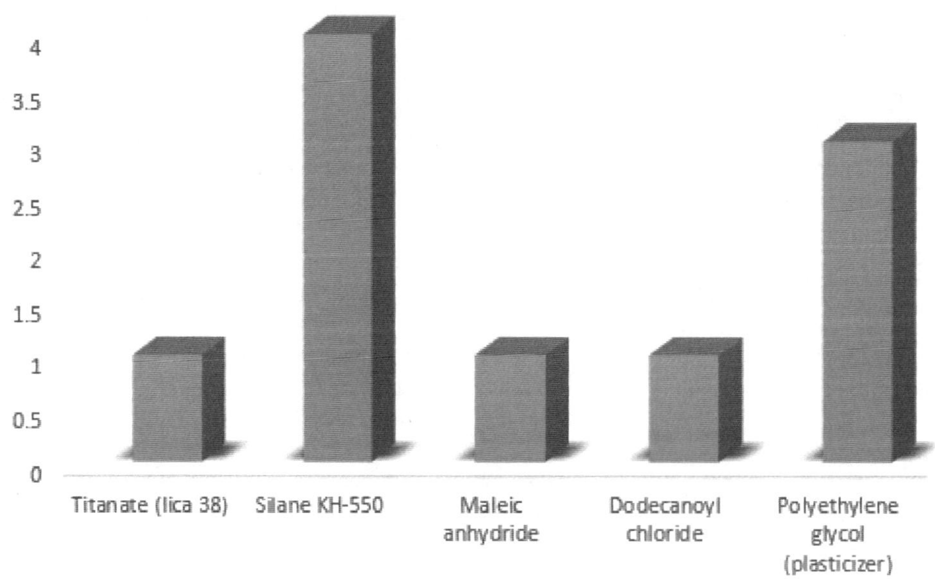


**Figure 3.** Coupling agent & Plasticizers.

From the studies conducted based on Figure 3, the commonly used coupling agent is silane KH-550 and followed by Polyethylene glycol, which is used as a plasticizer. Likewise, as a plasticizer, silane-coupling agent (KH-550) as well as polyethylene glycol (PEG) has been reported to work with each other to increase the sustainability and printability of 3D-based composite filaments [16]. Two techniques are primarily known in order to preserve the proper viscosity of the PLA-printing ink which are raising the manufacturing temperature or by using a plasticizer. It is more effective to use a suitable plasticizer because it helps the material to survive at low temperatures without triggering any thermal deterioration. In general, low molecular weight polyethylene glycol has been commonly accepted as an effective plasticizer for PLA. It is, therefore anticipated that the use of PLA/PEG blends could enhance the manufacturing process of 3D structures for tissue engineering [23].

PEG with strong interfacial compatibility was commonly used in industrial processing technology as an essential plasticizer [15]. In addition, PEG is a biocompatible hydrophilic, polymer utilized in many applications, ranging from industrial processing to biotechnology [23]. Cellulose micro/nanomaterials could be added to reinforced PLA based bio-composites only with modification. To improve its stability with the PLA matrix and microcrystalline cellulose, and to increase the crystallinity and storage modulus of the resulting bio-composites, the titanate coupling agent (Lica38) has been published [16].

## 3. Mechanical properties of PLA reinforced cellulose

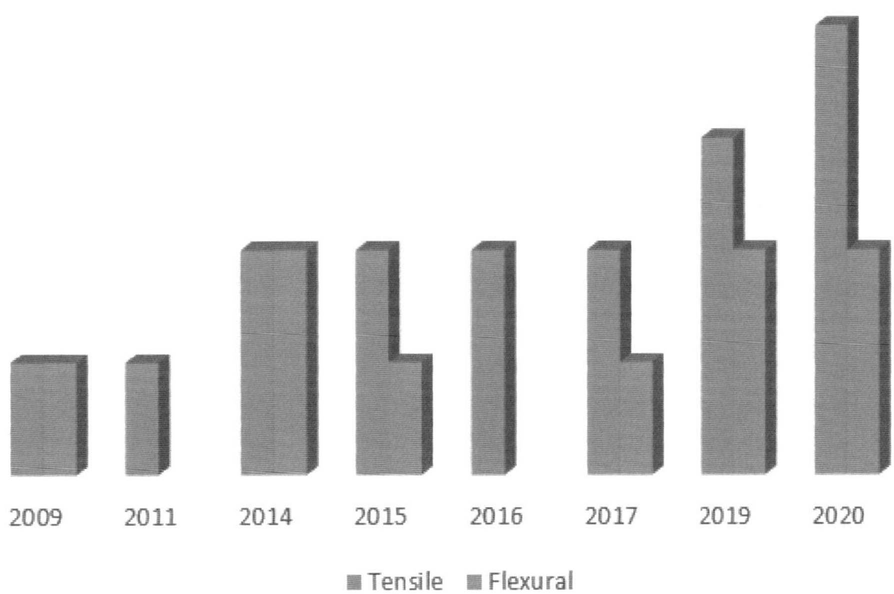


**Figure 4.** Types of mechanical properties to study PLA reinforced cellulose.

The evaluation of mechanical properties aids in the analysis and development of more effective and less expensive components and goods as they last longer. Thus, by determining the mechanical properties, it helps to develop and design items that are needed. Various mechanical testing experiments were undertaken to research the polymer composite mechanical properties, for example, tensile test, impact test and flexural test. Based on Figure 4, the tensile strength test is the most common test that is preferred by researchers when compared to other tests which are flexural, impact (Izod) and impact (Charpy) test. Awal (2015) reported that, with the addition of fibres and fillers, the mechanical properties of PLA bio-based composites might be strengthened. As an example, Awal (2015) stated that initially, the tensile strength of PLA was 62.7 MPa. However, the tensile strength was improved due to the fibre reinforcements as wood fibres were integrated into the bio-composites [24]. In addition, Bledzki (2009) stated that a boost in tensile strength by factor 1.45 could be accomplished with the addition of 30 % wt of man-made cellulose fibres [19].

## 4. Applications of cellulose in biomedical fields

The application of cellulose can be implemented in various medical fields such as pharmaceutical, dental and medical fields [25]. Therefore, the most significant fact, however is that cellulose is a polymer of natural origin and has a highly biocompatible character. These characteristics are one of the potential prospects relevant to the applications of cellulose derivatives in the biomedical field [20]. Recently, with many developments in the biomedical field, nanocellulose has been identified as an essential biomaterial. This application includes the growth of blood vessels, nerves, skin substitutes for burns and wounds, the mechanism of drug release, regeneration of gums, tissue engineering scaffolds and bone reconstruction. Nanocellulose mats were known to be very skilled in facilitating pain relief and accelerating granulation for successful wound healing. The structure and scale of such membranes could be manipulated to make them adequately efficient for the treatment of large and diverse body regions. Thus, with the help of nanocellulose, dental tissue reconstruction is being carried out. Crystalline nanocellulose is known to have several significant benefits as a drug delivery excipient. The advantage of having a considerably larger nanocrystalline cellulose surface area makes it possible to bind vast amounts of drugs to the surface of this cellulosic material with the promise of high payloads and optimal dosing power [25]. In many industries such as automotive, aerospace, maritime, electronic, packaging, and building, natural fibre reinforced polymer composites have been used due to its outstanding properties and benefits found over synthetic fiber reinforced polymer composites [21]. Table 2 summarizes the application of cellulose in the biomedical field.

**Table 2.** Application of cellulose in biomedical field.

| Field | Applications | References |
|---|---|---|
| Medical | Growth of blood vessels, nerves, skin substitutes for burns and wounds, bone reconstruction, tissue engineering scaffolds | [21,26] |
| Dental | Dental tissue reconstruction, regeneration of gums | [26] |
| Pharmaceutical | Drug delivery | [26] |

## 5. Conclusion

The types of commonly used biopolymers consisting in the group of biodegradable have been analysed in this article. Extensive usage of cellulose in various forms and sizes have been used as fillers. In order to improve the printability and sustainability of composite filament based on 3D printing, silane-coupling agent (KH-550) and polyethylene glycol are used as plasticizer. In general, based on the research that has been performed, the most widely used mechanical properties to measure the PLA/cellulose specimens are tensile strength compared to others. As PLA is a semi-crystalline polymer, the thermal annealing process is highly recommended to increase its crystallinity. Thus, due to its highly bio-compatible characteristics different forms of cellulose are being used in the biomedical while PLA are being incorporated in automotive, aerospace and packaging industry. Some research studies suggested that in order to obtain biomaterials, more work needs to be done using 3D printing when cellulose is involved.

**Acknowledgements**
The authors would like to thank Universiti Malaysia Pahang (www.ump.edu.my) for the facilities and resources provided for this research. This work was financially supported by Universiti Malaysia Pahang internal grants (RDU1903134, RDU190354), and Fundamental Research Grant Scheme (FRGS) FRGS/1/2019/TK03/UMP/02/15, provided by the Ministry of Higher Education, Malaysia.

## References

[1] Valvez S, Santos P, Parente J, Silva M and Reis P 2020 3D printed continuous carbon fiber reinforced PLA composites: A short review *Procedia Structural Integrity* **25** 394-9
[2] Gu S H, Nicolas V, Lalis A, Sathirapongsasuti N and Yanagihara R 2013 Complete genome sequence and molecular phylogeny of a newfound hantavirus harbored by the Doucet's musk shrew (Crocidura douceti) in Guinea *Infection, Genetics and Evolution* **20** 118-23
[3] Dhinakaran V, Manoj Kumar K P, Bupathi Ram P M, Ravichandran M and Vinayagamoorthy M 2020 A review on recent advancements in fused deposition modeling *Materials Today: Proceedings* **27** 752-6
[4] Solomon I J, Sevvel P and Gunasekaran J 2020 A review on the various processing parameters in FDM *Materials Today: Proceedings*
[5] Shunmugasundaram M, Maughal A A b and Ajay Kumar M 2020 A review of bio-degradable materials for fused deposition modeling machine *Materials Today: Proceedings* **27** 1596-600
[6] Ngo T D, Kashani A, Imbalzano G, Nguyen K T Q and Hui D 2018 Additive manufacturing (3D printing): A review of materials, methods, applications and challenges *Composites Part B: Engineering* **143** 172-96
[7] Mohd Pu'ad N A S, Abdul Haq R H, Mohd Noh H, Abdullah H Z, Idris M I and Lee T C 2020 Review on the fabrication of fused deposition modelling (FDM) composite filament for biomedical applications *Materials Today: Proceedings* **29** 228-32
[8] Dai L, Cheng T, Duan C, Zhao W, Zhang W, Zou X, Aspler J and Ni Y 2019 3D printing using plant-derived cellulose and its derivatives: A review *Carbohydr Polym* **203** 71-86
[9] Liu Z, Wang Y, Wu B, Cui C, Guo Y and Yan C 2019 A critical review of fused deposition modeling 3D printing technology in manufacturing polylactic acid parts *The International Journal of Advanced Manufacturing Technology* **102** 2877-89
[10] Penumakala P K, Santo J and Thomas A 2020 A critical review on the fused deposition modeling of thermoplastic polymer composites *Composites Part B: Engineering* **201**
[11] Zhang X, Fan W and Liu T 2020 Fused deposition modeling 3D printing of polyamide-based composites and its applications *Composites Communications* **21**
[12] Gopi S, Balakrishnan P, Chandradhara D, Poovathankandy D and Thomas S 2019 General scenarios of cellulose and its use in the biomedical field *Materials Today Chemistry* **13** 59-78
[13] Assessment U O o T and Assessment U S C O o T 1993 Biopolymers: Making Materials Nature's Way. Congress)
[14] Chandra R and Rustgi R 1998 Department of Polymer Technology and Applied Chemistry,“ *Biodegradable Polymers", Delhi College of Engineering, India* 1273-335
[15] Wang Z, Xu J, Lu Y, Hu L, Fan Y, Ma J and Zhou X 2017 Preparation of 3D printable micro/nanocellulose-polylactic acid (MNC/PLA) composite wire rods with high MNC constitution *Industrial Crops and Products* **109** 889-96

[16] Dong J, Mei C, Han J, Lee S and Wu Q 2019 3D printed poly(lactic acid) composites with grafted cellulose nanofibers: Effect of nanofiber and post-fabrication annealing treatment on composite flexural properties *Additive Manufacturing* **28** 621-8

[17] Dong J, Huang X, Muley P, Wu T, Barekati-Goudarzi M, Tang Z, Li M, Lee S, Boldor D and Wu Q 2020 Carbonized cellulose nanofibers as dielectric heat sources for microwave annealing 3D printed PLA composite *Composites Part B: Engineering* **184**

[18] Kowalczyk M, Piorkowska E, Kulpinski P and Pracella M 2011 Mechanical and thermal properties of PLA composites with cellulose nanofibers and standard size fibers *Composites Part A: Applied Science and Manufacturing* **42** 1509-14

[19] Bledzki A K, Jaszkiewicz A and Scherzer D 2009 Mechanical properties of PLA composites with man-made cellulose and abaca fibres *Composites Part A: Applied Science and Manufacturing* **40** 404-12

[20] Oprea M and Voicu S I 2020 Recent advances in composites based on cellulose derivatives for biomedical applications *Carbohydr Polym* **247** 116683

[21] Venkatarajan S and Athijayamani A 2020 An overview on natural cellulose fiber reinforced polymer composites *Materials Today: Proceedings*

[22] Graupner N, Herrmann A S and Müssig J 2009 Natural and man-made cellulose fibre-reinforced poly(lactic acid) (PLA) composites: An overview about mechanical characteristics and application areas *Composites Part A: Applied Science and Manufacturing* **40** 810-21

[23] Serra T, Ortiz-Hernandez M, Engel E, Planell J A and Navarro M 2014 Relevance of PEG in PLA-based blends for tissue engineering 3D-printed scaffolds *Mater Sci Eng C Mater Biol Appl* **38** 55-62

[24] Awal A, Rana M and Sain M 2015 Thermorheological and mechanical properties of cellulose reinforced PLA bio-composites *Mechanics of Materials* **80** 87-95

[25] Bhat A H, Khan I, Usmani M A, Umapathi R and Al-Kindy S M Z 2019 Cellulose an ageless renewable green nanomaterial for medical applications: An overview of ionic liquids in extraction, separation and dissolution of cellulose *Int J Biol Macromol* **129** 750-77

# A review of FDM and graphene-based polymer composite

**J Mogan[1], L Sandanamsamy[1], N A Halim[1], W S W Harun[2,*], K Kadirgama[3] and D Ramasamy[2]**
[1]Institute of Postgraduate Studies, Universiti Malaysia Pahang, 26300 Gambang, Kuantan, Pahang, Malaysia
[2]Department of Mechanical Engineering, College of Engineering, Universiti Malaysia Pahang, 26300 Gambang, Kuantan, Pahang, Malaysia
[3]Faculty of Mechanical and Automotive Engineering Technology, Universiti Malaysia Pahang, 26600, Pekan, Pahang, Malaysia

*Corresponding author's email: sharuzi@ump.edu.my

**Abstract.** Graphene is a carbon that has a unique structure that is excellent in enhancing mechanical, electrical and thermal properties. The fused deposition modelling (FDM) process is a widely used 3D printing method for its low investment and operating cost. Although the FDM process is cheaper and affordable, yet the printed parts are more fragile compare to other 3D printing methods. This paper covers about FDM process and the type of base materials and filler materials. However, the focus is mainly on ABS and graphene. The mechanical properties of ABS/Graphene polymer composite and application of ABS and graphene in the industry were also discussed. Hence, it proved that graphene enhances the properties of ABS. This study is done to improve polymer-based filaments for future references.

**Keywords:** Acrylonitrile Butadiene Styrene (ABS); Fused Deposition Modelling (FDM); Graphene; Additive Manufacturing; Mechanical Properties; Applications.

## 1. Introduction

In the 1980s, computer-aided design (CAD) was first introduced by rapid prototyping and this technology specifically provided the realization of user design without applying a mould or other machining preparation. This technology has since been developed exponentially, and in combination with various implementing materials, different forms of technology have been invented and developed. Due to the range of technologies available, additive manufacturing (AM) is referred to more frequently than rapid prototyping. AM has been applied for years, but this method has drawbacks, admittedly, and brings to the world a new age of manufacturing. [1]

Additive Manufacturing (AM) which also known as 3D printing is one of the techniques of adding material to build products from 3D model data, commonly layer after layer built through computer-aided design (CAD), as related to multiplying technology aspect [2]. The material is however melted (fused), cooled and also solidified in this process, fulfilling 3D geometries while not adopting complex models. AM technology is generally referred to as 3D printing [3]. Numerous AM methods depend on plastic lamination, such as laminated object manufacturing (LOM)[4], Stereolithography apparatus (SLA), which rely on the polymerization of images. The production of fused filaments (FFF) and selective laser sintering (SLS) depended on the plastic filaments being melted[5]. Due to the variation of cheap filaments and minimal printing device prices, the FDM technique is more popular than those in industrial applications. FDM 3D printers are commonly used in many such as manufacturing, automobile, medicine and aeronautics for rapid prototyping [6].

One of the most commonly used additive manufacturing techniques is fused filament manufacturing (FFF), which is mainly known as fused deposition modelling (FDM). In reality, the term 3D industries printing is generally used to refer to this technology[7, 8]. The most frequently used thermoplastic polymers are polylactic acid (PLA) and acrylonitrile-butadiene-styrene (ABS), as well as polycarbonate (PC) and polycaprolactone (PCL)[9]. The schematic diagram of the phase of FDM is shown in Figure 1. The FDM raw material will be in the form of a filament which is also heated in the liquefier head to a semi-solid state and then extruded to form a coating on the previously accumulated layer through the heated nozzle. The newly accumulated layer is often solidified due to the dissipation relation between layers retained to the previously accumulated layers[10].While Fused Deposition Modelling (FDM) is also well recognized mostly for operating costs and affordable investment costs, FDM's printed parts are more brittle compared to other typical plastic manufacturing processes, including moulding, injection, CNC, extrusion and blowing[11].

Although FDM produces materials of a reasonably machinable quality, today the world needs much good 3D printed materials than traditional techniques, and they can be obtained by using conventional blends of filament and carbon. Material durability and stability are the main factors in evaluating filament products for the manufacture of finished products using 3D printing. [12, 13]. Graphene nanoplatelets have recently been investigated as possible reinforcement fillers for polymers. Graphene nanoplatelets are tiny particles made up of sheets of graphene. This nanofiller can be used as a multifunctional reinforcement because it has excellent mechanical, thermal and electrical properties with a 2D graphene deformed structure. Therefore, thermoplastic nanocomposite packed with graphene greatly improves mechanical properties and thermal stability. [14-16]. This review paper aims to discuss graphene based polymer composite using FDM process. However, a particular focus will be given to ABS and graphene.

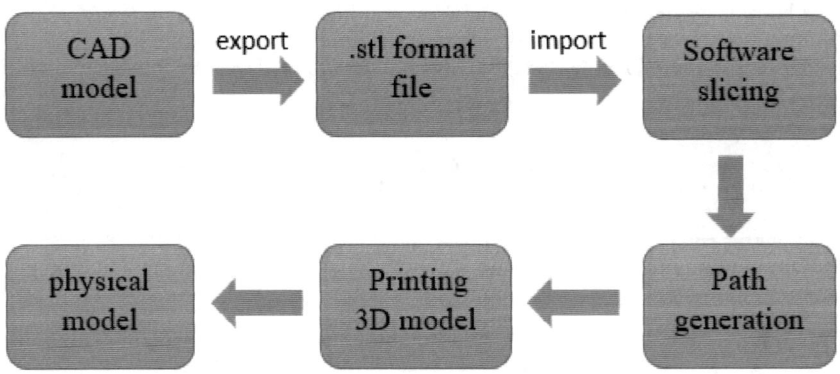


**Figure 1.** Basic steps for FDM process.

IOP Conf. Series: Materials Science and Engineering 1078 (2021) 012032 doi:10.1088/1757-899X/1078/1/012032

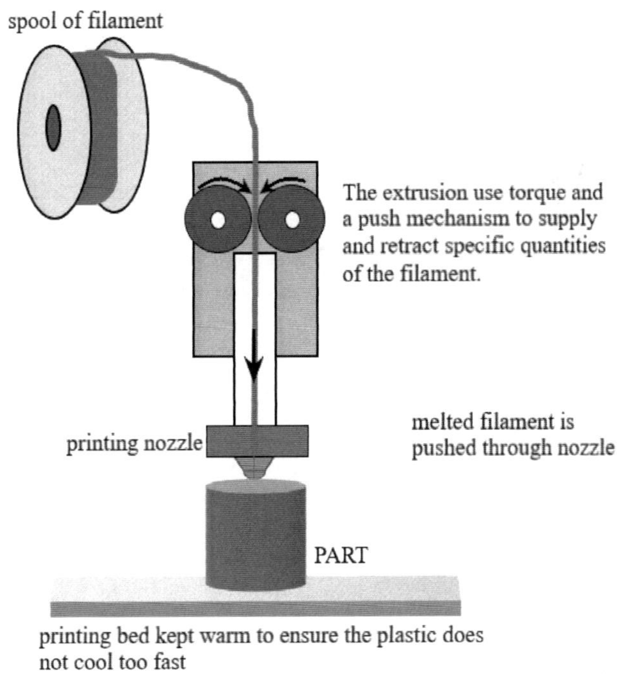


**Figure 2.** Schematic diagram of FDM process.

## 2. Type of material used

*2.1 Base materials used for FDM process*

Acrylonitrile, butadiene, and styrene are derived as ABS. Acrylonitrile is also a synthetic monomer formed from ammonia and propylene. Butadiene is a petroleum hydrocarbon obtained from the steam cracking C4 fraction and styrene monomer is produced using ethyl-a hydrocarbon obtained in the benzene and ethylene reaction dehydrogenation [17].

In general, ABS, which stands for Acrylonitrile Butadiene Styrene, is a thermoplastic that usually functions as a copolymer. The behaviour of thermoplastic materials includes physical, chemical, mechanical, thermal and morphological characteristics such as fracture resistive ability, amorphous vs crystalline, the temperature of melting and glass transformation, surface behaviour, length of the carbon chain, molecular structure, and molecular weight vs molecular density [18].

ABS comprises a system of slightly cross-linked polybutadiene (rubber) to increase the durability of its dispersion, containing styrene-acrylonitrile (SAN). The SAN ratios consist of polybutadiene in the blend and styrene and acrylonitrile in the SAN, the sum of inserted SAN, the size of rubber particles and the distribution of size, as well as the density of rubber cross-links [19]. ABS compositions are 15% to 35% acrylonitrile, 5% to 30% butadiene and 40% to 60% styrene, primarily used in household commercial products as well as Lego brick materials. [20].

Due to its reasonably good workability and outstanding mechanical efficiency, acrylonitrile-butadiene-styrene (ABS) is also commonly used in various electronic devices, household appliances, medical equipment, and others [21]. The ABS content is also known to be cheap, amorphous and manufactured at relatively low temperatures in terms of efficiency [22].

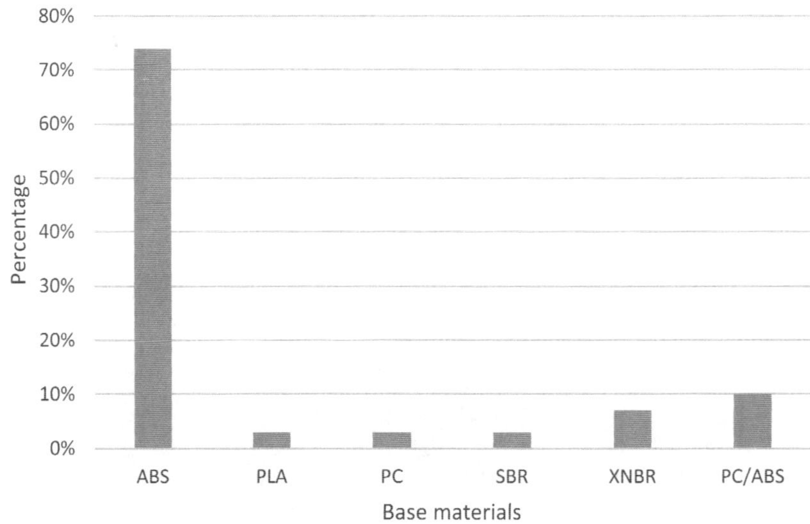


**Figure 3.** Type of polymers used as base materials.

Figure 3 shows the base materials that have been used. The base materials are Acrylonitrile Butadiene Styrene (ABS), Polylactic acid which is also known as polylactide (PLA), polycarbonate (PC), styrene-butadiene rubber (SBR), Carboxylate Nitrile (XNBR) and PC/ABS. From the figure above, ABS is the most used thermoplastic as the base materials. ABS plastic merges the strength and rigidity of styrene and acrylonitrile with the strength of rubber polybutadiene. ABS is superior for the hardness, shine, durability and electrical insulation properties, although the cost of manufacturing ABS is around 2 times the cost of producing polystyrene[17]. These ABS materials have been widely researched with different agents such as graphene, hollow glass bead filled, nano silica, poly methyl methacrylate (PMMA) and wood plastic to improve their strength.

**Table 1.** Physical properties of ABS.

| Classification | Value | Reference |
|---|---|---|
| Thermal Conductivity(K) | 0.1 to 0.15 W/*m-K* | [17, 23] |
| Density($\rho$) | 0.9 g/cm3 to 1.53 g/cm3 | [23, 24] |
| Specific Gravity | 1030 kg/m3 | [25] |
| Tensile modulus | 2 to 2.4 GPa | [23, 25] |
| Melting Point | 230 to 255 °C | [25] |
| Glass temperature | 110 °C | [23] |
| Solubility in Water | Insoluble | |

*2.2 Filler materials used for FDM process*

Graphene is an atomic carbon layer with minimal thickness and unique electronic structures, resulting in excellent mechanical and electrical properties[26]. It is also powerful, almost transparent, lightweight, and has good heat conductivity and electricity. Carbon nanotubes (CNTs) and graphene are both excellent materials for applications in electrochemistry. Compared to CNTs, graphene has gained popularity due to its economic ease and greater purity[27]. The use of constantly dispersed nanocomposites is motivated by the use of materials with improved mechanical properties required for different applications[28].

Graphene is a two-dimensional hexagonal carbon atom lattice and the fundamental building block for other allotropic carbon types, including nanotubes of carbon, fullerenes, graphite and diamonds. Graphene's remarkable properties are linked to the electronic structure and carbon atoms arranged in the hexagonal planar lattice. Therefore, graphene is an ideal reinforcement material to improve thermoelectrical efficiency and structure of the metal matrix composites[29].

Graphene has a two-dimensional geometric feature that can deliver very high aspect ratios and is responsible for the highest value over its surface-to-volume proportion. As the suitable candidate for a reinforcing material in polymer nanocomposites, its significant price of Young's Modulus and tensile strength enhances its mechanical properties. The mechanical strengths of polymer nanocomposites are affected by the huge increase of graphene. It not only does its incorporation as nanofiller transform the heat distortion temperature, hardness, stiffness, toughness, and also economic ease due to the unique nanostructures. This is subject to the aspect ratios and large surface-volume also polymer matrix connection[26].

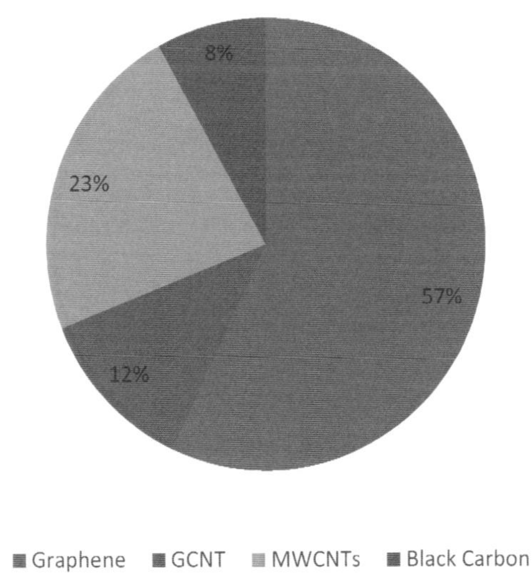


**Figure 4.** Type of carbon used as filler materials.

Figure 4 shows the type of carbon used as filler materials. The filler materials are graphene, graphene oxide-carbon nanotubes (GCNT), multiwalled carbon nanotubes (MWCNTs) and black carbon. Based on the studies conducted, graphene is the most used carbon filler compared to other carbon filler. This is because nanocomposites of the polymer network with graphene and its subordinates as fillers have shown exceptional potential for numerous important applications, such as gadgets, environmentally sustainable energy vitality, aviation and car companies. As stated some time recently, 2-D graphene has better electrical, mechanical and warm properties as well as other interesting elements, including a higher proportion of viewpoints and a greater specific surface area compared to various fortifications[30].

**Table 2.** Physical and mechanical properties for graphene.

| Classification | Value | Reference |
|---|---|---|
| Density | 1.06 g/cm3 | [29] |
| Surface area | 2,630 m2 /g | [29] |
| Thermal conductivity | 5300 W/m.K | [29, 31] |
| Tensile strength | 130 GPa | [29] |
| Young's modulus | 1.02 TPa | [29, 31] |

## 3. Mechanical properties of ABS/Graphene

Testing mechanical properties helps to evaluate and design materials and products so that they last longer are more efficient and less expensive. It also helps in designing products that are desired. Several mechanical test studies have been carried out in the literature to study the mechanical properties of polymer composites, such as tension test, compression test, impact test, bending test, hardness, creep and fatigue. Most mechanical properties studies of ABS/Graphene materials manufactured by the FDM were in tension scope in the papers read. There is some researcher has discovered to study impact, bending, flexure and hardness.

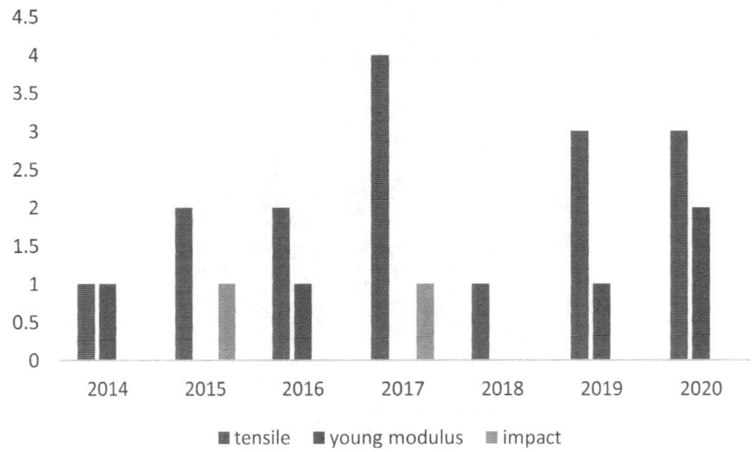


**Figure 5.** Type of mechanical properties tested from year 2014 to 2020.

Figure 5 shows the mechanical properties being testing from year 2014 to 2020. Based on the study conducted, the most preferred mechanical properties to be tested is the tensile test. Meanwhile, young modulus are second highly tested mechanical properties. Impact test is the least tested mechanical properties. In order to obtain the material parameter that includes ultimate tensile strength, yield stress, young's modulus, toughness and ductility, the tensile test is applied to a sample. In preparing a sample for the test, there are many options for shapes and designs. Even though there are many choices in the design of the test sample, those designs must follow the design of the ASTM standards.

One of the common designs used by most researchers was a flat dog-boned shape sample, [32]. Different forms, such as round specimens, were also used in tensile studies, according to ISO-7500/1 [33], CT specimens (ASTM E 647-08) [34], and non-standard shapes such as struts [35], and real part shapes [36].

Many researchers have stated that graphene has enhanced the mechanical properties of ABS. Hence, it is proved that ABS/graphene polymer composite material has better mechanical properties compared to ABS based materials. Sithiprumnea (2016) stated that samples are produced through different processes such as extrusion, compression moulding and also fused-deposition modelling, the thermo-mechanical properties of ABS and also its nanocomposites were compared. The existence of graphene nanoplatelets has increased the ABS tensile modulus[9].

The use of graphene nanoplatelets in ABS indeed increases ABS conductivity. The disadvantage is that the polymer matrix structure has been interrupted, thus reducing its impact strength, but the polymer elasticity somehow has increased. The images of morphology display strong dispersion and close interaction of the graphene platelet filler and polymer matrix, which has a positive effect on the electrical and mechanical properties of nanocomposites[37]. The fabrication of ABS/graphene composites will also potentially expand the applications and capabilities of 3D printed products.

## 4. Application of ABS and Graphene in industries

Acrylonitrile Butadiene Styrene (ABS) is a thermoplastic resin commonly used because of its excellent mechanical and processing properties in industrial sectors such as automotive, electronics, machinery, telecommunication instruments and construction [38, 39]. In addition, ABS is chosen as the manufacturing material in the automotive industry where it is utilized to manufacture bumpers and pillar trims[40].

Due to their outstanding electrical, mechanical and thermal properties, carbon nanofillers (carbon nanotubes, graphene oxides, carbon nanofibers, and reduction of graphene oxide) have been proved to be of great involvement in polymer composites. The relatively attractive materials used for different applications such as transparent electrodes, field-effect transistors, optoelectronics, energy storage materials, composites, and biosensors have been identified to be carbon nanofillers[41]. Besides that, graphene and carbon nanotubes (CNTs) are also amazing materials to be used in electrochemistry. Compared to CNTs, graphene has gained popularity due to its greater purity and economic ease[42]. Other than that, In addition, graphene has been used to coat metallic materials, indicating its potential to enhance tribological behaviour and extend the life span of these materials[43].

## 5. Summary

The types of commonly used thermoplastic and carbon have been analysed in this article. ABS has been used as the base material while the filler material is graphene. FDM printed products are easily breakable compare to other manufacturing processes due to their printing orientation and also material. Hence, based on the studies that have been conducted, graphene plays a significant role as a filler material in enhancing the mechanical properties of ABS. In general, based on the research that has been performed, the most commonly used mechanical properties to measure ABS/Graphene samples is tensile strength. The common application of graphene is in the electronic industry such as biosensors. Other than that, ABS has also been used in various industries where it is utilized to produce telecommunication instruments and electronic devices. Thus, due to its high compatiblity characteristics, different forms of graphene are being incorporated as electrodes, field-effect transistors, energy storage materials, and composites. Therefore, some research studies suggested implementing graphene as a filler in order to enhance the properties of polymers.

**Acknowledgments**
The authors gratefully acknowledge the support provided to the facilities and resources for this research by Universiti Malaysia Pahang (www.ump.edu.my). The authors would like to thank their support for the internal grants of Universiti Malaysia Pahang (RDU1903134, RDU190354) and the support is given by the Ministry of Higher Education, Malaysia, to the Fundamental Research Grant Scheme (FRGS.) FRGS/1/2019/TK03/UMP/02/15.

**References**
[1] Wang, Y., Zhou, Y., Lin, L., Corker, J. and Fan, M. 2020. Overview of 3D additive manufacturing (AM) and corresponding AM composites. Journal of Composites Part A. 139: 106114. doi:10.1016/j.compositesa.2020.106114
[2] Technologies, A.C.F.o.A.M. and A.C.F.o.A.M.T.S.F.o. 2012. Terminology, Standard terminology for additive manufacturing technologies. Astm International.
[3] Wang, J., Goyames, A., Gaisford, S. and Basit, A.W. 2016. Stereolithographic (SLA) 3D printing of oral modified-release dosage forms. International Journal of Pharmaceutics. 503(1):207-212. doi: 10.1016/j.ijpharm.2016.03.016
[4] Ahn, D., Kweon, J.H., Choi, J. and Lee, S. 2012. Quantification of surface roughness of parts processed by laminated object manufacturing. Journal of Materials Processing Technology. 212(2):339-346. doi: 10.1016/j.jmatprotec.2011.08.013
[5] Turner, B.N., Strong, R. and Gold, S.A.J.R.P.J. 2014. A review of melt extrusion additive manufacturing processes: I. Process design and modeling. Journal of Rapid Prototyping. 20/3: 192–204. doi: 10.1108/RPJ-01-2013-0012
[6] Heidari-Rarani, M., Rafiee-Afarani, M. and Zahedi, A.M. 2019. Mechanical characterization of FDM 3D printing of continuous carbon fiber reinforced PLA composites. Journal of Composites Part B: Engineering. 175:107147. doi: 10.1016/j.compositesb.2019.107147
[7] Ramanath, H. S., Chua, C.K., Leong, K.F. and Shah, K.D. 2008. Melt flow behaviour of polyε-caprolactone in fused deposition modelling. Journal of Material Science. 19(7): 2541-2550. doi:10.1007/s10856-007-3203-6
[8] Hill, N. and Haghi, M.J.R.P.J. 2014. Deposition direction-dependent failure criteria for fused deposition modeling polycarbonate. Journal of Rapid Prototyping. 20(3): 221-227. doi:https://doi.org/10.1108/RPJ-04-2013-0039
[9] Dul, S., Fambri, L. and Pegoretti, A. 2016. Fused deposition modelling with ABS–Graphene nanocomposites. Journal of Composites Part A: Applied Science and Manufacturing. 85: 181-191. doi:10.1016/j.compositesa.2016.03.013
[10] Rajpurohit, S.R. and Dave, H.K. 2018. Analysis of tensile strength of a fused filament fabricated PLA part using an open-source 3D printer. The International Journal of Advanced Manufacturing Technology. 101(5-8): 1525-1536. doi: 10.1007/s00170-018-3047
[11] Jiamjiroch, K. and Doungkam. P. 2019. Analysis of Printing Pattern and Infiltration Percent over the Tensile Properties of PLA Printed Parts by a Fuse Deposition Modelling Printer. Journal of Material Science and Engineering. 501:012028. doi:10.1088/1757-899X/501/1/012028
[12] Ashraf, M., Gibson, I. and Rashed, M.G. 2018. Challenges and prospects of 3d printing in structural engineering. 13th International Conference on Steel, Space and Composite Structures (Perth, WA).
[13] Gokhare, V.G. and Raut, D.N. 2017. A review paper on 3D-Printing aspects and various processes used in the 3D-Printing. 6(6): 953-958.
[14] Duguay, A.J., Nader, J.W., Kiziltas, A., Gardner, D.J. and Dagher, H.J. 2014. Exfoliated graphite nanoplatelet-filled impact modified polypropylene nanocomposites: influence of particle diameter,

filler loading, and coupling agent on the mechanical properties. Journal of Applied Nanoscience. 4(3): p. 279-291. doi: 10.1007/s13204-013-0204-2
[15] Jiang, X. and Drzal, L.T.J.P.c. 2010. Multifunctional high density polyethylene nanocomposites produced by incorporation of exfoliated graphite nanoplatelets 1: morphology and mechanical properties. Journal of Polymer Composites. 31(6): 1091-1098. doi:10.1002/pc.20896
[16] Pedrazzoli, D. and Pegoretti, A.J.C.P.A.A.S. 2014. Manufacturing, Expanded graphite nanoplatelets as coupling agents in glass fiber reinforced polypropylene composites. Journal of Composites Part A: Applied Science and Manufacturing. 66: 25-34. doi:10.1016/j.compositesa.2014.06.016
[17] Dimic, A., Miskovic, Z., Jelovac, D., Mitrovic, R., Ristivojevic, M. and Majstorovic, M. 2016. Application of rapid prototyping in maxillofacial surgery.
[18] Kumar, R., Singh, R. and Ahuja, I.P.S. 2018. Investigations of mechanical, thermal, and morphological properties of FDM fabricated parts for friction welding applications. Journal of Measurement. 120: 11-20. doi:10.1016/j.measurement.2018.02.006
[19] Peterson, A.M. 2019. Review of acrylonitrile butadiene styrene in fused filament fabrication: A plastics engineering-focused perspective. Journal of Additive Manufacturing. 27: 363-371. doi:10.1016/j.addma.2019.03.030
[20] Sagias, V. , Giannakopoulos, K. and Stergiou, C. 2018. Mechanical properties of 3D printed polymer specimens. Journal of Procedia Structural Integrity. 10:85-90. doi: 10.1016/j.prostr.2018.09.013
[21] Frascio, M., Avalle, M. and Monti, M. 2018. Fatigue strength of plastics components made in additive manufacturing: first experimental results. Journal of Procedia Structural Integrity 12: 32-43. doi: 10.1016/j.prostr.2018.11.109
[22] Hart, K.R. and Wetzel, E.D. 2017. Fracture behavior of additively manufactured acrylonitrile butadiene styrene (ABS) materials. Journal of Engineering Fracture Mechanics. 177:1-13. doi:10.1016/j.engfracmech.2017.03.028
[23] Kim, Y., Choi, E.S., Kwak, W. and Shin, Y. 2007. Analysis of the Thermal Distribution by Using Laser-Beam Irradiation. Journal of Korean Physical Society. 51: 503-508.
[24] Khaleed, H.M.T., Badruddin, I.A., Saquib, A.N., Addas, M.F., Kamangar, S. and Khan, T.M. 2019. Novel Approach to Manufacture an AUV Propeller by Additive Manufacturing and Error Analysis. Journal of Applied Sciences. doi:10.3390/app9204413
[25] Lipian, M., Kulak, M. and Stepian, M. 2019. Fast Track Integration of Computational Methods with Experiments in Small Wind Turbine Development. Journal of Energies. doi:10.3390/en12091625
[26] Sharma, V., Kapoor, S., Goyal, M. and Jindal, P. 2020. Enhancement of the mechanical properties of graphene based acrylonitrile butadiene styrene (ABS) nanocomposites. Journal of Materials Today: Proceedings. 28: 1744–1747. doi:10.1016/j.matpr.2020.05.155
[27] Pumera, M.J.C.A.E.J. 2009. The electrochemistry of carbon nanotubes: fundamentals and applications. 15(20): 4970-4978. doi:10.1002/chem.200900421
[28] Jindal, P., Sain, M. and Kumar, N. 2015. Mechanical characterization of PMMA/MWCNT composites under static and dynamic loading conditions. Journal of Materials Today: Proceedings. 2(4-5):1364-1372. doi:10.1016/j.matpr.2015.07.055
[29] Naseer, A., Ahmad, F., Aslam, M., Guan, B.H., Harun, W.S.W., Muhamad, N., Raza, M.R. and German, R.M. 2019. A review of processing techniques for graphene-reinforced metal matrix composites. Journal of Materials and Manufacturing Processes. 34(9): 957-985. doi:10.1080/10426914.2019.1615080

[30] Sreenivasulu, B., Ramji, B. R. and Nagaral, M. 2018. A review on graphene reinforced polymer matrix composites. Journal of Materials Today: Proceedings.5(1):2419-2428. doi: 10.1016/j.matpr.2017.11.021
[31] Lee, S.K., Kim, H. and Shim, B.S. 2013. Graphene: an emerging material for biological tissue engineering. Journal of Carbon Letters. 402-751. doi: 10.5714/CL.2013.14.2.063
[32] Zhang, B., Dembinski, L. and Coddet, C. 2013. The study of the laser parameters and environment variables effect on mechanical properties of high compact parts elaborated by selective laser melting 316L powder. Journal of Materials Science & Engineering A. 584: 21-31. doi:10.1016/j.msea.2013.06.055
[33] Yadroitsev, I., Pavlov, M., Bertrand, P.H. and Smurov, I. 2009. Mechanical properties of samples fabricated by selective laser melting.
[34] Riemer, A., Leuders, S., Thone, M., Richard, H.A., Troster, T. and Niendorf, T. 2014. On the fatigue crack growth behavior in 316L stainless steel manufactured by selective laser melting.Journal of Engineering Fracture Mechanics. 120: 15-25. doi: 10.1016/j.engfracmech.2014.03.008
[35] Li, P. 2015. Constitutive and failure behaviour in selective laser melted stainless steel for microlattice structures. Journal of Materials Science & Engineering A. 622: 114-120.doi:10.1016/j.msea.2014.11.028
[36] Delgado, J., Ciuranna, J. and Sereno, L. 2011. Comparison of forming manufacturing processes and selective laser melting technology based on the mechanical properties of products. Journal of Virtual and PhysicalPrototyping. 6(3):167-178. doi:10.1080/17452759.2011.613597
[37] Alauddin, S.M., Ismail, I., Zaili, F.S., Illias, N.F. and Aripin, N.F.K. 2018. Electrical and Mechanical Properties of Acrylonitrile Butadiene Styrene/Graphene Platelet Nanocomposite. Journal of Materials Today: Proceedings. 5: S125-S129.
[38] Nabi, G., Malik, N., Tahir, M.B., Raza, W., Rizwan, M., Maraj, M., Siddiqa, A., Ahmed, R. and Tanveer, M. 2020. Synthesis of graphitic carbon nitride and industrial applications as tensile strength reinforcement agent in red Acrylonitrile-Butadiene-Styrene (ABS). Journal of Physica B: Physics of Condensed Matter. doi:10.1016/j.physb.2020.412556.
[39] Scaffaro, R., Botta, L. and Benedetto, G.D. 2012. Physical properties of virgin-recycled ABS blends: Effect of post-consumer content and of reprocessing cycles. Journal European Polymer. 48: 637–648. doi:10.1016/j.eurpolymj.2011.12.018.
[40] K.Yadav, D., Srivastava, R. and Dev, S. 2020. Design & fabrication of ABS part by FDM for automobile application. Journal of Materials Today: Proceedings. 26: 2089–2093. doi:10.1016/j.matpr.2020.02.451
[41] Jyoti, J., Kumar, A., Dhakate, S.R. and Singh, B.P. 2018. Dielectric and impedance properties of three dimension graphene oxidecarbon nanotube acrylonitrile butadiene styrene hybrid composites. Journal of Polymer Testing. 68:456–466. doi:10.1016/j.polymertesting.2018.04.003
[42] Sharma, V., Kapoor, A., Goyal, M. and Jindal, P. 2020. Enhancement of the mechanical properties of graphene based acrylonitrile butadiene styrene (ABS) nanocomposites. Journal of Materials Today: Proceedings. 28:1744–1747 doi:10.1016/j.matpr.2020.05.155
[43] Mura, A., Adamo, F., Wang, H., Leong, W.S., Ji, X. and Kong, J. 2019. Investigation about tribological behavior of ABS and PC-ABS polymers coated with graphene. Journal of Tribology International. 134:335–340. doi:10.1016/j.triboint.2019.02.017

[1] Wang Y, Zhou Y, Lin L, Corker J, Fan M J C P A A S and Manufacturing 2020 Overview of 3D additive manufacturing (AM) and corresponding AM composites 106114
[2] Technologies A C F o A M and Terminology A C F o A M T S F o 2012 *Standard terminology for additive manufacturing technologies*: Astm International)
[3] Wang J, Goyanes A, Gaisford S and Basit A W 2016 Stereolithographic (SLA) 3D printing of oral modified-release dosage forms *International Journal of Pharmaceutics* **503** 207-12
[4] Ahn D, Kweon J-H, Choi J and Lee S 2012 Quantification of surface roughness of parts processed by laminated object manufacturing *Journal of Materials Processing Technology* **212** 339-46
[5] Turner B N, Strong R and Gold S A J R P J 2014 A review of melt extrusion additive manufacturing processes: I. Process design and modeling
[6] Heidari-Rarani M, Rafiee-Afarani M and Zahedi A M 2019 Mechanical characterization of FDM 3D printing of continuous carbon fiber reinforced PLA composites *Composites Part B: Engineering* **175**
[7] Ramanath H, Chua C, Leong K and Shah K J J o M S M i M 2008 Melt flow behaviour of poly-ε-caprolactone in fused deposition modelling **19** 2541-50
[8] Hill N and Haghi M J R P J 2014 Deposition direction-dependent failure criteria for fused deposition modeling polycarbonate
[9] Sithiprumnea Dul L F, Alessandro Pegoretti 2016 Fused deposition modelling with ABS–graphene nanocomposites
[10] Rajpurohit S R and Dave H K 2018 Analysis of tensile strength of a fused filament fabricated PLA part using an open-source 3D printer *The International Journal of Advanced Manufacturing Technology* **101** 1525-36
[11] Jiamjiroch P D a K 2019 Analysis of Printing Pattern and Infiltration Percent over the Tensile Properties of PLA Printed Parts by a Fuse Deposition Modelling Printer
[12] Ashraf M, Gibson I and Rashed M G 2018 Challenges and prospects of 3d printing in structural engineering. In: *13th International Conference on Steel, Space and Composite Structures (Perth, WA),*
[13] Gokhare V G, Raut D, Shinde D J I J o E R and Technology 2017 A review paper on 3D-Printing aspects and various processes used in the 3D-Printing **6** 953-8
[14] Duguay A J, Nader J W, Kiziltas A, Gardner D J and Dagher H J J A N 2014 Exfoliated graphite nanoplatelet-filled impact modified polypropylene nanocomposites: influence of particle diameter, filler loading, and coupling agent on the mechanical properties **4** 279-91
[15] Jiang X and Drzal L T J P c 2010 Multifunctional high density polyethylene nanocomposites produced by incorporation of exfoliated graphite nanoplatelets 1: morphology and mechanical properties **31** 1091-8
[16] Pedrazzoli D, Pegoretti A J C P A A S and Manufacturing 2014 Expanded graphite nanoplatelets as coupling agents in glass fiber reinforced polypropylene composites **66** 25-34
[17] <APPLICATIONOFRAPIDPROTOTYPINGINMAXILLOFACIALSURGERY.pdf>
[18] Kumar R, Singh R and Ahuja I J M 2018 Investigations of mechanical, thermal and morphological properties of FDM fabricated parts for friction welding applications **120** 11-20
[19] Peterson A M J A M 2019 Review of acrylonitrile butadiene styrene in fused filament fabrication: A plastics engineering-focused perspective **27** 363-71

[20] Sagias V, Giannakopoulos K and Stergiou C J P S I 2018 Mechanical properties of 3D printed polymer specimens **10** 85-90
[21] Frascio M, Avalle M and Monti M J P S I 2018 Fatigue strength of plastics components made in additive manufacturing: first experimental results **12** 32-43
[22] Hart K R and Wetzel E D J E F M 2017 Fracture behavior of additively manufactured acrylonitrile butadiene styrene (ABS) materials **177** 1-13
[23] <Analysis-of-the-Thermal-Distribution-by-Using-Laser-Beam-Irradiation.pdf>
[24] Khaleed H M T, Badruddin I A, Saquib A N, Addas M F, Kamangar S and Yunus Khan T M 2019 Novel Approach to Manufacture an AUV Propeller by Additive Manufacturing and Error Analysis *Applied Sciences* **9**
[25] Lipian M, Kulak M and Stepien M 2019 Fast Track Integration of Computational Methods with Experiments in Small Wind Turbine Development *Energies* **12**
[26] Sharma V, Kapoor S, Goyal M and Jindal P J M T P 2020 Enhancement of the mechanical properties of graphene based acrylonitrile butadiene styrene (ABS) nanocomposites
[27] Pumera M J C A E J 2009 The electrochemistry of carbon nanotubes: fundamentals and applications **15** 4970-8
[28] Jindal P, Sain M and Kumar N J M T P 2015 Mechanical characterization of PMMA/MWCNT composites under static and dynamic loading conditions **2** 1364-72
[29] Naseer A, Ahmad F, Aslam M, Guan B H, Harun W S W, Muhamad N, Raza M R, German R M J M and Processes M 2019 A review of processing techniques for graphene-reinforced metal matrix composites **34** 957-85
[30] Sreenivasulu B, Ramji B and Nagaral M J M T P 2018 A review on graphene reinforced polymer matrix composites **5** 2419-28
[31] Lee S K, Kim H and Shim B S 2013 Graphene: an emerging material for biological tissue engineering *Carbon letters* **14** 63-75
[32] Zhang B, Dembinski L, Coddet C J M S and A E 2013 The study of the laser parameters and environment variables effect on mechanical properties of high compact parts elaborated by selective laser melting 316L powder **584** 21-31
[33] Yadroitsev I, Pavlov M, Bertrand P, Smurov I J è A E d P and Rapide F 2009 Mechanical properties of samples fabricated by selective laser melting
[34] Riemer A, Leuders S, Thöne M, Richard H, Tröster T and Niendorf T J E F M 2014 On the fatigue crack growth behavior in 316L stainless steel manufactured by selective laser melting **120** 15-25
[35] Li P J M S and A E 2015 Constitutive and failure behaviour in selective laser melted stainless steel for microlattice structures **622** 114-20
[36] Delgado J, Ciurana J, Serenó L J V and Prototyping P 2011 Comparison of forming manufacturing processes and selective laser melting technology based on the mechanical properties of products: in this work, the superior property of the selective laser melting technology is presented by comparing four real parts manufactured using forming processes and selective laser melting technology and analysed for tension, compression and flexural **6** 167-78
[37] Alauddin S M, Ismail I, Zaili F S, Ilias N F and Aripin N F K J M T P 2018 Electrical and Mechanical Properties of Acrylonitrile Butadiene Styrene/Graphene Platelet Nanocomposite **5** S125-S9
[38] Nabi G, Malik N, Tahir M B, Raza W, Rizwan M, Maraj M, Siddiqa A, Ahmed R and Tanveer M 2020 Synthesis of graphitic carbon nitride and industrial applications as tensile strength reinforcement agent in red Acrylonitrile-Butadiene-Styrene (ABS) *Physica B: Condensed Matter*

[39] Scaffaro R, Botta L and Di Benedetto G 2012 Physical properties of virgin-recycled ABS blends: Effect of post-consumer content and of reprocessing cycles *European Polymer Journal* **48** 637-48

[40] Yadav D K, Srivastava R and Dev S 2020 Design & fabrication of ABS part by FDM for automobile application *Materials Today: Proceedings* **26** 2089-93

[41] Jyoti J, Kumar A, Dhakate S R and Singh B P 2018 Dielectric and impedance properties of three dimension graphene oxide-carbon nanotube acrylonitrile butadiene styrene hybrid composites *Polymer Testing* **68** 456-66

[42] Sharma V, Kapoor S, Goyal M and Jindal P 2020 Enhancement of the mechanical properties of graphene based acrylonitrile butadiene styrene (ABS) nanocomposites *Materials Today: Proceedings* **28** 1744-7

[43] Mura A, Adamo F, Wang H, Leong W S, Ji X and Kong J 2019 Investigation about tribological behavior of ABS and PC-ABS polymers coated with graphene *Tribology International* **134** 335-40

# Performance assessment of conceptual bifacial solar PV system in varying albedo conditions

**S Sreenath[1], K Sudhakar[2, 3,*] and Yusop A F[2]**

[1] Renewable Energy and Energy Efficiency Research Cluster, Universiti Malaysia Pahang, 26600 Malaysia
[2] Faculty of Mechanical and Automotive Engineering Technology, Universiti Malaysia Pahang, 26600 Malaysia
[3] Automotive Engineering Centre, Universiti Malaysia Pahang, 26600 Malaysia

*Corresponding author's email: sudhakar@ump.edu.my

**Abstract:** There is a rapidly growing interest in bifacial module based solar PV. Unlike conventional solar PV modules, the energy output of the bifacial PV module is affected by the reflectance of the ground surface (albedo). This paper aims to assess the performance of conceptual bifacial solar PV systems in varying albedo conditions. Besides, the comparison of energy generation between monofacial and bifacial solar PV system is carried out. In this regard, four sites on an educational campus having different albedo factor is chosen. The annual energy output, specific yield, and performance ratio are estimated using PVSyst simulation software. The energy output of 164.5 MWh/year is predicted for the minimum albedo factor of 0.15. While 175.8 MWh/year is expected for the maximum albedo factor of 0.85. The specific yield for Case 5 (bifacial) is 10.58 % higher than that estimated for Case 1 (monofacial). Based on annual energy generation, the site of Case 4 is the best location for the solar PV system among the studied cases. It is concluded that setting up of bifacial solar PV module on the ground surface with a high value of albedo factor is suitable to maximize energy generation.

**Keywords:** Albedo; Bifacial PV Module; Specific Yield; Performance Ratio; PVSyst.

## 1. Introduction

Solar PV systems are widely employed in renewable energy technology for the past decades. As per the IRENA report, the installed capacity of the solar PV system crossed 580.159 GW in 2019, and the share of solar PV system reached 22 % of the installed capacity of renewable energy systems [1]. The PV modules are the main component of the solar PV system and it performs the function of converting the sun's energy into electricity. The conventional PV modules can generate electricity from the surface facing the sun only and those are called mono facial PV modules. A bi-facial solar PV module can generate power from both sides i.e. the side facing the sun and the side facing away from the sun. The bifacial solar PV modules are relatively new in the solar market [2]. The current solar market is dominated by crystalline silicon (cSi) based monofacial PV modules. By 2023, it is expected that the demand for bifacial modules will reach as high as 40 GW [3]. The absorption of light occurs on both sides of bifacial solar PV modules. The solar irradiation reflected from the ground is also utilized for energy generation. The incident solar energy on the bifacial PV module varies with the value of ground reflectance (albedo). Hence, the value of albedo plays an important role in the amount of solar irradiation reflected on the backside of the bifacial PV module.

Albedo is defined as the ratio between solar irradiation reflected from the ground and the global solar insolation incident on the location. The value of albedo for a black surface is 0 as it absorbs all light rays falling on it. A perfect reflector (white surface) has an albedo value of 1. A bifacial solar PV system is fixed on different ground surfaces such as grass, bare soil, tarred road, snow-cover. Usually, the albedo varies between 0.2 (eg: urban locations, grasslands) and 0.8 for snow-covered ground [4]. The ground surfaces and its albedo factor for common surfaces are Asphalt (0.15), Grass (0.25), Concrete (0.35), and Aluminum (0.85). PVSyst software is capable to simulate a bifacial solar PV system for a selected site. Two pyranometers are used for the measurement of Albedo ( one pointing to the sun and the other facing the ground) [5]. The material selection and bifaciality are the two ways in which the energy output from solar PV cells can be enhanced [6]. In simulation software such as PVsyst, bifacial PV modules are characterized by a term called bifaciality factor. It is the ratio of the nominal efficiency at the rear side and the nominal efficiency of the front side. The ratio of the nominal power output of the PV module (kW) and the area of the PV module gives the nominal efficiency. So, the irradiance on the rear side can be added to the front irradiance based on the bifaciality factor during the simulation procedure.

Many literatures reported the performance of monofacial PV module based solar system using operational or simulated values. Kumar et al. analyzed the performance of conceptual off-grid solar PV system situated in educational institute using PVsyst modeling software and estimated that the performance ratio varied between 86 % in December and 64 % in April (due to the variation in ambient temperature) [7]. Sukumaran and Sudhakar assessed the performance of the 12 MW solar PV plant located in Cochin, India using PVSyst software and reported that the average performance ratio (PR) and final yield as 79.33 % and 4.31 kWh/kWp/day respectively [8]. Kumar and Sudhakar compared the real performance of the 10 MW solar PV power plant in Ramagundam, India with the predicted performance from PVsyst software. It was reported that the predicted energy generation (16.05 GWh/year) matched closely with the monitored energy generation (15.61 GWh/year) [9]. Arora et al. assessed the performance of 186 kW grid-connected solar PV system located in Haryana, India using PVsyst software and estimated a PR value of 85.6 % which is in close agreement with the estimated PR using operational values [10]. Boddapati and Daniel carried out the performance assessment of a single axis tracked 50 MWp solar PV power plant in Kurnool, India using monitored values and reported that the PR varied between 75.24 % (October) and 83.44 % (August) [11]. In the energy analysis carried out for a 10 MWp grid-interactive canal-top solar PV plant by Kumar et al. using multi-year operational data, it is reported that the monthly average values of final yield varied between 2.46 hours/day (Jul 2017) and 5.56 hours/day (May 2015) [12]. Ahmad et al. evaluated the energy performance of 250 kW grid-connected solar PV system (carport) using PVsyst software and disclosed that the performance ratio is 75 %, final yield is 3.80 kWh/kWp/day, and energy generation is 347 MWh/year [13]. Aryal and Bhattarai assessed the performance of 115.2 kW solar PV power plant situated in Kathmandu, Nepal using PVsyst simulation software and reported that the solar PV plant will inject 199 MWh of energy with a performance ratio of 83.5 %, and a specific yield of 1728 kWh/kWp respectively [14]. Abotaleb and Abdallah compared the performance of two bifacial-silicon solar PV modules, fixed at two different mounting configurations located in the desert climate of Qatar and reported that the energy yield of a typical latitude tilted bifacial PV module is about 14 % higher than the vertically tilted (east-west) configuration [15]. Katsaounis et al. compared the daily energy yield of a set of monofacial and bifacial PV module located on the western side of Saudi Arabia and reported that the gain of a bifacial PV module is 10 % and 15 % for a tilt angle of 25° and 45° respectively when compared to the monofacial PV module [16].

From past literature, it is observed that many authors reported the performance of conceptual solar PV systems using PVSyst software. However, only a handful of these literature studied the performance of bifacial solar PV modules. As per the author's knowledge, the performance of the bifacial solar PV system in varying albedo conditions of educational institutions has not been reported yet. The objectives of the present study are to evaluate the energy performance of monofacial solar PV system located in a university campus, to estimate the energy performance of bifacial solar PV system situated in the educational institution for different albedo conditions, to compare the energy generation between monofacial and bifacial solar PV system. The best location for solar PV among the selected locations based on annual electricity generation is also estimated.

## 2. Methodology

*2.1 Site selection*

A grid-connected solar PV system is proposed for the Pekan campus of Universiti Malaysia Pahang (UMP) which is situated in Pahang state, Malaysia. The co-ordinates of UMP Pekan are 3.54°N latitude and 103.43°E longitude respectively. It is located at about 12 meters from the sea level. The Pekan campus of UMP covers nearly 642 acres and has vacant areas for the installation of solar PV systems. The on-site solar energy generation supports reducing energy costs and green campus initiatives [17]. A 100 kW solar PV system is proposed for the educational institution. The sites for solar PV installation are chosen using Google map. Different foreground surfaces such as concrete rooftops, aluminum roofs, grasslands, and tarred surfaces (car parking) are selected (Table 2). Figure 1 shows the position of the chosen sites in the educational institution.

**Figure 1.** Position of selected sites for proposed solar PV in educational institution.

*2.2 System Description*

A 100 kW solar system is proposed in this study. The solar PV system mainly consists of PV modules, inverters, and the balance of the system. Since the proposed solar PV is grid-connected, battery banks are not included. This proposed solar PV system is assumed to be interconnected with the existing electrical facility. The chosen monofacial and bifacial solar PV modules have a power rating of 520 Wp. The conversion efficiency of the monofacial PV module is assumed as 20.39 % and the bifacial PV module is assumed as 22.48 %. The PV array is fixed at a 10° tilt angle and oriented towards the south direction [18]. An inverter with a nominal power of 100 kW is selected for the proposed system. The typical meteorological year climate data is collected from the database of PVGIS ( cell 3° to 4° latitude (111km) and 103° to 104° (111km) longitude). The selected location receives annual solar irradiation of 1830.4 kWh/m2 and observes the average annual temperature of 26.15 °C (Table 1).

**Table 1.** Monthly variation of climate parameters in the selected educational institution.

| Months | Global Horizontal Irradiation (kWh/m$^2$) | Ambient temperature (°C) | Wind Speed (m/s) |
|---|---|---|---|
| Jan | 141.7 | 24.69 | 4.65 |
| Feb | 149.7 | 25.32 | 4.10 |
| Mar | 147.7 | 25.04 | 3.45 |
| April | 186.0 | 26.70 | 3.66 |
| May | 159.2 | 27.12 | 2.19 |
| June | 149.5 | 27.27 | 2.13 |
| July | 170.1 | 27.13 | 4.30 |
| Aug | 171.1 | 26.49 | 2.97 |
| Sept | 164.0 | 26.67 | 2.84 |
| Oct | 160.8 | 26.36 | 2.28 |
| Nov | 113.2 | 25.79 | 3.35 |
| Dec | 117.5 | 25.17 | 4.41 |

### *2.3 PVsyst software*

PVSyst software is one of the widely used PV simulation software among energy professionals and academic researchers [8][10]. This software is capable of estimating the energy performance of solar PV systems such as grid-connected, stand-alone, pumping, and DC-grid. The simulation results from PVsyst are grouped in the form of a report which can be divided into three sections. All the main input parameters given for the simulation are provided in the first section. The second part includes the main results of the simulation, with a monthly table and graphs of normalized values. The third section contains an arrow loss diagram that showcases the system's behavior, with all detailed losses. PVSyst provides two different 2D models for estimating the energy performance of the bifacial solar PV system. The ''unlimited sheds'' model is suitable for fixed-tilt systems and the ''unlimited trackers'' model is appropriate for horizontal single-axis tracking systems. In the ''unlimited sheds'' 2D model, it is assumed that the bifacial PV array has a single orientation, equal distance with adjacent PV arrays, and a very large area (to exclude edge effect). To simplify the bifacial calculation, the mutual shading between PV arrays from a single direction is only considered [19]. In the simulation, the users are required to configure the bifacial PV module, inverter's specifications, number of modules in series, and number of strings. Also, there is an option to choose bifacial PV modules that are available in the databases. In the "General Simulation Parameter "dialogue box, the parameters such as albedo value, shading factor, mismatch loss, and other basic parameters for simulation can be configured. The shed parameters such as tilt & azimuth angle shed width, height above ground can be configured inside the dialogue box of ''unlimited sheds''.

### *2.4 Simulation procedure*

At first, the geographical location of the proposed solar PV plant is specified and the corresponding meteorological data is chosen from the built-in database. A basic solar PV system variant is defined by the user. It includes the orientation of the PV modules, the proposed power capacity, and the model of PV module & inverter [7]. Based on this, a basic configuration is proposed by PVsyst. In this study, this basic PV array configuration consists of 198 numbers of solar PV modules, 11 strings of 18 modules in series. Also, reasonable default values for all parameters that are required for the simulation are prefilled by the software. These parameters include far shadings, near shadings, specific loss parameters. The economic evaluation of the proposed solar PV system is not considered in the present analysis. The albedo values are adjusted deliberately based on the foreground surface of the selected solar PV site and its application. The simulation is carried out for 5 different cases as shown in Table 3. Except in case 1, bifacial PV modules are considered in all other cases. For each simulation, the albedo value is entered

IOP Conf. Series: Materials Science and Engineering 1078 (2021) 012033 doi:10.1088/1757-899X/1078/1/012033

depending on the type of ground surface on which the solar PV system is positioned. Ground reflectance values for the chosen ground surface is obtained from the webpage of PVSyst software [20].

**Table 2.** Different cases considered in the present study

| Studied Case | Site Name | PV Technology used | Type of sites | Site ground Surface | Albedo factor |
|---|---|---|---|---|---|
| Case 1* | NA | cSi-Monofacial | NA | NA | NA |
| Case 2 | Site I | cSi-Bifacial | Car park area | Asphalt | 0.15 |
| Case 3 | Site II | cSi-Bifacial | Open space | Grass | 0.25 |
| Case 4 | Site III | cSi-Bifacial | Concrete roof | Concrete | 0.35 |
| Case 5 | Site IV | cSi-Bifacial | Metal rooftop | Aluminum | 0.85 |

*Case 1 is considered as the base case, NA- Not Applicable

*2.5 Performance assessment*

In this study, the energy performance of bifacial and monofacial solar PV systems is assessed based on the values of energy generation, specific yield, performance ratio. These values are obtained through the PVSyst software simulations which are carried out for the five cases considered in the study [21].

(a) Annual Energy generation (Egrid): It accounts for the energy output from the solar PV system in a year (365 days). The total amount of electrical energy fed into the grid is the cumulative sum of daily generated energy (Ed). The yearly energy generation is given by equation 1.

$$E_{grid} = \sum_{d=1}^{365} E_d \tag{1}$$

(b) Specific yield (SYa): It is defined as the amount of electrical energy (kWh) that is generated for every kW of installed capacity over a period. Annual specific yield is the ratio of energy output from the solar PV system in a year (Egrid) and the installed capacity (Po) as given in equation 2.

$$SY_a = \frac{E_{grid}}{P_o} \tag{2}$$

(c) Performance ratio (PR): The ratio between the energy that is available for feeding to the grid and the theoretically possible energy generation from the solar PV array is termed as performance ratio. The theoretical energy generation is the product of solar irradiation (kWh/m2) on the PV array, area of PV array (m2), and PV module efficiency. It is also defined as the ratio between the final yield and the reference yield. PR is an important metric that is commonly used in the performance assessment of the solar PV system. Higher values of PR correspond to a higher amount of solar irradiation being transformed into electrical energy.

$$PR_a = \frac{E_{grid}}{\eta_{pv} \times H_t \times A_{pv}} \tag{3}$$

## 3. Results and discussions

The performance analysis of the proposed 100 kW solar PV system in five different cases is presented in this section. These results are obtained through simulation using PVsyst software.

*3.1 Energy analysis*

It is to be noted that the nameplate power of both the monofacial and bifacial PV module is intentionally chosen to be identical as 520 Wp. The monthly variation of energy generation and PR for Case 1 is depicted in Figure 2. The maximum energy generation is observed in April (15.62 MWh) and the minimum energy output is found in November (10.36 MWh). The annual energy generation in this base case is 159.01 MWh. The PR is almost constant with only a 2 % difference between maximum and minimum values. The variation of energy output and performance ratio for Case 2 is given in Figure 3. The energy generation varied between 16.14 MWh (April) and 10.7 MWh (November). The annual energy generation is estimated to be 164.45 MWh which is 5.44 MWh more than the base case value. This can be attributed to the enhanced solar energy input using bifacial solar PV modules. The average value of PR is 0.88 with the slightest monthly variation. Figure 4 depicts the monthly variation of energy generation and PR for Case 3. The maximum energy generation is observed in April (16.51MWh) and the minimum energy output is found in November (10.93 MWh). The PR reached 0.90 for most of the months with the least monthly variation. This value is higher than the base case PR value of 0.85. The variation of energy output and performance ratio for Case 4 is given in Figure 5. The annual energy generation is estimated to be 172.01 MWh which is 7.5 % more than the base case value. This can be attributed to improved performance using bifacial PV modules and increased ground reflectance (0.35). The value of PR varied between 0.91 (April) and 0.93 (June). Figure 6 depicts the monthly variation of energy generation and PR for Case 5. The maximum energy generation is observed in April (17.23 MWh) and the minimum energy output is found in November (11.39 MWh). The annual energy generation and average PR is 175.78 MWh and 0.94 respectively which is the highest value among the studied cases. This increase in values can be attributed to bifacial PV technology employed in high albedo conditions.

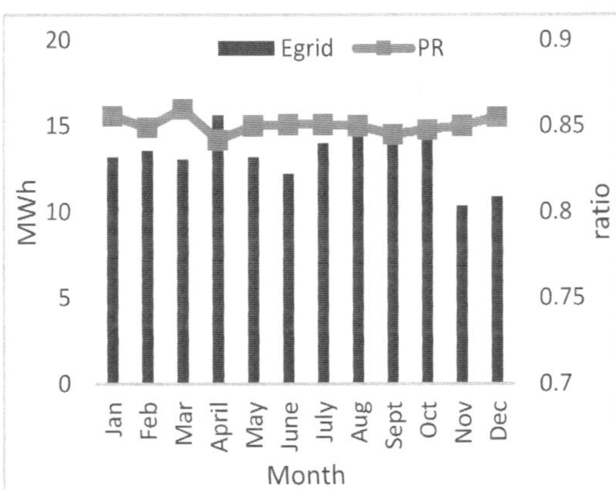


**Figure 2.** Monthly variation of energy generation and PR for case 1

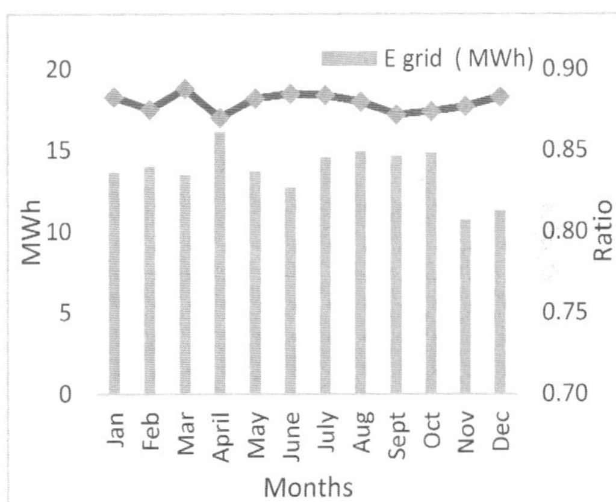


**Figure 3.** Monthly variation of energy generation and PR for case 2

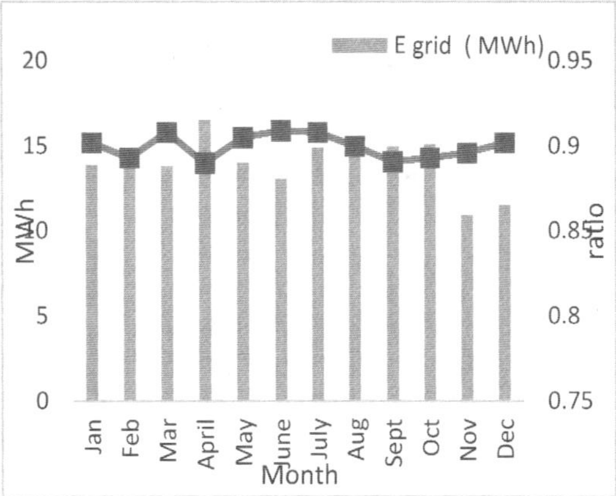


**Figure 4**. Monthly variation of energy generation and PR for case 3

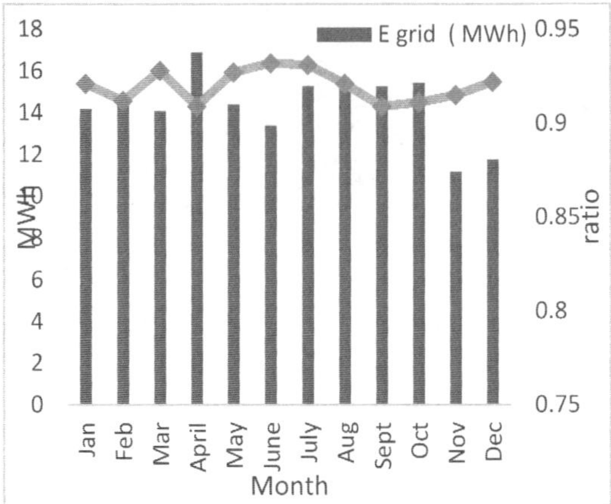


**Figure 5.** Monthly variation of energy generation and PR for case 2

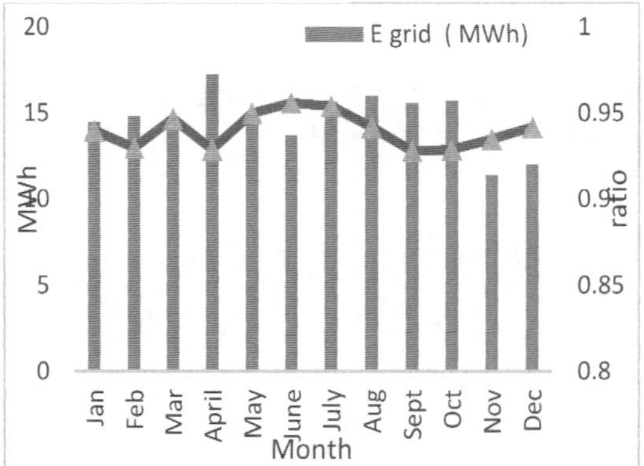


**Figure 6.** Monthly variation of energy generation and PR for case 5

*3.2 Discussions*

The comparison between 5 cases in terms of three technical parameters is provided in Figure 7. The monofacial solar PV system in Case 1 has the lowest value of energy generation, specific yield, and PR. Solar PV systems in all other sites have shown improvement in specific yield and PR which can be attributed mainly to the use of bifacial PV modules. Among these sites, a relative increase in energy generation is observed with the increasing value of albedo. Among the studied cases, it is observed that monthly variation in PR is minimal. Also, the peak value of energy generation occurred in April. This can be attributed to the least difference in the climatic condition and identical system design. Park et al. reported the specific yield for 1.5 kW bifacial solar PV system as 1351 kWh/kWp for an albedo of 0.21 (grey concrete) based on outdoor experiment [22]. Since higher energy output is possible for the same module area of bifacial PV modules, the cost needed for the balance of system (BOS) will be lesser than the conventional system [2]. The design and widespread implementation of bifacial solar PV is also influenced by the availability of an accurate reliable simulation model. Most of the PV simulation software is capable of energy estimation of monofacial solar PV only. Apart from PVSyst, software tools like SAM and Bifacial_Radiance can be used to design and predict the energy generation of bifacial solar PV system [23].

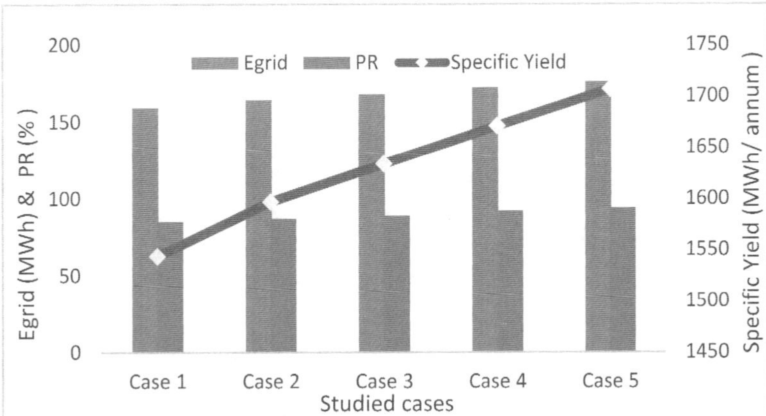


**Figure 7.** Comparison of performance of proposed solar PV system in the studied cases.

## 4. Conclusions

In this paper, the performance of bifacial solar PV systems in varying albedo is assessed using PVSyst software. The albedo of four common ground surfaces in an educational institution is considered. The energy generation and PR value for the monofacial solar PV system (Case 1) are 159 MWh/year and 0.85 respectively. The specific yield and PR at Case 5 (bifacial) is 10.58 % higher than that estimated for Case 1 (monofacial). An improvement in energy generation is observed with the increasing value of the albedo. The energy output for an albedo factor of 0.15 is 164.5 MWh/year. While 175.8 MWh/year is obtained for an albedo factor of 0.85. Based on annual energy generation, site 4 is the best location for the solar PV system among the studied cases. This can be attributed to two factors. Firstly, the bifacial solar PV module is used in Case 4. Secondly, the albedo value is highest for the ground surface of Site 4. The limitations of the study include the absence of analysis on the economic and social impact of the bifacial solar PV system in an educational institution. The selected sites receive an ample amount of rain. However, the impact of moisture on the albedo of selected ground surfaces is not considered in the present study. The share of the bifacial module in solar PV installation is expected to increase in near future. So, the present study will be beneficial to energy professionals, solar project developers, and investors. The future work is to carry out the economic analysis of such a solar PV system in Malaysian condition and compare it with monofacial solar PV plants. The effectiveness of the application of highly-reflective paints and/or building materials on foreground surfaces of the bifacial solar PV system can be studied further.

## Acknowledgments

The authors are grateful for the financial support provided by the Universiti Malaysia Pahang (www.ump.edu.my) through the Doctoral Research Scheme (DRS) and PGRS1903172. The assumptions and case studies reported in this article are only examples that are based on limited open-source information. The opinion, facts, findings, insights, and discussions in the document are solely of the authors and do not necessarily reflect the opinion of any organization involved directly or indirectly. Hence the authors are not responsible for any consequences thereof with the use of information presented in this work.

## Data availability

The data in this research can be shared upon request only due to the inclusion of third-party information.

## Conflict of interest statement

The authors declare that they have no known competing financial interests or personal relationships that could influence the work reported in the submitted manuscript.

## References


[1] "Renewable Energy Technologies," International Renewable Energy Agency (IRENA), 2019. www.irena.org/Statistics/View-Data-By-Topic/Capacity-and-Generation/Technologies (accessed Apr. 18, 2020).

[2] T. S. Liang et al. 2019 A review of crystalline silicon bifacial photovoltaic performance characterisation and simulation, *Energy and Environmental Science*, **vol. 12**, pp. 116–148

[3] M. Hutchins, "The long read: All about albedo," pv magazine, 2020.

[4] C. W. Thurston, "NREL Builds First Solar Database For Bifacial Albedo," CleanTechnica, 2019. https://cleantechnica.com/2019/02/18/nrel-builds-first-solar-database-for-bifacial-albedo/ (accessed Nov. 10, 2020).

[5] B. Marion, "Albedo Data to Facilitate Bifacial PV System Planning," 2019.

[6] A. Hasan and I. Dincer 2020 A new performance assessment methodology of bifacial photovoltaic solar panels for offshore applications, *Energy Convers. Manag.*, **vol. 220**, p. 112972

[7] R. Kumar, C. S. Rajoria, A. Sharma, and S. Suhag 2020 Design and simulation of standalone solar PV system using PVsyst Software: A case study, *Mater. Today Proc.*

[8] S. Sukumaran and K. Sudhakar 2017 Fully solar powered airport: A case study of Cochin International airport, *J. Air Transp. Manag.*, **vol. 62**, pp. 176–188

[9] B. Shiva Kumar and K. Sudhakar 2015 Performance evaluation of 10 MW grid connected solar photovoltaic power plant in India, *Energy Reports*, **vol. 1**, pp. 184–192

[10] R. Arora, R. Arora, and S. N. Sridhara 2019 Performance assessment of 186 kWp grid interactive solar photovoltaic plant in Northern India, *Int. J. Ambient Energy*

[11] V. Boddapati and S. A. Daniel 2020 Performance analysis and investigations of grid-connected Solar Power Park in Kurnool, South India, *Energy Sustain. Dev.*, **vol. 55**, pp. 161–169

[12] M. Kumar, S. S. Chandel, and A. Kumar 2020 Performance analysis of a 10 MWp utility scale grid-connected canal-top photovoltaic power plant under Indian climatic conditions, *Energy*, p. 117903

[13] O. A. Ahmad, W. H. Habeeb, D. Y. Mahmood, K. A. Jalal, and H. Sayed 2019 Design and performance analysis of 250 kw grid-connected photovoltaic system in iraqi environment using pvsyst software, *Indones. J. Electr. Eng. Informatics*, **vol. 7**, pp. 415–421

[14] A. Aryal and N. Bhattarai 2018 Modeling and Simulation of 115.2 kWp Grid-Connected Solar PV System using PVSYST, *Kathford J. Eng. Manag.*, **vol. 1**, pp. 31–34

[15] A. Abotaleb and A. Abdallah 2018 Performance of bifacial-silicon heterojunction modules under desert environment, *Renew. Energy*, **vol. 127**, pp. 94–101

[16] T. Katsaounis et al. 2019 Performance assessment of bifacial c-Si PV modules through device simulations and outdoor measurements, *Renew. Energy*, **vol. 143**, pp. 1285–1298

[17] C. Dondariya, D. Porwal, A. Awasthi, and A. Kumar 2018 Performance simulation of grid-connected rooftop solar PV system for small households: A case study of Ujjain , India, *Energy Reports*, **vol. 4**, pp. 546–553

[18] S. Sreenath, K. Sudhakar, and A. F. Yusop 2020 Technical assessment of captive solar power plant : A case study of Senai airport , Malaysia, *Renew. Energy*, **vol. 152**, pp. 849–866

[19] PVSyst, "PVsyst 7 Help," PVSyst, 2019. https://www.pvsyst.com/help/index.html?bifacial_module.html (accessed Dec. 10, 2020).

[20] PVSyst, "Albedo coefficient," PVSyst, 2020. https://www.pvsyst.com/help/albedo.htm%0A (accessed Nov. 10, 2020).

[21] K. A. Kumar, K. Sundareswaran, and P. R. Venkateswaran 2014 Performance study on a grid connected 20kWp solar photovoltaic installation in an industry in Tiruchirappalli (India), *Energy Sustain. Dev.*, **vol. 23**, pp. 294–304

[22] H. Park, S. Chang, S. Park, and W. K. Kim 2019 Outdoor performance test of bifacial n-type silicon photovoltaic modules, *Sustain.*, **vol. 11**

[23] C. Deline, S. A. Peláez, B. Marion, B. Sekulic, M. Woodhouse, and J. Stein 2019 *Bifacial PV System Performance: Separating Fact from Fiction*

# Review on thermal energy storage and eutectic nitrate salt melting point

**I Samylingam[1], K Kadirgama[1] , Navid Aslfattahi[2], L Samylingam[3], D Ramasamy[4,5,*], W S W Harun[4], M Samykano[4] and R Saidur[3]**

[1] Faculty of Mechanical and Automotive Engineering Technology, Universiti Malaysia Pahang, 26600, Pekan, Pahang, Malaysia
[2] Department of Mechanical Engineering, Faculty of Engineering, University of Malaya 50603 Kuala Lumpur, Malaysia
[3] Research Center for Nano-Materials and Energy Technology (RCNMET), School of Science and Technology, Sunway University, Bandar Sunway, Petaling Jaya, 47500, Selangor Darul Ehsan, Malaysia
[4]Department of Mechanical Engineering, College of Engineering, Universiti Malaysia Pahang, 26300 Gambang, Kuantan Pahang, Malaysia.
[5]Automotive Excellence Center, Universiti Malaysia Pahang, 26600 Pekan, Pahang, Malaysia

*Corresponding author's e-mail: deva@ump.edu.my

**Abstract.** In solar concentrates, thermal energy (TES) storage has a significant function (CSP). This article will discuss the forms of TES and TES content, focusing on the material for latent heat storage. Sensitive heat storage, latent heat storage and chemical reaction thermal storage classes can divide TES into three classes. Among the thermophysical properties for CSP is the latent heat storage content, which is used by more researchers. Dividing latent heat storage material into material for the organic, inorganic and eutetic phases change material (PCM). There are an advantage and downside to any form of storage material. Thermal stability at high temperature and low cost, however the specific heat capacity of the sensible heat storage material is very low compared to the latent heat storage materials..

**Keywords:** Thermal Energy Storage; Phase Change Material; Melting Point.

## 1. Introduction

Different renewable energy sources have been established since the leakage of toxic gas into the atmosphere and the death of fossil fuels have led. Unpredictable quantities of energy are required with the current population growth and global development. It has been established that solar energy is the highest abundance of renewable sources in the pursuit of alternative clean sources. Sun is providing us with 120 000 TW of energy per hour[1]. Several research studies have been performed with the aim of recognizing solar potential impacts at areas with high solar radiation such as North Africa, Southern Europe, the Middle East and South Asia; The Andosal solar energy facility, regarded as the largest CSP, provides 181.1 million kW/h of electricity up to 7.5 hours after sunset in Spain [2].

Two well-known technologies that use solar power to produce electricity are photovoltaic (PV) and CSP. A semiconductor unit turns sunlight into direct current in PV. The CSP plant consists of three main blocs that are the solar receptor, fluid transport system and power block for concentration and high temperature. The sun ray concentrates and coincides with a solar collecting field thermal fluid (HTF) which is subsequently deposited in a thermal storage tank and transported to the electricity generation block. CSP-TES plant is shown in Figure 1 [3]. HTF can also be stored as a TES unit to be used when sunlight is not available. The use of HTF as a TES substantially decreases fossil-supported fuel consumption, leading to a decrease in power costs [4]. PV is more suitable for low power demand at an affordable price. While CSP is more cost effective which is preferred to be used on a larger scale [5].

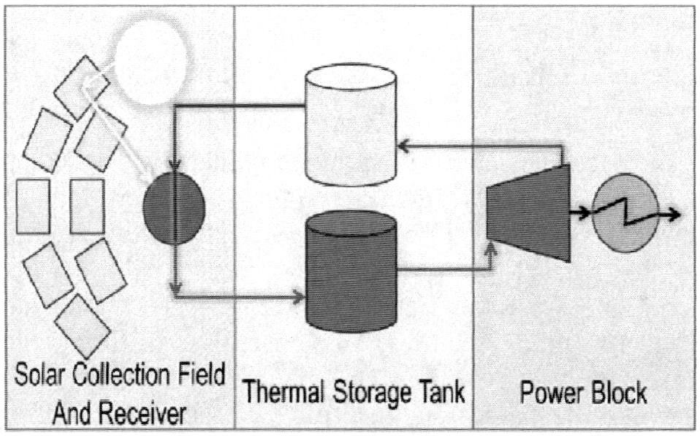


**Figure 1.** CSP-TES plant.

A big limitation of solar collection is that it is only available in the light of the sun. This is possible with HTF as TES. Thus, TES is an essential part of the CSP plant. For future CSP technologies, the architecture of the TES system plays a crucial role. When choosing TES sych as cost efficiency, high boiling point, high chemical stability, low melting point and good thermophysical properties, few parameters should be considered. The decrease in the cost of TES decreases CSP significantly. Improved CSP quality at the boiling point. TES that is high temperature chemical stable is important as they are less corrosive. TES is important. The smaller the TES melting point is the smaller the amount of heat energy is required to keep the HTF going, thereby reducing costs [4, 6, 7].

The choice of TES material has a vital part to be played by CSP technology, and therefore in this paper, the analysis will be discussed of different types of TES material for CSP applications.

## 2. Thermal energy storage

Three types of sensible heat storage, latent heat storage and chemical heat storage can be categorized as a TES system. The description of TES is shown in Figure 2.

### *2.1 Sensible Heat Storage*

In sensible heat storage (SHS) is stored by increasing the temperature of a solid or fluid thermal energy. Sensible heat can express thermal energy stored as [8]

$$Q = mC_p\Delta T \qquad (1)$$

Mass (kg) is represented by m, Cp is the specific heat capacity (kJkg$^{-1}$K$^{-1}$) and $\Delta T$ is the temperature raised during the charging process. Table 1 shows some of the solid and liquid materials used as sensible heat storage.

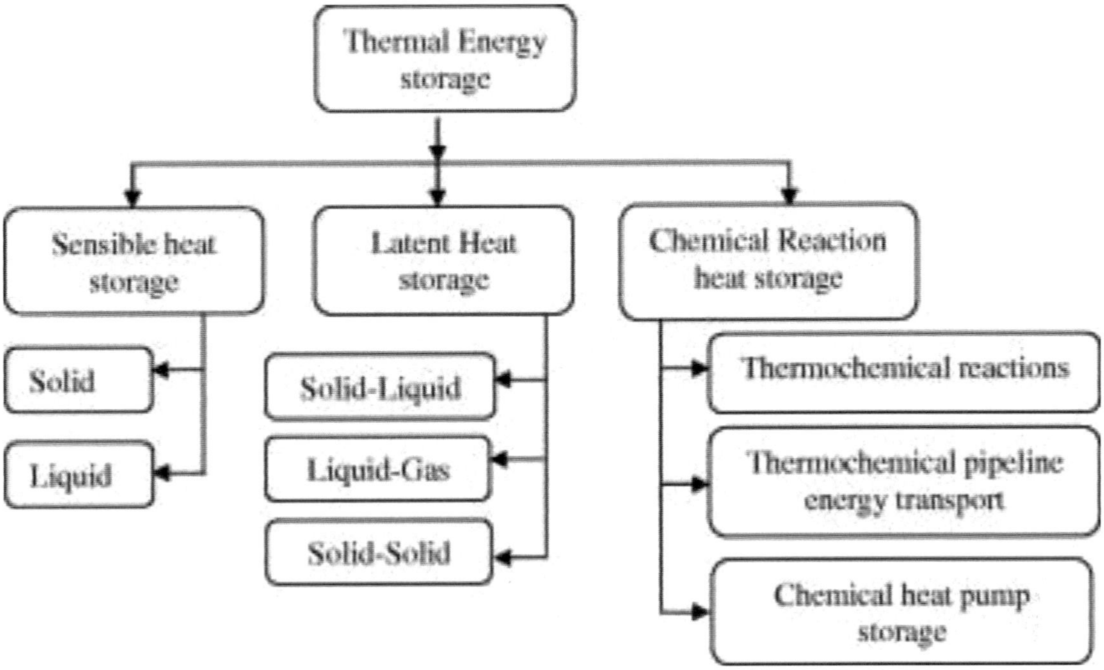


**Figure 2.** TES.

**Table 1.** List of solid and liquid materials.

| Medium | Fluid type | Temperature range (°C) | Density (kg/m$^3$) | Specific heat (J/kgK) |
|---|---|---|---|---|
| **Rock** | - | 20 | 2560 | 879 |
| **Concrete** | - | 20 | 1900-2300 | 880 |
| **Castable ceramic** | - | - | 3500 | 866 |
| **Water** | - | 0-100 | 1000 | 4190 |
| **Engine oil** | Oil | Up to 160 | 867 | 2200 |
| **Caloriea HT43** | Oil | 12-260 | 888 | 1880 |
| **Vegetable oil** | Oil | Up to 295 | 802 | 2509 |
| **Diphenyl oxide/ biphenyl** | Thermal oil | 12- 257 | 897 | 1630 |
| **Ethanol** | Organic Liquid | Up to 78 | 790 | 2400 |
| **Propanol** | Organic Liquid | Up to 97 | 800 | 2500 |

The heat storage sensitive increases with a higher storage medium temperature. Thus it is advisable that the medium has long-term thermal cycling stability, compatibility with containments, high specific heat capacity and above all low costs. Sensible heat storage can be defined as the storage of liquid media or solid media by way of thermal storage media.

*2.1.1 Liquid Media Storage*

The liquids in heat storage are abundant and efficient economically. Below are the advantages and drawbacks of some of the media chosen.

Water is one of the best storage media at low temperatures. It has more precise heat than other materials and is available cheaply and widely. Because of its high vapor pressure, however, expensive insulation and containment resistant pressure are required for applications at high temperatures. A large variety of temperatures can use water, say 25-90°C. Water stores 250 kJ/kg or 2.5 x $10^5$ kJ/m$^3$ when the temperature changes 60°C. Hot water for washing, bathing, etc is required, and is often used in space heating radiators. Water can be used as a storage device and a transport means of electricity, for example, in a solar power system. It is a widely used storage medium for warm water and solar heating applications [9, 10] . As the design requirements for water storage media have a great deal of data published [9-11], due to their simplicity. Due to the flood forces which ensure the highest temperature at the top and the lower temperature at the bottom of the tank thermal stretching or thermocline in a solar water heat storage tank may be created. Stratification is achieved by the absence of mixing during storage, obtaining a dual advantage[9, 10]: i) enhance energy efficiency by loading at a lower temperature instead of a mixed storage temperature; and ii) maximizing the energy produced if the inlet fluid of the collector is added for a charge.

A variety of materials such as steel, aluminum, reinforced concrete and fiber glass, form the water storage tanks. The tanks have glass wool, mineral wool or polyurethane insulation. The tanks used range in size between hundreds of liters and several thousand cubic metres. Underground natural aquifers were considered for large-scale storage applications. Aquifers contain groundwater and are geological formations that provide a possible way for heat storage for long periods. The medium for water storage includes gravel or sand saturated with water. There are large storage volumes as standard aquifer measurements vary from hundreds of thousands up to millions of cubic metres. For example, for every 10 K temperature difference, 105 $m^3$ of aquifer material can store approximately 3 MJ of heat. This form of storage is suitable for saisonal storage, i.e. heat transfer from summer to heating. Due to its large nature, small loads, e.g. individual homes, cannot be stored on the water. There has been widespread recognition of the utility of aquifer thermal energy storage and many experimental and theoresponsible studies have been carried out [12]. However, a more thorough computer simulation analysis is required to determine the optimal system configuration, because of the many different systems and physical parameters. Due to its low cost properties, high input/output rates and large volume, the attractiveness of aquifer storages.

Solar ponds are a simple and inexpensive way to absorb and store vast volumes of solar energy in the form of low-thermal energy (50-95°C). They have potential in room heating and cooling, in the supply of heat and power for the industrial process. Four specific factors can be used to identify solar ponds as a a) conveyance, (b) a gelled and non-gelled collector or storage, (c) multilayer, or non-partitioned and (d) separate storage or storage pond. Most research activities are currently, however, focused on the non-convective solar pond salt gradient [13]. A density gradient in this form of the solar pond is formed with the use of salt water (or seawater) whose concentration increases with surface depth. The salt most frequently used in this type of pond is sodium chloride (NaCl) and magnesium chloride ($MgCl_2$). The black or dark bottom of the salt gradient pool absorbs solar radiation with the consequent rise in water temperature to 95°C. The absorption of thermal energy from the lower layers of the pool will effectively be accomplished without disrupting the upper layer. The biggest solar pond in the United States is a 3,354$m^2$ experimental plant in El Paso, Texas, which has been running successfully since its launch in 1986. The pond operates an organic turbine generator of 70 kWe and a 5 000 gallons a day desalination plant while supplying a neighboring food processing company with process warmth. In its heat storage zone, the pond reached and sustained temperatures above 90°C, produced over 100 kWe during peak power output and produced over 80,000 gallons of potable water over 24 hours [13].

Petroleum-based oil and molten salts are the most often suggested water substitutes. The thermal power is 25 - 40% of the weight of water. These replacements, however have lower steam pressure than water and can work at temperatures greater than 300°C. Due to stability and safety purposes, oil is limited to less than 350 °C and can be fairly costly. Therminol and Caloria-HT are some of the oil candidates considered [9]. Some molten mixtures of inorganic salts for high temperatures have been proposed (300°C and above). The first element is sodium hydroxide that can be used for temperatures of up to 800°C and has a melting point of 320°C [14]. It is highly corrosive, however and difficult at high temperatures to contain. As potential sensible heat storage media, liquid metals are also listed.

*2.1.2 Solid Media Storage*

Strong materials such as rocks, metals, asphalt, sand, bricks and so on can be used for low or high-temperature thermal energy storage. The energy may then be retained at low temperatures or high temperatures because these materials do not freeze or boil. Thermal energy as a sensible solid thermal heat can be retained to prevent the difficulties of the high vapor pressure of the water and the disadvantages of other liquids. In addition, their jar should not spill the solids. Cast iron is the top product in the list of solid materials that are used in critical thermal storage [14]. Therefore, it is far more costly than brick or stone, and the payback period is also longer. As the storage material due to its low cost, pebble beds or rock piles are usually favored.

The rock bed or rock battery is made up of a bed of rock material that can be packed loosely and which can be used to transport heat. The heat energy in the packaged bed is retained by pushing warm air into the bed and is recirculated to the heated bed by the ambient air. In addition to the thermophysical properties of the stone, the energy contained in a packaged bed storage system depends on many parameters, including the rock size and shape, the density of packaging, heat transfer fluid etc. Solar energy can also be contained in soil or in rocks and is useful for use in buildings. This form of storage is mostly used in combination with solar air heaters for temperatures of up to 100°C [15]. The characteristic size of the rock pieces used usually varies between 1 and 5 cm. About 300 - 500 kg of rock per square meter of solar collector area for space warming applications is an approximated thumb rule followed for sizing. The rock and the concrete are stored at around 36 kJ/kg or 105 kJ/$m^3$ with a temperature change of 50°C. The storage of rock or steep beds can also be used up to 1000°C for much higher temperatures. To explain the geometric and thermal properties of packed litters, King and Burns [15] have several features such as particulate size, vacuum fraction, bed cross sectional surface and bed length, superficial air velocity and the Reynolds number. Fluidized beds like rock beds can be used for solar applications of medium, moderate and high temperature [16].The exchange rate of heat between the heat transfer fluid and the storage medium in fluidized beds is much faster than in rock beds, which can be of benefit to a variety of applications. The thermal stored fluidized bed can also be used for the recovery of waste heat.

Heat storage can be found both in new and existing buildings as well as in water and air systems. Ceramic brick composed of olivine, magnesite, microthermal, and feolite is the most common storage medium. Construction and structure cement with active or passive storage designs may also be used. Manufacturers continue to make changes in isolation and storage materials. Floor warming is the most common method of construction mass used in thermal storage. The ground becomes a large radiating surface at low temperatures where concrete serves as a medium for heat storage. The floor can be heated with water, direct wiring or air ducts with heat transfer fluid. Water is the most commonly used medium of heat transfer. Many floor warm buildings don't use on peak electricity to charge the storage, but they can be modified to use on peak heating through a control system. A Swiss producer with applications in the UK, Spain, Korea and Japan has designed a state-of-the-art phase-change material for floor storage heating [14]. Inertia building and hollow core construction are two other categories of building mass storage. Controls are built to reduce heating energy costs by adding the total building mass to heat decreases over peak periods. The use of building mass inertia is a technique that does not include additional HVAC equipment but involves precise controls. In Sweden, core ventilation systems with hollow air circulated via floor slabs, connecting it to the building mass, were developed[17].

Effects for energy stockpiling are mostly either inorganic salts or metals [18, 19], as suggested for high-temperature (120-1400°C). Examples of these metals included aluminum, magnesium and zinc. When high thermal conductivity is priortised compared to the cost, the use of a metal medium can be advantageous. The use of solid industrial waste as a store for energy storage may include copper slag, iron slag, cast iron slag, aluminum slag and copper chips.

In general, sensible heat storage materials gain from high thermal stability and inexpensive materials with exception of thermal oils and liquid metals. Although the heat-storage materials are responsive, the thermal energy storage density is 50-100 times smaller than the latent heat. The stability of the temperature during discharge is its major downside. Over time the release phase is constant, the outlet temperature of HTF decreases steadily.

*2.2 Thermochemical energy storage*

In a chemical reaction, the chemical heat storage mechanism depends on the energy released and consumed by the reform and breakup of molecular bonds. The heat stored is governed by the amount of storage content, the heat reaction and the conversion scope.

$$Q = a_r m \Delta h_r \tag{2}$$

$a_r$ is fraction reacted and $\Delta h_r$ is endothermic heat of reaction,

The significant benefit of such energy storage is that the energy density is theoretically high. The chemical storage heat content is shown in table 2 [20-25]

**Table 2.** Chemical heat storage material.

| Compound | Reaction | Energy storage density | Reaction Temperature ($^0$ C) |
|---|---|---|---|
| **Ammonia** | $NH_3 + \Delta H \leftrightarrow 1/2N_2 + 3/2\, H_2$ | 67 kj/mol | 400-500 |
| **Calcium Hydroxides** | $Ca(OH_2) \leftrightarrow CaO + H_2O$ | 3 GJ/m$^3$ | 500 |
| **Calcium carbonate** | $CaCO_3) \leftrightarrow CaO + CO_2$ | 4.4 GJ/m$^3$ | 800-900 |
| **Metal hydrides** | $Metal\, xH_3 \leftrightarrow yH_2 + (x-y)H_2$ | 4 GJ/m$^3$ | 200-300 |
| **Magnesium oxide** | $MgO + H_2O \leftrightarrow Mg(OH_2)$ | 3.3 GJ/m$^3$ | 250-400 |

The maximum TES density and capacity to store TES with low thermal loss is the chemical thermal energy storage device. But the TES material, like $Mg(OH)_2$, is charged and the grain grows as decomposition takes place, leading to low porosity.

Thermochemical storage (TCS) production is undergoing a basic laboratory stage, and is far from validated design and material to be commercially transferred. However, TCS is being examined in detail because when the chemical reaction partners are separately stored and because it may provide outstanding volumetric energy density[26], it is almost uncommon to store energy. Gas-solid reversible reactions are especially ideal for heat storage due to the easy isolation of the gas emitted during heat absorption. Diverse types of reaction systems for gas-solid are being explored: metal hydrides (80–400°C), decarboxylation of metal carbonates (100–950°C), dehydration of metal hydroxides (250-800°C), thermal desoxygenation of metal oxides (600–1000°C). While reaction enthalpy (400-1100kWhm$^{-3}$ depending on temperature) is generally extremely high, a feasible energy density is between 200 and 500 kWh. There are primarily two reasons for this: one is that the solid must be in powder or pellets; second, the reactants (solids and gas) must in most cases, be contained in different tanks (closed systems).

In the EU FP7, CaO/H2O responses and open redox cycles are studied by the projects TCS-Power and RESTRUCTURE (2011-2014) [27]. Particularly important in steam power generation applications are the heat storage systems based on metal hydroxides, where water (steam) reacts with metal oxide (e.g. CaO). However many issues have been reported and remain unresolved for the majority of them (e.g. tendency to agglomerate solid, weak heat transport properties, low reactive kinetics, potential dehydration crystallizations, and high-temperature sintering). As it was named, thermochemical storage has a significantly large storage density that accommodates restrictions in plant space and a loss-free and long-term storing option.

Unfortunately, since the technology was in its infancy, its cyclic stabilization and its simple incorporation with solar power concentration are still doubtful. It shows promise: a large variety of temperatures, from 50°C to more than 1000°C, are used to provide a thermochemical storage environment at which sensible and latent heat will waste substantial quantities of energy in the form of radiation. This means, thermochemical storage with solar tower receptors that can reach very high temperatures between 1300 and 1500°C can be implemented in an effective manner [27].

### *2.3 Latent Heat Storage*

Latent heat storage (LHS) materials maintain heat in their latent temperature during a constant temperature operation, such as phase shift. Solid-liquid is the most common phase transition. While special latent thermal energy is lower for solid-solid phase transformation, it does not require leakage nor encapsulation. Phase change of liquid-gas has the highest latent phase heat, but it is not always used as a result of the large shift in storage material [28]. TES stored latent heat is expressed as

$$Q=mL \tag{3}$$

m represents mass and L is specific latent heat.

Latent heat storage materials are commonly known as phase change materials (PCM). PCM can be classified into organic, inorganic and eutectic.

#### *2.3.1 Organic PCM*

The organic PCM contains compounds parrafin and non-parrafin. The organic PCM has temperatures of phase change between 18°C and 30°C close to the normal thermal range of the human body. Although chemically stable and non-toxic, non-corrosive and easy to use, they decay at high temperatures and have very low thermal conductivity.

Parrafin is a widely used organic PCM and consists of a combination of primarily CH3-(CH2)-CH3 straight n-alkanes. Upon crystallization of the chain (CH3), a huge amount of latent heat is released. The melting point also increases as the number of atoms. Parrafin is cheaper, odorless and non-corrosive to supercool. They are chemically stable below 500°C with changes in small volumes after melting. Their vapor pressure in the melting form is low and is not separated in phase. Parrafin has a low thermal conductivity and low density, the main disadvantage. [29, 30]. The most critical group of materials for PCM storage is the non-parrafins. Research on organic material and the ester, alcohol, fatty acid and glycol that are suitable for energy storage was carried out by Sauhney and Buddhi [31] and Abhat et.al [32], Organic materials are generally heat-fused, flammable, low thermal conductivity, low flammable, different toxicity levels and not stable at high temperatures. Table 3 shows some latent heat and non-parrafin material.

The dependence of the heat potential of n-alcanes on the number of atoms C between 19 and 48 was assessed by C.M.L. Atkinson et al [33]. Detailed analyses of the available liquid heat-capacity data for the alkanes can be found in the following publications: M. In 1996 and 2001 Zábransky'et others were focused on IUPAC-supported Projects (IUPAC) [33-35]. A number of researchers have currently been widely involved in the production of different heat storage materials on the thermal properties of paraffin waxes. All these studies have shown the high latent heat storage capacity between 200 kJ $kg^{-1}$ to 250 kJ $kg^{-1}$ and a great number of thermal temperatures with a wide range of melts with thermal stability up to 250 °C is absorbed, stored and released by waxes during phases transitions between the solid and the liquids. During repeated phase transitions, they show no segregation. They are inert chemically, odorless, durable, cheap, ecologically harmless, and not poisonous. They are not harmful. Different enthalpic fusion and melting points for the same alkanes were recorded in different studies. For instance, n-octadacane has been measured with fusion enthalpy 233.8 kJ $kg^{-1}$ (35.6 °C) [36], 228.2 kJ $kg^{-1}$ (28.1°C) [37] and 222 kJ $kg^{-1}$ (31.0°C). These fluctuating values can reflect changes in physical characteristics of various crystalline structures and polymorphism according to n-alkan molecular parameters and heating and cooler conditions during DSC analysis [38].

The melting point for paraffin wax increases and the inducted dipole attractions between n-alkane chains, e.g. n-dodecan melting point (12 C atoms), increase to -9.6°C for each while n-octacanosane melting point (28 C atoms) increases by 62°C for each. Due to the conformational differtances of specific specimens, the measured Hfusion values do not display such a frequent increase in magnitude. Reseacrhs have shown that each n-alcane crystallization temperature, which corresponds to a minute of overcooling during freezing, is slightly lower than its melting temperature. As well as supercooling, the poor thermal conductivity (k) of the paraffin waxes (approximately 0.2$Wm^{-1}$ $K^{-1}$, which requires improvement in order to increase their skills in thermal application [39-42], is another drawback.

**Table 3.** Non parrafin material.

| Material | Melting point ($^0$ C) | Latent heat (kJ/kg) |
|---|---|---|
| **Isomalt** | 147 | 275 |
| **Pentaerythritol** | 187 | 255 |
| **Phenol** | 41 | 120 |
| **Glycerin** | 17.9 | 198.7 |
| **Cetyl alcohol** | 49.3 | 141 |
| **Diphenyl amine** | 52.9 | 107 |
| **Bees Wax** | 61.8 | 177 |
| **Hydrophosporic acid** | 55 | 213 |
| **Benzoic acid** | 121.7 | 142.8 |
| **Quinone** | 115 | 171 |
| **Stearic acid** | 69 | 209 |
| **Polyethylene glycol 600** | 12.5 | 129.1 |
| **Polyethylene glycol 35000** | 68.7 | 166.9 |

Fatty acids represent organic PCMs and have an enormous TES potential for widely used thermal storage applications [43]. Because of their reducibility from vegetable and animal oils, fatty acids have many TES benefits and are environmentally friendly [44, 45]. On the other hand, their principal nuisance is characterized as low corrosiveness, bad smell and high rates and their capacity to use these properties is restricted. The other downside is the problem of leakage during the melting process with regard to fatty acids [46]. The highest melting points and latent heat in organic PCMs are sugar alcohols. Due to temperature changes, the thermal storage systems for applications such as solar or waste heat recovery for medium temperature (90°C-250°C) are suitable. [47]. The cost and non-toxicity of alcohols are minimal. They also exhibit polymorphism and they can occur in two or more crystalline states. The intake of alcohol in TES due to polymorphism is seriously impaired. Polymorphs of various types may have important physiochemical properties variations. The stability of alcohols is low due to oxidation of the environment by oxygen, causing changes in phase temperature and fragile material with reduced storage capabilities [47]. The change in temperature in the Polyethylene glycol (PEG) process is near to the ambient temperature, increasing with the molecular weight of the polymer. PEG is unique to other organic PCMs because it dissolves in water while the hydrophobic ones in nature. One of PEG's big issues is the highest supercooling of all organic PCMs. The difference between the smelting and freezing point can reach as high as 30°C-40°C, which is not ideal.

### *2.3.2 Inrganic PCM*

Metallic and salt hydrates are known as inorganic products. Usually this material is used as organic matter decomposed at hiogh temperatures in high-temperature operations. As a PCM the advantages of metals and alloys are high heat storage capacity and maximum thermal conductivity per unit volume. Metals and alloys are expensive, but the TES criterion is the best choice. Metals and alloys have limited heat energy storage capacity per unit weight. They therefore have the issue of excess weight [48]. Cárdenas and León have identified several high-temperature potential PCM metals and metal alloys that melt temperatures above 300°C that several other researchers have proposed and investigated. As with other authors, scientists neglect the use of metal alloys as a PCM despite their benefitial properties such as high thermal conductivity, high resistance to corrosion and minor changes in volume due to changes in phases and no subcooling. Therefore in certain applications, metal and metal alloys have more thermophysical properties and can compete with salts, particularly if weight is not a deciding factor [28]. Metal and alloys are experiencing changes in their microstructure due to precipitation, oxidation and separation. Inert atmosphere is important to prevent oxidation when using these materials. [49].

The high latent fusion heat, high thermal conductivity, high availability, low volume shift and low cost hydrates make salt hydrates one of the most attractive inorganic PCM in low temperature applications (-20-140°C). [50]. Inorganic salts with n water molecules forming a regular crystalline solid with the general formula AB*nH2O are salt hydrates similar to [51]. The type of bond formed usually is ion-dipole or hydrogen. Although in some systems the water is more directed toward the anion and salt cation in others the molecules of the system are correctly located and oriented [52]. Salt hydrates are slightly acidic and plastic compliant, taking into account the distribution and storage. The most popular salt hydrates available are CaCl2*6H2O and Na2SO4*10H2O. The salt hydration process may be influenced by the crystallisation/melting process. A hydrated crystal could dissociate from moles in water to a lower hydrate during melting (Eq 4) (Eq 5) [53]:

$$AB{*}nH_2O \rightarrow AB{*}mH_2O + (n\text{-}m)\ H_2O \quad (4)$$
$$AB{*}nH_2O \rightarrow AB + nH_2O \quad (5)$$

Low-specific heat, subcooling, insulation and fusion heat reduction through incongruous melting is among the disadvantages of salt hydrates. Mechanical turmoil, PCM encapsulation, incorporated thickeners and additional water use can improve the incongruous melting process. [54-57]. Mechanical turbulence, PCM encapsulation, integrated thickeners and additional water can improve the incongruous melting. Table 4 shows certain inorganic substances that may be used as PCM [58].

**Table 4.** Inorganic material.

| **Material** | **Melting Point ($^0$ C)** | **Latent Heat (kJ/kg)** |
|---|---|---|
| $LiClO_3.3H_2O$ | 8 | 155-253 |
| $CaCl_2.6H_2O$ | 29 | 190.8 |
| $LiNO_3.2H_2O$ | 30 | 296 |
| $LiNO_3.3H_2O$ | 20 | 189 |
| $CoSO_4.7H_2O$ | 40.7 | 170 |
| $Ni(NO_3)_2.6H_2O$ | 57 | 169 |
| $NaOH.H_2O$ | 64.3 | 273 |
| $MgCl_2.6H_2O$ | 117 | 167 |
| $LiNO_3$ (Lithium nitrate) | 253 | - |
| $KNO_3$ (Pottasium nitrate) | 333 | 266 |
| $NaNO_3$ (sodium nitrate) | 307 | 172 |
| $Ca(NO_3)_2$ (Calsium nitrate) | 561 | - |
| $MgCl_2$ (Magnesium Chloride) | 714 | 452 |
| $Na_2CO_3$ (Sodium Carbonate) | 854 | 275.7 |

*2.4 Eutectics PCM*

Eutectic is a minimum melting composition of two or more components that during crystallisation melts and freezes each to form a component crystal mixture. The inorganic, organic and/or organic combination may be used in this component. [59].

The use of individual composite fatty acids also has a practical constraint on achieving comfort requirements such as home heating / cooling, textiles and indoor air conditioning, because of their phase-in-transition temperature. By combining two or more fatty acid components with eutetic mixtures, the correct temperature can be mitigated until each component has a sharp, freezing peak temperature and eutectic temperatures [60]. Zhao et al. produced a phase chart and showed their transitional properties of the phase shift for a series of binary and ternary eutestic fatty acids derived from the four typical bioPCM (stearic acid, palmic acid, myric acid and lauric acid) [61]. Li et al. [62] also documented a series of eutectic binary fatty acids, showing that the TES applications were good thermal efficiency. It is noticeable that the eutectical mixtures do not address issues like supercooling and phase separation.

Organic-organic eutectics are ideal for TES applications of low to moderate temperature thanks to their advantage of congruent melting, eco-friendly and non toxic bio-design, minute degree of supercooling, low vapor pressure during phase transformation, compatibility and durability with building material. In addition a large range of eutectic PCMs can be adapted to enhance the adaptability of PCMs for TES applications in the light of the requested thermophysical properties of LHTES. [63]. Cai et al. investigated the thermophysical properties of the prepared form of stable, acid-based Eutectic Fiber (PET). Eutectic PCMs with different quantities of PET has a working temperature range of 35-53°C [64]. Kant et al. have produced ternary eutectics of fatty acids for low-temperature applications with an experimental temperature range between 14 and 21°C [65]. Sharma et al. [66] indicated that myrtical acid and palmic acid were thermally reliable for long term use after the thermal analysis of these fatty acids following 1,500 melting and freezing cycles.

Huang K et al. [67] reported that the desired floor heating PCMs must have an ideal melting temperature of 26 to 32°C, given that the heating system usually has a heating surface temperature of 22 to 26°C. Furthermore, the convenient indoor temperature for the thermal storage materials in energy efficient building fields like wallboards for the phases change is usually allowed in the range of about 16-25°C. In other words, it is possible to consider and apply wallboards to decrease the temperature variance inside a room and to increase comfort inside the indoor atmosphere [68, 69] if PCMs have a melting temperature of about 16-25°C. Boulard Tet al [70] also summarized that PCMs with a melting temperature of approximately 20-28°C can by way of encapsulation in flate plastic bags or packaging into specially built polyethylene bags, considered as promising materials for the solar greenhouse heating system. Literature from Mondal S [71] and Sarier N et al [72] suggests the most productive and acceptable candidates for temperature control and isolation of cloth and textiles in PCMs with a melting temperature between 15 and 35°C. These PCMs may store or released redundant energy to adjust humans' temperature and by means of phase change processes, improve thermal comfort when the temperature of the environment is higher or lower than the temperature of human comfort.

Literature have successfully studied preparation, phase diagrams, eutetic mass ratios and thermal energy storage propagations for the binary fatty acids eutectics (e.g capric acid-lauric acid, capric acid-myristic acid , lauric acid-stearic acid and myristic acid-stearic acid, capric acid-palmitic acid , lauric acid- palmitic acid) and part of ternary fatty acid eutectics (e.g capric acid-lauric acid- palmitic acid, capric acid-myristic acid-palmitic acid, lauric acid-myristic acid-stearic acid) but also some of these have been successfully investigated in a lot works of literature. The following summarizes the cooling peak temperature ranges of binary and ternary fatty acid eutectics investigated by the researchers according to literature data and research papers [67-69, 73-79]: (1) 20-25°C: including eutectics of the capric acid-lauric acid, of the capric acid-lauric acid- palmitic acid and of the capric acid-myristic acid-palmitic acid; (2) 25-30°C: including eutectics of the capric acid-myristic acid, of capric acid-palmitic acid and of the capric acid-palmitic acid-stearic acid; (3) 30-35°C: including eutesctics of the lauric acid-myristic acid-palmitic acid, of the myristic acid-myristic acid-stearic acid and capric acid-stearic acid, of which (4) 35-40°C: including eutectics of the lauric acid-palmitic acid-stearic acid, of the lauric acid-myristic acid and of the lauric acid-palmitic acid; (5) 40-45°C: including eutectics of the lauric acid-stearic acid, (6) 45-50°C: including eutectics of stearic acid-myristic acid-palmitic acid, myristic acid-palmitic acid and stearic acid-myristic acid. (7) >50°C. Objective energy consumption criteria can obviously be considered and extended for three kinds of fatty acid eutectics namely capric acid-lauric acid, capric acid-lauric acid- palmitic acid and capric acid-myristic acid- palmitic acid eutectics. In addition, there is also very few candidates for use as PCMs in this temperature range for thermal-regulated fibers and textile applications.

$Mg(NO_3)_2.6H_2O$ (MNH) and glutarin acid (GA) which have a melting enthalpy of 189.0 J/g higher than the raw material are produced by Suimin Li et al [80] organic-inorganic eutetic PCM. Due to the characteristics of the structure and the crystallisation behaviour, latent heat increases are associated with crystalline eutectic MNH-GA processing. Moreover, eutectic MNH-thermal GA's stability has been improved as the melting temperature decreased by less than 66.7°C dehydration. Karthick et al. [81] developed a PCM eutectic mixture to improve copper indium diselenide PV System efficiency. The module's temperature decrease was noted up to 9°C compared to the reference module. Liu, Yang, and $Na_2SO_4 \cdot 10H_2O$-$Na_2HPO_4$-$12H_2O$ and $Na_2CO_3.10H_2O$ $Na_2HPO_4.12H_2O$ have been characterized as eutectics of salts hydrated to find their use in the thermal storage of various solar systems. The eutectics showed no phase separation and the supercooling level was reduced significantly[82].

Figure 4 offers a description of the storage density, operating temperature temperature, longevity and trade status of the TES. Owing to high variations in enthalpy during the phase transition, the storage density of the LHS medium is typically higher than that of the SHS, while the thermo-chemical energy density is greater than that of all other modes. Different thermochemical energy storage materials are currently under research and no such device is available on the market. The commercial feasibility of the LHS is constrained by material characteristics and is thus mainly commercialized during the production process as opposed to SHS. Figure 4 demonstrates that in comparison to LHS materials which constitute one quarter of their lifetime and thermochemical energy storage, the life cycle of the materials is about 20 years, which is approximately 1/10 of the lives of SHS materials [83]. The LHS range of operating temperatures among those different types of TES is most versatile, with operating temperatures significantly lower for sensible and thermo-chemical energy storage [84]. The benefit of latent thermal energy over sensible thermal storage is that it has a density of approximately 10 times greater than TES, an output of 50-150 kWh/t, and an efficiency of ~75%-96% [85, 86].

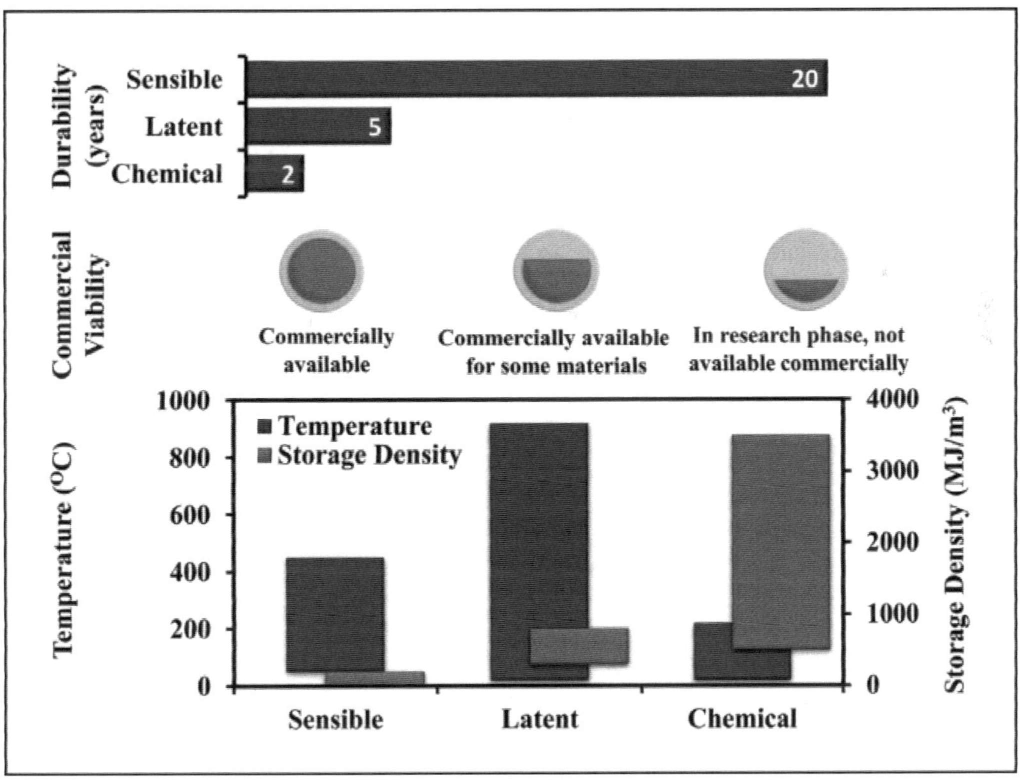


**Figure 3.** Overview of PCMs for TES.

## 3. Molten salt

The term 'molten salt' refers usually to a substance produced by the melting of inorganic salt or the mixture of salt. A number of salt mixtures have been produced to achieve unique purposes such as fluorides, carbonates, nitrates, etc. For example, molten salt fluids are needed in the CSP plant in order to get a lower freeze point in the salt tunnel. Solar salt, a combination of $NaNO_3$ and $KNO_3$ with a melting point of 236°C is used in a commercial system. Equally significant for optimum construction and cost-effective operations for use in a melting salt reactor are properties such as high temperature stability (up to 750°C), radiation stability and neutronical properties. Table 5 shows some commercially used molten salt [87].

**Table 5.** Molten salt and applications.

| Molten Salt | Eutectics | Applications |
|---|---|---|
| Fluorides | $LiF–BeF_2$<br>LiF-NaF-KF | TES , HTF,Molten Salt Reactors (MSR), |
| Carbonates | $NaBF_4–NaF$<br>$Li_2CO_3$-$Na_2CO_3$ | High temperature Fuel Cells, Molten Salt Oxidation |
| Chlorides | $KCl–MgCl_2–NaCl$<br>KCl–NaCl | TES, HTF, MSR |
| Nitrates | $NaNO_3–KNO_3–NaNO_2$<br>$KNO_3–NaNO_3$ | HTF and TES |

Molten salts, poor thermal conductivity and liquid leakage have the two major drawbacks of reducing their use for heat storage. A very successful way of halting the leakage of the liquid was a chemically inert stable composite PCM formation, which spread through porous support. Many porous supports, including metal foams, porous carbon materials like expanded graphite (EG), graphite foams, carbon nanopodes, porous ceramics, montmorillonites, vermiculite and diatomite have been successfully used for PCMs [88]. EG is one of the best possible PCM supports due to its excellent thermal conductivity, good corrosion resistance and low density. Xiao et al [89] considered EG to help solve the issue of low thermal molten salt leakage. A cascading latent heat storage (CLHS) system has been recently recorded to serve as an efficient TES alternative that ensures the best use of the storage material. In the CLHS system, the temperatures of phase change were ordered from the lowest to the highest according to their melting temperatures.

Zhong et al [90] synthesized by means of impregnation process porous heterogeneous composite PCMs from EG and binary molten salts. The binary salts of this substance were $LiNO_3$-KCl, $LiNO_3$-$NaNO_3$ and $LiNO_3$-Nacl. The double percentage of molten salt was between 77.8 and 81.5% and the percentage of encapsulative efficiency was between 72.8 and 78.8%. Via EG impregnation, binary molten salts have increased their thermo-conductivity 4.9-6.9 times. The porous heterogeneous composite's phase change properties displayed high temperature stability that can retain their properties even after 100 cycles. Solar power plants can retain their properties at high working pressure in closed systems via the high temperature heat transfer fluid. Due to its better thermal properties than synthetic mineral oil, molten salts may be used as direct HTF in CSP plants etc. Real, due to their high freezing temperature points, the direct applications as thermal transmission fluids of molten salts are inhibited.

Sang et al [91] have researched $K_2CO_3$-$Li_2CO_3$-$Na_2CO_3$ as PCM and MgO as ceramic supporting material for ternary carbonate-based composite phase change material (CPCM) as (CSM). The composite composed of 50 wt percent salts was found to be 28 MPa highest mechanical power, 1.48 J/g·K special thermal heat capacity and 158.7 kJ/kg latent heat. These materials showed strong cycling efficiency and a wider range of operating temperatures and thus a broader prospect of application to mid and high temperatures A $NaNO_3$ CPCM was manufactured by Ye et al. [92]. Using the MgO as CSM and the multicarbon nanotubes as thermal conductivity enhancement material (TCEM), the addition of carbon nanotubes greatly improved the thermal conductivity of the composite and increased the degree of its rise, as the charge of carbon and the operating temperature increased. Qin et al. [93] investigated the use of diatomite as the same CSM process of the sodium composite, and the findings suggested that excellent chemical and physical stability was obtained at the insulation of 880°C and that a combination of mechanical strength and energy is preferred by the diatomite of 45% in the composite.

For some time, chloride salts were presented as LHTES media. The TES device accompanying the solar thermal power plant design study from Clayton suggested two ternary chloride salt pulses as PCMs: KCl-NaCl-$MgCl_2$ and KCl-NaCl-$CaCl_2$ [94]. As a promising LHTES medium, Kamimoto recommended two pure chloride salts, $FeCl_3$ and the double salt, $KMgCl_3$ [95]. However as nitrates currently are, nothing has been done to fix their possible use as critical heat-storage media in the molten state. Marcus recommended chlorides to be potential LHTES while also neglecting their use as sensible liquid storage media [96]. Tanaka's study identified the use of the eutetic system KCl-LiCl as a pcm in a TES dual-stage superheating device where pressurized water is the main stage of sensible heat storage [97]. The TES media can be called various chlorides. There are main categories of an alkaline metal chloride, alkaline earth, transition metal, p-bloc metal, and rare earth metal [98]. The general categories of content are: For the purposes of this analysis we are limited to salts that have higher melting temperatures – more than 400 degrees Celsius (Ferric Chloride, for example, is ignored since its melting temperature is 307.6 degrees Celsius [99]). In combination with eutectic mixtures, the resulting salt systems could melt in temperatures below 400°C; these materials are expected to continue to be used as fuse-sensitive storage media at elevated temperatures. Furthermore many of the pure salts assumed to be displayed cost well beyond that of other candidate salts (e.g., rare earth metals chlorides); as such, salts costing up to a minimum of kg.

### *3.1 Nitrates Salt*

Molten alkali nitrates, especially mixtures of sodium and potassium nitrate with other additives, have been used very successfully as fluids for energy storage or heat transfer in some cases. In this section , we will study the thermophysical and thermochemical properties of binary, ternary and quarternary molten nitrates. Eutectic points occur in nitrate salt mixtures where the system solidifies with some other composition at a low temperature at a particular chemical composition. Two components of the liquid mixture are still in the balance at the eutectic point and each of the components is crystals, so if the temperature is lower than the eutectic temperature, each component will begin to crystallise out of the mixture [100].

Probably the most studied is the binary mixture method, but the detailed phase diagram is not yet accepted. In order to understand the processes for the improvement of the working of the nitrate salt mixture it is important to define thermal properties of the pure components and binary mixtures, known as binary systems and molten salt mixtures for commercial and essential applications, such as fluid for transport, which functions as a hot storage medium in the molten salt TES tank. $NaNO_3$ shrinks to 307°C and $KNO_3$ shrinks to 337°C.[101]. It is widely accepted that, the composition of 50 mol % $KNO_3$ and 50 mol % $NaNO_3$ or 54wt % $KNO_3$ and 46wt % $NaNO_3$ has always been accepted that the minimum melting point is near 220°C [102]. As for the binary mixture of 60-40 weight percentage for $NaNO_3$-KNO3, commo9nly known as solar salt's melting point is around 221°C [103]. While this salt mixture does not have the lowest melting point, its low investment cost also emphasises it. There are some disadvantages to this binary nitrate mixture in which this molten salt can quickly freeze and obstruct the pipeline in the evening or winter. Therefore, to deal with this issue, some auxiliary costs can be added and the expenditure will increase. The molten salt mixture of binary $KNO_3$-$NaNO_3$ is stable up to temperatures of 500°C and the melt varies only slightly in weight, although some NOx gases have formed over this temperature. Another of the most studied heat transfer fluid additives is $LiNO_3$, which is believed to boost the efficiency of molten salts, extend the working range of temperatures with a low melting point and greater thermal stability. Experimentally identified that at a composition of 44 mol percent $NaNO_3$ and 56 mol percent $LiNO_3$, the eutectic temperature for this binary system was $196^0$ C.

At a composition of 46 mol percent $NaNO_3$, the measured eutectic point was defined. The melting point is $183^0$ C [101].

Ternary $NaNO_3$ and $KNO_3$ mixtures have significantly smaller melting point than binary salt and alkaline earth-nitrates. For a more dynamic combination of salt, such as eutectc ternary, a more dramatic reduction of melting points can be seen. Several experiments and patents in the melt chemistry literature were conducted in order to lower $KNO_3$ and $NaNO_3$ eutectical temperature [104]. In addition, by adding the other salt or regulating the nitrate-nitrite ratio, the upper working temperature is extended [105]. Bradshaw and Siegel mentioned that the inclusion or replacement of $LiNO_3$ in the ternary system of $KNO_3$-$NaNO_2$-$NaNO_3$ is an acceptable function for improving the thermal stability range of salt, although its price is the crucial problem related to this additive [106]. The thermal power of the ternary mixture of $LiNO_3$-$NaNO_3$-$KNO_3$ salt, according to researcher Reddy [107]. The heat potential in the liquid state is very stable and, with little slope, increases linearly with temperature. $Ca(NO_3)_2$ shows the highest potential as an additive in solar energy. Its low cost and ability to minimise the melting point of alkaline nitrates show the tremendous potential of the additive to be integrated into new energy storage formulations of molten salts and make it a primary candidate for the substitution of binary solar salts [108].

Some molten salt has been used as heat transfer fluid, including binary mixture, ternary mixture, and more dramatically is a quarternary mixture. The potential for salt development is to enhance its physiochemical characteristics, in particular its melting point, thermal stability and heat power, to create new quaternary mixtures or to integrate new components. Quaternary salt, which contains different $NaNO_3$, $LiNO_3$, $KNO_3$, and $Ca(NO_3)_2$ weight ratios, has stronger chemical and physical properties than the commonly used binary solar salt (60 wt % $NaNO_3$-40 wt % $KNO_3$) [109]. Bradshaw et al. [106] are known to reveal the mixture of anhydrous compositions belonging to the quaternary system of $LiNO_3$-$NaNO_3$-$KNO_3$-$Ca(NO_3)_2$, said that these compositions have a melting point <$95^0$ C and thermal stability up to $500^0$ C. The melting point of a salt mixture based on sodium nitrate and potassium nitrate is reduced by molten salt mixtures that include calcium nitrate and lithium nitrate. The transition observed in the thermal analysis did not melt the $KNO_3$ content properly with the other components at temperatures above $300^0$ C and the signal agrees with values recorded in the literature and experimental values obtained for potassium nitrate melting [110]. For this mixture, the maximum stability is $580.36^0$ C with decomposition beginning at $469.56^0$ C. Table 6 shows some of the binary and ternary nitrate salt compositions and their melting point.

**Table 6.** binary and ternary nitrate salts' melting point.

| $LiNO_3$ | $NaNO_3$ | $KNO_3$ | $Ca(NO_3)_2$ | Melting point | Reference |
|---|---|---|---|---|---|
| 56 mol% | 44 mol% | | | 196 $^0$ C | [103] |
| | 60 wt% | 40 wt% | | 221 $^0$ C | |
| 25.9 wt% | 20 wt% | 54.1 wt% | | 118 $^0$ C | [111] |
| 30 wt% | 52 wt% | 18 wt% | | 120 $^0$ C | [112] |
| 25.9 wt% | 54.1 wt% | 20.06 wt% | | 118 $^0$ C | [109] |
| | 15 wt% | 43 wt% | 42 wt% | 175 $^0$ C | [113] |
| | 10 wt% | 27 wt% | 63 wt% | 130 $^0$ C | [114] |
| | 9 wt% | 30 wt% | 61 wt% | 133 $^0$ C | [108] |
| | 13 wt% | 40 wt% | 47 wt% | 160 $^0$ C | |
| | 12 wt% | 44 wt% | 44 wt% | 130 $^0$ C | |
| | 7-34 wt% | 43-50 wt% | 16-48 wt% | 120-190 $^0$ C | [112] |

**Table 7.** Quaternary Nitrate salts' melting point.

| Nitrate Salt System | Melting point |
|---|---|
| **$LiNO_3$ $NaNO_3$ $KNO_3$ $NaNO_2$** | 99 |
| **$LiNO_3$ $NaNO_3$ $KNO_3$ $KNO_2$** | 96.4 |
| **$LiNO_3$ $KNO_2$ $KNO_3$ $NaNO_2$** | 100.3 |
| **$LiNO_3$ $NaNO_3$ $KNO_3$ $Ca(NO_3)_2$** | 109 |

## 4. Conclusion

It's evident from this review that choosing a good TES material benefits the CSP process. An ideal TES material has good thermophysical properties such as high energy storage, good thermal stability, non toxic, non corrosive, non flammable, high thermal conductivity, low melting point, compatible with container and etc. Although it's difficult to identify a perfect TES many researchs have been conducted to improvise current TESs to match the requirement. Water media storage is a better SHS compared to solid media storage as the temperature range and heat storage is higher. Molten salts are more suitable PCM (especially eutectic nitrate salts) compared to other LHS as it has high thermal stability, high energy storage (compared to SHS), less corrosive and lower melting point. Among the eutectic nitrate salt quaternary salt has the lowest melting point compared to a single, binary and ternary molten salt. The addition or increasd weight percentage of $LiNO_3$ proven to be reducing the melting point of molten salt. For future studies, a review on other thermophysical properties such as specific heat, thermal stability and thermoconductivity of the eutectic nitrate salt can be compared.

## Aknowledgements

The author would like to thanks the Minister of Higher Education and Universiti Malaysia Pahang (UMP) to provide the financial assistant under grant no. FRGS/1/2019/TK07/UMP/02/3 and RDU192207

## References

[1] Glaser P E 1968 Power from the sun: Its future *Science* **162** 857-61

[2] Cavallaro F 2010 Fuzzy TOPSIS approach for assessing thermal-energy storage in concentrated solar power (CSP) systems *Applied Energy* **87** 496-503

[3] Lorentei L 2011 lvarez JL A, Blanco D. Performance model for parabolic trough solar thermal power with thermal storage: comparison to operating plant data *Sol Energy* **85** 2443-60

[4] Kearney D, Herrmann U, Nava P, Kelly B, Mahoney R, Pacheco J, Cable R, Potrovitza N, Blake D and Price H 2003 Assessment of a molten salt heat transfer fluid in a parabolic trough solar field *J. Sol. Energy Eng.* **125** 170-6

[5] Quaschning V and Muriel M B 2002 Solar power-photovoltaics or solar thermal power plants? *VGB POWERTECH-INTERNATIONAL EDITION-* **82** 48-52

[6] Statistics R C 2016 International Renewable Energy Agency (IRENA). ed)

[7] Kearney D, Herrmann U, Nava P, Kelly B, Mahoney R, Pacheco J, Cable R, Potrovitza N, Blake D and Price H 2002 Evaluation of a molten salt heat transfer fluid in a parabolic trough solar field. In: *International Solar Energy Conference,* pp 293-9

[8] Lane G A 1983 Solar heat storage: latent heat materials

[9] Duffie J A and Beckman W A 1974 Solar energy thermal processes. University of Wisconsin-Madison, Solar Energy Laboratory, Madison, WI)

[10] Wyman C, Castle J and Kreith F 1980 A review of collector and energy storage technology for intermediate temperature applications *Solar Energy* **24** 517-40

[11] Abhat A 1980 Short term thermal energy storage *Revue de physique appliquee* **15** 477-501

[12] Kangas M and Lund P 1994 Modeling and simulation of aquifer storage energy systems *Solar energy* **53** 237-47

[13] Lodhi M 1996 Solar ponds in alkaline lake and oil well regions *Energy conversion and management* **37** 1677-94
[14] Hasnain S 1998 Review on sustainable thermal energy storage technologies, Part I: heat storage materials and techniques *Energy conversion and management* **39** 1127-38
[15] King R and Burns A 1981 Sensible heat storage in packed beds. In: *Proc. Int. Conf. on Energy Storage, Brighton, UK,* pp 231-45
[16] Weast T and Shannon L 1980 Thermal energy storage systems using fluidized bed heat exchangers
[17] Anderson L, Bernander K, Isfält E and Rosenfeld A 1979 Storage of heat and coolth in hollow-core concrete slabs. Swedish experience, and application to large, American-style buildings. RLI Byggdata AB, Saltsjobaeden (Sweden); AB Strangbetong, Stockholm (Sweden ...)
[18] Shitzer A and Levy M 1983 Transient behavior of a rock-bed thermal storage system subjected to variable inlet air temperatures: Analysis and experimentation
[19] Geyer M 1991 *Solar power plants*: Springer) pp 199-214
[20] Hadorn J-C 2015 *Solar and heat pump systems for residential buildings*: John Wiley & Sons)
[21] Lovegrove K, Luzzi A and Kreetz H 1999 A solar-driven ammonia-based thermochemical energy storage system *Solar energy* **67** 309-16
[22] Hahne E 1986 Thermal energy storage some views on some problems. In: *International Heat Transfer Conference Digital Library*: Begel House Inc.)
[23] Kubota M, Yokoyama K, Watanabe F and Hasatani M 2000 Heat releasing characteristics of CaO/CaCO3 reaction in a packed bed for high temperature heat storage and temperature up-grading. In: *Proceedings of the 8 th International Conference on Thermal Energy Storage (Terrastock 2000),*
[24] Visscher K 2004 SolarCombi systems, heading for 100% solar fraction *ECN presentation given at VSK Utrecht*
[25] Kato Y 2004 Personal notes on solar chemical storage chapter. June)
[26] Cot-Gores J, Castell A and Cabeza L F 2012 Thermochemical energy storage and conversion: A-state-of-the-art review of the experimental research under practical conditions *Renewable and Sustainable Energy Reviews* **16** 5207-24
[27] Prieto C, Cooper P, Fernández A I and Cabeza L F 2016 Review of technology: thermochemical energy storage for concentrated solar power plants *Renewable and Sustainable Energy Reviews* **60** 909-29
[28] Cárdenas B and León N 2013 High temperature latent heat thermal energy storage: Phase change materials, design considerations and performance enhancement techniques *Renewable and sustainable energy reviews* **27** 724-37
[29] He B and Setterwall F 2002 Technical grade paraffin waxes as phase change materials for cool thermal storage and cool storage systems capital cost estimation *Energy conversion and management* **43** 1709-23
[30] Stamatiou A, Obermeyer M, Fischer L J, Schuetz P and Worlitschek J 2017 Investigation of unbranched, saturated, carboxylic esters as phase change materials *Renewable Energy* **108** 401-9
[31] Conference B W E A 1994 *Wind Energy Conversion: Proceedings of the... British Wind Energy Association Conference* vol 16: Mechanical Engineering Publications)
[32] Abhat A, Heine D, Heinisch M, Malatidis N and Neuer G 1981 Development of a modular heat exchanger with integrated latent heat energy store *ikes*
[33] Atkinson C, Larkin J and Richardson M 1969 Enthalpy changes in molten n-alkanes and polyethylene *The Journal of Chemical Thermodynamics* **1** 435-40
[34] Zábranský M, Ruzicka Jr V, Domalski E S and Majer V 1996 *Heat capacity of liquids: critical review and recommended values* vol 1: Amer Inst of Physics)

[35] Zábranský M, Růžička Jr V and Domalski E S 2001 Heat capacity of liquids: critical review and recommended values. Supplement I *Journal of Physical and Chemical Reference Data* **30** 1199-689
[36] Sarier N and Onder E 2007 The manufacture of microencapsulated phase change materials suitable for the design of thermally enhanced fabrics *Thermochimica acta* **452** 149-60
[37] Parks G S, Moore G E, Renquist M L, Naylor B F, McClaine L A, Fujii P S and Hatton J A 1949 Thermal data on organic compounds. XXV. Some heat capacity, entropy and free energy data for nine hydrocarbons of high molecular weight *Journal of The American Chemical Society* **71** 3386-9
[38] Chickos J S, Acree Jr W E and Liebman J F 1999 Estimating solid–liquid phase change enthalpies and entropies *Journal of Physical and Chemical Reference Data* **28** 1535-673
[39] Farid M M, Khudhair A M, Razack S A K and Al-Hallaj S 2004 A review on phase change energy storage: materials and applications *Energy conversion and management* **45** 1597-615
[40] Tyagi V, Kaushik S, Tyagi S and Akiyama T 2011 Development of phase change materials based microencapsulated technology for buildings: a review *Renewable and sustainable energy reviews* **15** 1373-91
[41] Khudhair A M and Farid M M 2004 A review on energy conservation in building applications with thermal storage by latent heat using phase change materials *Energy conversion and management* **45** 263-75
[42] Agyenim F, Hewitt N, Eames P and Smyth M 2010 A review of materials, heat transfer and phase change problem formulation for latent heat thermal energy storage systems (LHTESS) *Renewable and sustainable energy reviews* **14** 615-28
[43] Sarı A and Karaipekli A 2012 Fatty acid esters-based composite phase change materials for thermal energy storage in buildings *Applied Thermal Engineering* **37** 208-16
[44] Yan Q 2011 The thermal properties of shape-stabilized fatty acid mixtures used for wallboard *International Journal of Sustainable Energy* **30** 47-54
[45] Fauzi H, Metselaar H S, Mahlia T and Silakhori M 2014 Thermo-physical stability of fatty acid eutectic mixtures subjected to accelerated aging for thermal energy storage (TES) application *Applied thermal engineering* **66** 328-34
[46] Tumirah K, Hussein M, Zulkarnain Z and Rafeadah R 2014 Nano-encapsulated organic phase change material based on copolymer nanocomposites for thermal energy storage *Energy* **66** 881-90
[47] Solé A, Neumann H, Niedermaier S, Martorell I, Schossig P and Cabeza L F 2014 Stability of sugar alcohols as PCM for thermal energy storage *Solar energy materials and solar cells* **126** 125-34
[48] Liu M, Saman W and Bruno F 2012 Review on storage materials and thermal performance enhancement techniques for high temperature phase change thermal storage systems *Renewable and Sustainable Energy Reviews* **16** 2118-32
[49] Fernandez A I, Barreneche C, Belusko M, Segarra M, Bruno F and Cabeza L F 2017 Considerations for the use of metal alloys as phase change materials for high temperature applications *Solar Energy Materials and Solar Cells* **171** 275-81
[50] Milian Y E, Gutierrez A, Grageda M and Ushak S 2017 A review on encapsulation techniques for inorganic phase change materials and the influence on their thermophysical properties *Renewable and Sustainable Energy Reviews* **73** 983-99
[51] Hyun D C, Levinson N S, Jeong U and Xia Y 2014 Emerging applications of phase- change materials (PCMs): teaching an old dog new tricks *Angewandte Chemie International Edition* **53** 3780-95
[52] Mehling H and Cabeza L F 2007 Thermal energy storage for sustainable energy consumption. In: *Conference Proceedings,* pp 257-77
[53] Sharma S D, Kitano H and Sagara K 2004 Phase change materials for low temperature solar thermal applications *Res. Rep. Fac. Eng. Mie Univ* **29** 31-64

[54] Safari A, Saidur R, Sulaiman F, Xu Y and Dong J 2017 A review on supercooling of Phase Change Materials in thermal energy storage systems *Renewable and Sustainable Energy Reviews* **70** 905-19
[55] Vickers N J 2017 Animal communication: when i'm calling you, will you answer too? *Current biology* **27** R713-R5
[56] Cabeza L F, Svensson G, Hiebler S and Mehling H 2003 Thermal performance of sodium acetate trihydrate thickened with different materials as phase change energy storage material *Applied Thermal Engineering* **23** 1697-704
[57] Letcher T M, Law R and Reay D 2016 *Storing energy: with special reference to renewable energy sources* vol 86: Elsevier Oxford)
[58] Sharma A 2009 Tyagi, VV Chen, CR Buddhi, D *Renewable and Sustainable Energy Reviews* **13** 318-45
[59] Adams E 1947 Ceramic Abstracts Section
[60] Cai Y, Gao C, Zhang T, Zhang Z, Wei Q, Du J, Hu Y and Song L 2013 Influences of expanded graphite on structural morphology and thermal performance of composite phase change materials consisting of fatty acid eutectics and electrospun PA6 nanofibrous mats *Renewable energy* **57** 163-70
[61] Ayotamuno M, Kogbara R, Ogaji S and Probert S 2006 Petroleum contaminated ground-water: Remediation using activated carbon *Applied Energy* **83** 1258-64
[62] Li M, Kao H, Wu Z and Tan J 2011 Study on preparation and thermal property of binary fatty acid and the binary fatty acids/diatomite composite phase change materials *Applied energy* **88** 1606-12
[63] Zalba B, Marın J M, Cabeza L F and Mehling H 2003 Review on thermal energy storage with phase change: materials, heat transfer analysis and applications *Applied thermal engineering* **23** 251-83
[64] Cai Y, Ke H, Lin L, Fei X, Wei Q, Song L, Hu Y and Fong H 2012 Preparation, morphology and thermal properties of electrospun fatty acid eutectics/polyethylene terephthalate form-stable phase change ultrafine composite fibers for thermal energy storage *Energy conversion and management* **64** 245-55
[65] Kant K, Shukla A and Sharma A 2016 Ternary mixture of fatty acids as phase change materials for thermal energy storage applications *Energy Reports* **2** 274-9
[66] Sharma R, Ganesan P and Tyagi V 2016 Long-term thermal and chemical reliability study of different organic phase change materials for thermal energy storage applications *Journal of Thermal Analysis and Calorimetry* **124** 1357-66
[67] Yuan Y, Zhang N, Tao W, Cao X and He Y 2014 Fatty acids as phase change materials: a review *Renewable and Sustainable Energy Reviews* **29** 482-98
[68] Sarı A and Karaipekli A 2008 Preparation and thermal properties of capric acid/palmitic acid eutectic mixture as a phase change energy storage material *Materials Letters* **62** 903-6
[69] Shilei L, Neng Z and Guohui F 2006 Eutectic mixtures of capric acid and lauric acid applied in building wallboards for heat energy storage *Energy and Buildings* **38** 708-11
[70] Boulard T, Razafinjohany E, Baille A, Jaffrin A and Fabre B 1990 Performance of a greenhouse heating system with a phase change material *Agricultural and forest meteorology* **52** 303-18
[71] Mondal S 2008 Phase change materials for smart textiles–An overview *Applied thermal engineering* **28** 1536-50
[72] Sarier N and Onder E 2012 Organic phase change materials and their textile applications: an overview *Thermochimica Acta* **540** 7-60
[73] Sharma A, Shukla A, Chen C and Wu T-N 2014 Development of phase change materials (PCMs) for low temperature energy storage applications *Sustainable Energy Technologies and Assessments* **7** 17-21
[74] Karaipekli A and Sari A 2007 Capric acid and palmitic acid eutectic mixture applied in building wallboard for latent heat thermal energy storage

[75] Yanping Y, Wenquan T, Xiaoling C and Li B 2011 Theoretic prediction of melting temperature and latent heat for a fatty acid eutectic mixture *Journal of Chemical & Engineering Data* **56** 2889-91
[76] Sarı A 2006 Eutectic mixtures of some fatty acids for latent heat storage: thermal properties and thermal reliability with respect to thermal cycling *Energy conversion and management* **47** 1207-21
[77] Inoue T, Hisatsugu Y, Ishikawa R and Suzuki M 2004 Solid–liquid phase behavior of binary fatty acid mixtures: 2. Mixtures of oleic acid with lauric acid, myristic acid, and palmitic acid *Chemistry and physics of lipids* **127** 161-73
[78] Yang X, Yuan Y, Zhang N, Cao X and Liu C 2014 Preparation and properties of myristic–palmitic–stearic acid/expanded graphite composites as phase change materials for energy storage *Solar Energy* **99** 259-66
[79] Zhang N, Yuan Y, Wang X, Cao X, Yang X and Hu S 2013 Preparation and characterization of lauric–myristic–palmitic acid ternary eutectic mixtures/expanded graphite composite phase change material for thermal energy storage *Chemical Engineering Journal* **231** 214-9
[80] Li S, Lin S, Ling Z, Fang X and Zhang Z 2020 Growth of the Phase Change Enthalpy Induced by the Crystal Transformation of an Inorganic–Organic Eutectic Mixture of Magnesium Nitrate Hexahydrate–Glutaric Acid *Industrial & Engineering Chemistry Research* **59** 6751-60
[81] Karthick A, Ramanan P, Ghosh A, Stalin B, Vignesh Kumar R and Baranilingesan I 2020 Performance enhancement of copper indium diselenide photovoltaic module using inorganic phase change material *Asia- Pacific Journal of Chemical Engineering* e2480
[82] Liang L and Chen X 2018 Preparation and thermal properties of eutectic hydrate salt phase change thermal energy storage material *International Journal of Photoenergy* **2018**
[83] H Abedin A and A Rosen M 2011 A critical review of thermochemical energy storage systems *The open renewable energy journal* **4**
[84] Gil A, Medrano M, Martorell I, Lázaro A, Dolado P, Zalba B and Cabeza L F 2010 State of the art on high temperature thermal energy storage for power generation. Part 1—Concepts, materials and modellization *Renewable and Sustainable Energy Reviews* **14** 31-55
[85] Sarbu I and Sebarchievici C 2018 A comprehensive review of thermal energy storage *Sustainability* **10** 191
[86] Singh D, Zhao W, Yu W, France D M and Kim T 2015 Analysis of a graphite foam–NaCl latent heat storage system for supercritical CO2 power cycles for concentrated solar power *Solar Energy* **118** 232-42
[87] Kuchibhotla A, Banerjee D and Dhir V 2020 Forced convection heat transfer of molten Salts: A review *Nuclear Engineering and Design* **362** 110591
[88] Acem Z, Lopez J and Del Barrio E P 2010 KNO3/NaNO3–Graphite materials for thermal energy storage at high temperature: Part I.–Elaboration methods and thermal properties *Applied thermal engineering* **30** 1580-5
[89] Xiao J, Huang J, Zhu P, Wang C and Li X 2014 Preparation, characterization and thermal properties of binary nitrate salts/expanded graphite as composite phase change material *Thermochimica acta* **587** 52-8
[90] Zhong L, Zhang X, Luan Y, Wang G, Feng Y and Feng D 2014 Preparation and thermal properties of porous heterogeneous composite phase change materials based on molten salts/expanded graphite *Solar energy* **107** 63-73
[91] Sang L, Li F and Xu Y 2019 Form-stable ternary carbonates/MgO composite material for high temperature thermal energy storage *Solar Energy* **180** 1-7
[92] Ye F, Ge Z, Ding Y and Yang J 2014 Multi-walled carbon nanotubes added to Na2CO3/MgO composites for thermal energy storage *Particuology* **15** 56-60
[93] Qin Y, Leng G, Yu X, Cao H, Qiao G, Dai Y, Zhang Y and Ding Y 2015 Sodium sulfate–diatomite composite materials for high temperature thermal energy storage *Powder Technology* **282** 37-42

[94] Clayton J 1978 Research on energy storage for solar thermal conversion. Final report. Xerox Electro-Optical Systems, Pasadena, CA (USA))
[95] Kamimoto M, Tanaka T, Tani T and Horigome T 1980 Investigation of nitrate salts for solar latent heat storage *Solar Energy* **24** 581-7
[96] Marcus Y 1982 Thermal energy storage (in Molten Salt Technology, editor Lovering, DG). Plenum Press, London)
[97] Tanaka T 1980 Solar thermal electric power systems in Japan *Solar Energy* **25** 97-104
[98] Myers Jr P D, Bhardwaj A, Goswami D Y and Stefanakos E 2015 Chloride salt systems for high temperature thermal energy storage: properties and applications. In: *Energy Sustainability*: American Society of Mechanical Engineers) p V001T03A3
[99] Lide D R 1995 *CRC handbook of chemistry and physics: a ready-reference book of chemical and physical data*: CRC press)
[100] Prigogine I and Defay R Chemical Thermodynamics; Longmann: London, UK, 1973 *There is no corresponding record for this reference.[Google Scholar]*
[101] Bauer T, Laing D and Tamme R 2011 Recent progress in alkali nitrate/nitrite developments for solar thermal power applications *Molten Salts Chemistry and Technology* **9**
[102] Berg R W and Kerridge D H 2004 The NaNO 3/KNO 3 system: the position of the solidus and sub-solidus *Dalton Transactions* 2224-9
[103] Bauer T, Laing D and Tamme R 2010 Overview of PCMs for concentrated solar power in the temperature range 200 to 350 C. In: *Advances in Science and Technology*: Trans Tech Publ) pp 272-7
[104] Kramer C and Wilson C 1980 The phase diagram of NaNO3—KNO3 *Thermochimica Acta* **42** 253-64
[105] Cordaro J G, Rubin N C and Bradshaw R W 2011 Multicomponent molten salt mixtures based on nitrate/nitrite anions *Journal of solar energy engineering* **133**
[106] Bradshaw R W and Siegel N P 2009 Development of molten nitrate salt mixtures for concentrating solar power systems *SolarPACES, Berlin*
[107] Reddy N and Yang Y 2011 Potential of plant proteins for medical applications *Trends in biotechnology* **29** 490-8
[108] Bergman A, Rassonskaya I and Shmidt N 1955 unknown title (Russian) Izvest *Sektora Fiz. Khim Anal., Akad. Nauk SSSR* **16** 156
[109] Wang T, Mantha D and Reddy R G 2013 Novel low melting point quaternary eutectic system for solar thermal energy storage *Applied energy* **102** 1422-9
[110] Fernández A, Lasanta M and Pérez F 2012 Molten salt corrosion of stainless steels and low-Cr steel in CSP plants *Oxidation of metals* **78** 329-48
[111] Reddy R G 2011 Molten salts: thermal energy storage and heat transfer media. Springer)
[112] Bradshaw R and Meeker D 1990 High-temperature stability of ternary nitrate molten salts for solar thermal energy systems *Solar energy materials* **21** 51-60
[113] Menzies A W and Dutt N 1911 THE LIQUIDUS SURFACE OF THE TERNARY SYSTEM COMPOSED OF THE NITRATES OF POTASSIUM, SODIUM AND CALCIUM *Journal of the American Chemical Society* **33** 1366-75
[114] Jänecke E 1942 The quaternary system Na, K, Ca, Mg *NO3 and its subsystems (in German), Z. für Elektrochemie und angewandte physikalische Chemie* **48** 453-512

# Smart solution for low humidity problems using automatic ultrasonic humidifier (AUH)

**Md Nazmus Sakib[1,2], Md Mahmudul Hasan Sohel[3], Shahanul Islam[3] and Minarul Islam[4,*]**

[1] Faculty of Manufacturing and Mechatronic Engineering Technology, University Malaysia Pahang (UMP), 26600, Pekan, Pahang, Malaysia.
[2] Department of EEE, Faculty of Science & Engineering, International Islamic University Chittagong (IIUC), Chittagong-4318, Bangladesh.
[3] Department of Mechanical Engineering, Faculty of Mechanical Engineering, Khulna University of Engineering & Technology, Khulna, 9203, Bangladesh.
[4] Faculty of Electrical and Electronic Engineering Technology, University Malaysia Pahang (UMP), 26600, Pekan, Pahang, Malaysia

*Corresponding author's e-mail: sumankhancsejust@gmail.com

**Abstract.** Humidity has a substantial impact on biological and environmental factors. Improper relative humidity (too high or very low) can create discomfort for individuals and affect several systems and apparatus. On the other hand, the correct type of humidifying machinery and functions may facilitate reliable, economic, and trouble-free control of moisture. Humidity in the winter region is always too low, which causes anxiety. This study aims to develop a microcontroller-based system to identify and eliminate such low humidity problems. A control system is implemented to generate air humidity according to the need to decide whether the unit is working. A wireless sensor was used to measure the room's relative humidity and temperature. The Humidifier is activated whenever the relative humidity is less than that of the setting's value according to the standard humidity value. A humidifier is developed using vaporizing ultrasound, which can still be considered the automatic ultrasonic Humidifier (AUH). The setting is suitable in the winter zone or where the relative humidity is below the standard comfort value. In this study, data were recorded in Khulna, Bangladesh, between April and May of 2018. Most of the sensor's recorded values were greater than the standard comfort Relative humidity, so the Humidifier was disabled. Humidifier activated while the sensor recorded lower relative humidity value with comparison to the standard set value.

**Keywords:** Humidity; Smart Solutions; Automatic Ultrasonic Humidifier (AUH).

## 1. Introduction

Humidity is the concentration of water vapor present in the air. Water vapor, the gaseous state of water, is generally invisible to the human eye. Humidity indicates the likelihood for precipitation, dew, or fog to be present. Humidity depends on temperature and the pressure of the system of interest. The same amount of water vapor results in higher humidity in cool air than warm air. A related parameter is the dew point [1]. However, there are also a number of variables and individuals, which overlap, making it difficult and challenging to construct a proper humidifier unit [2]. Many instances may also be associated with breathing disorders, including hypertension, and some may result from depression. Besides, patients often will hyperventilate, triggering symptoms of, among many others, loss of feeling, light-headedness, and lack of concentration [3]. Air conditioning also decreases the temperature and reduces humidity, decreasing discomfort in the summer [4]. Studies suggest because when temperature rise is preserved around 35%-55%, individuals typically feel relaxed. Water evaporates very quickly from the body whenever the weather is clean, creating a sensation of coldness at temperatures of 23.89 ° C or even more [5]. Because the psychological health of relative humidity (RH) is sometimes viewed as a difference in temperature, through adequate humidity regulation throughout cooler altitudes, pleasant environments can be achieved [6]. During winter, once cold outdoor weather is warmed inside, the moisture can decrease as low as 10-20 % [7]. The dry air can cause health problems through dried away respiratory tract, including the nasal passages and mouth, and heart failure. That since the body's low humidity dampens moisture, the convection of tissues can arise [8]. The moist equilibrium may also be disrupted by low humidity and causing irritation and even droopy eyelids [9]. Dry air influences hardwood floors, allowing the joint to weaken and detach or parts to break. If moisture is insufficient, textbooks, documents, and works of art can even be compressed or warped and become fragile [10]. Lower temperatures can cause various problems, namely health issues, widespread destruction, and machinery harm [11]. The effect of low humidity on the human body is shown in Figure 1.

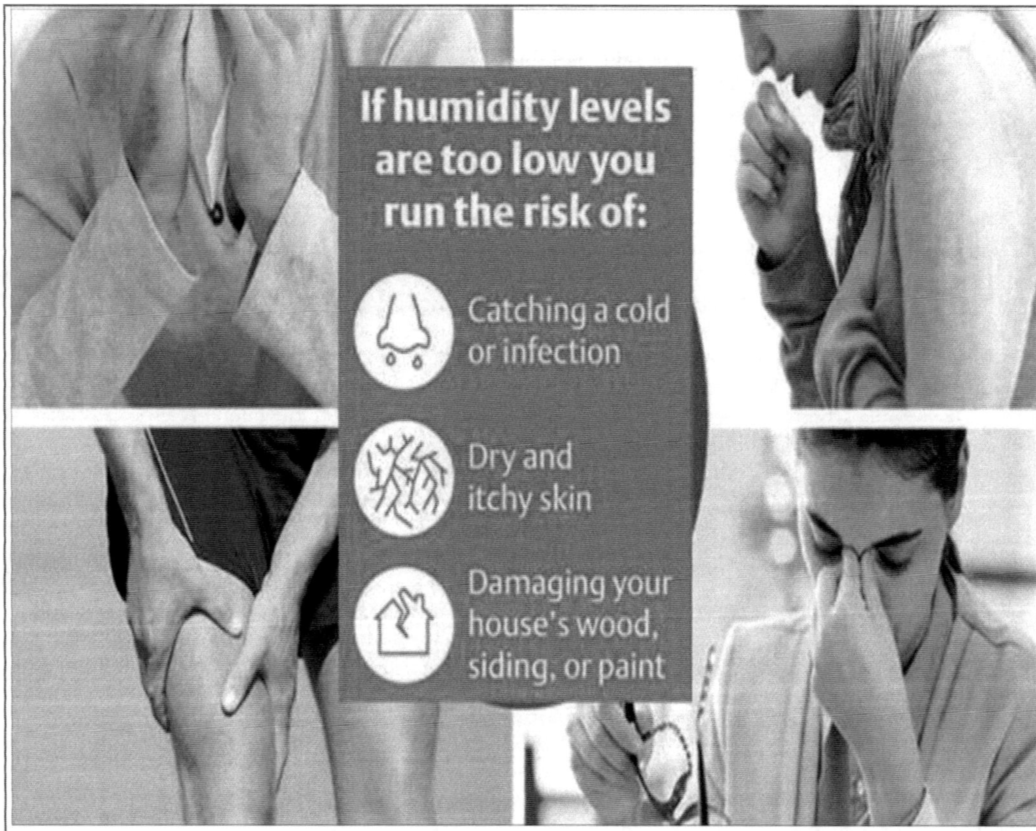


**Figure 1.** Effect of low humidity on the human body [12].

To avoid spikes that can destroy information technology (IT) machinery, the American Society of Heating, Refrigerating, and Air Conditioning Engineers (ASHRAE) has proposed a range of 45-55 % RH in data centres [13]. In museums and art galleries, humidifiers are available to protect fragile artworks and museum rooms at which inflammatory acne induced by warming is minimized for tourists' convenience and during winter [14]. Among several other exogenous conditions, weather, hygiene, wind speed, and heat flux, the need for moisture is known to become very necessary to note that humidity is probably the least apparent to sensory consciousness. Most individuals respond faster to fluctuations in climate, odors, or dense particles throughout the wind, draughts, or thermal gradient. During which humidity levels interact with all these factors, this becomes an integral component of the climate's complete regulation. In terms of developing reliable, inexpensive, and trouble-free thermal management, appropriate humidification devices, and devices may well be beneficial [15]. The developed system can also be used to remove low humidity observed in different ways, maintaining the climate conditions at the necessary level of human health [16,17]. For those countries where humidity is slightly lower than the comfort zone, the system is more useful [18]. The system is much more critical for maintaining continuous thermal comfort throughout the winter regional countries to remove such problems [19]. Its most humidity levels rise, the harder it becomes to raise moisture steadily [20]. In Bangladesh, as humidity is higher in any season than in the winter countries, it is less useful for Bangladesh than other nations. And it can be used in certain areas, such as labs, raising livestock, clinics, etc.

The above studies show acceptable performance to detect humidity but still has some limitations discussed. The objective of this research is to detect humidity in the weather automatically. Humidity problems in weather envelopes can develop from six different sources: use of wet materials in construction, the capillary rise of ground moisture, leakage of rainwater or melted snow, condensation of vapor, persistent high humidity conditions, and leakage of piped water. This paper is concerned with condensation caused by dew point temperatures occurring on or within the weather and closely related phenomena.

## 2. Methodology

### *2.1. Automatic ultrasonic humidifier*

And use a concrete esophagus pulsing at an ultrasonic level, an ultrasonic humidifier produces water molecules leaving the Humidifier in the form of cold mist. A small fan blows the fog typically, but there are no filters for specific ultra-mini models. To produce a high relevant acts amplitude in a liquid droplet, ultrasonic air purifiers use a piezoelectric material. This has an incredibly thin layer of particles about such a micrometer in size that evaporates rapidly into air circulation. Many attachments like Grove Temperature and Humidity sensor, Arduino Uno R3, Grove Water Atomizer, Matrix Membrane Keypad, Breadboard, LCD module, Buzzer, Potentiometer, hop wire and therefore on it should be operational efficiency to control the ultrasonic Humidifier [21]. Initially, the humidity sensor, keyboard, LCD monitor, Arduino UNO, and ultrasonic transducer must be linked through a hop cable to a circuit board. First, the proposed system has to set the input value and then need to read the humidity value. After reading the humidity value, it should check whether it is lower than the setting value or not. If no, then again, need to read the humidity value (besides, it will show not activated). If yes, then the system will be activated. The Arduino algorithm/code has been used to develop the system. Developing and running the Arduino algorithm/code needs to download Arduino software (which is open-source software) and need USB cable to connect between PC and Arduino to upload the code to Arduino. On USB boards, the wire between the USB and the power plugs selects the power source. Place the wire on the two pins closest to the USB plug to power the USB port's board (good for controlling low power devices like LEDs). Place the wire on the two pins closest to the power plug to power the board from an external power supply (needed for drive systems and other high current devices). Connect the board to a computer's USB port the other way. The proposed control system workflow of a humidifier is shown in Figure 2.

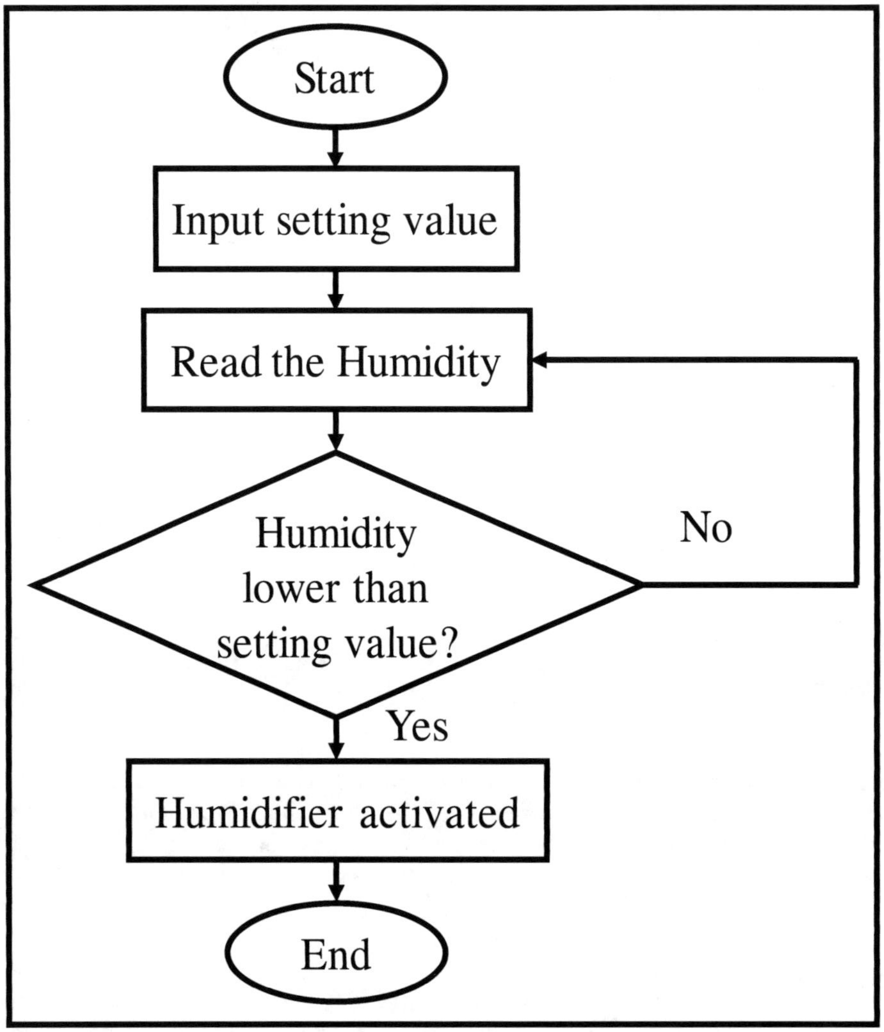


**Figure 2.** The proposed control system workflow of a humidifier.

A pre-designed computer program (Arduino IDE 1.8.13 software has been used to develop the code) for the microprocessor must be used to automate the computer to operate the machine correctly. The keyboard will define that settings function since turning mostly on the computer. The surrounding air's existing moisture is determined by a humidity sensor and shown on a liquid crystal display (LCD). The sensor produces an alarm condition to the Arduino unless the humidity is lower than that of the establishing level of humidity initially signed mostly on storage [22]. The sensor is accessed, and the procedure of humidifying begins the sensor, after which recognizes the consistent moisture performance of the total climate periodically. Unless the spatial moisture attains the developing design, the Humidifier will immediately shut down. Finally, the proposed experimental system setup of the humidifier control system is shown in Figure 3.

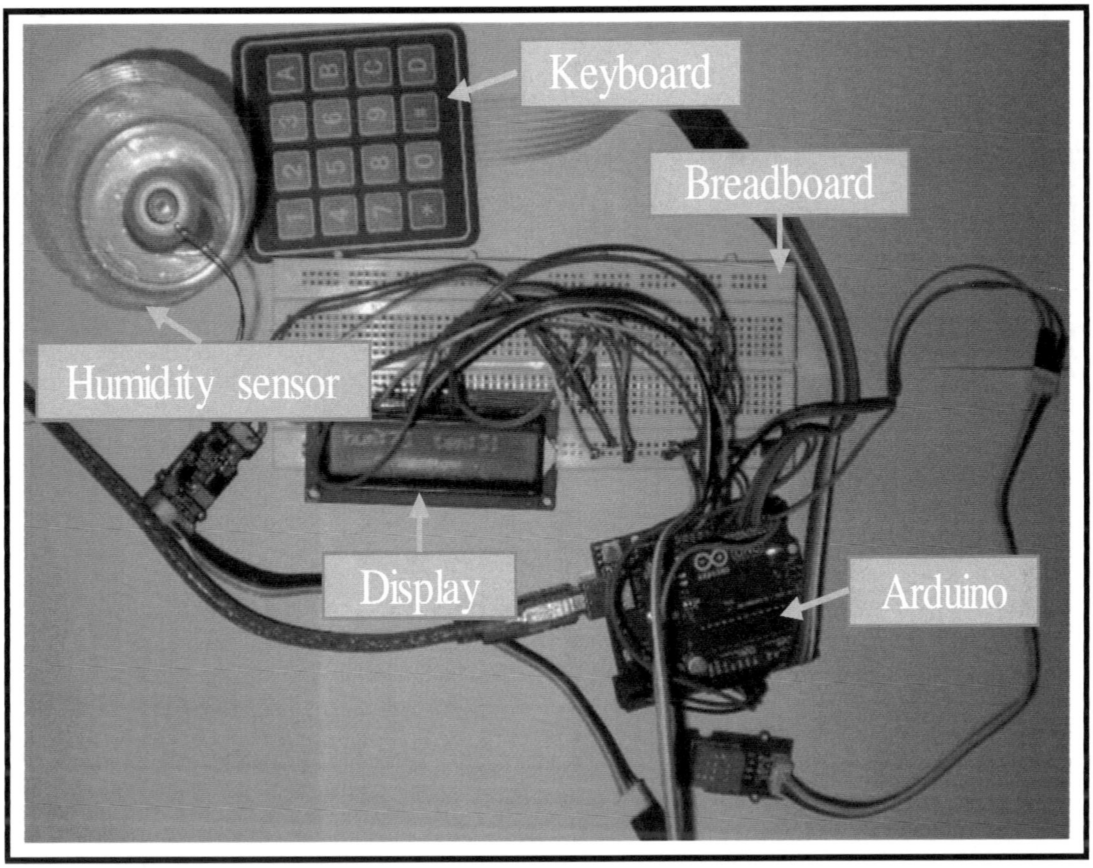


**Figure 3.** The proposed experimental system setup of the humidifier control system.

### 3. Results and discussion

All the data related to the project and paper were taken for a couple of months in April and May of 2018, at Khulna University of Engineering and Technology, Thermal Lab. Data has been taken inside the room to avoid the air motion outside. Due to the weather change, it has been recorded 5 to 10 days interval. From the thermal lab, with the help of a Fluke 971 Handheld hygrometer, we found dry bulb temperature (DBT) °C, actual relative humidity (RH) in %. At the same time, we recorded the sensor RH. Table 1 is formed by the different values of the various parameters from 10th April to 16th of May 2018. Data were taken at different times of the day has shown in table 1, due to the relative humidity change with the change of sunlight intensity and position. It was helpful to train the system for all conditions. The last column of the table showed the sensor's deviation and the actual value of relative humidity. The sensor showed a deviation range from 0.5 to 2.70. The average value of the deviation is 1.38, which justifies that the sensor is quite acceptable to the operation. Due to manual setup and frequent air temperature reading changes, the sensor produces some deviation values than actual. Sensor RH value depends on the motion of the air and atmospheric pressure. So, it can deteriorate due to the absence of a controlled atmosphere in the lab.

IOP Conf. Series: Materials Science and Engineering 1078 (2021) 012035 doi:10.1088/1757-899X/1078/1/012035

**Table 1.** Data for calibration of temperature and humidity with and without sensor.

| Obs. | Date of test | Time | DBT °C | Sensor Reading %RH | Actual %RH | Difference between actual VS sensor reading %RH |
|---|---|---|---|---|---|---|
| 1 | | 11:00 pm | 30.00 | 54.80 | 57.50 | 2.7 |
| 2 | 10th April | 11:10 pm | 32.90 | 54.00 | 56.70 | 2.7 |
| 3 | | 11:25 pm | 33.80 | 57.30 | 58.00 | 0.7 |
| 4 | 17th April | 12:25 am | 32.00 | 63.00 | 65.00 | 2 |
| 5 | | 12:35 pm | 33.50 | 63.40 | 64.90 | 0.5 |
| 6 | | 12:40 pm | 34.40 | 61.00 | 62.00 | 1 |
| 7 | 2nd April | 3:05 pm | 35.00 | 61.50 | 63.00 | 1.5 |
| 8 | | 3:15pm | 34.90 | 62.00 | 63.00 | 1 |
| 9 | | 3:30 pm | 35.20 | 61.50 | 62.00 | 0.5 |
| 10 | | 3:00 pm | 34.50 | 76.50 | 77.50 | 1 |
| 11 | 3rd May | 3:10pm | 34.20 | 74.50 | 76.50 | 2 |
| 12 | | 3:30 pm | 34.00 | 74.30 | 76.50 | 2.2 |
| 13 | | 1:00 pm | 35.10 | 73.20 | 74.50 | 1.3 |
| 14 | 10th May | 1:10 pm | 34.70 | 71.30 | 73.00 | 1.7 |
| 15 | | 1:25 pm | 34.95 | 72.00 | 73.50 | 1.5 |
| 16 | | 11:00 am | 35.40 | 78.30 | 79.00 | 0.7 |
| 17 | 16th May | 12:15 pm | 35.50 | 78.60 | 80.00 | 1.4 |
| 18 | | 12:30 pm | 35.50 | 80.00 | 80.50 | 0.5 |

Some experiments were conducted in various situations to evaluate the Humidifier's efficiency. The Humidifier's overall efficiency can be enhanced by using a useful humidity sensor and comprehensive airflow control. The system operated adequately from the tests performed, and the system raised the humidity of the atmosphere to the anticipated level. Figure 4 represents the relationship between actual and sensor %RH value. Due to six seasons in Bangladesh, the relative humidity value changes 45% to 83% round the year. We recorded 57.50% to 80.50% and changed the RH value very sharply. But the actual and sensor vale of RH so good consistency. The consistency of the humidity curve proves the sensor's effectivity, which enhances humidifier efficiency.

In Bangladesh, the average annual relative humidity is 65.8%, and the average monthly relative humidity ranges from 45% in March to 80% in June. The American Society of Heating, Refrigerating, and Air Conditioning Engineers, Inc (ASHRAE) provides guidelines intended to satisfy most building occupants wearing an average amount of clothing while working at a desk. The ASHRAE guidelines recommend a relative humidity (RH) of 30 to 60 percent [23]. For the project, we collected data in April and May 2018, shown in Table 2, and we set the standard value of 60% RH in April, while in May, we selected 62% RH. Few values showed lower than expected value than Humidifier active, but for the higher value, it was disabled.

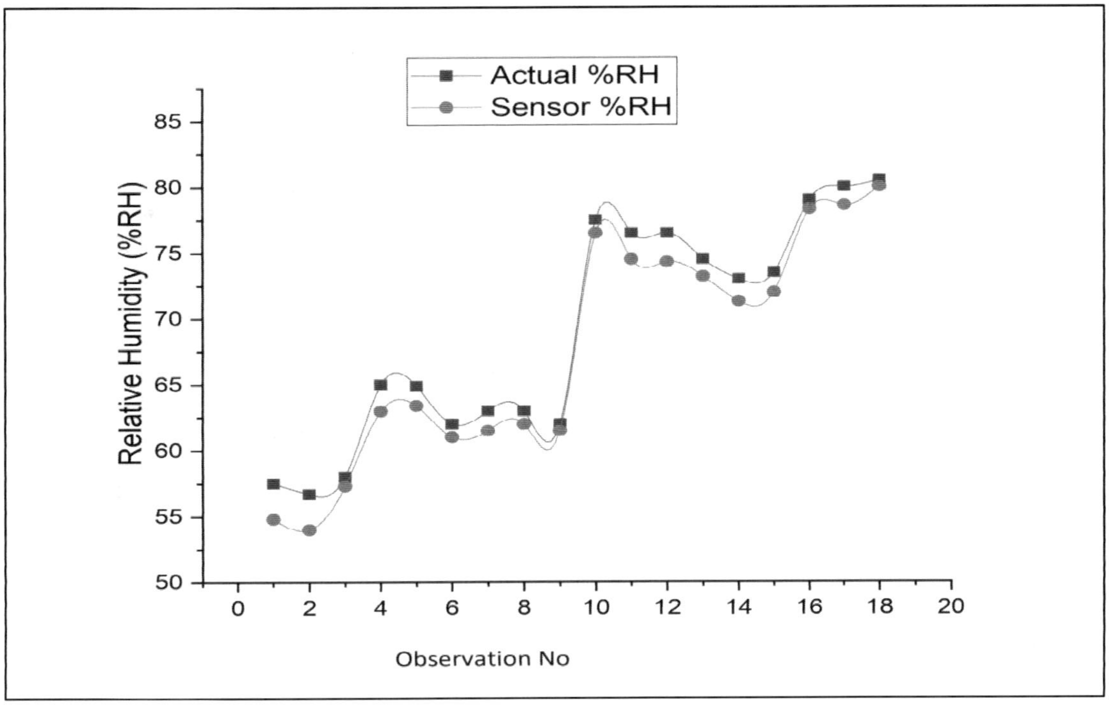


**Figure 4.** Difference between actual and sensor reading of %RH.

**Table 2.** Data for Humidifier activity with respect to setting humidity.

| Observation No | Date for Test | Setting value Relative humidity (%RH) | Sensor Reading Relative humidity (%RH) | Humidifier Action |
|---|---|---|---|---|
| 1 | | 60 | 54.80 | Activated |
| 2 | 10th April | | 54.00 | Activated |
| 3 | | | 57.30 | Activated |
| 4 | 17th April | 62 | 63.00 | Not Activated |
| 5 | | | 63.40 | Not Activated |
| 6 | | | 61.00 | Activated |
| 7 | 2nd April | | 61.50 | Activated |
| 8 | | 62 | 62.00 | Not Activated |
| 9 | | | 61.50 | Activated |
| 10 | | 62 | 76.50 | Not Activated |
| 11 | 3rd May | | 74.50 | Not Activated |
| 12 | | | 74.30 | Not Activated |
| 13 | | 62 | 73.20 | Not Activated |
| 14 | 10th May | | 71.30 | Not Activated |
| 15 | | | 72.00 | Not Activated |
| 16 | | | 78.30 | Not Activated |
| 17 | 16th May | 62 | 78.60 | Not Activated |
| 18 | | | 80.00 | Not Activated |
| 19 | | | 75.50 | Not Activated |
| 20 | 21st May | 62 | 73.00 | Not Activated |
| 21 | | | 73.50 | Not Activated |

This study's proposed aims were to develop a microcontroller-based system to identify and eliminate such low humidity problems. We developed an ultrasonic Humidifier with attachments of Grove Temperature and Humidity sensor, Arduino Uno R3, Grove Water Atomizer, Matrix Membrane Keypad, Breadboard, LCD module, Buzzer, Potentiometer, and hop wire. The Arduino algorithm/code has been used to operate the system. Initially, we try to measure the sensor efficiency shown in table1 and figure 1. The sensor was suitable to perform the relative humidity reading according to the efficiency standard. Our project was in the middle of April and May 2018; data were taken at the Khulna region of Bangladesh, where the temperature and relative humidity are relatively high. So very few data were collected below the setting value and which activated the system. The authors proposed that this system is suitable for the winter region where temperature and relative humidity are below the standard value. It will reduce the low damage humidification problems like human skins, electronics, and so on.

## 4. Conclusion

The system helps increase the amount of humidity inside predetermined significant stress. Computers, document suppliers, microelectronics companies, patient hospital beds, cold storage areas, museums and art galleries, exhibition galleries, and where a specific level of humidity are to be preserved must be considered. During the winter and then when the humidity is smaller than the sufficient amount, the system is often useful. A country like Bangladesh recorded a higher value of relative humidity rather than standard. Our study is appropriate in the winter season or areas where humidity is lower than that of standard. Using sprinklers rather than atomizers and incorporating features for continuous operation, using specific temperature and humidity sensors can be improved.

**Acknowledgements**

This work is supported by Khulna University of Engineering and Technology. The authors would also like to thank Faculty of Electrical & Electronics Engineering Technology, Universiti Malaysia Pahang (http://www.ump.edu.my) for providing the facilities to complete this work.

## References

[1] Imam S A, Choudhary A and Sachan V K 2016 Design issues for wireless sensor networks and smart humidity sensors for precision agriculture: A review *International Conference on Soft Computing Techniques and Implementations, ICSCTI 2015* (Institute of Electrical and Electronics Engineers Inc.) pp 181–7

[2] Khan N, Khattak K S, Ullah S and Khan Z 2019 A low-cost IoT based system for environmental monitoring *Proceedings - 2019 International Conference on Frontiers of Information Technology, FIT 2019* (Institute of Electrical and Electronics Engineers Inc.) pp 173–8

[3] Vinothkannan M, Kim A R, Gnana Kumar G and Yoo D J 2018 Sulfonated graphene oxide/Nafion composite membranes for high temperature and low humidity proton exchange membrane fuel cells *RSC Adv.* **8** 7494–508

[4] Rahmah N and Wonorahardjo S 2020 Critical Review of Advanced Material for Transit-Oriented Development in a Hot-Humid Climate *IOP Conference Series: Earth and Environmental Science* vol 532 (IOP Publishing Ltd) p 012015

[5] Castano L M and Flatau A B 2014 Smart fabric sensors and e-textile technologies: A review *Smart Mater. Struct.* **23** 053001

[6] Zougmore T W, Malo S, Kagembega F and Togueyini A 2018 Low cost IoT solutions for agricultures fish farmers in Afirca: A case study from Burkina Faso *ICSCC 2018 - 1st International Conference on Smart Cities and Communities* (Institute of Electrical and Electronics Engineers Inc.)

[7] Bucci G, Faccio M and Landi C 2000 New ADC with piecewise linear characteristic: case study - implementation of a smart humidity sensor *IEEE Trans. Instrum. Meas.* **49** 1154–66

[8] Sheth M and Rupani P 2019 Smart Gardening Automation using IoT with BLYNK App *Proceedings of the International Conference on Trends in Electronics and Informatics, ICOEI 2019* vol 2019-April (Institute of Electrical and Electronics Engineers Inc.) pp 266–70

[9] Boutsika L G, Enotiadis A, Nicotera I, Simari C, Charalambopoulou G, Giannelis E P and Steriotis T 2016 Nafion® nanocomposite membranes with enhanced properties at high temperature and low humidity environments *Int. J. Hydrogen Energy* **41** 22406–14

[10] Gozuoglu A, Ozgonenel O and Karagol S 2019 Fuzzy Logic Based Low Cost Smart Home Application *ELECO 2019 - 11th International Conference on Electrical and Electronics Engineering* (Institute of Electrical and Electronics Engineers Inc.) pp 64–8

[11] Molina-Lopez F, Briand D and De Rooij N F 2012 All additive inkjet printed humidity sensors on plastic substrate *Sensors Actuators, B Chem.* **166–167** 212–22

[12] Anon The Effects of Low Humidity and How to Deal With it | ThermoPro

[13] Feng Y, Cabezas A L, Chen Q, Zheng L R and Zhang Z Bin 2012 Flexible UHF resistive humidity sensors based on carbon nanotubes *IEEE Sens. J.* **12** 2844–50

[14] Iacovidou E, Purnell P and Lim M K 2018 The use of smart technologies in enabling construction components reuse: A viable method or a problem creating solution? *J. Environ. Manage.* **216** 214–23

[15] Khan F A, Ibrahim A A and Zeki A M 2020 Environmental monitoring and disease detection of plants in smart greenhouse using internet of things *J. Phys. Commun.* **4** 55008

[16] Halder S and Sivakumar G 2018 Embedded based remote monitoring station for live streaming of temperature and humidity *International Conference on Electrical, Electronics, Communication Computer Technologies and Optimization Techniques, ICEECCOT 2017* vol 2018-Janua (Institute of Electrical and Electronics Engineers Inc.) pp 284–7

[17] Chekired F, Houtti S, Bouroussis C A, Rahmani A, Tilmatine A and Canale L 2020 Low Cost Automation System for Smart Houses based on PIC Microcontrollers *Proceedings - 2020 IEEE International Conference on Environment and Electrical Engineering and 2020 IEEE Industrial and Commercial Power Systems Europe, EEEIC / I and CPS Europe 2020* (Institute of Electrical and Electronics Engineers Inc.)

[18] Ticci S, Less B, Walker I and Sherman M 2015 Development of a Seasonal Smart Ventilation Controller to Reduce Indoor Humidity in Hot-Humid Climate Homes *Effective ventilation in high performance buildings* pp 851–61

[19] Bencheikh M A and Boukhenous S 2019 A low Cost Smart Insole for Diabetic Foot Prevention *Proceedings of the 2018 International Conference on Applied Smart Systems, ICASS 2018* (Institute of Electrical and Electronics Engineers Inc.)

[20] Tang R, Wang S and Shan K 2018 Optimal and near-optimal indoor temperature and humidity controls for direct load control and proactive building demand response towards smart grids *Autom. Constr.* **96** 250–61

[21] Zhang Y, Wu Z, Zhang M, Mai J, Jin L and Wang F 2018 Smart indoor humidity and condensation control in the spring in hot-humid areas *Build. Environ.* **135** 42–52

[22] Mekonnen Y, Namuduri S, Burton L, Sarwat A and Bhansali S 2020 Review—Machine Learning Techniques in Wireless Sensor Network Based Precision Agriculture *J. Electrochem. Soc.* **167** 037522

[23] Sharmin S, Glass K, Viennet E and Harley D 2015 Interaction of Mean Temperature and Daily Fluctuation Influences Dengue Incidence in Dhaka, Bangladesh ed M Kasper *PLoS Negl. Trop. Dis.* **9** e0003901

# Heat transfer analysis in compact heat exchanger with fibre reinforced plate

**M Hamood[1,*], A N Oumer[1], Azizuddin Abd Aziz[2], M Ishak[2], J P Siregar [1] and O Mamat[3]**

[1]Collage of Engineering, Department, Universiti Malaysia Pahang, Malaysia, 26300, Gambang, Pahang, Malaysia
[2]Faculty of Mechanical and Automotive Engineering Technology, Universiti Malaysia Pahang, Malaysia, 26600, Pekan, Pahang, Malaysia,
[3]Department of Mechanical Engineering, Universiti Teknologi PETRONAS, 32610, Seri Iskandar, Perak, Malaysia

[*]Corresponding author's email: mhhalsham@gmial.com

**Abstract.** Due to their lightweight and ease of manufacturing, fibre-reinforced composites are replacing metals in many heat exchanger applications. However, their heat transfer dissipation performance is not investigated in detail. This paper aims to study experimentally the heat transfer performance of fins made from fibre reinforced composites in compact heat exchangers. Thin fins were manufactured from reinforced fibre composites by the hot pressing technique. The fins have dimensions of 11.5 cm wide and 11.5 cm long. Five different airflow velocities with fixed water flow rates were considered in the analysis. The temperature distribution inside the tube was measured at eight different points using a K-type thermocouple.
The results showed that the water temperature at the outlet section was the highest at the lowest airflow velocity. Moreover, the air-side heat transfer coefficient increased along the tube length at all airflow velocities. The comparable heat transfer performance results make the fibre reinforced fins a good choice in the compact fin-and-tube heat exchanger.

**Keyword:** Compact Heat Exchangers; Fiber-Reinforced Composites; Heat Dissipation; Heat Transfer Coefficient.

## 1. Introduction

Heat exchangers are considered as one of the most important industrial and processing equipment used worldwide. As it is one of the applications of thermal physics (thermodynamics) and it is also considered one of the applications of heat transfer science. Heat exchangers are widely used in heating, refrigeration, air conditioning, and in power plants, chemical and petrochemical plants, in addition to oil refineries and gas processing plants. There are many classifications of heat exchangers and they vary depending on the location and the system adopted [1]. However, heat exchangers can be classified according to the type of flow arrangement. There are three arrangements for the flow of fluids in heat exchangers; the first is the parallel flow heat exchangers, the two fluids enter the heat exchanger from one side and transfer parallel to the other end of the exchanger. The second type is heat exchangers with a counter flow, the two fluids enter from the opposite sides and go in the opposite direction. Third, the heat exchangers with the crossflow, the two fluids move in a vertical position so that one of the fluids is perpendicular to the other, which this research will focus on. There are many types of heat exchangers [1], the most prominent of which is compact heat exchanger, fin and tube heat exchangers are characterized by their excellent efficacy in heat dissipation and are characterized by lightweight and space-saving, which made them used extensively in most applications engineering, for example, car coolers, heat energy conservation, air conditioners, and in aircraft and spacecraft. Heat transfer is carried out in this heat exchanger by conduction.

In the last decade, heat exchangers made of the reinforced polymer have spread, and have received great interest due to the many properties that the polymer possesses. For example, it is light in weight, easy to manufacture, and most importantly, has good corrosion resistance. On the contrary, metal heat exchangers have a lot of defects such as heavy weight and ease of corrosion, which greatly reduces its life and efficiency. In addition, polymeric heat exchangers are distinguished by their ability to insulate electricity, which makes them ideal for use in electronic devices. Finally, polymers are more resistant to chemicals than metals, which makes them more durable as a material for heat exchangers[2].

Many studies have invented the need to develop the metal heat exchangers into exchangers made from alternative materials. Therefore, attention was focused on making compact reinforced polymer heat exchangers as a first step, due to the ability of the polymer composite material to handle liquids and gases, and its resistance to corrosion. More importantly, the polymer composite is inexpensive, light in weight, and small in size to save space. All of these advantages give the polymer material a competitive advantage over the metal heat exchangers [3]. Polymer matrix composites are among the most common advanced composites. These composites consist of a polymer thermoplastic or thermosetting reinforced by fibre (natural carbon or boron). A variety of shapes and sizes can be moulded from these materials. Along with resistance to corrosion, they provide great strength and stiffness. The reason why these are most prevalent is their low cost, high strength and simple production principles. The fibre reinforced polymer does not sufficiently dissipate heat, thus accumulating heat stress. However, its many distinctive properties such as lightweight, rust resistance and ease of manufacture, which make it a suitable choice for designing a high-efficiency heat exchanger. Eunbi Lee mentioned that some additions and improvements can be made to the reinforced polymer to increase the thermal conductivity, through-thickness using a layer-by-layer covering of inorganic crystals. Three types of inorganic crystal fillers consisting of aluminium, magnesium, and copper were used to prepare the highly thermally conductive CFRP composites through the layer-by-layer coating process. The vertical thermal conductivity of pure CFRP at a very low content of 0.01 wt percent was increased by up to 87 percent while using magnesium filler. It was also confirmed that the higher the thermal conductivity enhancement was, the better were the mechanical properties. In order to increase the thermal conductivity of composites, many different materials with high thermal conductivity have been used as fillers, such as carbon nanotubes (CNTs), boron nitride (BN), aluminium oxide, diamond, and graphene. Graphene has received more interest due to its unique structure and some of its exciting properties, such as its high thermal conductivity (Li et al., 2017).

Teweldebrhan et al. stated that the thermal conductivity measurement of suspended single-layer graphene was one of the highest thermal conductivities of currently known materials, about 5000 $Wm^{-1}K^{-1}$ [4]. A promising method of advancing the thermal conductivity of composites has been considered to be the alignment of hBN using an electric field. This allows small quantities of hexagonal boron hBN filler to be used to achieve high thermal conductivity. In addition, it was suggested that by adding electric field alumina, the thermal conductivity value of hBN / polymer composites can be increased by 15 times for 15 vol % plate-like alumina [5]. Experimental research was conducted by [6] to study the performance effect of fin types for compact finned flat-tubing of HEs. Three shape types of fins connected to three out of three flat tube bank arrays are smooth, wavy and rectangular grooved fins were used in that research. Compared to the wavy and plain fin, the rectangular fin was found to have the highest heat transfer performance where the wavy fin is greater than plain perfect. The impact of fin spacing on the fin temperature and the heat transfer coefficient of plain fin-and-tube HEs was studied by Chen and Hsu [7]. They found that the greater the distance between the fins, the higher the heat transfer coefficient. In contrast, the temperature decreased along the length of the fin. [8] performed two-dimensional numerical simulations to assess the efficiency of circular and elliptical tubes with one and two rows of tubes at the lower number of Reynolds. Their findings reveal that, under equal operating conditions, the heat exchanger with the elliptical tube provides greater overall efficiency than the heat exchanger with the circular tube. The fine thickness and tube diameter were explored by Lu et al. [9].

From his work, it is found that the heat transfer is improved by decreasing tube diameter with a set frontal inlet velocity. Thus, the purpose of this research are to investigate the heat transfer performance of compact fine-and-tube heat exchangers (CFTHE) made from reinforced polymer composites and to optimize the geometrical and process parameters for enhanced heat dissipation of fibre reinforced plate. The air of this study is to investigate the heat transfer performance of carbon fibre reinforced polymer composite fins and copper tube compact heat exchangers at various airflow velocities. The fluid medium flowing inside the tubes is water at one fixed temperature inlet value.

## 2. Methodology

### *2.1. Material*

The materials used to make a rectangular mold for the manufacturing of flat fins, a mild still was used. The drawing for the machining was done using Solidworks 2016 software. The finalized dimension of the mold is 23.5cm x 23.5cm with 2mm. After the CAD drawing, the drawing has been transferred to Mastercam software to create detailed instructions (G-code) and to automate the manufacturing process of the mold using a Computer Numerical Control (CNC) machine. After creating detailed instructions (G-code) from Mastercam software, the mold was fabricated from the mild steel plate using the CNC machine.

### *2.2 Fin preparation*

The fins were prepared using VORAFUSE method. It is a technique that has been developed by Dow Automotive Systems. It is the combination of epoxy resin with carbon fibre for prepreg (combination of fibres and uncured resin) applications in order to improve the handling of materials and the cycle time during the molding of composite structures. The hand lay-up process is the most commonly used manufacturing process. The fibre preforms are initially placed in a mold where an anti-adhesive coat (thin layer) is used to easily remove them. The resin material is then poured for reinforcement. The roller is used to enhance the interaction between the reinforcement and matrix materials ' successive layers. Previously, no analysis was carried out on the material form of the final shape; the parameters were therefore set at 12 cm x 12 cm and 1 mm with inline setup. With constant fin spacing, a total of 13 woven carbon fibre fins were fabricated. Similar work documented by [10].

### *2.3 Experimental setup and procedure*

The full experimental setup with the measurement instruments used in the experiment and the position of the temperature sensors in the inline fin is shown in Figure 1. Air is used as the working fluid, and in order to obtain the thermal-hydraulic performance of the fin and tube heat exchangers, the experiments were performed in an open wind tunnel. The system of the experiment consists of an inlet section, a test section, and measuring tools for air temperature and water, pressure drop, velocity of air, and controller device for water temperature. The system starts up by operating the suction fan so that air flows into the test section over the fin and tube. By using the coil heater, the water is heated between 40 to 50 °C in a basin, then pumped through the tube to the inlet of HE by using hot water pump (with power 22W and water flow 800L/h), and then this water returns to the basin and heated again and continues in a circular motion at the same temperature.

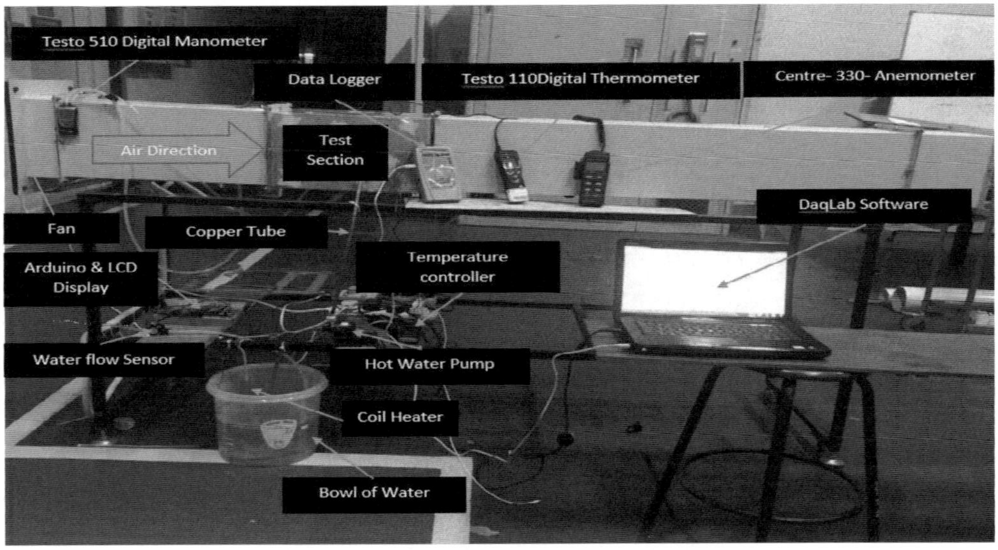


**Figure 1.** Experimental setup with measuring.

The wind tunnel is square in shape, each side is 26 cm. A straightener device has been placed at the inlet to avoid the velocity of the air flows any distortion. The test section is 1.2 meters away from the straightening device, in order to filter the air before it reaches the test section. Two holes were made for the bottom wall of the wind tunnel one before the test section and the other after it, for fixing the pitot tubes of the digital manometer. The back side of the wind tunnel has also been drilled, where a thermometer device could be installed to measure the velocity and temperature of the airflow before entering the test section. After that, air flows through the finned-and-tube HEs and then discharged to the adjacent places. The wind tunnel is also installed 50 cm high on a steel stand to reduce noise and vibration during system operation.

The exhaust fan was operated with a 50 W AC current, by means of a frequency converter the fan speed was controlled to five different speeds as shown in Table 1. Since the frequency converter is not accurate to control the desired air velocity, a Digital Anemometer was used where the frequency converter is tuned Moreover, the Digital Anemometer can also measure the temperature, so the temperature of the air entering the test section is measured by it. Then the Testo 110 Digital Thermometer was used to measure the air temperature after it left the test section. The REX C-100 was also used to keep the temperature constant on the surface of the circular tube. TESTO 510 is a model of digital differential pressure manometer, and it was used to measure the pressure of air at the inlet and the outlet of the test section. Figure 2 shown the test section made of 13 reinforced polymer fins with 2 cm distance between each other and a copper tube by 5mm diameter. Eight locations were selected on the test section and connected to the K-type sensor, so that six points were distributed on the test section inside the wind tunnel (point 2, 3, 4, 5, 6, and7) and two points outside the wind tunnel, one of them in the inlet (point 1) and the other at the outlet (point 8). Then by using OM-DAQPRO-5300 data logger, the temperature was measured for the eight points for each second during two minutes, then this process was repeated for the five different velocities of the airflow, Figure 2 shows the location of the sensors in the test section.

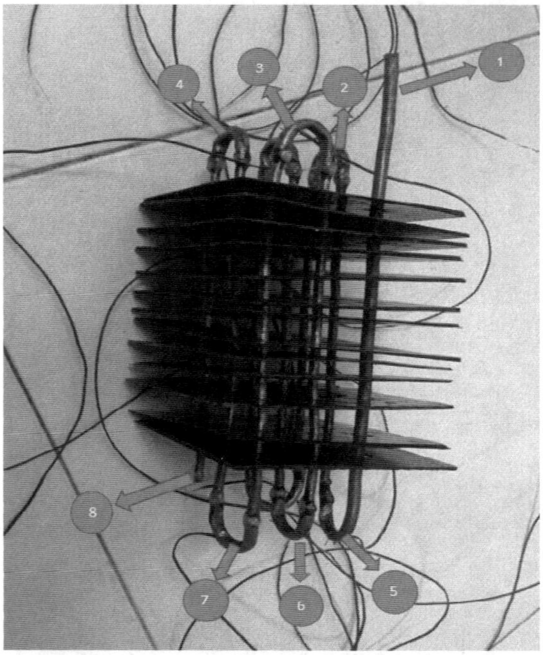


**Figure 2**. Test section and the 8 positions of K-type.

*2.4 Data reduction*

In this section, the data reduction is identified as a primary purpose for evaluating the heat transfer performance of heat exchangers by finding the heat transfer coefficient. The heat transfer rate transferred from the hot water to the tube is calculated using water properties and temperature differences.

$$Q^{\cdot} = c_p \dot{m}^{\cdot} (T_{in} - T_{out}) \tag{1}$$

Where, $C_p$, $\dot{m}^{\cdot}$ , $T_{in}$ and $T_{out}$ are the specific heat capacity of water, the water mass flow rate, hot water inlet temperature and water outlet temperature, respectively. The heat transfer coefficient was calculated using Equation (2)

$$h = \frac{Q}{A\,\Delta T} \tag{2}$$

$$\Delta T = \frac{(T_i - t_o) - (T_o - t_i)}{\ln\left[\frac{(T_i - t_o)}{(T_o - t_i)}\right]} \tag{3}$$

$$\Delta T = T_{wall} - T_{mean} \tag{5}$$

$$T_{mean} = \frac{T_{air\ inlet} + T_{air\ outlet}}{2} \tag{6}$$

Where ΔT can be calculated from the log mean temperature difference or the change between the wall temperatures and meat temperature, *h* is the heat transfer coefficient, and *A* is the surface area of the circular tube and the fin surfaces.

## 3. Result and dissection

Two random airflow velocities are selected and presented here in Figs. 3 and 4 to prove the flow and temperature distributions are at steady-state conditions at every point during the experiment. As shown in both figures, the temperature values at various thermocouple locations for the airflow velocities of 2.3 and 3.3 m/s show straight horizontal curves at all times. At the hot water inlet location (T1), there are slight temperature variations between 25-30, 50-55 and 100-105 seconds, however, these variations are insignificant. This indicates that the temperature distribution is in a stable condition at all locations. Moreover, the results indicated that as the flow advances from the inlet to the outlet section (from T1 to T8), the temperature of the working fluid decreased as expected. For both velocity cases, the temperature variation is significant between T1 & T2 and T7 & T8 while the temperature change is small for the other thermocouple locations. The temperature distribution of the working fluid measured at the fixed water flow rate of 0.22L/s are shown in Table 1. The temperature variation along the tube length was measured at eight positions at different airflow velocities which were 1.8, 2.3, 2.6, 2.8, 3.2 m/s. As can be seen from the table, a gradually decreasing trend of temperature from point T1 to T2 for all airflow velocities. The temperature drop was around 10°C from the inlet to the outlet of the test section. This is due to the fact that the hot water loses energy to the air flowing over the surface of the tube when it advances toward the outlet section. Moreover, at one specific thermocouple location, at the outlet section (T8) for instance, the temperature variations with increased airflow velocity are not significant. However, the higher the airflow velocity, supposed to be the greater the amount of heat absorbed during the passage of the air through the test section as was reported by A. Y. Adam et al. for the case of fin-and-flat-tube compact heat exchangers made from aluminium [11]. The discrepancies might be due to low flow rate of heated water flowing through the circular tube and usage of heaters by A. Y. Adam et al. which made the wall temperatures of the tube constant at all locations.

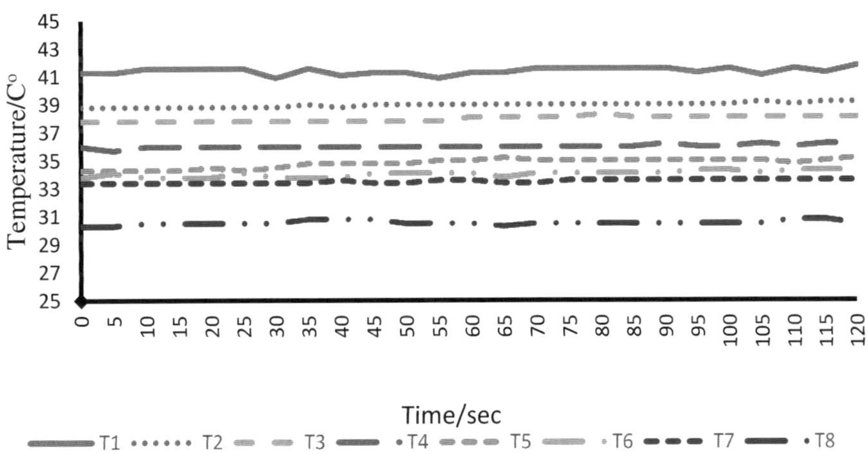


**Figure 3.** Variation of temperature with time at various thermocouple locations for the air velocity of 2.3 m/s.

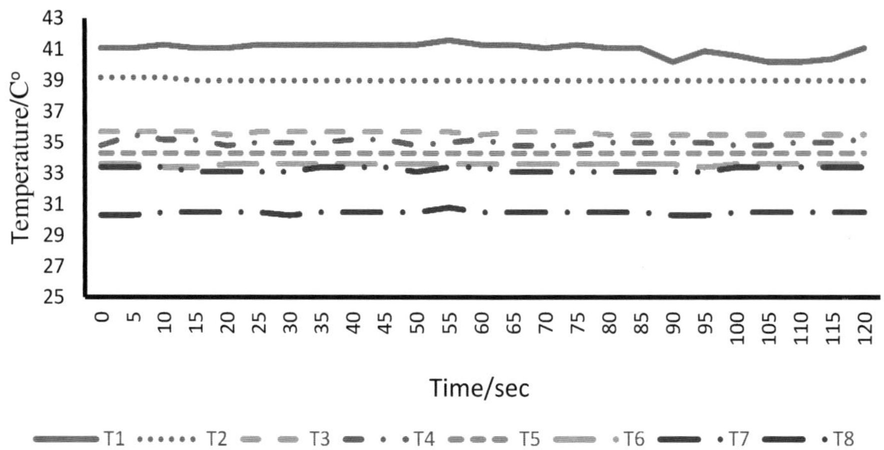


**Figure 4**. Variation of temperature with time at various thermocouple locations for the air velocity of 3.2 m/s.

**Table 1.** Temperature for each position during the five different velocity.

| Temperature position | Temperature C | | | | |
|---|---|---|---|---|---|
| | 1.8 m/s | 2.3 m/s | 2.6 m/s | 2.8 m/s | 3.2 m/s |
| T1 | 41.3 | 41.8 | 41.6 | 40 | 41.1 |
| T2 | 38.5 | 39.2 | 39.5 | 39.2 | 39 |
| T3 | 37.8 | 38.1 | 38.3 | 38.1 | 35.5 |
| T4 | 36 | 36.2 | 36.2 | 36 | 35.2 |
| T5 | 34.3 | 35.2 | 35 | 35 | 34.3 |
| T6 | 33.6 | 34.3 | 34.3 | 33.3 | 33.6 |
| T7 | 33.4 | 33.6 | 33.6 | 34.3 | 33.4 |
| T8 | 30.3 | 30.5 | 30.5 | 30.5 | 30.5 |

Figure 5 depicts the effects of airflow velocity on the local air-side heat transfer coefficient ($h$). This figure reveals that the local air-side heat transfer coefficient has increased significantly when the hot water travelled from the inlet to the outlet section of the tube at all airflow velocities. In general, the effect of airflow velocity on the heat transfer coefficient is negligible. However, However, as the water flow advances towards the outlet ($x$ increases), it can be noticed a trend characterized by the increase of $h$ with decreasing air flow rate. These different behaviours are associated with the time required by the air to absorb more heat from the hot water when travelling at low speed. It is difficult to generalize the overall trend as it might change when the waterside mass flux is changing instead of kept constant.

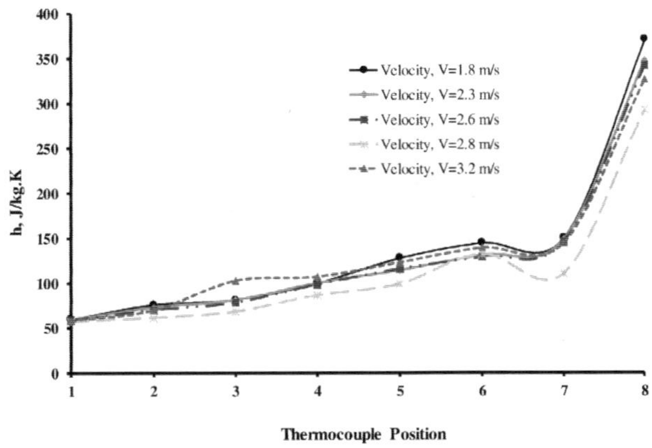


**Figure 5.** Heat Transfer Coefficient Variation of the Air-Side with Thermocouple Positions

## 4. Conclusion

From the heat transfer results, it was found that the fins made of the fibre-reinforced composite can be used in compact heat exchangers due to the heat dissipation performance is comparable with the conventional heat transfer materials. Through the experiment, which was for five different velocities of airflow, it was observed an average of 10 °C decrement in the water-side temperature. However, at the constant water-side flow rate, increasing the air-side flow rate has an insignificant effect on the average air-side heat transfer coefficient at all tube locations. Hence, fibre-reinforced fins can be used as an alternative material for heat dissipation in the compact heat exchanger.

### Acknowledgements

The authors would like to thank Universiti Malaysia Pahang and the Ministry of Higher Education Malaysia for the financial support given under FRGS/1/2019/TK03/UMP/02/18 and RDU190319, respectively.

### References

[1] Kuppan Thulukkanam, "Classification of Heat Exchangers," 2013, pp. 1–77.

[2] M. A. Arie, A. H. Shooshtari, R. Tiwari, S. V. Dessiatoun, M. M. Ohadi, and J. M. Pearce, "Experimental characterization of heat transfer in an additively manufactured polymer heat exchanger," *Appl. Therm. Eng.*, vol. 113, no. November, pp. 575–584, 2017, doi: 10.1016/j.applthermaleng.2016.11.030.

[3] L. Zaheed and R. J. J. Jachuck, "Review of polymer compact heat exchangers, with special emphasis on a polymer film unit," *Appl. Therm. Eng.*, vol. 24, no. 16, pp. 2323–2358, 2004, doi: 10.1016/j.applthermaleng.2004.03.018.

[4] D. Teweldebrhan *et al.*, "Superior thermal conductivity of single layer graphene," *Nano Lett.*, vol. 8, p. 902, 2008.

[5] X. Zheng, S. Kim, and C. W. Park, "Enhancement of thermal conductivity of carbon fibre-reinforced polymer composite with copper and boron nitride particles," *Compos. Part A Appl. Sci. Manuf.*, vol. 121, no. August 2018, pp. 449–456, 2019, doi: 10.1016/j.compositesa.2019.03.030.

[6] H. U. Choi and R. Fatkhur, "Experimental Investigation on Effect of Fin Shape on the Thermal-Hydraulic Performance of Compact Fin-and-Tube Heat Exchangers Experimental Investigation on Effect of Fin Shape on the Thermal-Hydraulic Performance of Compact Fin-and-Tube Heat Exchangers," 2018, doi: 10.1088/1757-899X/318/1/012070.

[7] H. T. Chen and W. L. Hsu, "Estimation of heat transfer coefficient on the fin of annular-finned tube heat exchangers in natural convection for various fin spacings," *Int. J. Heat Mass Transf.*,

vol. 50, no. 9–10, pp. 1750–1761, 2007, doi: 10.1016/j.ijheatmasstransfer.2006.10.021.

[8] L. A. O. Rocha, F. E. M. Saboya, and J. V. C. Vargas, "A comparative study of elliptical and circular sections in one- and two-row tubes and plate fin heat exchangers," *Int. J. Heat Fluid Flow*, vol. 18, no. 2, pp. 247–252, 1997, doi: 10.1016/S0142-727X(96)00063-X.

[9] C. W. Lu, J. M. Huang, W. C. Nien, and C. C. Wang, "A numerical investigation of the geometric effects on the performance of plate finned-tube heat exchanger," *Energy Convers. Manag.*, vol. 52, no. 3, pp. 1638–1643, 2011, doi: 10.1016/j.enconman.2010.10.026.

[10] P. Moorthy, A. N. Oumer, and M. Ishak, "Experimental Investigation on Effect of Fin Shape on the Thermal-Hydraulic Performance of Compact Fin-and-Tube Heat Exchangers," *IOP Conf. Ser. Mater. Sci. Eng.*, vol. 318, no. 1, 2018, doi: 10.1088/1757-899X/318/1/012070.

[11] A. Y. Adam, A. N. Oumer, A. Alias, M. Ishak, and M. M. Noor, "Investigation of thermal-hydraulic performance in flat tube heat exchangers at various tube inclination angles," *Int. J. Automot. Mech. Eng.*, vol. 14, no. 3, pp. 4542–4560, 2017, doi: 10.15282/ijame.14.3.2017.12.0359.

# Laser cladding process to enhanced surface properties of hot press forming die: A review

**B Norhafzan[1,2], C M Khairil[1] and S N Aqida[2,3,*]**

[1] Politeknik Muadzam Shah, Lebuhraya Tun Razak, 26700 Muadzam Shah, Pahang, Malaysia
[2] College of Engineering, University Malaysia Pahang,26600 Pekan, Pahang, Malaysia
[3] Automotive Engineering Centre, Universiti Malaysia Pahang, 26600 Pekan, Pahang, Malaysia

*Corresponding author's email: norhafzan23@gmail.com

**Abstract.** Laser cladding is one of the advance processes in laser surface treatments. The process involved a laser beam to combines another material that has different metallurgical properties on a substrate, whereby a very thin layer of the substrate has to be melted for it to achieve metallurgical bonding with minimal dilution of added material and substrate. The resulted properties were characterized by surface topography, subsurface microstructure, hardness, and residual stresses. The objective of this paper is to review the factors that affected of cladding process to get the best cladding, suitable to enhance hot press forming die surface and subsurface. The parameter control, metallurgical bonding between coating and substrate, and effect of powder size were discussed.

**Keywords:** Laser Cladding; Surface Properties; Hardness; Dilution.

## 1. Introduction

Surface properties for many engineering applications especially in the hot press forming industry are very important to endure high temperature, wear, and friction. Physical properties include time consumption, practicality and cost are among the factor that needs to be considered for engineering materials application. The most important criteria for the selection of die material are its resistance to plastic deformation, wear, and fatigue during operation. To provide resistance to wear, fatigue, and plastic deformation, the die hardness should be as high as possible. During hot forming, the dies soften and lose their hardness quickly if the temperature on the dies exceeds the tempering temperature [1].

Implementation of surface modification, treatments, or coatings is important to reduce the lubricant requirements and increase tool life. The heating and cooling processes in laser processing depend on the processing parameter, which affects the chemical composition, microstructures, and properties of materials [2]. The laser cladding was found to deliver excellent bonding, improve hardness, have no flaws, and produce new mechanical properties and high temperature stability of the cladding layer, which improved the lifetime of die. The laser cladding process provides and enhanced the surface hardness of the materials. The clad layer is stronger, due to higher dilution of the substrate [3]. Besides that, laser cladding also an effective method to produce microstructure refinement. A study in surface treatment shown grain refinement in cobalt alloy coating when Nano particles are added [2]. This paper presents a review on surface modification using the laser cladding technique. The effect of the clad layer is used to modify the surface behavior of various materials to improved surface hardness and wear resistance. The metallurgical bonding between coating and substrate, and laser parameter control were presented.

## 2. Laser cladding

One of the techniques to modified or improved surface properties in advanced manufacturing is the laser cladding process. Laser cladding is one of the laser surface treatments. The laser surface treatments are based on a change of the material compositions or grain structure of the sub-surface layer, due to a heating and cooling cycle by a laser source [4]. The advantages of implement laser cladding are improved grain size, increased hardness, minimal dilution, and produce new surface properties. Laser cladding uses a laser heat source to produce the desired clad layer during laser processing. The deposited material can be transferred to the substrate by several methods such as pre-placed powder, powder injection, paste powder on the substrate, or wire feeding as shown in Figure 1. Laser cladding with powder injection is proven to be the most effective method for the produced clad layer. During the cladding process, the laser beam melts the powder particles and produces thin layer on the substrate [5]. Laser cladding can be used as a safe technology to repair die surfaces, especially on critical contact surfaces. Laser cladding also increases tool life and saves manufacturing costs.

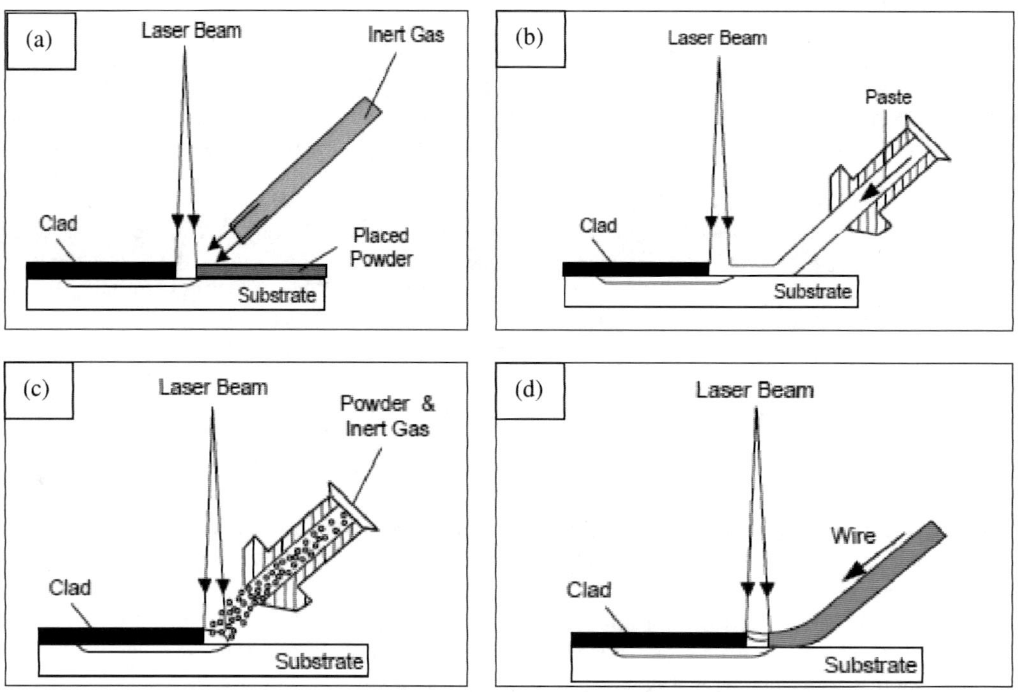


**Figure 1.** Different methods of laser cladding: (a) placed powder laser cladding, (b) paste laser cladding, (c) powder injection laser cladding and (d) wire feeding laser cladding. [5]

## 3. Effects of laser parameters on cladding properties

Laser processing parameters affect the variation of hardness, microstructure and geometry of the clad layer. It is measured by metallurgical characterization and microhardness testing [6]. Important features in laser cladding are clad height and dilution. The parameters used to control the setting are laser power (P), velocity of moving laser beam (v), diameter of the laser beam (D), and preheating temperature of the substrate (T) [7]. Processing parameters are one of the methods to improve the cladding quality is using a different substrate, powder, and type of laser will gain a different parameter. In order words, the different experimental parameters will led to different microstructure [8]. Combining appropriate processing parameters such as peak power, overlapping factor, and pulse duration during the laser cladding process using a pulse laser is crucial to produce independent and uniform results. Other effects such as overlapping factors for the width and depth layer can be seen throughout the molten pool [9]. The selection of proper processing parameters is required to avoid high residual stress concentration [10]. Also, this parameter can help determine the clad profile, dilution of the cladding metal, a fusion between layer, homogeneity, and surface finish defect such as porosity, crack, etc. [11]. The clad characteristics such as height and width can be designed to get optimum metallurgical characteristics with maintain its maximum deposition rate [12]. The hardness of clad layer is much influence by the amount and type of powder used, and also involves the wear resistance of its clad layer which affects the hardness and wear resistance [13]. In pulse energy lasers, the formations of clad layer are mostly influenced by pulse energy itself. Pulse duration increased laser pulse and overlap are factors that affect the surface roughness of clad layer. It has been observed that a rougher surface is obtained with a large laser spot increment and overlap with shorter pulse duration. Smoother surface can be obtained in using small laser spot increment and overlap with the use of longer pulse duration [14]. Many researchers do research on preheating the substrate before cladding process. This is because many of them find that this is the most effective method that can reduce the cooling rate significantly. Without using this method first, the clad layer will most probably produce crack in the end result [15].

## 4. Metallurgical bonding and dilution in cladding

Other important factors in the performance of laser cladding are the dilution rate of single-track cladding. The dilution rate is large and a large amount of substrate material will be mixed to the coating thus affecting the properties of the coating. But, when a small dilution rate is applied, a good metallurgical bonding cannot be achieved in the interface between the clad layer and substrate. Approximately, the appropriate dilution rate of the laser cladding layer should be around 10% to 25% [15]. During the laser cladding process, the temperature in the middle of a single track is higher than on both sides of the track. Thus, the thickness in the middle of the layer is greater than both sides of the layer, as shown in Figure 2.

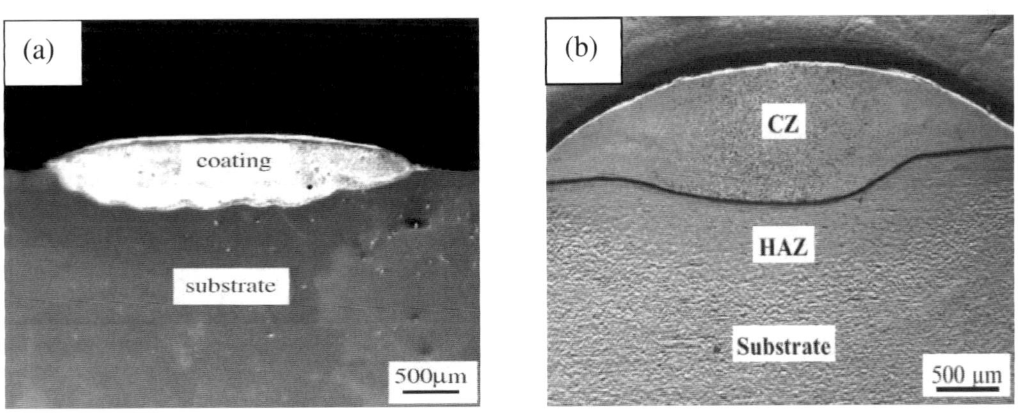


**Figure 2**. SEM micrograph (a) and (b) showing cross section morphology of one coating track. [10, 11].

The morphology of the cross-section layer can be divided into three regions, clad zone (CZ), heat affected zone of the substrate (HAZ) and the substrate [16]. The diluted zone reduces surface hardness and expandability between the clad layer and the substrate, which is expected to strengthen the adhesion of the layer to the substrate. The clad layer is free from pores and cracks, with low dilution, homogeneous microstructure and good metallurgical bond to the interface region [11].

## 5. Hardness properties

Laser claddings are proven to increase surface hardness up to 4 times of the substrate due to secondary carbides, grain refinement, and hard non-equilibrium microstructure formation. During cladding process, the molten pool rapidly solidifies and produces a fine grain refinement [17]. From the previous study, the main reasons for the increase in microhardness in the hardness of the AISI 304 stainless steel/ $Al_3O_3$ composite coating on AISI 1045 steel are due to the strength solution effect of the Al element and also due to presence of martensite in the heat affected zone [18]. The increased surface hardness of the clad surface is due to the presence of martensite and fine unsaturated carbides in the microstructure [19]. The high hardness value of the substrate near the intermixed area is due to the rapid melting and solidification of steel resulting in martensite in these steels [20]. Reducing the particle size of powder will increase the hardness of the copper matrix composite by 47% compared to pure copper [21]. Figure 3 shows, various microhardness distribution on the cross-section of the laser cladding layer. Surface hardness increase in clad layer and rapidly decrease through HAZ to the substrate and constant low hardness in the base substrate. Higher hardness in clad layer due to rapid heating and cooling, where grain sizes were improved. While delay solidification in HAZ causes latent heat and produce low hardness in HAZ.

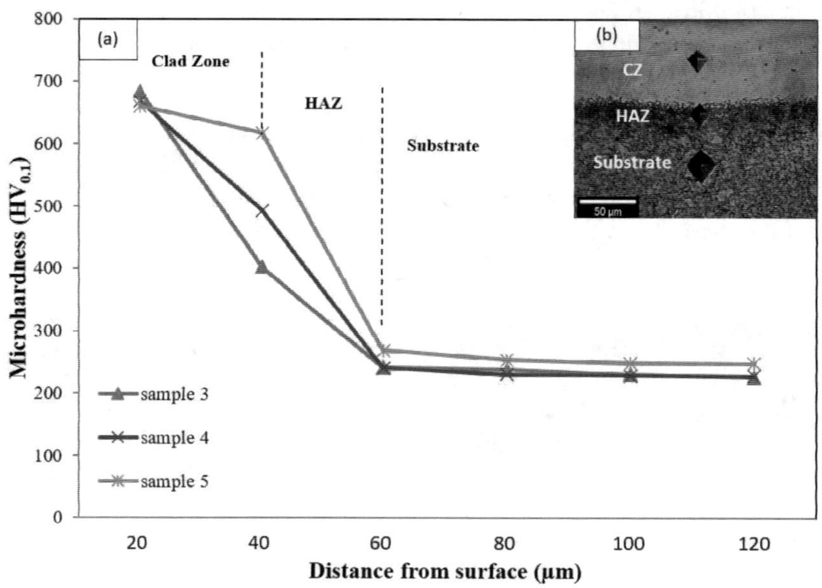


**Figure 3**. Various microhardness distribution curves on the cross section of the laser cladding layer. [22]

## 6. Effects of powder sizes on cladding morphology

Different sizes and types of powder give some effect on the microstructure of the clad layer. Table 1 shows the type of various powder and effect on claddings morphology. The main methods in the preparation of powder are to use high-energy ball milling technique. Powder size, morphology, uniformity as well as microstructure and properties of cladding property mainly been influenced by the milling time in the preparation of the powder itself [23].

**Table 1.** Type of powders used in various laser cladding and their effect on clad morphology.

| Reference | Type of powder | Effect |
|---|---|---|
| [24] | Chromium (Cr) | Good cladding layer without defects (cracking and porosity), improve hardness and corrosion resistance |
| [14] | Tungsten (WC) | Improve wear resistance and decreased porosity in the coating |
| [25] | Titanium (Ti) | Microstructure becomes refine and uniform |
| [13] | Copper, Molybdenum, Silicon (Cu, Mo, Si) | Improve hardness and wear resistance |

The magnified cross-sectional SEM morphologies of laser cladding coatings near the top surface after polishing with powders pre-milled with different time shown in Figure 4**,** the figure shows some effect on the microstructure of the coating. The powder was fabricated with a pre-mill for 1 hour shows course cellular and uniaxial phase. For the powder that pre-mill for 12 hours, present the dendrites as well as relatively slander cellular and uniaxial microstructure. The result from sample show that similar structure gain from pre-milled powder for 24 hours to that 12 hours but with smaller grain sizes.

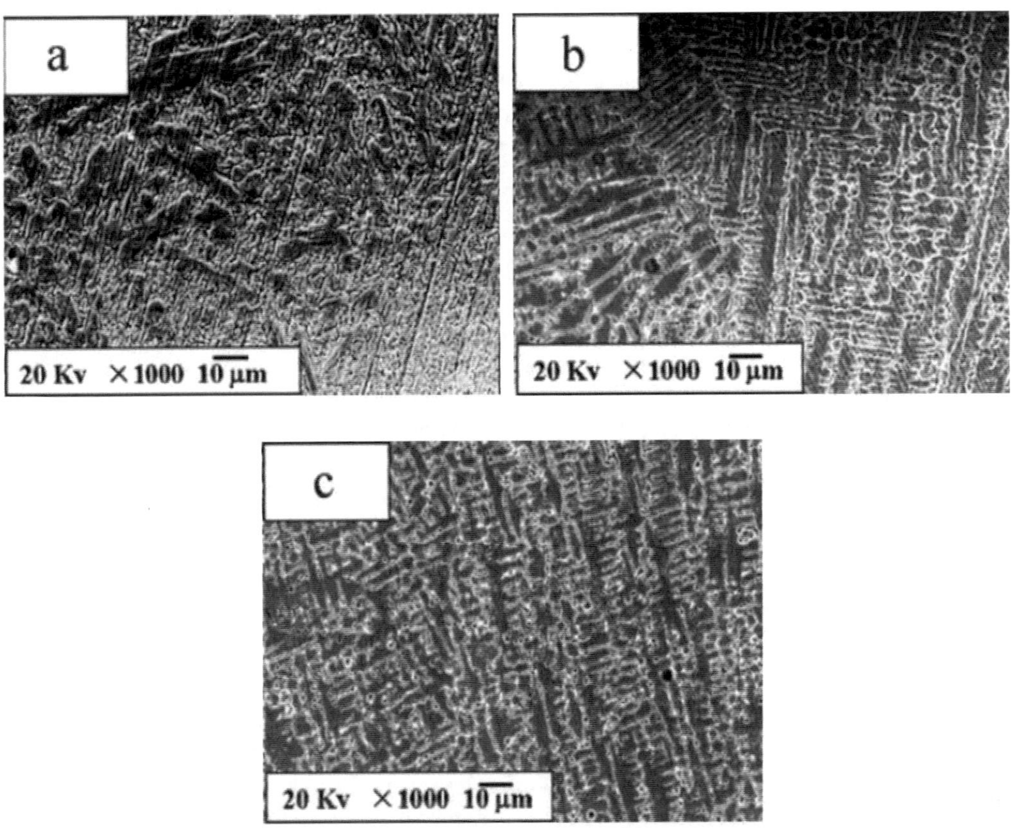


**Figure 4**. The magnified cross-sectional SEM morphologies of laser cladding coatings near the top surface after polishing with powders pre-milled for (a) 1 h, (b) 12 h and (c) 24 h [23].

It can be concluded that different size of the powder to make the coating are composed of grain with different shape and size. It shows that different size of the powder produced clad layer with consisting of grain with different shape and size.

## 7. Conclusions

The study of various works in the laser cladding process indicates that parameter control, substrate preheating, sizes of the powder, type of powder and dilution rate affect the metallurgical bonding, wear-resistance and hardness of the coating layer. Therefore, to enhance surface properties of the HPF die material, the above factors should be emphasized. Using suitable parameters can produce hard layer with excellent metallurgical bonding to the substrate material, uniform composition and preferable coating thickness.

**Acknowledgments**

The authors would like to acknowledge the financial support from University Malaysia Pahang under RDU182203 UMP Flagship research grant and Politeknik Muadzam Shah Pahang.

## References

[1] Altan T and Deshpande M 2011 Selection of die materials and surface treatments for increasing die life in hot and warm forging *ERC for Net Shape Forming, Paper* 1-32
[2] Montealegre M, Castro G, Rey P, Arias J, Vázquez P and González M 2010 Surface treatments by laser technology *Comtemporary Materials, I–1* 19-30
[3] da Silva M D, Partes K, Seefeld T and Vollertsen F 2012 Comparison of coaxial and off-axis nozzle configurations in one step process laser cladding on aluminum substrate *Journal of Materials Processing Technology* **212** 2514-9
[4] Schneider M F and Schneider M F 1998 Laser cladding with powder
[5] Toyserkani E, Khajepour A and Corbin S F 2004 *Laser cladding*: CRC press)
[6] Santhanakrishnan S, Kong F and Kovacevic R 2011 An experimentally based thermo-kinetic hardening model for high power direct diode laser cladding *Journal of Materials Processing Technology* **211** 1247-59
[7] Bax B, Pauly C, Leibenguth P, Woll K and Mücklich F 2012 Synthesis of B2-RuAl coatings on mild steel by laser cladding *Surface and Coatings Technology* **206** 3931-7
[8] Lin P, Zhang Z and Ren L 2014 The mechanical properties and microstructures of AZ91D magnesium alloy processed by selective laser cladding with Al powder *Optics & Laser Technology* **60** 61-8
[9] Sahoo C K and Masanta M 2015 Effect of pulse laser parameters on TiC reinforced AISI 304 stainless steel composite coating by laser surface engineering process *Optics and Lasers in Engineering* **67** 36-48
[10] Farahmand P and Kovacevic R 2014 An experimental–numerical investigation of heat distribution and stress field in single-and multi-track laser cladding by a high-power direct diode laser *Optics & Laser Technology* **63** 154-68
[11] Boddu M R, Landers R G and Liou F W 2001 Control of laser cladding for rapid prototyping—A review. In: *Proceedings of the Solid Freeform Fabrication Symposium,* pp 6-8
[12] Farayibi P, Folkes J, Clare A and Oyelola O 2011 Cladding of pre-blended Ti–6Al–4V and WC powder for wear resistant applications *Surface and Coatings Technology* **206** 372-7
[13] Peilei Z, Xiaopeng L, Yunlong L, Hua Y, Zhishui Y, Chonggui L and Qinghua L 2014 Microstructure and wear behavior of Cu–Mo–Si coatings by laser cladding *Applied surface science* **311** 709-14
[14] Huang S, Nolan D and Brandt M 2003 Pre-placed WC/Ni clad layers produced with a pulsed Nd: YAG laser via optical fibres *Surface and Coatings Technology* **165** 26-34
[15] Wang D, Tian Z, Wang S, Shen L and Liu Z 2014 Microstructural characterization of $Al_2O_3$–13 wt.% $TiO_2$ ceramic coatings prepared by squash presetting laser cladding on GH4169 superalloy *Surface and Coatings Technology* **254** 195-201

[16] Lei Y, Sun R, Tang Y and Niu W 2015 Microstructure and phase transformations in laser clad CrxSy/Ni coating on H13 steel *Optics and Lasers in Engineering* **66** 181-6
[17] Cheng H, Yi J, Fang Z, Dai S and Zhao X 2013 Tribology property of laser cladding crack free Ni/WC composite coating *Materials Transactions* M2012237
[18] Xu P, Lin C, Zhou C and Yi X 2014 Wear and corrosion resistance of laser cladding AISI 304 stainless steel/Al2O3 composite coatings *Surface and Coatings Technology* **238** 9-14
[19] Telasang G, Majumdar J D, Padmanabham G, Tak M and Manna I 2014 Effect of laser parameters on microstructure and hardness of laser clad and tempered AISI H13 tool steel *Surface and Coatings Technology* **258** 1108-18
[20] España F A, Balla V K and Bandyopadhyay A 2010 Laser surface modification of AISI 410 stainless steel with brass for enhanced thermal properties *Surface and Coatings Technology* **204** 2510-7
[21] Uddin S M, Mahmud T, Wolf C, Glanz C, Kolaric I, Volkmer C, Höller H, Wienecke U, Roth S and Fecht H-J 2010 Effect of size and shape of metal particles to improve hardness and electrical properties of carbon nanotube reinforced copper and copper alloy composites *Composites Science and Technology* **70** 2253-7
[22] Norhafzan B, Aqida S, Chikarakara E and Brabazon D 2016 Surface modification of AISI H13 tool steel by laser cladding with NiTi powder *Applied Physics A* **122** 1-6
[23] Wang L, Zhou J, Yu Y, Guo C and Chen J 2012 Effect of powders refinement on the tribological behavior of Ni-based composite coatings by laser cladding *Applied Surface Science* **258** 6697-704
[24] Zhang H, Zou Y, Zou Z and Shi C 2014 Effects of chromium addition on microstructure and properties of TiC–VC reinforced Fe-based laser cladding coatings *Journal of alloys and compounds* **614** 107-12
[25] Gao W, Zhao S, Wang Y, Zhang Z, Zhou C and Lin X 2015 Refinement of Fe-based alloy doped Ti cladding layer *Surface and Coatings Technology* **270** 16-23

# Hydroinformatics based technique for leak identification purpose – An experimental analysis

**M Y Hanafi[1,2], M F Ghazali[1,2,3,*], W H Azmi[1], M F M Yusof[1,2] and M A PiRemli[1,2]**

[1]Faculty of Mechanical & Automotive Engineering Technology, University Malaysia Pahang, 26600 Pekan, Pahang, Malaysia
[2]Advance Structural Integrity and Vibration Research (ASIVR), Faculty of Mechanical Engineering, University Malaysia Pahang, Pekan, Pahang, Malaysia.
[3]Centre of Excellence for Advanced Research in Fluid Flow (CARIFF) University Malaysia Pahang Lebuhraya Tun Razak 26300 Gambang Kuantan Pahang.

*Corresponding author's email: fairusham@ump.edu.my

**Abstract.** Analysis of single point pressure transient signal has the potential of providing information on the pipeline system. The technique uses the frequency response and experiments are satisfactory for low-frequency bandwidth. The concept behind this approach is that reflections are detected by leak-induced disturbances. Such reflections are often very difficult to identify, particularly because of excessive noise. The common approach to boost the analysis is to use a data denoising approach. The wavelet transforms for leak detection in water pipelines are proposed in this article. The wavelet filter is optimally calibrated using maximal kurtosis values and is chosen for leakage detection. The suggested approach is tested using high-noise laboratory test signals. The findings show that the method is superior to current signal processing methods for the conditions used. The system permits not only the detection of leakage but also the location and evaluation of its magnitude with errors of less than 7%.

**Keywords:** Hydroinformatics; Pressure Transient; Wavelet; Leak; Non Revenue Water.

## 1. Introduction

The differential within the quantity of water produced in between distribution is referred to as non-revenue waters. This is linked to many causes, such as system leakage through the pipe. Water failure is a common concern and a critical issue for water management [1]. Statistical research showed that Malaysia's usual non-revenue water level was in 2010, at 29.4 per cent, leading to large currency, weight, and supply crises and abuse of living. [2]. In 2017, the average grew to 35.3% [3]. The effect can therefore refer us to the negative consequences and actions of governments. High non-revenue water amount makes a negative reputation for water supply authorities [4]. It reflects the delicacy of the water distribution network [5]. The consequence will also have negative implications on infrastructure and the ecosystem. [6]. The operators of water supplies have a negative thought about strong non-income sewage. This is the issue of the instability of the water supply [7] [8].

There are several ways to reduce and manage non-revenue water [9]. A very economical approach is responsive monitoring as the concern is normally verified by the customer or noted by the own operations of the companies. The procedure can be tolerated in locations in which the supplies are sufficient or great value for money. In comparison, the effective maintenance of the leakage consists of two techniques, a regular survey and leakage monitoring. Continuous review is a reason to watch for leaking pipes and adjust and interpret the metered flow into the temporary zone to track night traffic in large volumes. Flow control is the tracking of leakage in areas or regions and optimises leak detection activities [10]. However, traditional leak detection systems are often expensive and must be chosen for more advanced techniques in a developing nation. The sales plan would also be drawback.

The leak detection analysis can be pioneered in finding the leak and taking measures for business. The dilemma we have to solve, though is the desire to benefit from the industry. The running expenses must be smaller to save revenue. Knowledge must be provided to more performance and accuracy. Method review can be replicated to track the pipe network system. The transient pressure technique may indeed evaluate leakage detection [2]. For example, transient analysis may provide critical leak detection, including an exact transient model for a device. [11]. The benefits of transient analysis and the number of problems in a single method can be catastrophic further explanations. [11]. In all these studies, the identification of signal reflection is an important element. The most generally employed signal processing techniques used in transmission pipelines are correlation, cepstrum and wavelet analysis [12]. Most leak detection work is directly linked to water pipelines. This is because the flow of water is also correlated with excessive quantities of noise. Weak Signal Noise Ratios (SNR) along with reflection produced by geometric pipeline properties, such as joints or valves, make it increasingly hard to detect water leakages in the process.

It is necessary to convert the time domain signal into the frequency domain in order to reveal the consequence of the signal pulse. Previously there have been several transformation techniques such as Fourier (FT), Short Time Fourier (STFT), and Hilbert Transform (HT). Wavelets Transform (WT) is a sort of study of ways to measure signal time and frequency. It can evaluate with different distinguish ratios; however, it also has the capacity to represent the sectional characteristics of the signal in terms of time domain and frequency domain. This is a sectional analysis approach that is ideal for consistent signal analysis, time and frequency identification, anomaly monitoring and denoising with a fixed-size window but changeable appearance and a growing time and frequency range [13]. Wavelet Transform in this study is used to analyse reflection signal generates from the leak on pipeline and remove the signal noise with the use of a morlet as a mother wavelet.

## 2. Wave propagation in pipes

Multiple sensors, include pressure, flow metres, and valve sensors are needed for existing monitoring procedures. The motion of the wave may be identified at a fixed point below the surface with a change of pressure [14]. Consequently, improvements in the structure of the pipeline such as the valve, intersection, block and leaks would reflect. This reflection is called a pressure wave, which passes through the pipeline at the sound speed [10, 15]. The pressure wave is simulated with the time domain based on modelling techniques in the transmission line. Wave can also be observed by fluid particle movement. The wave time can be captured by forming a pressure wave with a single remote sensor at a certain spot. Speed is the distance that the wave travels to that distance from the time the wave travels. Where $l$ = length , $t$ = time and $a$ = speed of sound. The pipeline length can be determined by several times in pipelines with sound speed.

$$l=ta \quad (1)$$

The speed of sound is formulated from multiplying of fluid properties which are $\gamma$, $Ru$, and $M$ and also the temperature in degree Kelvin.

$$a = \sqrt{\frac{\gamma RT}{M}} \quad (2)$$

The tube is supposed to be rigid since flexible tubes slow down the system speed. The waves of pressure only travel in the fluid.

## 3. Experimental design

A pressure sensor is used in this research to obtain the signal reaction produced by the solenoid valve. The solenoid valve is normally closed during the operation and irregularities in the water hammer are caused by the opening and closing of the valve for three consecutive periods. The test data was obtained using MATLAB tools.

The research was conducted in a 152.3-m (Fig. 1) platform circular loop experimental test rig. The device consisted of Medium High Density Polyethylene (MDPE), an external 60-mm cross section, medium diameter within the 55-mm and 2.6-mm cross-section of thickness. A round hole with different sizes in diameter was fitted on 49.7-m (Fig. 2) from the pipeline sample act as the leak. A water surface storage is connected to the tube exit from the tube locks. This protects fast expansion phenomena of the pressure waves and decreases the adverse pressure wave, since data are affected in the transducer. The testing process has a driven sound speed of 524.3 ms-1. Fig. 1 shows the pressure sensor and the solenoid valve that is positioned 58.9-m (Fig. 2) metres away from the electrical motor to prevent disturbance during the data acquiring. The product of friction is caused by the flow of the water and humidity from the channel. This approach is frequently used in dam research. The concept was taken from the use of pressurised water progression in the pipeline system. The pressure transient reaction occurs hypothetically when the water movement is interrupted suddenly, by shutting or opening the valve in the pipeline system. In combination with the pipeline system, the wave propagation of water will be produced, as are the wave propagation characteristics as a leak detection method, and the pipeline network. The wave of the pipeline network travels with several other signals according to the size of the leak, leak duration, tube-style function and pipe diameter.

Fig.1 illustrated the experimental test platform in the Fluid Mechanic Laboratory at the University of Malaysia-Pahang. The fire hydrant cap was designed and manufactured to fitted the solenoid valve and the pressure sensor for the test run. The solenoid valve generates water hammer phenomena in both components of the device and a pressure sensor is used for signal acquisition. In certain situations, the system may be shut down with the abrupt opening or closing of the valve and the pump failure. This produces a "water hammer" on the pipeline. This phenomenon also affected both pressurise pipe structures and caused sudden pressure change and pipe collapse [16]. This effect is generated by the solenoid valve (Fig. 1) for creation of a water hammer in pipeline system. The same situation is developed in actual situations in a pressure condition in the underground pipe [17].

**Figure 1**. Experimental test rig in laboratory scale.

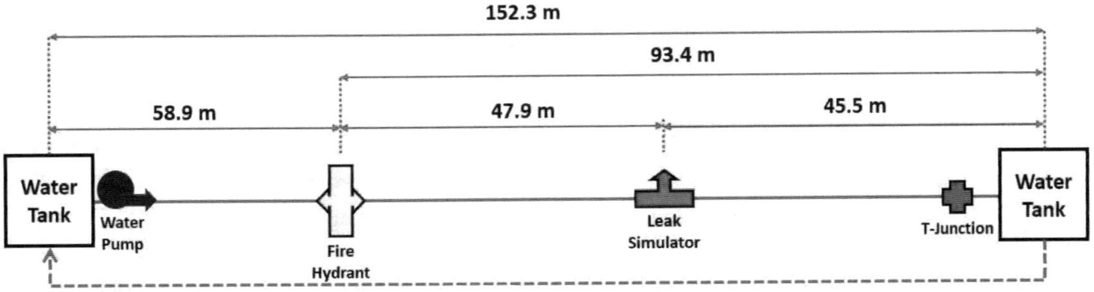


**Figure 2**. Experimental test rig sketch.

## 4. Results and discussion

The test starts by obtaining the transient pressure signal from a test rig using MATLAB. The acquired signal regains the properties of the pipeline overall. Water hammer waves dispersed in both directions along with the pipeline system away from the outburst source via the speed of sound in the water pipeline system. The tests were performed with the MDPE pipe. 10kHz of sampling frequency were sample at 1 second. The NI-9234 card and DASYLAB were used in the data acquisition phase then the signal further analysed using CWT through MATLAB platform. A clear transient (fig. 3a and 3b) can be seen at the start of the response signal indicates the opening and closing the solenoid valve. In the raw response, no other important features were available since the reflection signal was masking in the signal noise. As illustrated in Fig. 3(a) and 3(b), the raw signal failed to detect clear reflection signal generates from the leak and any disturbances in the pipes. Therefore, the further analysis is needed to extract the information contained in the signal response. The reflected signal was combined with the Morlet as a mother wavelet for every data set. The signal then analysed by continuous wavelet transformation (CWT). The small scale of the mother wavelet recognises that the frequency is high [18]. Wavelet compressed the signal and vice versa.

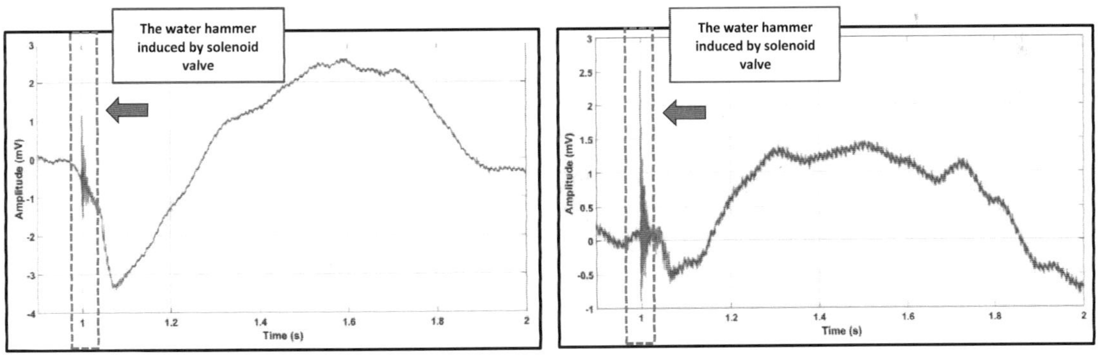


**Figure 3**. Raw signal (a) No-leak pipe (b) Raw signal for leak pipe.

The step of signal analysis began with the signal denoising through the wavelet-based filter by tuning the maximum value of kurtosis. This is because the background noise and modulations of the reflected signal are too strong in the analysed data. The denoised signal then transformed into time-frequency representation and time-frequency (TF) spectrum through CWT as shown in Fig. 4,5,6 and 7. The plots had uniform time $t_r=t / Tc$, where: $t_c = 2l / c$ is the characteristic time of the pipe; l is the length of a pipe; c is the speed of the wave of pressure. In Fig. 3, the rising spikes at $t$ = 1 sec are equal to the start of the solenoid valve shock wave. The pressure wave then travels along the pipe and reflects the upstream reservoir. At $t_r$ = 1 the pressure wave enters the valve and reflects and induces a rapid pressure drop. The shock wave proceeds to pass from the valve into the tank, smoothing and attenuating it in time vice versa [19].

The repetitive pattern and the rapid changes in the pressure signal were obvious in the wavelet maps where absolute intense values were almost consistently located in time (Fig. 4 to Fig. 7). Irregularities in the pressure signal that fit the start of the open-closing solenoid valve and the locations of the reflected waves are illustrated by the presence of instant frequency and energy (Fig. 4 to Fig. 7) of the reflected signal. These phenomena were more visible at a larger scale, as the wavelet coefficients are very small for smaller scales for all time.

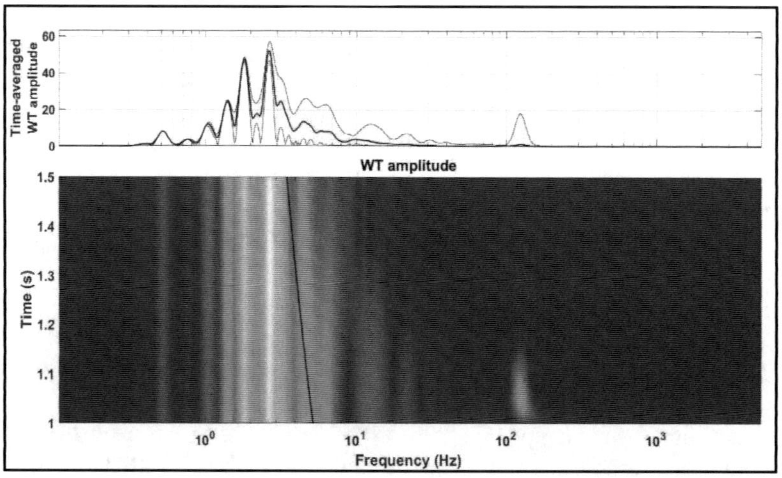


**Figure 4.** Healthy pipeline system (a) Time -frequency representation (b) Wavelet spectrum.

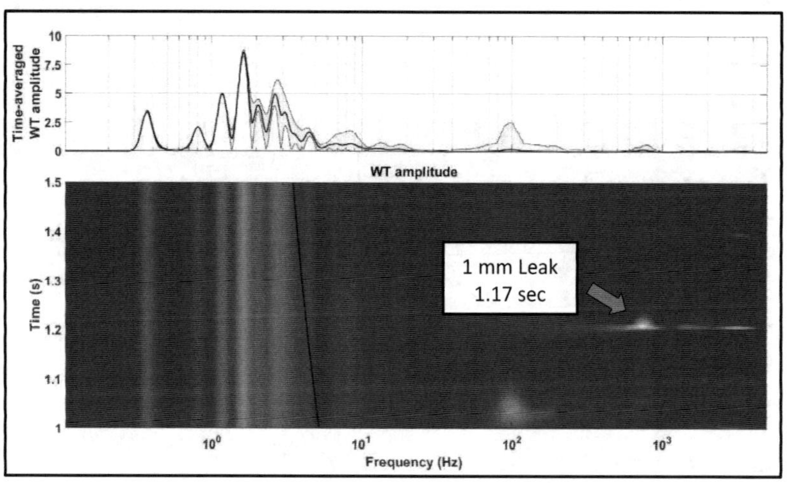


**Figure 5.** 1 mm leak (a) Time -frequency representation (b) Wavelet spectrum.

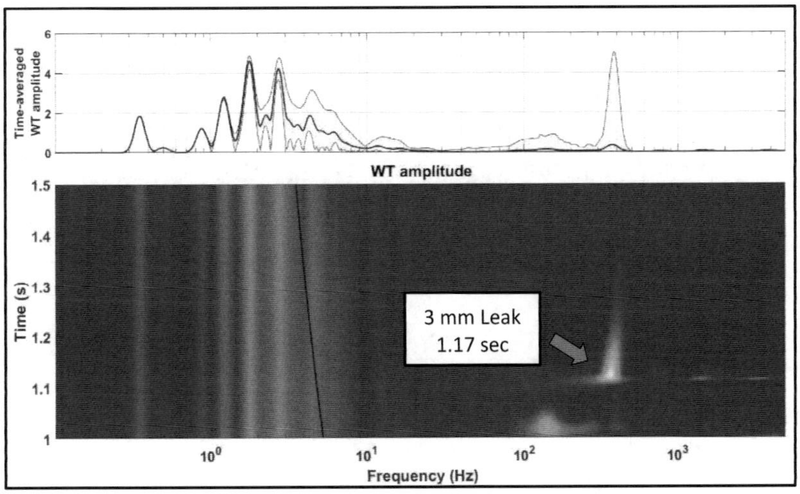


**Figure 6.** 3 mm leak (a) Time -frequency representation (b) Wavelet spectrum.

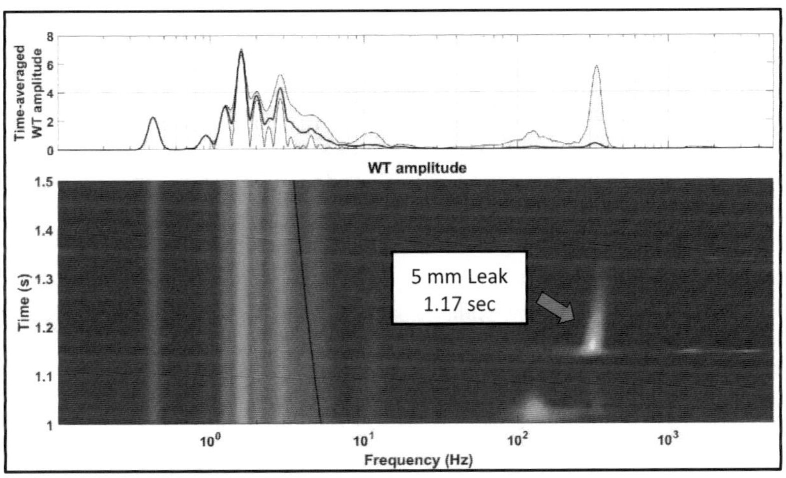


**Figure 7.** 5 mm leak (a) Time-frequency representation (b) Wavelet spectrum.

Referring to the Fig. 4-7, the enlarge time scale (1 sec - 1.5 sec) results of wavelets transform for the signal with the various condition of the size of the leak. The Fig. 4 shows the result for a healthy condition pipeline without any leakage tapping on the pipe. From the observation (Fig. 4), on both TF representation and spectrum, at all along with the position of time, $t$ =1 to $t$ = 1.5 (sec), there was no strange frequency developed and energy (amplitude) of the signal was consistent. The result was very contrasting when compared to the Fig. 5 – 7 where the leak are simulated at the position 47.9-m ( $t$ = 1.18 sec). Fig. 5 – 7 illustrate the TF representation and wavelet spectrum of the signals contain leak 1-mm, 3-mm, a 5-mm in diameter. As shown on Fig. 5 – 7, at the distanced where the leak was simulated on the pipeline system, the energy (amplitude) and instant frequency were appeared. The amplitude of the reflection signal slightly increases align with the increasing the leak size. The measured pressure signal provides much more information on the system than is purely necessary for the identification of leaks. While it was chosen intentionally for the experimental apparatus to prevent changes in the pipe characteristics, some other unwanted reflections can occur. In addition, as with any experimental signal, a certain noise rate is inevitable. In the wavelet diagram, both at larger and smaller scales, local maximum lines are present outside the tank and leakage. The experimental data was tabulated on the table as below.

**Table 1.** Comparison of the position of pipeline features between measured and experimental position.

| Pipe Feature | Measured Distanced (m) | Measured Times (sec) | Analysed Distanced (m) | Analysed Times (sec) | Error % |
|---|---|---|---|---|---|
| No-Leak | - | - | - | - | - |
| 1-mm circular hole | 48.1 | 1.19 | 49.3 | 1.17 | 2.5 |
| 3-mm circular hole | 48.1 | 1.19 | 49.3 | 1.17 | 2.5 |
| 5-mm circular hole | 48.1 | 1.19 | 51.1 | 1.21 | 6.2 |

## 5. Conclusion

To conclude, the early identification of leaks will prevent waste and harm to users. Any suggested method for early leak detection should be efficient, quicker and cheaper. For low reliability leak signal, CWT displays the ability to detect the leak in the water pipeline as the decomposition signal. The transformation of the wavelet has been shown to effectively eliminate the noise from the leak signal. Conversely, the study of the wavelet transforms presented in this article capable of detecting local singularities due to a leak. In addition, the use of a proper wavelet transformation enables the selection of information on the existence of a leak from other peculiarities. For better accuracy and consistency, the results outlined in this study should also be checked by varying the environmental and field condition during the experimentation.

## Acknowledgements

The author wants to thank University Malaysia Pahang for the support, facilities and sources for this research. The acknowledgement is also to University Malaysia Pahang for providing internal grant RDU190346, Postgraduates Research Grant Scheme (PGRS) PGRS1903174 and Ministry of Higher Education (MOHE), Malaysia grant FRGS/1/2017/TK03/UMP/02/1( RDU170121).

## References

[1] Yusop H M, Ghazali M, Yusof M and Hamat W 2017 Improvement of Cepstrum Analysis for the Purpose to Detect Leak, Feature and Its Location in Water Distribution System based on Pressure Transient Analysis *J. Mech. Eng.* **4** 103-22

[2] Yusop H M, Ghazali M, Yusof M, Remli M P and Kamarulzaman M 2017 Pipe leak diagnostic using high frequency piezoelectric pressure sensor and automatic selection of intrinsic mode function. In: *IOP Conference Series: Materials Science and Engineering*: IOP Publishing) p 012091

[3] Ministry of Water L a N R K 2018 Non-Revenue Water (NRW) Statistics - Datasets - MAMPU. (http://www.data.gov.my/data/en_US/dataset/non-revenue-water-nrw-statistics: http://www.data.gov.my/data/en_US/dataset/non-revenue-water-nrw-statistics)

[4] See K F and Ma Z 2018 Does non-revenue water affect Malaysia's water services industry productivity? *Utilities Policy* **54** 125-31

[5] Alkasseh J M, Adlan M N, Abustan I, Aziz H A and Hanif A B M 2013 Applying minimum night flow to estimate water loss using statistical modeling: a case study in Kinta Valley, Malaysia *Water resources management* **27** 1439-55

[6] Liemberger R and Wyatt A 2019 Quantifying the global non-revenue water problem *Water Supply* **19** 831-7

[7] Jang D and Choi G 2018 Estimation of non-revenue water ratio using MRA and ANN in water distribution networks *Water* **10** 2
[8] Pickard B D, Vilagos J, Nestel G K, Fernandez R, Kuhr S and Lanning D 2008 Reducing non-revenue water: a myriad of challenges *Florida Water Resour. J* 26-32
[9] Farley M 2003 Non-Revenue Water–International Best Practice for Assessment, Monitoring and Control. In: *12th Annual CWWA Water, Wastewater & Solid Waste Conference,*
[10] Taghvaei M, Beck S and Staszewski W 2006 Leak detection in pipelines using cepstrum analysis *Measurement Science and Technology* **17** 367
[11] Yusop H M, Ghazali M, Yusof M, Remli M P and Samykano M 2018 Fault Identification in Pipeline System Using Normalized Hilbert Huang Transform and Automatic Selection of Intrinsic Mode Function. In: *MATEC Web of Conferences*: EDP Sciences) p 02002
[12] Urbanek J, Barszcz T, Uhl T, Staszewski W, Beck S and Schmidt B 2012 Leak detection in gas pipelines using wavelet-based filtering *Structural Health Monitoring* **11** 405-12
[13] Tang X, Liu Y, Zheng L, Ma C and Wang H 2009 Leak detection of water pipeline using wavelet transform method. In: *2009 International Conference on Environmental Science and Information Application Technology*: IEEE) pp 217-20
[14] Lighthill J 2001 *Waves in fluids*: Cambridge university press)
[15] Beck S and Staszewski W 2004 Cepstrum analysis for identifying reflection points in pipeline networks. In: *International conference on pressure surges,*
[16] Amin M, Ghazali M, PiRemli M, Hamat A and Adnan N 2015 Leak detection in medium density polyethylene (MDPE) pipe using pressure transient method. In: *IOP Conference Series: Materials Science and Engineering*: IOP Publishing) p 012007
[17] Yusop H M, Ghazali M, Azmi W, Yusof M, PiRemli M and Rahim M 2019 A Study of Ikaz and Normalized Hilbert transform for solving faulty in pipeline distribution system using transmission line modelling. In: *IOP Conference Series: Materials Science and Engineering*: IOP Publishing) p 012083
[18] Hafizi Z, Kahandawa G, Epaarachchi J, Lau K, Canning J and Cook K 2013 NIR fibre Bragg grating as dynamic sensor: an application of 1D digital wavelet analysis for signal denoising. In: *Fourth International Conference on Smart Materials and Nanotechnology in Engineering*: International Society for Optics and Photonics) p 87930B
[19] Ferrante M and Brunone B 2003 Pipe system diagnosis and leak detection by unsteady-state tests. 2. Wavelet analysis *Advances in Water Resources* **26** 107-16

# Influence of polymer coated aggregate on the aggregate impact and crushing value

**A F Ahmad[1], A R Razali[1,*], F R M Romlay[1] and I S M Razelan[2]**

[1]Faculty of Mechanical and Automotive Engineering Technology, Universiti Malaysia Pahang, 26600 Pekan, Pahang, Malaysia
[2]Faculty of Civil Engineering Technology, Universiti Malaysia Pahang, Lebuhraya Tun Razak, 26300 Gambang, Kuantan, Pahang, Malaysia

*Corresponding author's email: akhtar@ump.edu.my

**Abstract.** Polyethylene Terephthalate (PET) and latex belong to the family of the polymer. PET type of plastic bottle is the most commonly found in the household waste stream. Recycling this type of plastic into another form is the cheapest and the best way of recycling. However, this may not become a green waste management solution when energy is used to transform this plastic into another useable form. In this study, PET was used to discover the potential for improvement of asphalt mixture properties. This study examines the plastic waste used as an aggregate coating. PET plastic bottle was shredded and sieved to obtain size 0.425 to 0.075mm then was blended with hot aggregate to form a thin layer of plastic coating on the aggregate surface. The comparison between unmodified aggregate properties and modified properties was studied. 1%, 3%, 5%, 7%, 9%, 11%, 13% and 15% of plastic by weight of aggregate were used as aggregate coating. The aggregate impact value test (AIV) and aggregate crushing value test (ACV) were done on the plastic coated aggregate. The test result had shown an improvement in the properties of plastic coated aggregate. From the experiment results conducted on the PET and aggregate, we can conclude that the lowest percent of aggregate loss on the crushing value test is 16.34% when PET content is 9% while 8.00% when PET content is 13% for the impact value test. The lowest percent of aggregate loss for impact value is then compared with the percent loss of PET Latex coated aggregate. The presence of latex and PET has caused the strength of the aggregate to increase where the percentage loss value when performed impact test is very low that is can achieve 3.48%. The quality of flexible pavement can be improved when the surface of the aggregate is coated with the polymer.

**Keywords:** Aggregate Test; Polyethylene Terephthalate; Plastic Coating; Road Performance.

## 1. Introduction

The performance of a road depends on the quality of aggregate used during the paving process. Crushed rock, gravel, slag, dust and sand are the example of aggregate that was used in pavement. Hot Modified Aggregate (HMA) can be formed by mixing the selected aggregate with asphalt binder [1]. The weight of the aggregate is one of the main support of HMA pavement component. This is because almost 95% of HMA is made up of aggregate. Hence, the characteristic of the aggregate can affect the performance of HMA pavement [2]. The characteristic of aggregate can be reinforced by the presence of plastic coating. Therefore, the presence of too much plastic waste in Malaysia can be used for the road construction industry.

Nowadays, there is a worldwide study on the use of waste material as secondary material. The most produced waste is plastic bottles [3]. Since the last decade, plastic bottles were so popular in the food industry because of safe, durable and good for packaging [4]. PET is a semi-crystalline thermoplastic polymer with high strength and transparency properties [5].

The soft blend can be achieved at low temperature and a stiffer blend is obtained at high temperature when PET is added which proves that it is a good modifier [6]. Another reason PET is considered as a good modifier is because it does not emit any harmful gas when heated to temperature range of 30°C to 180°C [7]. PET has good binding properties [8], hence it can be mixed with bitumen to enhance it properties. Malaysia has tropical climate temperature which is around 22°C to 32°C [9]. At this range of temperature, PET are remain stable and would not decomposed or degraded [10].

This study is focus to evaluate the properties of the aggregate by comparing the coating aggregate with unmodified aggregate. The different percent of PET is used to coat the aggregate and this process also known as dry process. Aggregate Impact Value Test and Crushing Value Test have been done to the unmodified and coated aggregate. The hypothesis for these test are aggregate percentage loss for crushing value and impact value of aggregate should be decrease as the amount of PET coated increase. Present of Latex can enhance the strength of aggregate by decreasing the aggregate percent loss.

## 2. Materials

PET, latex and aggregate is the main material for this study. Mineral water bottles were collected and shredded into small sizes. PET is defined as a polymer that happens by polymerization reaction between acid (terephthalic acid) and alcohol (ethylene glycol) [5]. PET is one of the polyester which easy to handle and chemically stable [11]. For this research, PET plastic bottles were collected and shredded. These shredded PET are sieved to get the size of 0.425 to 0.075mm. PET has an average molecular weight between 30000 - 80000 $gmol^{-1}$ and can be melt at temperatures 255°C to 265°C. The density of PET is 1.41 $gcm^{-3}$ and has a glass transition temperature between 69°C to 155°C [12].

High ammonia natural rubber (HANR) latex was provided by Lembaga Getah Malaysia. Field latex contains 64% water content, 36% rubber and a small number of impurities. Field latex can only last for 3 hours in liquid form. It needs to be preserved for storage and will be used later in different applications.[13].Therefore, ammonia is used to preserve the latex and to keep latex in liquid condition.

The aggregate used for this study is obtained from quarry in Berkelah, Pahang. HMA consists of three type of aggregate which are coarse aggregate, fine aggregate and mineral filler[14]. The size for coarse aggregate is 2.36mm or larger [15, 16] while the size for fine aggregate is between 2.36mm to 0.075mm[17]. Mineral filler is composed of dust or slag that is less than 0.075mm [16, 18]. 10mm size of aggregate was used in this research.

## 3. Methodology

Aggregate Impact Value Test and Crushing Value Test have been done to the selected aggregate. Based on British Standard where this standard is also used as a guideline in Malaysia, the limit of average aggregate percentage loss for crushing value test and impact value test is must less than 30% [19-21].

For impact value test, the experiment shows how to know the impact value of road aggregate in the laboratory and to evaluate its suitability in road construction based on impact value. This experiment complied with the SSRW in Malaysia [22] and British Standard BS 812-112:1990 [20]. The test is initiated by heating the aggregate at 160°C and then coat the aggregate with PET at the designated percentage. After cold the aggregate at room temperature, it then undergoes impact value test which according to the method stated in the British Standard BS 812-112:1990. The experiment is repeated to obtain the average of the result. Figure 1 shows the schematic image and apparatus of the impact value test.

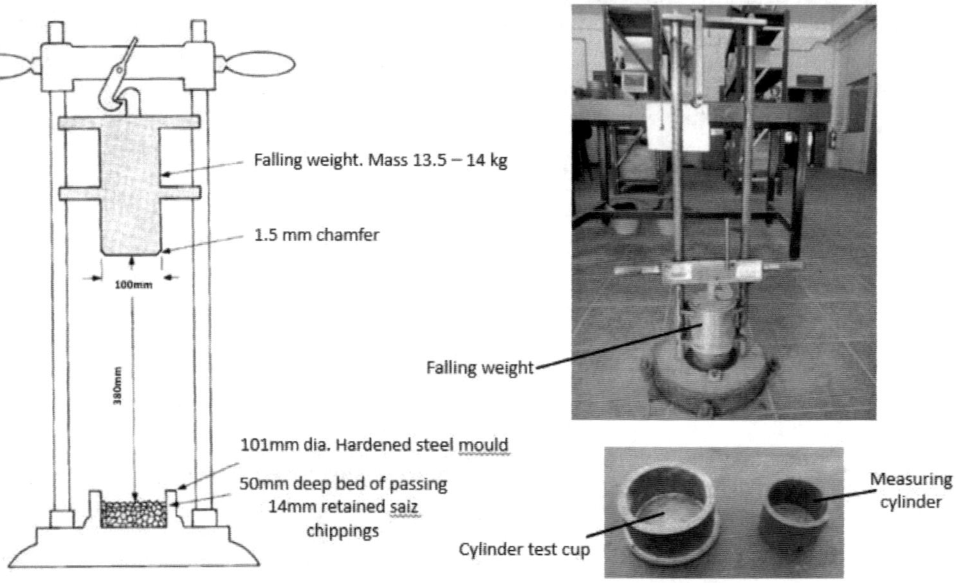


**Figure 1.** Schematic image and apparatus of the impact value test.

Aggregate crushing value is a percentage of the weight of the crushed aggregate obtained from the test where the specific load is applied to the aggregate tested. The best aggregate quality to be used in the surface course should not more than 30% aggregate lost. This experiment was based on British Standard BS 812-110 : 1990 [21] which on par with Standard Specification of Road Work (SSRW) in Malaysia [22].

The first step of this test is the same as the aggregate impact value test which heats the aggregate at 160°C and then coat the aggregate with PET at the designated percentage. Figure 2 shows the shredded PET plastic used for this research. After the aggregate was cold at room temperature, it then undergoes a crushing value test. The experiment is repeated to obtain the average of the result. Figure 3 shows the schematic image and apparatus of the crushing value test.

**Figure 2.** Plastic size 0.425 to 0.075mm used to coat the aggregate.

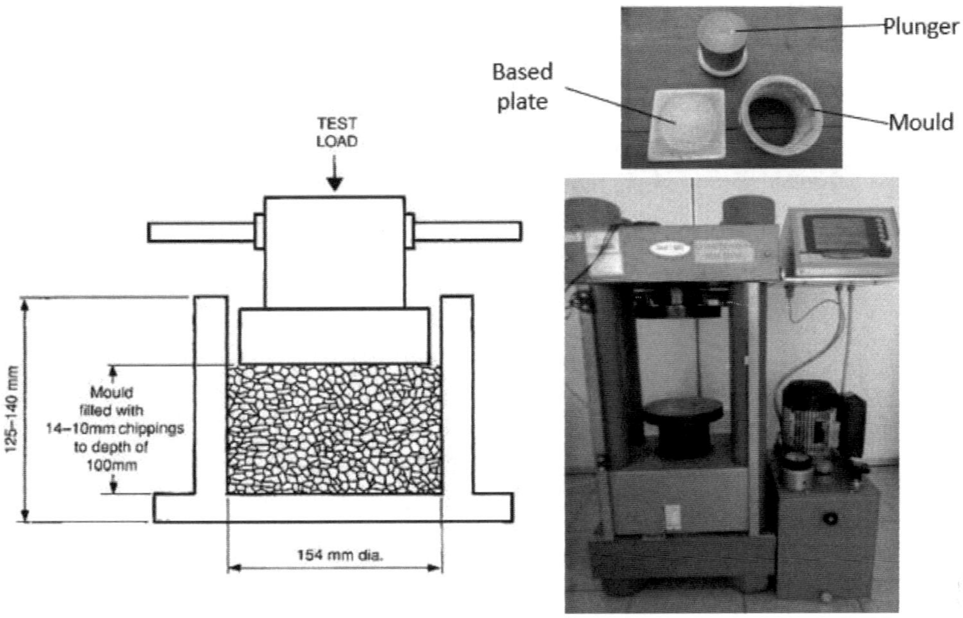


**Figure 3**. schematic image and apparatus of the crushing value test.

## 4. Results and discussions

From this experiment, the aggregate test result shows the aggregate quality value .The aggregate quality value can be used as the guideline to classify the stones according to their toughness property. The toughness property can be evaluated by looking at the aggregate loss after done the impact and crushing test. Percent of aggregate loss can be calculated by dividing the weight of coated aggregate on pan after test with the weight of coated aggregate before test. The answer is then multiple with one hundred to get the percentage of aggregate loss [21]. The result of aggregate impact value tests is shown in Table 1 while the result of aggregate crushing value test is shown in Table 2. This table shows the comparison between PET coated aggregate and unmodified aggregate (no PET).

**Table 1.** Impact value tests.

| % of plastic | Weight of Crushed Agg. (g) | | % Loss |
|---|---|---|---|
| | M1 | M3 | |
| 0 | 312.94 | 53.32 | 17.04 |
| 1 | 300.36 | 46.52 | 15.49 |
| 3 | 312.50 | 40.61 | 13.00 |
| 5 | 323.94 | 40.38 | 12.47 |
| 7 | 327.65 | 40.52 | 12.37 |
| 9 | 334.11 | 40.01 | 11.98 |
| 11 | 342.67 | 34.43 | 10.05 |
| 13 | 343.36 | 37.56 | 8.03 |
| 15 | 346.15 | 39.32 | 11.36 |

**Table 2.** Aggregate crushing value.

| % of plastic | Weight of Crushed Agg. (g) | | % Loss |
|---|---|---|---|
| | M1 | M3 | |
| 0 | 2660.62 | 677.72 | 25.47 |
| 1 | 2743.99 | 511.35 | 18.64 |
| 3 | 2491.21 | 430.06 | 17.26 |
| 5 | 2591.47 | 439.65 | 16.97 |
| 7 | 2604.79 | 436.67 | 16.76 |
| 9 | 3646.39 | 432.15 | 16.33 |
| 11 | 2780.10 | 469.69 | 16.89 |

Unmodified aggregate shows a higher percent of aggregate loss compared to PET coated aggregate. The trend was proven by whenever the amount of PET is decreased, the aggregate percent loss was an increase. Figure 4 shows that the percent loss of aggregate was influenced by the amount of PET coated. Based on Standard Specification of Road Work (SSRW) in Malaysia [22], the road is in the category weak on surface when the impact value is more than 35% and it will consider being exceptionally strong when the value below 10%. However, when the impact value is between 20% to 30%, the road is consider in category satisfactorily for road surfacing while the value is between 10% to 20% the road are in strong category. From the graph in figure 4, the lowest aggregate loss was 8 % when 13% of PET is used to coat the aggregate. The aggregate loss starts to increase when 15% of PET is used to coat the aggregate. Without any plastic coating, the aggregate is still in in good condition for road surfacing to comply with the SSRW which the aggregate loss is 25.47%. However, coating the aggregate with PET was increase the category of aggregate from satisfactorily for road surfacing to strong and exceptionally strong. Based on the result, 13% of PET coated aggregate was achieved the lowest aggregate lost which is 8%. From that indicator, the combination of latex and PET was used to coat the aggregate. The total amount combination of latex and PET is 13%. Table 3 shows the result of the impact value test for latex and PET coated aggregate. From that results, combination of 7% PET and 6% Latex seems most suitable wish it achive 3.48% of aggreagte lost. Figure 5 shows the PET+Latex coated agrregate. Figure 6 shows that the percent loss of aggregate was influenced by the amount of 13% PET coated and PET+Latex coated.

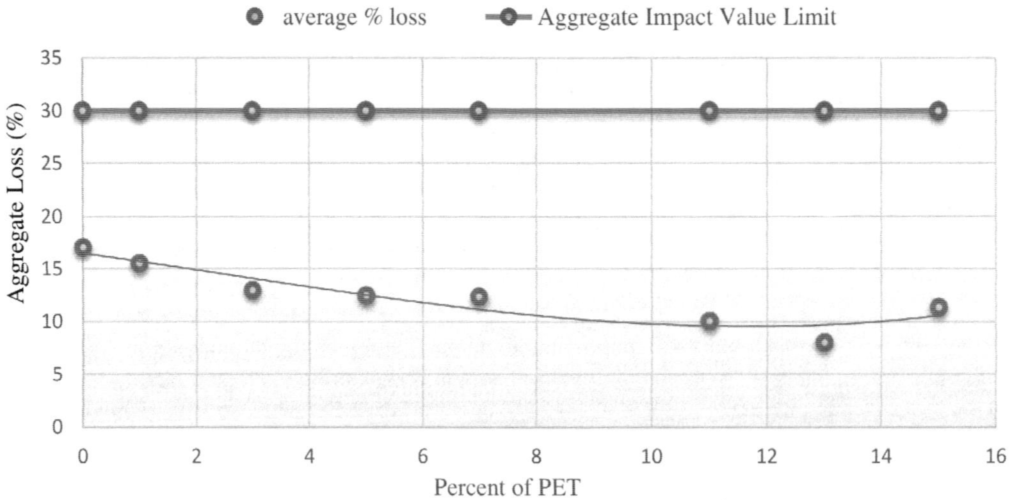


**Figure 4.** Impact value test.

**Table 3.** Result of aggregate impact value test for PET+Latex coated aggregate.

| Amount of binder (PET+Latex) | | % Loss |
|---|---|---|
| PET (%) | Latex (%) | |
| 5 | 8 | 4.34 |
| 7 | 6 | 3.48 |
| 9 | 4 | 4.17 |

**Figure 5.** PET+Latex coated aggregate.

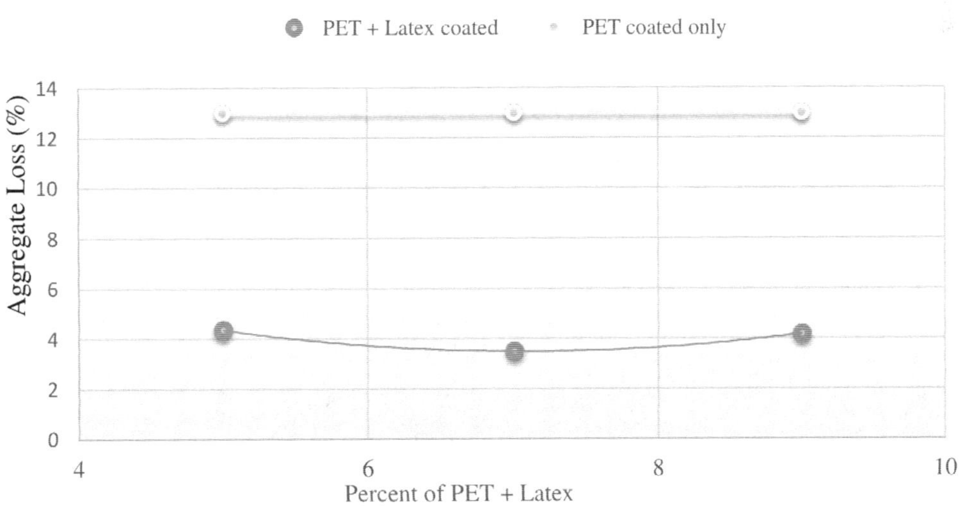


**Figure 6.** Percent loss of aggregate was influenced by the amount of 13% PET coated and PET+Latex coated.

Figure 6, it shows that the combination of PET and Latex can increase the strength of aggregate. The same result also achieves by Wayal and Wagal which state that coating of plastics and rubber over the stone aggregate improves the quality of the aggregate [23]. This is because the presence of latex can make the aggregate elastic and able to protect aggregate from strong impact [24]. The indicator of this experiment result could be accepted when the percent loss of aggregate impact lies in the range of strong for road surfacing. Based on SSRW in Malaysia, the percentage of aggregate loss for impact value for wearing course should be below 30% [19, 22, 25]. This test is important to adjudge the suitability of aggregate whether it strong enough to be used in road construction. The ability of the road stone to resist the impact and pounding action due to traffic load can be determined by this test. In order to produce high quality pavement, this test also helps to determine the durability of stone that has the ability to break into smaller pieces if repeated impact is applied.

The PET coated aggregate has lower percent of aggregate loss than the unmodified aggregate. 16.34% of aggregates loss was found in 9% of PET plastic coating. Figure 7 shows that when the amount of PET increase, the percent of aggregate loss decrease.

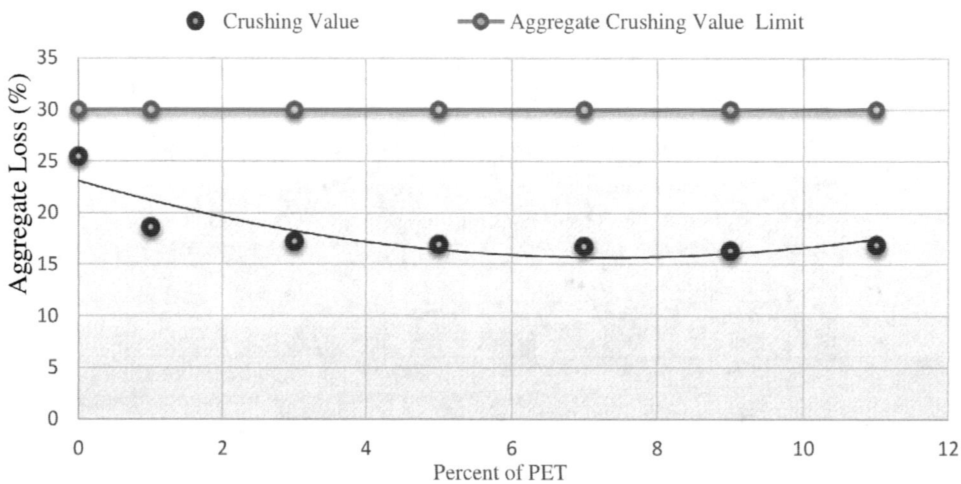


**Figure 7.** Crushing value test.

The ability of an aggregate to resist crushing can be measured through the aggregate crushing value. Based on the test results, it showed that all results are adhering to the SSRW in Malaysia which the aggregate loss recorded is 25.47%, 18.64%, 17.26%, 16.97%, 16.76%, 16.33% and 16.89%. Low aggregate crushing value is more preferred to increase the quality of the pavement. The best quality of aggregate for surface layer pavement should have lower than 30% of aggregate loss. Based on all results, this kind of aggregate was strong enough for this study as it will be applied to the surface course layer. It was proved that coating of aggregate with PET can improve the strength of properties of aggregate. This is because the PET coated aggregate can help to fill the pores at the surface. The brittleness of the aggregate is measured as an impact value. Void sand air cavity in the aggregate can be reduced by coating the aggregate with polymer. Cracking can be prevented when the polymer layer formed. Coating aggregate with PET can increase the strength of the aggregate. This can be proved by comparing the percentage of aggregate loss on a plain aggregate and modified aggregate. To compare with the study by Kumar and Vikranth on 2017 [26], they used Polypropylene (PP) as their modifier. Their result shows that as the amount of PP is increase, the percent of aggregate loss was decreased. Gurpreet Singh and Rajiv Chauhan [27] studied about coating the aggregate by using the PET. The trend of their results is also the same with this study. The results comparison of aggregate impact value is presented in figure 8 and the comparison of aggregate crushing value is presented in figure 9. The impact value test of Kumar and Vikranth study shows lower percent of aggregate loss compared to this research. This might be due to the use of a different type of plastic. PP has a melting point of 164°C while PET is 252°C [28]. Therefore, PP can coat the aggregate better because it melt easily compared to the PET.

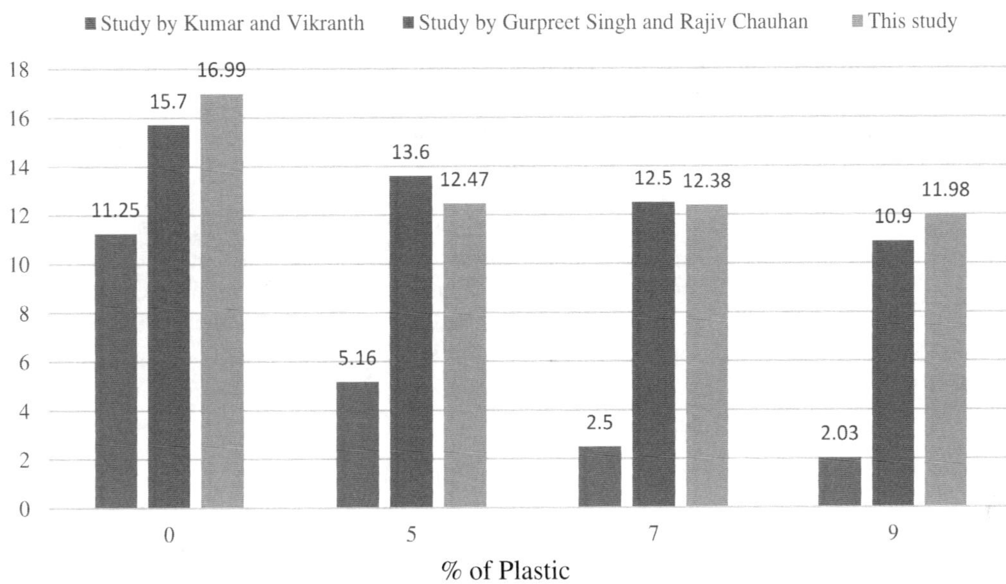


**Figure 8.** Comparison of Aggregate Impact Value between researchers.

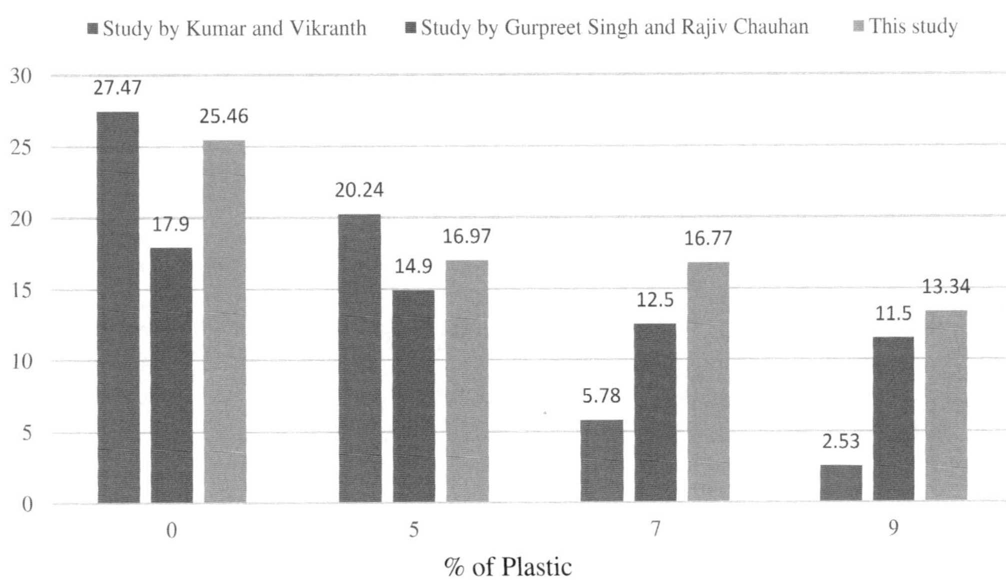


**Figure 9.** Comparison of Aggregate Crushing Value between researchers.

## 5. Conclusions

Coating aggregate with polymer (PET and latex) can produce new raw material for flexible pavement which this also known as dry process. The results allows the following conclusion to be drawn:

i. The lowest percent of aggregate loss on the crushing value test is 16.34% when PET content is 9% while 8% when PET content is 13% for the impact value test.

ii. The presence of latex has caused the strength of the aggregate to increase where the percentage loss value when performed impact test is very low that is can achieve 3.48%.
iii. The quality of aggregate can be improved when the surface of the aggregate is coated with a polymer.
iv. The quality of the pavement can be improved as the aggregate strength increases.
v. Using waste plastic in road construction is able to help the waste polymer disposal process in an eco-friendly method. With this eco-friendly method, the waste that disposes of through incineration and land filling can be avoided. It also able to help to create jobs as rag pickers.

**Acknowledgments**
The authors would like to acknowledge Universiti Malaysia Pahang for funding through the Doctoral Research Scheme (DRS) scholarship. In addition, further appreciation was also given to the Ministry of Education Malaysia for providing the Fundamental Research Grant Scheme FRGS/1/2019/TK03/UMP/02/14 @ RDU1901132. Finally, big thanks to the UMP Research and Innovation Department for providing the Postgraduate Research Grants Scheme (PGRS) under grant no PGRS1903173.

**References**
[1] Iqbal A, Yousuf M, Ullah K, Adnan M, Ahmad M, Ashiq M, Shehzad U, Munir M, Rehman S U and Akhtar M 2020 Utilization of Waste Plastic Polymers to Improve the Performance of Modified Hot Mix Asphalt *Pakistan Journal of Engineering and Technology* **3** 162-71
[2] Almadwi F S and Assaf G J 2020 Determination of the permanent deformation of hot mix asphalt with natural desert sand *Construction and Building Materials* 121255
[3] Stelluto G A, Plum L, Pollock V, Moll A, Stieha V and Salzman N 2020 Recycled PET Plastic
[4] Subramani T and Rahman A F 2017 An Experimental Study On The Properties Of Pet Fibre Reinforced Concrete *International Journal of Application or Innovation in Engineering & Management (IJAIEM)* **6** 058-66
[5] Alzuhairi M, Khalil B and Hadi R 2017 Nano ZnO Catalyst for Chemical Recycling of Polyethylene terephthalate (PET) *Engineering and Technology Journal* **35** 831-7
[6] Sulyman M, Haponiuk J and Formela K 2016 Utilization of recycled polyethylene terephthalate (PET) in engineering materials: a review *International Journal of Environmental Science and Development* **7** 100
[7] Dawale S 2016 Use of waste plastic coated aggregates in bituminous road construction *International journal of advancement in engineering technology, management and applied science* **3** 118-26
[8] Bhatnagar A, Kumar G, Pal G S, Singh P P and Gupta V 2019 To Enhance The Strength Of Bituminous Roads Using Waste Plastic
[9] Lei J, Yang J and Yang E-H 2016 Energy performance of building envelopes integrated with phase change materials for cooling load reduction in tropical Singapore *Applied Energy* **162** 207-17
[10] Dankwah J, Koshy P and Sahajwalla V 2014 Reduction of FeO in EAF steelmaking slag by blends of metallurgical coke and end-of-life polyethylene terephthalate *Ironmaking & Steelmaking* **41** 401-9
[11] Zhou J, Fei X, Li C, Yu S, Hu Z, Xiang H, Sun B and Zhu M 2019 Integrating Nano-Cu2O@ ZrP into in situ polymerized polyethylene terephthalate (PET) fibers with enhanced mechanical properties and antibacterial activities *Polymers* **11** 113
[12] Awaja F and Pavel D 2005 Recycling of PET *European Polymer Journal* **41** 1453-77
[13] Azahar N, Hassan N A, Jaya R P, Kadir M, Yunus N and Mahmud M Z H 2016 An overview on natural rubber application for asphalt modification *Int. J. Agric. For. Plant* **2** 212-8

[14] Abd Shareef R and Al-Hdabi A 2020 Hot Mix Asphalt Characteristics with Sugar Industry Waste Materials as Mineral Filler. In: *IOP Conference Series: Materials Science and Engineering*: IOP Publishing) p 012005
[15] Abhijith B and Krishnan J M 2020 Influence of the Aggregate Gradation on the Rutting Resistance of Bituminous Mixtures. In: *Proceedings of the 9th International Conference on Maintenance and Rehabilitation of Pavements—Mairepav9*: Springer) pp 3-12
[16] Sutradhar D, Miah M, Chowdhury G J and Sobhan M A 2015 Effect of using waste material as filler in bituminous mix design *American Journal of Civil Engineering* **3** 88-94
[17] Lambebo B A 2020 Mix Design for Wearing Course of Flexible Road Pavement by Marshall Method *International Research Journal of Engineering and Technology (IRJET)* **07**
[18] André F P, Miceli H, Moura L C, Neumann R and Tavares L M 2019 Upgrading a manufactured fine aggregate for use in concrete using dry rare-earth magnetic separation *Minerals Engineering* **143** 105942
[19] Adanikin A, Olutaiwo A and Ashamo A 2018 Mechanical Strength Determination Of Crushed Stone Aggregate Fraction For Road Pavement Construction (Case Study: Selected Quarries In Western Nigeria)
[20] British Standard B 1990 Testing aggregates. Method for determination of aggregate impact value (AIV) *Vol. BS* 812-112
[21] Standard B 1990 BS 812-110: 1990 Testing Aggregates-Part 110: Methods for Determination of Aggregate Crushing Value (ACV) *BSI, UK*
[22] Raya J K 2008 Standard Specification For Road Work, Section 4: Flexible Pavement Jkr *Spj* S4
[23] Wayal A S and Wagle M D 2013 Use of waste plastic and waste rubber in aggregate and bitumen for road materials *Polymer* **40** 36
[24] Atahan A O and Yücel A Ö 2012 Crumb rubber in concrete: static and dynamic evaluation *Construction and Building Materials* **36** 617-22
[25] Irior K, Abdullahi M and Aguwa J 2020 The capitalized title is Compressive and Flexural Strength of Concrete Using Periwinkle Shell as Partial Replacement to Coarse Aggregate. EasyChair)
[26] Kumar T K and Vikranth J 2017 A Study on use of Plastic Waste (Polypropylene) in Flexible Pavements *International Journal of Engineering and Management Research (IJEMR)* **7** 554-60
[27] Singh G and Chauhan R 2020 A Study on using Plastic Coated Aggregate for evaluation of modified Bituminous Concrete Mix. In: *IOP Conference Series: Materials Science and Engineering*: IOP Publishing) p 012052
[28] Wu C-M, Murakami R-I, Lai S-G, Lin P-C, Koinkar P, Lee K-C, Lai C-C and Rwei S-P 2019 Investigation on the interface modification of PET/PP composites *Modern Physics Letters B* **33** 1940019

# AUTHOR INDEX